Field Guide to Some
Carbonate Rock Environments:
Florida Keys

Field Guide to Some

Carbonate Rock Environments

Florida Keys and Western Bahamas

New Edition

H. Gray Multer
Fairleigh Dickinson University

KENDALL/HUNT PUBLISHING COMPANY
2460 Kerper Boulevard, Dubuque, Iowa 52001

Cover Photographs

Front cover (outside): Patch reef community consisting of *Montastrea annularis* (M), *Porites asteroidies* (P), *Halimeda opuntia* (H), and several acylonarians (A) such as sea whips and sea fans. Constituent particles of the loose sand surrounding this reef contain skeletal elements of these carbonate producing organisms. Compare with outside back cover showing photo of Pleistocene outcrop of lithified equivalent patch reef coral (*M. annularis*) surrounded by similar (cemented) skeletal sand.

Front cover (inside): Index map of Florida-Bahama area.

Back cover (outside): Exposed bedrock section showing *Montastrea annularis* coral in growth position surrounded by calcarenite. Key Largo Canal Cut, Key Largo. Time needed for formation of this exposed reef rock along canal is estimated by Hoffmeister and Multer (1964) to be between 250 and 500 years. Coral is approximately 2 meters high. Note: Compare with front cover.

Back cover (inside): Aerial photograph of the Florida Keys taken from Gemini IV, courtesy of NASA. Compare with figure 3.1. Numbers indicate: (1) Florida Straits, (2) Molasses outer reef, (3) Mosquito Bank back reef, (4) Rodriguez Key mud bank, (5) Tavernier, (6) Crane Key, (7) U.S. Route #1, (8) Carysfort outer reef.

Copyright © 1977 by Kendall/Hunt Publishing Company

Library of Congress Catalog Card Number: 76-50173

ISBN 0—8403—1646—1

Printed in the United States of America

Contents

Preface

This book describes both modern and Pleistocene carbonate environments found in the southern Florida and western Bahama region—an area unique and world-famous for investigations concerning its great variety of readily accessible carbonate sediments and rocks. The book is designed not only for persons who will use it as a guide in the field but also for those who will use it as a reference book or text in the classroom or laboratory.

The purpose of this Field Guide is to serve the students of carbonate rocks by helping them locate, examine, discuss and ponder both modern and Pleistocene carbonate environments. These objectives are accomplished by (1) dividing the book into chapters, each of which covers a general area (i.e., Lagoonal Areas, Outer Reef Areas, etc.) containing a variety of specific environments, (2) describing and illustrating features of each of these environments and indicating with questions (at the end of each chapter) some of their special characteristics, and (3) summarizing, quoting and noting the existence of pertinent literature.

Classification of natural environments has always presented challenges for geologists. Thus organization of a guidebook which contains as wide a variety of environments as are found in the Florida Keys-western Bahamas region has been especially difficult. Readers may disagree with some boundaries and definitions used, but it should be understood that the author has tried to define boundaries and use terms that will be most helpful to geologists who are interested in interpreting ancient analogs of the south Florida-western Bahama environments.

Some chapters include environments which would also be appropriate for discussion in other chapters (i.e., blanket sediments are found in both the Inner and Outer Shelf Areas). In such cases discussion is confined primarily to one chapter with only reference or brief discussion made to it in another. The chapter entitled "Transitional Areas" may cause initial confusion since "transitions" in sediments and biota occur in all areas covered by this Guidebook. The "Transitional Areas" chapter, however, was specifically designed for petroleum geologists interested in those usually localized transitional and often interfingering carbonate lithofacies helpful in interpreting paleoenvironmental settings and sometimes displaying good sorting and channel characteristics.

As a discipline, the field of "carbonate rocks" has accumulated a large amount of objective data and many theories. However, in many ways, this subfield of geology is in a state of evolution and is suffering "growing pains." Therefore *questions* and as yet *untested ideas* are often more appropriate than flat statements and generalizations. For this reason the "Field Questions" at the end of the chapters suggest ideas to ponder as well as important features to observe. **Note:** After a thorough reading of the text, many of these questions can be answered in the classroom or laboratory without the student having been in the field (although a good set of colored 35 mm slides and samples from the instructor would be an additional asset).

The completion of this Guidebook is the result of the cooperation of many persons who have allowed extensive use of their published material, participated in helpful discussions and provided encouragement during the writing. Although it is impossible to list them all, public recognition is certainly due to the following: W. D. Bock, H. K. Brooks, P. Enos, G. M. Friedman, C. D. Gebelein, R. N. Ginsburg, G. M. Griffin, I. G. Macintyre, J. D. Milliman, A. C. Neumann, N. D. Newell, E. A. Shinn and M. P. Weiss.

The amount of significant *unpublished* material furnished the author for successive editions of this Guidebook creates a unique dimension for readers interested in the Holocene of the Florida-Bahama area.

These thirty-nine contributions, when woven into the fabric of the published literature, provide a truer picture of the current "state of the art" in the world of carbonate investigations of the region. The quality of the present edition is particularly enhanced by the extensive contributions of unpublished text and illustrative material by D. L. Kissling and J. A. Zischke. Exciting new research directions are noted in the illustrated summaries of some unpublished theses which will in the years to come be further expanded by the individual authors involved. Examples are those submitted by M. Boardman, P. M. Harris, V. Jindrich, S. M. Landon, J. P. Manker and E. R. Warzeski. Others who have provided unpublished material (text and/or illustrations) which enhance the Guidebook and for which the author is grateful are: A. Antonius, R. G. Bader*, R. J. Bain, M. M. Ball, R. C. G. Bathurst, W. D. Bock, A. H. Coogan, R. J. Cuffey, J. R. Dodd, R. J. Dunham, P. Enos, C. D. Gebelein, R. N. Ginsburg, G. W. Gray, G. M. Griffin, J. P. Howard, D. W. Kirtley, J. C. Kraft, J. A. Lineback, I. G. Macintyre, G. Müller, A. C. Neumann, J. C. Ogden, L. C. Pray, M. A. Roessler, E. A. Shinn, D. C. Steinker, J. W. Teeter, A. Thorhaug, W. J. Turney*, G. L. Voss and H. R. Wanless.

Permission to reproduce text and/or illustrative material from published literature is gratefully acknowledged from the following:

American Association of Petroleum Geologists
Bulletin of the Geological Society of America
Bulletin of Marine Science
Bulletin of the Museum of Natural History
Geology
Journal of Geology
Journal of Paleontology
Journal of Sedimentary Petrology
Marine Geology
National Geographic Magazine
Nature
Reinhold Book Corporation
Science
Shale Shaker
Society of Economic Paleontologists and Mineralogists
University of Miami Press

The writer's long and valued association with J. E. Hoffmeister as a colleague as well as coinvestigator on numerous National Science Foundation-supported projects has made possible much of the information presented herein. Support for these projects by the National Science Foundation is gratefully acknowledged.

Photography of the corals in Appendix III was done by S. J. Simonds of specimens mostly loaned by the Museum of the Rosenstiel School of Marine and Atmospheric Science. All figures in this Guidebook that are not credited to other sources are those of H. G. Multer. L. H. Collinvaux provided identification of *Halimeda* in figure A.15.

A bibliography is provided at the end of the text giving complete citation for all reproduced material as acknowledged on these pages.

The writer is particularly indebted to his wife, Susan, whose constant encouragement and help in compiling, typing and editing have made the present edition possible.

Contribution No. 77-1, Department of Earth Sciences, Fairleigh Dickinson University, Madison, N.J.

*Deceased.

INTRODUCTION

LOCATION OF AREA

Carbonate environments described in this Field Guide are located from Miami south along the arc of the Florida Keys to Loggerhead Key in the Dry Tortugas and in the Bahamas, including Bimini, Cat Cay, Grand Bahama Bank, Andros Island and Little Bahama Bank (see inside front cover and index maps at the back of the Guidebook).

GENERAL STATEMENT

The Florida Keys and the Bahamas both represent shallow platforms capped with Pleistocene bedrock which protrudes only rarely above the present sea level. In the Florida Keys the emergent bedrock is represented by a crescentic chain of small, low islands extending some 376 kilometers from Miami to Dry Tortugas. Emergent bedrock in the Bahamas consists of low, marginal lips of flat, saucerlike platforms, mostly underwater. On the submergent platforms of both areas are a great variety of modern carbonate environments, each of which are characterized by certain types of bottom sediments, textures, and structures.

In one sense, the Florida Keys-Bahama area can be pictured as a vast "carbonate rock factory" which is actively producing many types of limestones. Some raw materials are Ca^{++}, Mg^{++}, CO_3^{--}; the manufacturing processes include evaporation, supersaturation, precipitation, photosynthesis; and the parts which are produced from these raw materials by these processes are spicules, ooids, crusts and skeletons. Such parts can be assembled by certain (diagenetic) processes which may include compaction, subaerial cementation, submarine cementation and dolomitization, yielding final end products which in the geologic literature have been referred to as oolitic limestones, coralline limestones, calcarenites, calcilutites or dolomites.

The above-mentioned "carbonate factories" have undergone many fluctuations in production rates during the late Pleistocene, due to glacial control of eustatic sea level resulting in fluctuations up to 130 meters (Milliman and Emery, 1968). The present picture is one of a steady, slow rise in sea level at a decreasing rate over the last 6,500 calendar years (see figure 1.1) according to Scholl et al. (1969). They also cite evidence for the generalization that submergence rates in South Florida largely determine as well as limit rates of coastal sedimentation in lagoon and estuarine areas.

As the sea gradually rose above the level of the continental shelf edge, the karst rock floor of both the Florida and Bahama platforms became covered with shelly mud and mangrove peat. At the same time coral-algal development on the windward edges of these same platforms produced an intermittent barrier or rim (which still sporadically persists). As sea level continued to rise producing more open circulation, skeletal sand covered the mud in many deeper water platform areas. Flooding of adjacent bedrock floors (i.e., Florida Bay during the past 4,000 years) also provided widespread conditions for deposition of carbonate mud as can be seen today.

A final and ever-present source of change in production rates for the various "carbonate rock factories" on both the Florida and Bahama platforms is local modification by tropical storms and hurricanes. If such storms occur at particular tide levels, they alter the usual products and produce characteristic features ("horns" of reef rubble on the back side of outer reefs, or layers of supratidal mud on supratidal islands).

In summary, Holocene sediments of the Florida Keys and Bahamas today present a vast array of textures and constituent particles characteristic of certain types of environments which have been subjected to fluctuating sea levels and storm action. Such sediments and textures may be used in interpreting ancient environments.

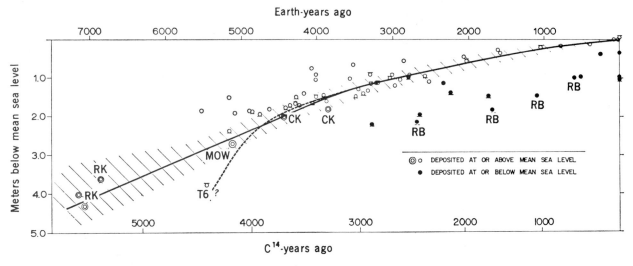

Figure 1.1. Unbroken curve is the revised submergence curve for south Florida; dashed curve is a previously published version (*1*). Short bar above or below data points identifies original data used to construct the dashed curve. Data accuracy has been discussed in detail elsewhere (*1, 11*) and is briefly described in reference (*7*). Width of hatched zone indicates the vertical uncertainty in the position of any portion of the revised submergence curve. The accuracy of the data justifies only the construction of a smooth or generalized submergence curve, which should not, therefore, be construed to mean the absence of minor "oscillations" in the late Holocene eustatic rise in sea level. Doubly circled points are dated peat deposits underlying Crane Key (*CK*), Man-O-War Key (*MOW*), and Rodriquez Key (*RK*); dated marine deposits from Crane Key and Rodriquez Bank (*RB*) are also identified (*8, 14*). Upper time scale represents approximate true earth (sidereal) years (*4*), lower scale is in standard radiocarbon years. From Scholl et al. (1969, p. 563). Copyright 1969 by the American Association for the Advancement of Science.

Figure 1.2. Aerial view looking southwest at various "carbonate factories" of the Florida Keys. Florida Bay (B) produces large quantities of mud, shoal areas (S) of the back-reef area produce a variety of muddy sand blankets and mounds, and patch reefs (P) (Mosquito Banks) are the site of skeletal sand production. The edge of the Florida Platform (upper left corner of photo) displays intermittent development of outer reefs which produce wide blankets of carbonate sand over the platform surface. Tavernier is at "T." Photo courtesy of Paul Enos.

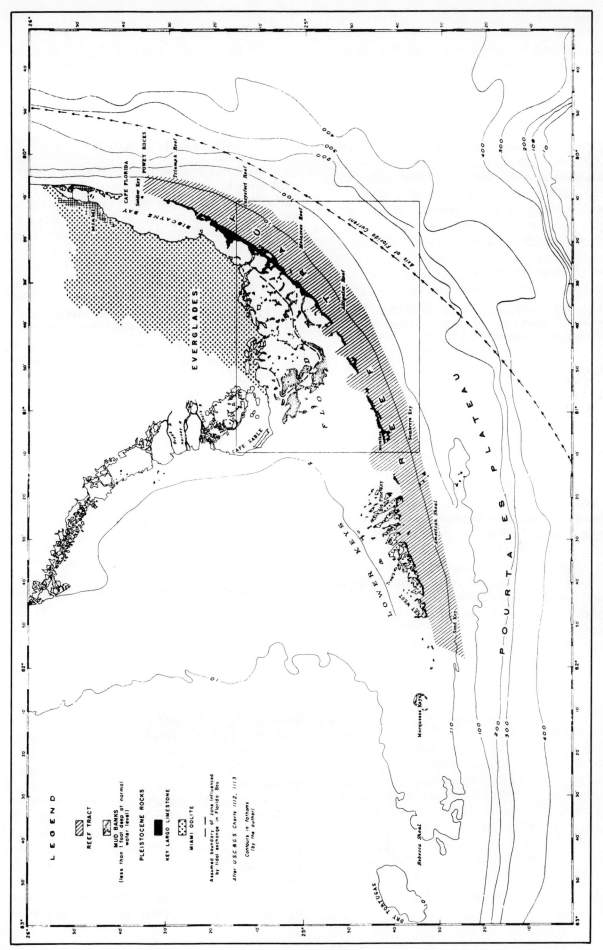

Figure 1.3. Index map of the Florida Keys area after Ginsburg (1956); see figure 3.1 for inset map area. Note: Subsequent mapping has changed distribution outline of above "Miami oolite" and caused it to be renamed Miami Formation consisting of an oolite facies and bryozoan facies; see chapter 9.

Indurated Pleistocene carbonate rocks making up the emergent portions of the Florida Keys and Bahamas represent a transitional type of limestone which in many ways spans the gap between the Holocene and the ancient carbonates. For example, ancient carbonates with a porosity of 0 to 10 percent stand in sharp contrast with the high porosities of the Holocene unconsolidated carbonate sediments of the Florida-Bahama area. However, the Pleistocene indurated rocks represent a lithology that spans the porosity spectrum (5 to 60 percent porosity). By studying these limestones, one is thus better able to understand diagenetic processes (such as cementation, hole-filling, and recrystallization) in the ancient rocks. In other words, because of the minimum amount of alteration in these Pleistocene limestones there is an unusual opportunity to study various diagenetic processes.

A review of the pertinent literature is highly recommended for carbonate students uninitiated to the Florida-Bahama area before their visit.

FIELD CONDITIONS

. . . or "fun in the sun"? Despite rumors circulated to the contrary, the weather in Florida is always unpredictable and can be nasty without warning. Precipitation and wind data as well as local advice indicates that June is usually a good month for fieldwork with rain and hurricane potentialities increasing towards September-October. Strong winds disturb the bottom sediment and often make the water opaque for several days and small boat handling difficult. The best weather (low winds) should be used for outer reef observations; the more protected Florida Bay sometimes can be observed even on windy days.

Besides rain and wind, Florida has lots of sun for which the field trip participant should be properly prepared. The following items are listed to make the initial venture more comfortable and hence more productive:

- hat with brim
- mask, snorkel, flippers, weight belt
- suntan oil and/or lotion, sun glasses
- sun preventative paste (white, opaque)
- lightweight, long-sleeved shirt to wear in water
- lightweight, long pair of slacks to wear in boat
- one pair of old canvas tennis sneakers to wear in water (when flippers are not appropriate)
- seasickness pills if believed possibly necessary
- rubber wetsuit if swimming in the winter months

Other accessory items which might prove valuable would be:

- plastic bucket, plastic sample bags, waterproof black felt marker pens, gloves for working in coral areas
- underwater camera

PROTECTED AREAS

Visitors should note the boundaries of various **protected areas** in the Florida Keys and abide by rules and regulations governing their use. These areas include the Everglades National Park (Crane Key is in this park), Ft. Jefferson National Monument (Dry Tortugas) and John Pennekamp Coral Reef State Park. In such areas the removal or destruction of natural features and marine life is prohibited. Collecting of scientific specimens is by permit only. For further information concerning the Everglades Park and Ft. Jefferson areas write to: Superintendent, Everglades National Park, P.O. 279, Homestead, Fla. 33030. **Note:** There are abundant opportunities for sampling Key Largo/Miami limestones from huge quanities of dredged rock piles (often of perfect specimen size!) along canals and construction sites. Visitors are urged to use these sources of supply and not to deface vertical walls of bedrock in quarries or along canals (i.e., Windley Key quarry or Key Largo Canal cut).

LITERATURE

Geologic literature concerning the area of this Guidebook is voluminous. Selected literature pertinent to local environments is noted within the text under appropriate heading (see table of contents) with full bibliographic citations at the end of the Guidebook.

Selected references concerning regional geology include Presseler (1947), Illing (1954), Parker (1955), Newell and Rigby (1957), Dubar (1958a), Baars (1963), Ginsburg et al., (1963), Gorsline (1963), Purdy (1963), Banks (1964), Puri and Vernon (1964), Scholl (1964), Chen (1965), Traverse and Ginsburg (1966), Perkins et al., (1968), Uchupi (1968, 1969), Scholl et al., (1969), Malloy et al., (1970), Multer (1971), Dubar (1974), and Ginsburg and James (1974a).

A summary of ideas concerning tectonic control of the Bahama area is given by Ball (1967b), and a review of seismic-refraction studies and related bibliography is given by Sheridan et al., (1966). See also Uchupi and Emery (1967); Lynts (1970); Ballard and Uchupi (1970). Gough (1967) describes magnetic anomalies. Paleocene and Eocene rocks of Florida are described by Chen (1964), and Puri and Vernon

(1964) give a summary of the geology of Florida and a guidebook to the classic exposures found in the state. For a description of a 14,585 foot test hole drilled on Andros see Spencer (1967) and figure A.7 (appendix).

A critical review of various hypotheses and types of evidence for the origin of the Florida-Bahama area together with a brief survey of Bahama wells and seismic characteristics is given by Meyerhoff and Hatten (1974). The latter paper also includes a summary of the subsurface stratigraphy of the Florida-Bahama region with special reference to the oil and gas production potentials; a comprehensive bibliography is included. See Dietz et al., (1970, 1973), Glockhoff (1973) and d'Argenio (1975) for a discussion of geotectonic evolution and subsidence of the Bahama Platform. See Olson (1974) for structural history and oil and gas potential of the offshore Florida-Bahama area. See Gebelein (1974) and Walper (1974) for a summary of the origin and history of the Bahama area.

For general bibliographic references the interested student is referred to Foster (1956), Bathurst (1971) and Milliman (1974); for an annotated bibliography of papers on corals and coral reefs see Milliman (1965); for a natural history bibliography of the Bahamas see Gillis et al., (1975).

Basic references for students of organic limestones are Johnson (1951) and Lowenstan (1963). Basic carbonate background information can be gotten from five carbonate symposium volumes by Le Blanc and Breeding (1957), Ham (1962), Pray and Murray (1965), Friedman (1969a) and La Porte (1974). See also Bass (1963), Taft and Harbaugh (1964), and Bricker (1971).

For a review of sea level changes in the Florida-Bahama area the reader is referred to the following papers and their included bibliographies: Milliman and Emery (1968), Scholl et al., (1969), Alexander (1974), Fairbridge (1974), Neumann and Moore (1975) and Missimer (1976).

Papers dealing with hurricanes and their effect on the south Florida region include: Pray (1965), Ball et al., (1967), Perkins and Enos (1968), Jindrich (1972), and Gentry (1974). For a summary of the effect of various types of storms on sediments of south Florida see Warzeski (1976b) and page 219 of this Guidebook. For hurricane effects along the coast of Great Abaco, Bahamas see Raphael (1975). A summary of climatological and hydrological data for south Florida is given by Thomas (1974). For a summary of climatic changes in south Florida see Brooks (1974, p. 309-311).

Data from the three carbonate field trip Guidebooks issued for the 1964 G.S.A. annual meeting in Miami are mostly covered in the text of this present Guidebook. Persons wishing to read these original editions should see Ginsburg (1964), Hoffmeister et al., (1964) and Purdey and Imbrie (1964). For guidebooks used on pertinent field trips during the 1974 G.S.A. annual meeting in Miami see Gebelein (1974), Kirtley (1974), Rodis et al., (1974) and Multer (1975).

The outstanding "classic" paper which set the stage for much future work in Holocene carbonates of the Florida Keys is by Ginsburg (1956). For summaries of various aspects of Florida-Bahama Holocene carbonates the reader is referred to two excellent texts by Bathurst (1971) and Milliman (1974). A general review of Caribbean coral reefs—including the Florida-Bahama region—is given by Milliman (1973) who also compares these reefs with those of other areas. See also Glynn (1973) for western Atlantic coral reef ecology. Cuffey (1974) describes the role of bryozoans in living (and fossil) reefs. See Ginsburg et al., (1971) for illustrated outlines covering the major groups of calcareous algae, their classification, reproduction, ecology and geologic occurrence. Woelkerling (1976) provides a well-illustrated guide and key to south Florida benthic marine algae including pertinent glossary and literature references. The latter two publications are a part of the "Sedimenta Series" of the Comparative Sedimentology Laboratory, Fisher Island Station, Miami Beach, Fla. 33139. Updated publication lists are available upon request.

Extensive use has been made of the Florida Keys-Bahamas area for interpretive models in dealing with ancient carbonate rocks. Examples of just a few of such comparisons are Beales (1958), Ginsburg (1960), Fischer (1964), Hoffmeister, Stockman and Multer (1967), La Porte (1967), Roehl (1967), Bishop (1968), Gutstadt (1968) and Shinn (1968b). See also article by Kraft in the appendix of this Guidebook. Interpreters of ancient carbonate rocks who use Holocene sediments as models should note certain possible limitations cited by Chave (1967). Hoffmeister (1974) reviews the evolution of the Pleistocene bedrock beneath the Florida Keys and Everglades with many illustrations. Gebelein (1974) discusses interpretation of ancient carbonates in terms of what may be seen in the field today in the Holocene of the Bahaman platform. Wilson (1974) recognizes three types of carbonate-platform margins in ancient rocks and relates them to some specific modern examples and general environmental conditions observable in the Holocene. See Abbott (1975) for a comparison of Florida coral banks with Silurian limestones.

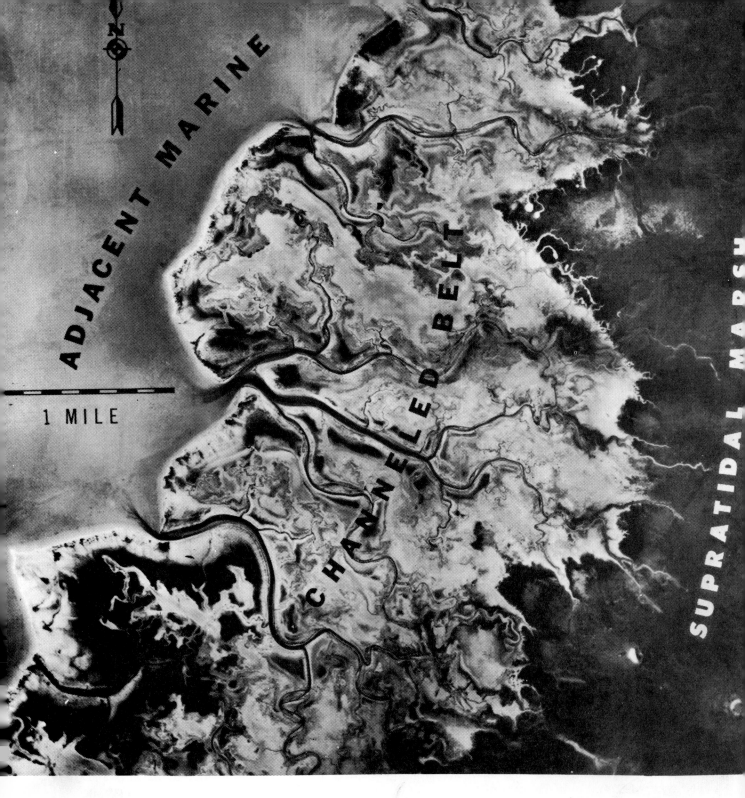

Figure 2.1. Aerial view of Loggerhead Point area showing subdivisions. Within channeled belt are (1) beach ridges, (2) levees, (3) channels, (4) intertidal flats and (5) ponds. Note large channel in lower left which is abandoned, filled and outlined by mangrove growth. From Shinn *et al.*, (1969).

INTERTIDAL AND SUPRATIDAL AREAS

The more obvious "action zone" of the Florida Keys-Bahama region lies between the tides and on the supratidal flats. Portions of the intertidal and supratidal environments may feature abundant plant and animal life, tidal flushing action, frequent diurnal changes in pH, temperature, salinity and intense burrowing, fragmentation, reworking and cut-and-fill action.

For the sake of completeness and to emphasize the importance of these specific environments, each will be discussed separately although some may represent specific parts of larger areas discussed elsewhere in this Guidebook. Specific environments discussed in this chapter include:

1. intertidal organic reef environments
2. rocky shore environments
3. carbonate sand beach environments
4. beachrock environments
5. tidal and supratidal mud flat environments.

Specific parts of these environments have characteristics which aid in the identification of their counterparts in ancient carbonate rocks. Roehl (1967) describes both modern and ancient representatives of these environments and tabulates frequency of certain fabrics (see figure 2.2).

Many examples of intertidal and supratidal action exist in the area coved by this Guidebook. Specific localities have been selected and are described below as typical occurrences that have been studied by geologists and are often readily accessible.

INTERTIDAL ORGANIC REEF ENVIRONMENTS

WORM REEFS (Bear Cut, Miami): The tube-building, marine polychaete *Phragmatopoma lapidosa* Kinberg produces thick, wave-resistant lit-toral reefs and encrustations along both shores of Bear Cut and along the eastern shore of Key Biscayne (see figure 2.3). The most readily available single large accumulation of this "worm reef" is found on the northeastern corner of Virginia Key where a 35-centimeter thick tabular (5 × 18 meters) reef is found exposed at low tide, resting on peat (fig. 2.4). These reefs can also be found as elongate bands parallel with the shoreline often cut by numerous surge channels or as clusters of low, rounded mounds (fig. 2.5). Multer and Milliman (1967) describe them in the Miami area at depths as great as 5 to 10 meters with most of the concentration, however, confined to depths of several meters.

These worm reefs are important in that they tend to stabilize the shoreline and sort local beach sands by construction of their tubes. Preferential selection of medium-size sand grains and flat, platey particles by the tube-building polychaete provides localization of these grains in the worm reef. In addition, these worm tubes contain considerably more silt-sized (62 microns) grains than do the adjacent beaches (see figure 2.6). The grains are extracted from the turbulent near-shore waters by small marine worms which agglutinate the grains, forming individual tubes often featuring an imbricate structure (see figure 2.7).

Kirtley and Tanner (1968) describe the distribution of sabellariid worms and reefs along the Florida coast (see figure 2.8) and note "inasmuch as the animals must depend upon turbulence for tube-building materials, food, and removal of metabolic wastes, the size and shape of the reefs is dependent upon prevailing local wave energy levels and water mass circulation properties." They also note (p. 76) the replacement of the protein cement by calcium carbonate, converting it into a type of beach rock which builds belts 100 meters or more in width and

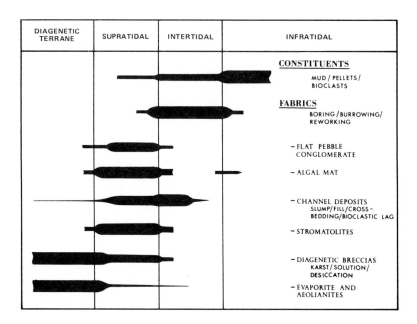

Figure 2.2. Estimate of relative frequency of occurrence of potential carbonate fabrics in epeiric-tidal flat and subaerial terraces. From Roehl (1967).

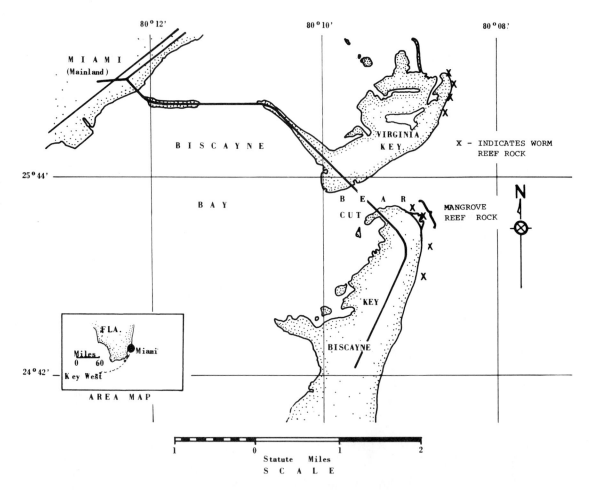

Figure 2.3. Index map of the Virginia Key and Key Biscayne (Miami) area showing location of Bear Cut, worm reef and mangrove reef rock. Rickenbacker Causeway connects these islands. From Hoffmeister and Multer (1965).

Figure 2.4. *Phagmatopoma lapidosa* reef at low tide resting on peat, northeast corner of Virginia Key (Miami) Florida. Broken slabs are from undermining and impact action during recent hurricane Inez. From Multer and Milliman (1967).

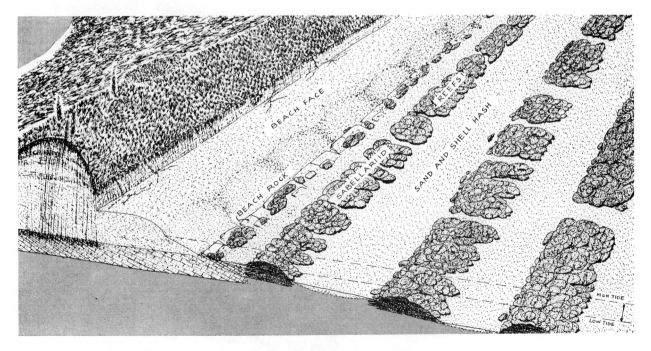

Figure 2.5. Schematic drawing showing shape and distribution of Sabellariid reefs. From Kirtley and Tanner (1965).

Size (in mm)	Sample 1			Sample 2			Sample 3		
	Beach sand	Worm tube sand	Conc. factor	Beach sand	Worm tube sand	Conc. factor	Beach sand	Worm tube sand	Conc. factor
4.0	–	–	–	–	–	–	–	–	–
2.0	–	0.2	–	tr	tr	1	0.6	–	0
1.0	1.1	0.1	0.1	0.6	0.4	0.7	1.0	0.2	0.2
0.5	10.7	29.0	2.7	8.7	10.4	1.2	4.7	5.1	1.1
0.25	21.8	35.0	1.6	35.5	52.5	1.5	22.3	42.7	1.9
0.12	55.2	25.4	0.5	46.3	28.5	0.6	61.5	44.0	0.7
0.06	11.1	8.4	0.8	8.7	5.5	0.6	9.6	5.2	0.5
< 0.06	0.1	1.8	18.0	0.3	2.6	8.4	0.2	2.8	12.6

Total sediment distribution (per cent)

Size statistics of total sample

Mean size (mm)	0.22	0.31		0.24	0.27		0.21	0.24	
Sorting (phi units)	0.80	0.92		0.76	0.78		0.68	0.71	
Skewness	−0.31	0.21		−0.13	0.18		−0.16	0.09	

Size (in mm)	Carbonate distribution (per cent)								
4.0	–	–	–	–	–	–	–	–	–
2.0	–	100	–	100	100	1.0	100	–	0
1.0	98.6	84	0.9	100	100	1.0	95.7	100	1.0
0.5	98.7	97.2	1.0	98.2	99.3	1.0	79.6	89.1	1.1
0.25	84.9	94.5	1.1	97.4	99.0	1.0	58.3	78.7	1.3
0.12	60.6	69.2	1.1	80.7	89.6	1.1	51.6	78.6	1.5
0.06	49.5	63.1	1.3	53.3	77.1	1.4	45.8	85.2	1.9
< 0.06	60.8	94.3	1.5	42.0	99.0	1.5	46.1	98.7	2.1

Figure 2.6. Size distribution of sediments from Key Biscayne, Florida. From Multer and Milliman (1967).

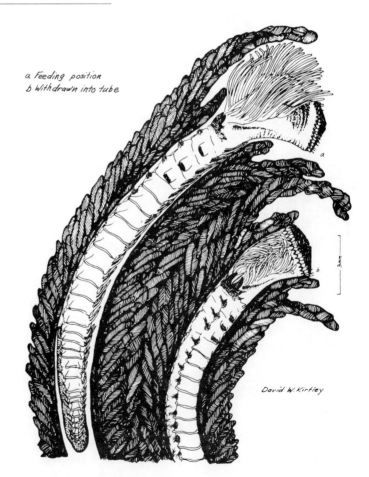

a Feeding position
b Withdrawn into tube

3 mm

David W. Kirtley

Figure 2.7. Semi-schematic diagram of *Phragmatopoma lapidosa* Kinberg. From Kirtley, (1966).

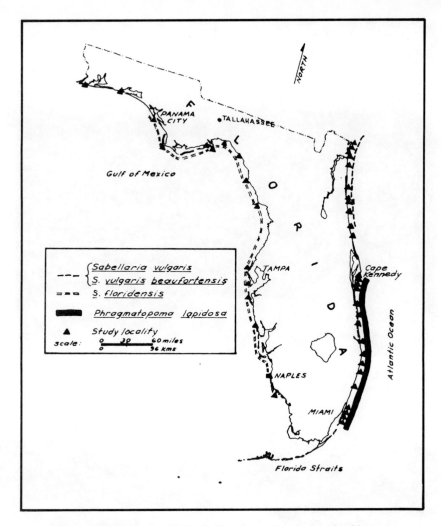

Figure 2.8. Distribution of sabellariid worms and reefs along Florida coast. From Kirtley (1966).

which may serve as the base for perched barrier islands. These same authors stress the importance of worm reefs for providing effective protection against beach erosion.

Ancient carbonate-rich sands, displaying elongate tabular zones of anomalous sorting, or even traces of tubular structure, may be due in some cases to the selective action of ancient littoral animals such as found in these sabellariid reefs. The fact that these organisms can be found today between the latitude 72° north and 53° south indicates their widespread stabilizing potential along both modern and ancient coastlines.

The interested student should refer to the following literature for other descriptions of worm reefs in various parts of the world: Galaine and Houlbert (1916), Herdman (1920), Richter (1920, 1921, 1927, 1928), Hartman (1944-47), Scholl (1958), Dollfus (1960) and Kirtley (1968).

MANGROVE REEFS (Bear Cut, Miami): A unique occurrence of fossilized mangroves can be found along the south side of Bear Cut on the northeast shore of the north sand beach of Key Biscayne from the parking lot which adjoins Rickenbacker Causeway. This wave-resistant reef rock has been described by Hoffmeister and Multer (1965) as occurring for a distance of about 360 meters along the shore and extending seaward in horizontal sheets for some 100 meters (fig. 2.3). The above writers note that:

The fact that woody material of the roots of some mangroves is capable of being converted into a rock substance opens up the possibility that other marine and brackish water plants of similar organic structure might do likewise. For example, rock structures which are of doubtful origin and which have tentatively been assigned to worm tubes and burrows of various animals may actually have originated in some type of marine vegetation.

Figure 2.9. Close-up of *Phragmatopoma lapidosa* tubes showing rasplike exterior rim. From Multer and Milliman (1967).

Figure 2.10. Intertidal mound of *Phragmatopoma Iapidosa*, Bear Cut, Miami, Florida.

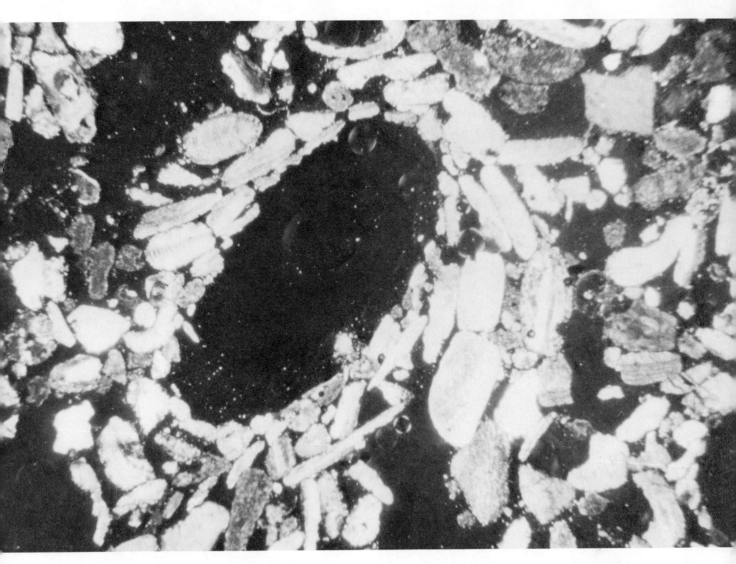

Figure 2.11. Photomicrographs showing cross section view of inclined worm tube. Note imbricate structure. Magnification approximately 15x, plain light.

The structural similarities between the roots of a living black mangrove (*Avicennia nitida*) (fig. 2.13) and the shapes and arrangement of the rods (figs. 2.14, 2.15) of the reef are very striking. The fossilized rods consist of a hard rim composed of subangular, fine-to-very-fine quartz grains with a limited amount of calcite grains, unidentified dark brown organic material, mollusk and algal fragments in a cryptocrystalline calcite bond. Insoluble residues show an average of 55 percent quartz in addition to organic scum. Thin section study indicates that small voids within the hard rim are lined with crystals of sparry calcite. Within the hard rim of the rods is a "paste" composed of calcium carbonate with a trace of fine-grain quartz and organic material. Radioactive dating indicates the age of the fossilized rods of this reef rock to be between 1,000 and 2,000 years.

The following is a description of the sequence of events believed by Hoffmeister and Multer (1965) to have led up to the formation of this fossil mangrove reef rock.

It is believed that the exposed reef rock as described in this paper is a remnant of a swamp of black mangroves which at one time extended seaward beyond the present edge of the reef and landward over at least the northern part of what is now Key Biscayne.

The substrate in which the roots were embedded was a calcareous-quartzitic sand of the intertidal zone. The trees were subsequently destroyed, and the woody material of the buried roots subjected to slow decomposition. This action released CO_2 which combined with available water, forming H_2CO_3. This in turn dissolved calcite and produced calcium carbonate-bearing solutions which percolated through the pore spaces of the sand.

Reprecipitation of the $CaCo_3$ in the sand immediately adjacent to the rotting root cemented quartz grains

Figure 2.12. General view of mangrove reef at low tide, Bear Cut, Key Biscayne. From Hoffmeister and Multer (1965).

Figure 2.13. Living black mangrove (*Avicennia nitida*) showing lateral horizontal roots and vertical pneumatophores arising from them, Bear Cut, Key Biscayne. From Hoffmeister and Multer (1965).

Figure 2.14. Edge of mangrove reef showing upper platform and exposed rods after sand has been removed by wave erosion, Bear Cut, Key Biscayne. From Hoffmeister and Multer (1965).

Figure 2.15. Looking down on surface of mangrove reef platform showing horizontal rods (former lateral roots) and vertical rods (former pneumatophores) arising from them, Bear Cut, Key Biscayne. From Hoffmeister and Multer (1965).

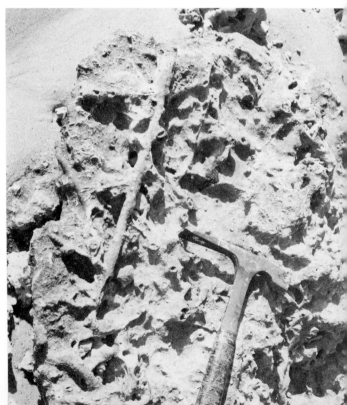

Figure 2.16. Flat-topped mangrove reef rock (M) exposed at low tide with encrusting layer of worm reef rock (W). Bear Cut, Miami.

together, forming a hard cylindrical rim around the root. This slowly grew outward as the action continued and resulted in a coating considerably thicker than the original periderm. At the same time continued decay of the organic material, surrounded by the hard but still porous rim, provided an environment for calcification within the woody structure and for replacement of the tissue itself by $CaCO_3$.

It is difficult to determine the exact cause of precipitation of the $CaCO_3$. It is believed, however, that microbial action was a significant factor in the process. Sisler (1962, p. 68, 69) discussed the various hypotheses for bacterial precipitation of $CaCO_3$. He endorsed (p. 69) the views of Bavendamm (1932) that no single species of bacteria could be solely responsible for $CaCO_3$ precipitation but ". . . that bacteria may be considered as biochemical agents which influence the carbon dioxide and calcium equilibria in a variety of ways, depending on the environment."

He also inclined (p. 68) toward Bavendamm's (1932) conclusion

. . . that if sulphate-reducing bacteria should play a key role in the precipitation of $CaCO_3$ their main scene of action would be the mangrove swamps.

Oppenheimer (1965, personal communication) inspected the fossil mangrove reef of Key Biscayne and found that the H_2O saturated sediment is anaerobic a few millimeters below the surface and that H_2S and black hydrotoilite are present. It is possible that microbial activity during the decomposition of the roots altered the chemical condition of the area surrounding the roots, producing the necessary change in carbonate equilibrium through pH transition which resulted in the precipitation of the $CaCO_3$. Such pH changes have been shown to occur between aerobic and anaerobic environments (Oppenheimer and Kornicker, 1958).

The importance of mangroves in the formation of new land by their baffle effect cannot be overemphasized. Vaughan (1910a) estimates that from one-third to one-half of the total area of the Florida Keys are occupied by them and that they are among the most important of constructional geologic agents. The thick tangle of roots which catch and hold sediment and both floating and storm debris can be readily observed by growths of the black mangrove (*Avicennia nitida*) (fig. 2.13), the white mangrove (*Laguncularia racemosa*) and the red mangrove

(*Rhizophora mangle*) exposed along the south shore Bear Cut. The interested reader is referred to Bowman (1917) and Davis (1940) for authoritative papers on mangroves. Persons interested in the flora and fauna of the littoral zone of Bear Cut should be aware of the publication by Bayer (1964) which also provides a checklist of representative plants and animals occurring at Bear Cut with pertinent references. See White (1970, p. 59-68) for a discussion of relic mangrove islands in Florida and Wanless (1974) for mangrove sedimentation in geologic perspective.

VERMETID REEFS (Ten Thousand Islands, Southwest Florida): Although outside the geographic area covered by this particular Guidebook, another type of intertidal reef rock has been described in the southwestern Florida area. Shier (1969) notes vermetid reefs of considerable extent in the Ten Thousand Island area. These intertidal reefs are made up of fused masses of the sessile gastropod *Vermetus (Thylaeodus) nigricans* Dall in a matrix of calcareous silt. They are underlain by slightly peaty basal lagoonal sand. Shier compares these reefs with the ancient rudist reefs from the Edwards limestone of Texas.

ROCKY SHORE ENVIRONMENTS

SPANISH HARBOR KEYS (South of Bahia Honda Key, Florida Keys): Along U.S. Route 1, after passing over Bahia Honda bridge (see "N" on index map 1) there is a chain of Keys called the Spanish Harbor Keys whose eastward shore is mostly rockbound and typical of such coasts in the Florida Keys (fig. 2.17).

A study of the causes and effects of intertidal erosion which may produce characteristic features such as smooth and irregular bedrock surfaces and nips is an important feature of intertidal environments and is included in this Guidebook. The measurement, classification, dating and correlation of marine terraces, however, is not the subject of this present Guidebook and the reader should refer to Newell (1963) and other voluminous literature on this subject for such information.

Tropical limestone shores are distinctive in many ways from their siliceous counterparts. Two of the principle reasons for their distinctiveness are (1) the more soluble nature of limestone, and (2) the intensive, destructive, and constructive role that organisms play along carbonate coasts. Revelle and Emery (1957) summarize the chemical action along limestone coasts. Some solution features such as they describe can be observed (see figure 2.18), especially

above the high tide mark in the Florida Keys. Other workers, describing chiefly physical and chemical breakdown of carbonate coastlines, are Fairbridge (1952), Russell (1962) and Emery (1946). Neuman (1966) reviews the literature describing the rock-boring and browsing organisms along limestone shorelines and indicates that wave cutting is ruled out in favor of "bioerosion" as responsible for 4- to 5-meter intertidal notches along the Bermuda limestone coast. He further notes (p. 92) that along such coasts the boring sponge *Cliona lampa* is "capable of removing as much as 6 to 7 kg of material from 1 m^2 of carbonate substrate in 100 days."

A very significant portion of today's erosion on the rocky coast of the Florida Keys is being caused by the direct action of organisms. Careful observations of most intertidal rocks will provide evidence for this statement (figs. 2.19 and 2.20). Ginsburg (1953b) considers physiochemical solution of bedrock of only local importance and suggests that the activities of boring and burrowing organisms are a major factor in intertidal bedrock erosion of the Florida Keys. He lists (p. 61) some of the responsible boring and burrowing organisms in the intertidal zone today as follows:

I. Forms which penetrate the rock:
    Phylum Porifera
        Family Clionidae
    Phylum Annelida, Class Gephyrea, Order Sipunculoidea
        Physcosoma antillarum
        Physcosoma varians
        Aspidosiphon speciosus
    Phylum Arthropoda, Class Crustacea, Order Cirripedia
        Lithotrya dorsalis

Several specimens of a stalked boring barnacle, *Lithotrya dorsalis,* have been collected, and at one locality, Indian Key, they were abundant. The sipunculoids are abundant in that portion of the intertidal zone where the rocks are usually covered with an algal mat. The bores are a few millimeters in diameter and several centimeters in length. Boring sponges are very common in this zone below the algal mat.

II. Forms which have exposed, generally shallow burrows:
    Phylum Mollusca, Class Pelecypoda
        Arca barbata
        Mytilus (Brachidontes) exustus
    Phylum Mollusca, Class Amphineura
        Acanthopleura granulata
    Phylum Echinodermata, Class Echinoidea
        Echinometra lucunter

Arca is found almost buried in the rock, and Otter (1937, p. 328-9) considers it an active borer. The hoof shell, *Hipponix*, is a ques-

Figure 2.17. Typical carbonate rock/bound coast showing color zones (Y) yellow intertidal, (B) black, covered by spring tides and (W) white, wet only during storms. Spanish Harbor Keys, Florida, at about midtide.

Figure 2.18. (Left) Marine solution basins with rills and undercut tunnels, etched in beach rock 1.5 m above low-tide level, Western Australia. From Revelle & Emery (1957, plate 226). Compare with marine solution basin with rills (right) found 1.5 m above low-tide level at Key Largo, Florida.

Figure 2.19. Holes in carbonate, littoral zone rocks are commonly formed by boring marine worms and mollusks. Worms (left of the hammer) in photograph crawled out from freshly broken rock surface.

Figure 2.20. Intense organic action riddles the intertidal zone in the Florida Keys.

tionable borer; Fisher (1950, p. 132) places it with *Patella* as excavating shallow depressions in the rock with its foot. Chitons occupy depressions in the intertidal zone. The facies of the local species are calcareous, as was the case with those collected during the Barrier Reef Expedition (Otter 1937, p. 332). Rasping with the radula is the mode of excavation. *Echinometra lucunter* has not been found out of water at normal low tide. It occupies roughly spherical cavities and can only be removed by breaking the rock. Otter (1937, p. 333) has reviewed the literature on the mode of excavation; it is thought to be a combination of the pick-like action of the teeth and abrasion by rotation of the spines. *Brachidontes* is tremely abundant locally and often covers rock surfaces completely in the intertidal zone (see figure 8). It is a byssus-attached form which is generally attached with its long dimension perpendicular to the rock surface. Commonly the umbo projects into the rock several millimeters. The manner in which these pits are produced is not known.

III.   Grazing forms:
      Phylum Mollusca, Class Gastropoda
          Hipponix antiquata
          Siphonaria spp.
          Littorina spp.
          Narita spp.
      Phylum Mollusca, Class Amphineura
          Acanthopleura granulata
The several grazing forms of the intertidal zone remove loose rock particles as they rasp the surfaces for food.

Constructional work along the littoral zone by organisms includes the building of worm reefs, mangrove reefs, and vermetid reefs, all of which tend to stabilize and often prograde the shorelines. Chemically precipitated, protective sheets of hard, microcrystalline, calcium carbonate "rind" may be found coating rocks of the upper littoral and supratidal zone (see Multer and Hoffmeister, 1968, p. 188-189).

A basic reference paper for intertidal study is Stephenson and Stephenson (1950) who break the intertidal environments into three color zones related to the tide levels (see figure 2.17). These are in descending order (1) a white, smooth-domed, bedrock surface wet only during storms, (2) a black zone covered by spring tides and showing some effects of intertidal erosion and burrowing, and (3) an intertidal yellow zone of intense undercutting and dissection with common, pinnaclelike remnants of bedrock. Two general platform levels can be observed in the Keys—the lower surface is usually found toward the base of the yellow zone and the upper platform surface is .3 to .6 meters above it in the black and white zone. At some places the white can be subdivided into a lower (gray) zone also.

Students interested in rock-boring organisms found in the intertidal and back-reef areas of south Florida should read Robertson (1963) who describes and illustrates both boring animals and plants of the area. It is interesting to note that Robertson believes that the effect of boring plants may be equal to, or greater than, that of the animal borers. A partial list of conclusions by Robertson is given below:

1. A study was made of the marine rock-boring fauna at Key Biscayne, Margot Fish Shoal, and West Summerland Key in southeast Florida.
2. The rock-boring fauna, as a whole, is typical of the West Indian faunal province and is comparable to that of other tropical regions.
3. The boring sponge *Cliona truitti* is recorded for the first time from southeast Florida. It is extremely common in the intertidal zone.
4. *Cliona caribboea* apparently is the dominant boring sponge in the back-reef environment.
5. The sipunculid genus *Lithacrosiphon,* represented by a single species, four species of the genus *Aspidosiphon,* and *Phascolosoma dentigerum* are recorded for the first time from Florida waters.
6. *Phascolosoma dentigerum* is by far the most common sipunculid in the intertidal zone.
7. Species of the genus *Aspidosiphon* are the most prominent sipunculids on the reef patches.
8. The boring barnacle *Lithotrya dorsalis,* although not encountered in large colonies, is a common member of the rock-boring community in southeast Florida.
9. The contribution of boring animals to the erosion of intertidal rock is slight above the midtide level, but appears to be considerable in a zone extending for several centimeters above mean low water.
10. Boring lamellibranchs are not prominent in the intertidal zone, but they are the most conspicuous boring animals in the back-reef environment.
11. *Lithophaga nigra, Lithophaga antillarum,* and the gastrochaenids are the most prominent of the boring mollusks at Margot Fish Shoal.

SOUTH BIMINI (Bahamas): An excellent example of the effectiveness of browsing and burrowing organisms in the littoral zone can be observed along the exposed, southwest rockbound shore of Nixson's Harbor area in the very southwest corner of South Bimini. In the littoral zone below the hotel and along the shore of the little uninhabited island offshore there can be seen all stages of the breakdown in the bedrock predominately by organic action. Figures 2.21 and 2.22 illustrate four stages in this sequence according to Multer (unpublished) with the initial stage brought about by depressions in the supratidal zone which collect rain and salt spray water. Browsing gastropods tend to widen these supratidal pools as they eat the algae which inhabit areas on the bottom and especially around the perimeter of individual pools (fig. 2.23). Within the middle and upper littoral zone the activity of echinoids and of chitons is particularly prominent

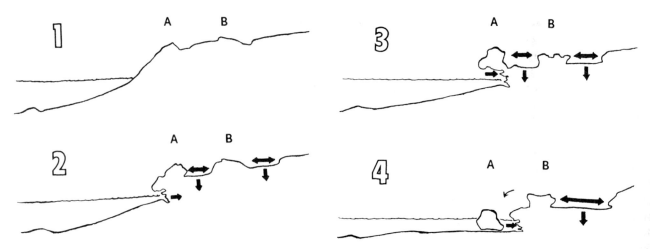

Figure 2.21. Schematic drawing illustrating various stages in development of supratidal pools and erosion of intertidal zone as shown at Nixson's Harbor.

Figure 2.22. Lateral widening and vertical deepening of supratidal pools combined with the mechanical undercutting and organic boring from the intertidal zone results in ultimate destruction of this Nixson's Harbor rocky shore at South Bimini.

with the microscopic teeth of individual chitons scraping off a layer of bedrock as they eat away at algae. Such action produces chiton-shaped holes to a depth of several centimeters in the bedrock (fig. 2.24). Continuation of these processes along with others, such as echinoid burrowing and scratching, boring algae and sponges, accounts for the final stage seen in figure 2.21 in which the undermined, exposed bedrock meets the lowered, flat-floored spray pools. The final process involves the storm action of waves breaking the remaining weakened rock apart. Large blocks of rock broken off in a similar manner can be found along the front of the exposed shoreline. This type of "modified littoral sinkhole" formation illustrates the dominant role of organisms in the intertidal area. **Note:** The formation of the flat-bottomed spray pools may be influenced by diurnal temperature and pH changes (Oppenheimer and Master, 1965, and Schmaltz and Swanson, 1969). Effective physical abrasive action can also be observed in the smooth-sided potholes found in the low tide area and filled with boulders rounded during high tide and storm periods.

Newell *et al.* (1959, p. 206-209) describe the white, gray, black and yellow zones (in descending order) of the Bimini rocky shore and notes the close, general similarity of shore zones of Florida with those on the opposite side of the Gulf Stream along the Bahama platform edge. He notes several exceptions to the above statement, however, including an encrusting coralline algae (*Porolithon pachydermum*) lip immediately below the lower yellow intertidal zone which is present along the Bahama coast but absent on the Florida coast.

The rocky shore profile given by Newell *et al.,* (1959, p. 207) is reproduced in figure 2.25, and a list of main indicator organisms given by him (p. 209) is noted below. Species marked with an asterisk are especially abundant.

White zone
  Gastropods
    Tectarius muricatus
    Littorina ziczac
    Nodolittorina tuberculata
    Echininus nodulosus

Gray Zone
  Gastropods
    Tectarius muricatus
    Littorina ziczac
    Nodolittorina tuberculata
    Echininus nodulosus

Black zone
  Algae
    Perforating green and bluegreen (e.g., *Entophysalis deusta*)
  Gastropods
    Not in tide pools

    Littorina ziczac
    Littorina mespillum
    Littorina meleagris
    Tectarius muricatus
    Nodilittorina tuberculata
    Echininus nodulosus
  Tide pools
    Puperita pupa
    Nitidella ocellata
    Planaxis lineatus
    Cerithium biminiense
    Purpura patula
  Lowest part of zone
    Nerita versicolor
    Nerita peloronta
    Nerita tessellata
  Crab
    Grapsus grapsus

Upper yellow zone
  Alga
    Entpromorpha sp.
  Chitons
    Chiton squamosus
    Acanthopleura granulata
  Gastropods
    Littorina ziczac
    Littorina mespillum
    Littorina meleagris
    Nodilittorina tuberculata
    Echininus nodulosus
    Livona pica
    Purpura patula
    Nerita versicolor
    Nerita peloronta
    Nerita tessellata
    Spiroglyphus "irregularis"
    Siphonaria alternata
    Acmaea jamaicensis
    Diodora listeri
  Sponge
    Clione sp.
  Barnacles
    Chthamalus angustitergum
    Tetraclita squamosa stalactifera (lower part of zone)

Lower yellow zone
  Alga (green)
    Valonia ocellata
  Chitons
    Chiton viridis
    Chiton squamosus
    Acanthopleura granulata
  Gastropods
    Nerita tessellata
    Spiroglyphus "irregularis"
    Petaloconchus aff. nigricans
    Thais rustica
    Livona pica
    Purpura patula
    Acmaea jamaicensis
    Fissurella barbadensis
  Pelecypods
    Isognomon bicolor
    Brachidontes exustus

Figure 2.23. Browsing gastropods widen flat-bottomed supratidal pool.

Figure 2.24. Chiltons are one of the destructive elements along the rocky shore of Nixson's Harbor, South Bimini.

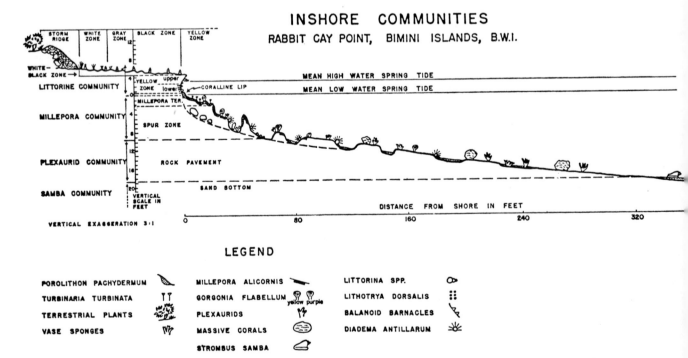

Figure 2.25. Shore profile and communities according to Newell, *et. al.* (1959, p. 207) for Rabbit Cay Point, south of Nixson's Harbor, South Bimini, showing vertical distribution of some of the most distinctive indicator organisms. Dashed line indicates base of grooves in spur zone.

Sipunculid
  Phaseolosoma antillarum
Barnacles
  **Tetraclita squamosa
  stalactifera*
  *Lithotrya dorsalis
Coralline lip zone
  Alga (encrusting coralline)
    Porolithon pachydermum
  Chiton
    Chiton viridis
  Gastropods
    *Spiroglyphus ''irregularis''
    Thais deltoidea
    Livona pica
  Echinoid
    Echinometra lacunter

Voss and Voss (1960) give an ecological survey of the invertebrates of the Bimini area and include a resume of previous workers and description of topography and plant life. They compare organisms found at Bimini with those found in similar habitats along the Florida Keys. One of the reasons for difference in the two areas cited is the fact that the Florida Keys are flanked by a 6 kilometer wide flat, shallow platform which reduces wave energy and affects intertidal fauna. See also Seibold (1962) and Bathurst (1967) for comments on morphology and biological observations at Bimini. See also Zahl (1952). See page 33 and 100 in this Guidebook for further description of the Bimini area.

PIGEON KEY (Along ''7-Mile Bridge,'' 5 Kilometers West of Marathon, Florida Keys): Pigeon Key, site of the University of Miami Field Station, represents an excellent example of a shallow water rocky shore environment covered in places with sand or loose rock rubble. Extensive shallow water rock-floored platforms provide intertidal and subtidal flats which lend themselves to the production of a series of inshore animal and plant communities readily available for study. Fluctuations in temperature, salinity, tides and turbidity produce fluctuating stresses on organisms surrounding Pigeon Key. Corals found in the Pigeon Key area have a relatively low-diversity and high tolerance for adverse physical conditions. These ''carbonate bank corals'' are described by Abbott (1975) as being analogous in many ways to the Silurian Wenlock limestone in England.

Professor James A. Zischke (1973; personal communication) and his students from St. Olaf College have compiled an unpublished ecological Guide to Pigeon Key, major portions of which are reproduced with permission on the following pages. See also pages 338 to 347 in the appendix for some common invertebrates of Pigeon Key.

AN ECOLOGICAL GUIDE TO THE SHALLOW-WATER MARINE COMMUNITIES OF PIGEON KEY, FLORIDA
by
James A. Zischke

Surface water conditions. Monthly surface water temperature and salinity records for Key West are given on page 318. The shallow waters around Pigeon Key tend to undergo rapid cooling and warming with fluctuations in the temperature of the overlying air. In the past extremely low water temperatures have caused extensive mortalities among fishes and some benthic invertebrates (e.g., during January 1970, low water temperature 10.5°C). Surface water salinity also undergoes marked fluctuation, varying from 28.0 to 38.8%; with periods of low salinity and high salinity occurring about two to three months after periods of high and low precipitation, respectively. The rather considerable fluctuations in salinity reflect the fact that Florida Bay is an estuary, receiving freshwater out-flow from the Everglades.

The Tide Tables for 1972 give the mean tide range for Pigeon Key as 0.7 feet with a spring range of 0.9 feet. Although the tidal range in the Keys is very low, extensive intertidal areas may exist because of the generally gentle slope of the shore and the extremely shallow water on the reef flat.

Because the water surrounding the Key is very shallow (generally under four feet for 50 yards off shore) the shore is normally not exposed to extreme wave action. However, moderate wave action may occur on any shore facing high winds.

Periods of high winds are accompanied by the churning up of fine sediment (marl) from the bottom causing a great increase in turbidity and the water may become milky white in some areas. After periods of high winds sedimentation rates as high as two to three inches in 24 hours have been reported. The high sedimentation rate on the reef flat effectively excludes from the area many benthic species which are unable to cope with heavy silting (e.g., many reef coral).

Inshore Communities of Pigeon Key

Stephenson and Stephenson (1950) made a detailed study of the life in the intertidal region of the Florida Keys. They considered the rocky platform or ledge that commonly fringes the shore, to be the most typical intertidal substrate in the area. Even casual observation, however, reveals that almost any type of substrate, including also: loose rock, sand or mud (marl), may occur and may even be more common than the rocky platform on some Keys. The zonation of life in the intertidal region is not as distinct in the Keys as it is on more northern shores where the tidal range is greater.

Voss and Voss (1955) described the subtidal shallow-water habitats at Soldier Key in Biscayne Bay. They recognized four different ''Zones,'' *Echinometra, Porites*—coralline algae, *Thalassia* and *Alcyonaria* (at increasing distance from shore), on the basis of the dominant organisms.

This guide employs the terminology introduced by Stephenson and Stephenson (1950) and Voss and Voss (1955).

Figure 2.26 is a map of the inshore habitats at Pigeon Key and figure 2.27 is a bottom profile (including the most common macroinvertebrates in each habitat) drawn from a representative transect line across a section of shore with a rocky platform.

Intertidal Zone—Physical Structure

The physical nature of the intertidal area on Pigeon Key is extremely variable. A well-developed rocky platform is present on the east and northwest shores but the platform on the southwest shore is heavily eroded with a large number of loose rocks lying on the surface (fig. 2.26). The *upper platform* is a gently sloping rocky ledge extending seaward from the edge of the terrestrial vegetation. The surface is pitted with numerous depressions, the larger ones forming tide pools up to 10 feet in diameter. At the seaward edge of the upper platform is the *platform face* which is marked by irregular pinnacles of algal covered rock projecting one to two feet above the *lower platform* below (fig. 2.27). The intertidal substrate on the south shore consists of loose rock on limestone pavement with a sandy surface dominating on the west and sand mixed with rock on the north shore (fig. 2.26).

Intertidal Zone—Plants and Animals

The species of intertidal organisms present on the shore vary depending on the type of substrate. Stephenson and Stephenson (1950) described three color zones (white, grey and black) on the upper platform. The zones are the result of differential weathering and algal growth on the limestone surface. On Pigeon Key all three of these color zones are present, although not prominent, wherever a well-developed platform occurs.

Upper platform. At the landward edge of the upper platform (white zone) lives a group of maritime plants the commonest of which is *Batis*, a fleshy herbaceous species which may extend out for some distance onto the face of the upper platform. A few animal species are normally limited in distribution to this maritime community, e.g., the small snails *Truncatella scalaris* and *Pedipes mirabilis* and the grapsoid crabs *Sesarma cinereum* and *Cyclograpsus integer* (all of which are commonly found under rocks or debris at low tide). The beach wrack, an assortment of plant material (marine grasses, *Sargassum* and other algae) deposited by the waves at the high tide line, harbors thousands of crustaceans (mainly amphipods and isopods) and insects.

Some animal species occupy only a narrow vertical range in the intertidal area while others may be distributed over two or more of the color zones. Among those species generally occurring high on the upper platform are: the snails *Littorina ziczac* (fig. 2.27), *L. lineata, L. lineolata, Tectarius muricatus* (fig. 2.27), *Nodilittorina tuberculata* and *Echininus nodulosus.*

Species that have a more general distribution on the upper platform include: the snails *Neritina virginea, Nerita tessellata, N. peloronta* and *N. versicolor* (fig. 2.27) and the isopod crustacean *Ligia baudiniana* (fig. 2.27). Other species may be distributed across the upper platform but are generally restricted to tidal pools formed by depressions in the platform surface. Some of these species are: the snails *Batillaria minima* (the most abundant intertidal mollusc, may be over 2000 per square meter, figure 2.27, *Nitidella ocellata* and *Planaxis lineatus,* the grapsoid crab

Pachygrapsus transversus (fig. 2.27) and the hermit crab *Clibanarius tricolor* (usually in *Batillaria* shells, figure 2.27).

Platform face. The yellow zone (structurally the platform face) represents the midtidal region and is inhabited by a large number of plant and animal species. Dense mats of algae covering the rock surface give this zone its characteristic color. The commonest of these are: the filamentous red algae *Bostrychia tenella* and *Ceramium* (several species). Also prominent here are the small nodules of the green alga *Valonia ocellata.*

Among the animals in the yellow zone are a large number of sessile species including: the large ribbed barnacle *Tetraclita squamosa* (fig. 2.27), the mussel *Brachidontes exustus* (fig. 2.27), the tree oyster *Isognomon bicolor* and the vermetid snail *Spiroglyphus annulatus,* the intertwined shells of which form masses on the rock surface. These sessile species serve as food for several predatory gastropods: *Thais haemastoma, T. deltoidea* (fig. 2.27), *T. rustica* and *Leucozonia nassa.*

Other forms commonly found in the yellow zone are: the large chiton *Acanthopleura granulata* (fig. 2.27), the false limpet *Siphonaria alternata* (restricted to the southwest shore, figure 2.26) and burrowed in the rock the sipunculid worms *Aspidosiphon brocki, Phascolosoma antilarum, P. perlucens* and *Themiste alutaceum.*

Intertidal areas without rocky platform. In addition to rocky platform the other types of intertidal substrate on Pigeon Key are: loose rock over limestone pavement, loose rock over sand and sand (fig. 2.26). Where the upper surfaces of the loose rocks are exposed at low tide, as on the south shore, they may support species similar to those found on the rocky platform, for example, the grazing gastropods such as *Nerita, Littorina* and *Tectarius.* Loose rock also provides many microhabitats, in crevices and on the under surfaces, which tend to increase the diversity of species found in such an area. The protection of these microhabitats allows certain species (e.g., sponges, anemones, ploychaetes, crabs and brittle stars), which are normally subtidal, to penetrate into the intertidal region just as tide pools allow a similar extension of range onto the rocky platform.

On intertidal sand such rocky shore species as the grazing snails are absent and the diversity of species is relatively low. However, certain sand burrowing forms, e.g., the sipunculid worm *Phascolion cryptus,* may be found in addition to such generally present intertidal species as the crab *Pachygrapsus transversus.*

Alternating submergence and emergence by the tide make the intertidal area an extremely demanding environment. The regularly changing water level produces an environmental gradient on the shore in relation to exposure to water or air. Relatively few species are adapted to live under conditions that alternate from aerial to aquatic as occurs on the rocky platform. They must be highly tolerant to desiccation and changes in temperature. The diversity of species in the high stress region of the intertidal is, therefore, comparatively low.

Primary productivity (plant production) on the upper platform is largely in the form of microscopic algae that encrust the rock surfaces and a few macroscopic algae that are able to live in the tidal pools. On the platform face and lower platform the dominant primary producers are the few species of filamentous red algae that form matlike growths on the rock surface. Much of the nutrient material

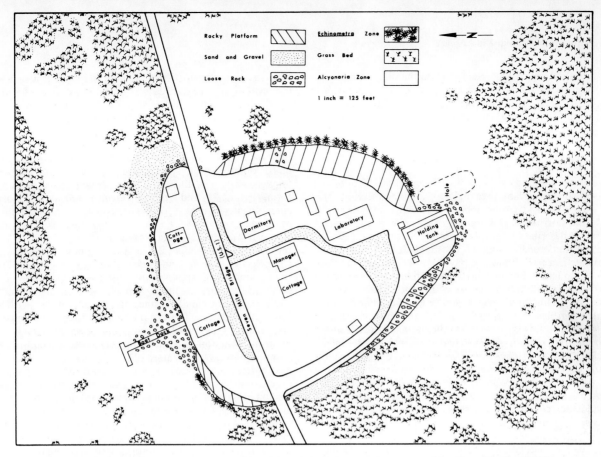

Figure 2.26. Map of Pigeon Key showing the location of the inshore marine communities. From Zischke (1973).

Figure 2.27. Bottom profile showing inshore communities on east side of Pigeon Key (not drawn to scale). From Zischke (1973).

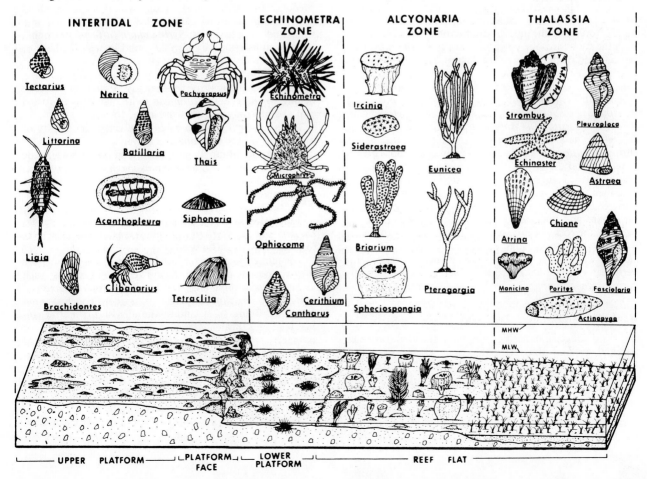

for intertidal animals is obtained from the plankton when the tide covers the area or it is imported in the form of leaves from terrestrial plants (e.g., mangrove) and marine grasses or as floating algae (e.g., *Sargassum*) carried in by waves. To become available for use by many animals this material must first be partially broken down by bacteria, fungi and protozoa to form detritus.

Lower platform. The region of the shore that is uncovered only during extremely low water of spring tides is an ecologically variable area on Pigeon Key. On sections of the shore where a well-developed rocky platform is present (east and northwest) this region corresponds to the lower platform described by Stephenson and Stephenson (1950) and the *Echinometra* zone of Voss and Voss (1955). The substrate is a hard limestone pavement which may be pitted or undercut by the boring sea urchin *Echinometra lucunter* (fig. 2.27). The surface of the rock is covered by mats of filamentous red algae (e.g., *Ceramium* and *Bostrychia*). In addition to the urchins the commonest invertebrates in this zone are: the colonial anemones *Zoanthus* and *Palythoa* (in matlike clusters about one-half inch thick), the chiton *Ischnochiton floridanus,* the herbivorous snails *Cerithium eburneum* and *C. litteratum* (fig. 2.27), the predator snails *Pisania tincta* (fig. 2.27) and *Morula nodulosa* (which replace *Thais* in the subtidal region), the crested oyster *Ostrea equestris* and the spider crabs *Microphrys bicornutus* (fig. 2.27) and *Macrocoeloma trisponsum.*

Loose rock habitat. On sections of the shore where the intertidal rocky platform is absent (e.g., south and north shores) the substrate in the region just seaward of the main intertidal zone is composed of loose rock or loose rock mixed with sand on the limestone pavement. This is the area of the highest species diversity of any of the shallow-water habitats around the Key. Some of the species commonly found in the numerous microhabitats associated with the rocks are: the polychaete annelids *Amphitrite ornatus* (with long unbranched tentacles), *Hermodice carunculata* (large, red-brown, with numerous setae) and *Eunice longicirrata,* the chitons *Calloplax janeirensis, Ischnochiton floridanus* and *I. limaciformis,* the gastropods *Pisania tincta, Cerithium eburneum, Columbella mercatoria, Morula nodulosa* and *Lucapina sowerbyi,* the small bivalve *Arcopsis adamsi,* the mantis shrimp *Pseudosquilla ciliata,* the snapping shrimp *Alpheus formosus* and the brittle stars *Ophiocoma echinata* (large, red to black, figure 2.27) and *Ophiothrix oestedii* (blue-green with glassy spines). Covering the rock surfaces are a number of species of algae as well as encrusting sponges, bryozoans and tunicates. Found on the pavement surface between the rocks in this zone are several species of algae, a variety of sponges, some stony corals, e.g., *Siderastrea siderea,* (fig. 2.27) and a large number of motile invertebrates including annelids, spider crabs and mantis shrimp (see accompanying list for species names).

Reef Flat

Two distinct communities dominate the shallow-water subtidal area of the reef flat. On limestone pavement substrate is the soft coral (alcyonarian) and sponge community and on soft sediment substrates are thick beds of sea grasses.

Alcyonaria-sponge community. Immediately seaward from the lower platform or loose rock area on Pigeon Key is generally a region of relatively strong current and low sedimentation. The substrate is limestone pavement with usually less than one inch of sediment and the most prominent animals are the large loggerhead sponges, *Spheciospongia vesparia* (fig. 2.27), and a variety of soft corals. This is the region designated the *Alcyonaria* zone by Voss and Voss (1955). In addition to *Spheciospongia* many other sponge species are found here including: *Spongia graminea,* three species of *Ircinia* (often in cup-form, figure 2.27) and *Halicondria melanadocia.* The most common alcyonarians around Pigeon Key are *Briareum asbestinum, Eunicea knighti, Pterogorgia anceps,* all shown in figure 2.27, and *Pseudopterogorgia acerosa.* A few species of stony coral are also common in this area, e.g., *Porites furcata* and *Siderastrea siderea* (fig. 2.27). Living among the sponges and soft corals are numerous small invertebrates including: annelid worms, brittle stars and a variety of small crustaceans (see accompanying list for species names). The canal systems within the sponges also house an overwhelming number of small organisms the most common of which are polychaete annelids and snapping shrimp (see Pearse, 1932).

The *Alcyonaria* zone is an area of relatively low primary productivity. Only a few scattered species of attached algae are found here. The dominant animal species in this zone are either carnivores (alcyonarians) or suspension feeders (sponges). Most of their nutrients are obtained from the plankton or imported from the grass beds (primarily as detritus). Little is known concerning the interrelationships of the animals in this habitat.

Sea grass community. Beds of marine grass cover vast areas of the reef flat. Although turtle grass, *Thalassia testudinum* (with flat leaves about 12 mm wide), is the predominant species in the region, two other species, shoal grass, *Diplantheria wrightii* (with flat leaves about 2 mm wide) and manatee grass, *Syringodium filiforme* (with thin cylindrical leaves), may also be found in separate beds or intermixed with *Thalassia.*

In addition to the grass species, the beds also contain a large number of algal species the most prominent of which belong to the genera *Halimeda* (with flat lime-encrusted plates) and *Penicillus* (a brush-shaped form).

The grass beds contain a highly diverse community of organisms of both epifauna and infauna.

Among the infaunal species in the grass beds are: tude dwelling annelids such as *Onuphis magna* (often with bits of shell cemented in the membranous tube), many burrowing bivalves including the pen shell, *Atrina rigida* (fig. 2.27), the cross-barred venus, *Chione cancellata* (fig. 2.27) and several species of arc shells belonging to the genera *Arca, Anadara* and *Barbatia* and numerous interstitial species (e.g., the tiny snail *Caecum*).

The numerous filter feeding bivalves of the infauna are preyed upon by certain surface dwelling gastropods, such as *Fasciolaria tulipa* (fig. 2.27) and *Pleuroploca gigantea* (which also feeds on the herbivorous queen conch, *Strombus gigas,* figure 2.27), and by the starfish *Echinaster sentus* (fig. 2.27) and *Oreaster reticulata* (large, red or orange). Besides the starfish the other prominent echinoderms of the grass beds are the primarily herbivorous sea urchins *Lytechinus variegatus* (usually covered with leaves and shell fragments), *Eucidaris tribuloides* (with heavy, blunt primary spines) and *Tripneustes ventricosus* (with short white spines) and the deposit feeding sea cucumbers *Actinopyga agassizi* (large, brown, with five anal teeth, figure 2.27) and *Holothuria floridana* (large, usually grey or black). Two snails lacking external shells which are commonly found in the *Thalassia* zone are: the small green and white nudibranch *Tridachia crispata* (seasonal) and the large spotted "sea hare" *Aplysia dactylomela* (a tectibranch gastropod). The crustaceans of the grass beds include: large forms such as the blue crab *Callinectes ornatus* and many smaller species such as the shrimp *Palaemon tenuicornis, Latreutus fucorum, Penaeus brasiliensis* and *Tozeuma carolinensis* and the mantis shrimp *Pseudosquilla ciliata.*

On the surface of the *Thalassia* beds are also a large assortment of sessile forms such as the sponges *Tethya* (small, spherical, usually green or orange with warty surface), *Cinachyra* (small, spherical, usually yellow or orange, interior with splinterlike spicules) and *Ircinia* and the stony corals *Manicina areolata* (fig. 2.27) and *Porites furcata* (fig. 2.27). Numerous species of small invertebrates are found among the sponges, corals, and bases of the grass blades. Some of these animals are: turbellarian flatworms, errant polychaetes, crabs and brittle stars (see accompanying list for species names).

Grazing on the epiphytic algae on the grass blades are several species of snails including: *Tegula fasciata, Modulus modulus* and two species of *Astraea* (fig. 2.27).

Thalassia beds cover extensive areas of the bottom in the shallow marine waters bordering southern Florida and along the coast of the Gulf of Mexico. The leaves of the grass act as a sediment trap and the rhizomes, which may extend nearly two feet into the substrate, serve to stabilize the bottom deposits. The beds contribute enormously to the primary productivity in this area. Because few animals (e.g., the sea urchin, *Lytechinus variegatus,* and parrot fish) feed directly on the grass its main contribution to the marine ecosystem is in the form of detritus produced when the leaves break off and decompose. In addition, the surfaces of the leaves support large populations of epiphytes (including many species of filamentous algae and diatoms), foraminifera, tube dwelling annelids and molluscs. Grass beds also provide nursery grounds for the early stages in the life histories of many species of shrimp, the spiny lobster, *Panulirus,* and numerous fish species: sea trout, jacks, pompano, barracudas, mullet, cowfish and snappers.

Variations in Communities

In the report of Voss and Voss (1955) on the ecology of Soldier Key the *Thalassia* zone was found just landward of the *Alcyonaria* zone, the reverse of the general situation at Pigeon Key. However, there are regions, especially on the west side of the island, where the grass beds are found close to shore. Thus the position of the grass beds with respect to the *Alcyonaria* zone is governed by substrate type and current velocity. Voss and Voss (1955) also described another zone, the *Porites*-coralline zone, between the *Echinometra* and *Thalassia* zones. No such zone is present in the shallow waters close to Pigeon Key. However, such an assemblage, composed mainly of the stony coral *Porites furcata* and the coralline algae *Janea, Lithothamnium* and *Goniolithon,* is present in the extremely shallow waters of nearby Molasses Key. A narrow band dominated by *Porites* also fringes many of the grass beds. Variations, therefore, exist in the types of inshore bottom communities encountered on different Keys. The major factors con-

tributing to these variations are: (1) nature of the intertidal substrate, i.e., rocky platform, platform plus loose rock, sand, sand plus loose rock or mud, (2) degree of mangrove growth on the shore, e.g., a rocky platform may, with time, become overgrown with trees transforming the shore into a typical mangrove community, and (3) human activity, e.g., clearing vegetation from shoreline, dredging and filling.

Other Shallow-Water Communities

Mangrove Community

Only a few scattered mangroves are present on the east shore of Pigeon Key. However, because they are so common in the Florida Keys the composition and importance of the mangrove community will be briefly considered.

The three species of mangrove common in south Florida display a characteristic zonation on the shore. *Rhizophora mangle* (red mangrove), with its characteristic prop roots, is found on the outermost edge of mangrove forests with *Avicennia nitida* (black mangrove), recognized by its pneumatophores sticking upright from the substrate, and *Laguncularia racemose* (white mangrove) located in the more landward portions of the stand.

On the stems and roots of the mangroves are many attached forms such as filamentous algae (e.g., *Bostrychia* and *Ceramium*), hydroid colonies and tunicates. More prominent on the aerial portions of the plants are: several gastropods, *Tectarius muricatus, Littorina angulifera* and *Melampus coffeus;* tree oysters, *Isognomon alatus;* and crabs, *Goniopsis cruentata.* In the shallow water at the base of the mangroves live a large assortment of animals including flatworms, annelids, jellyfish *(cassiopeia xamachana),* fiddler crabs *(Uca sp.),* horseshoe crabs *(Limulus polyphemus),* and small forage fish.

Stands of mangrove cover approximately 700 square miles of the coast in south Florida (including the Keys). These forests and their associated animal communities are extremely important to the biological energetics of the region (Odum, 1971). It has been estimated that mangroves loose about three tons of leaves per acre per year. These leaves undergo physical breakdown through the action of a variety of forces including fragmentation by crabs and amphipods. Bacteria, fungi and protozoa accumulate on the surfaces of the fragments and provide the nutrient material utilized by the animals feeding on this detritus. A fragment may be passed through the digestive tracts of a series of the bottom dwelling organisms providing energy with each passage. The many small detritus feeding animals (worms, crabs, shrimp and forage fish) later provide food for young fishes such as gray snapper, lady-fish, red drum, sheepshead, snook and tarpon.

CHECKLIST OF
MACROFLORA, INVERTEBRATES AND
FISHES OF PIGEON KEY

The following is a list of the species of algae, grasses and invertebrates that have been collected by James A. Zischke, Department of Biology, St. Olaf College, Northfield, Minn. from the inshore waters surrounding Pigeon Key. The list does not include all species present, nor does it include species that are exclusively found in mangrove and coral reef communities or forms generally restricted to deeper water.

The numbers to the right indicate the habitats from which the species have been collected according to the following code:

Code of Habitat Numbers (See also figures 2.26 and 2.27.)
1—Intertidal Region
2—*Echinometra* Zone
3—Loose Rock
4—*Alcyonaria*—Sponge Zone
5—Grass Beds
6—Free Floating

Algae

Greens (Chlorophyceae)

Acetabularia crenulata	2,4
Avrainvillea nigricans	2,4,5
Caulerpa racemosa	1,2
Cladophoropsis macromeres	3,4
Cympolia barbata	3
Dasycladus vermicularis	2,3
Dictyosphaeria cavernosa	1,2
Entermorpha ligulata	1,3
Halimeda discoidea	3,4,5
Halimeda monile	3,4,5
Halimeda opuntia	3,5
Halimeda tuna	5
Penicillus capitatus	3,4,5
Penicillus dumetosus	3,4,5
Penicullus lamourouxii	4,5
Sargassum sp.	6
Udotea conglutinata	3,5
Valonia aegagropila	1
Valonia ocellata	1
Valonia ventricosa	2,4

Brown (Phaeophyceae)

Dictyota dichotoma	3,4
Lobophora variegata	2
Padina perindusiata	2
Padina sanctae-crucis	3
Sargassum sp.	6
Turbinata turbinata	6

Reds (Rhodophyceae)

Acanthophora spicifera	2
Amphiroa fragilissima	3,5
Bostrychia tenella	1,2
Centroceras clavulatum	2,3
Ceramium sp.	1,2
Champia parvula	4
Chondria dasyphylla	3,4
Chondria sp.	3,4
Eucheuma isiforme	2,4
Goniolithon sp.	4, 5
Gracilaria sp.	3,4
Heterosiphonia gibbesii	2,3,4
Laurencia poitei	3,4,5
Lithothamnium sp.	3,4
Wrangelia argus	2

Marine Grasses (Angiospermae)

Diplantheria (Halodule) wrightii	5
Syringodium filiforme	5
Thalassia testudinum	5

Porifera

Anthosigmella varians	3,4,5
Callyspongia vaginalis	3,4
Chondrilla nucula	2,3,4,5
Chondrosia collectrix	3,4,5
Cinachyra cavernosa	3,4
Cliona sp.	2,3,4,5
Dysidea etheria	3,4,5
Geodia gibberosa	3,4,5
Haliclona rubens	3,4
Haliclona viridis	3,4
Halicondria melanadocia	4,5
Ircinia campana	3,4,5
Ircinia fasciculata	3,4,5
Ircinia strobilina	3,4,5
Neopetrosia longleyi	3,4,5
Spheciospongia vesparia	3,4,5
Spongia graminea	4
Tedania ignis	3,4,5
Tethya sp.	3,4,5
Verongia longissima	3,4

Cnidaria

Hydrozoa

Lytocarpus philippinus	3
Millepora alcicornis	3,4
Physalia physalis	6
Sertularia inflata	3
Velella velella	6

Scyphozoa

Aurelia aurita	6
Cassiopeia xamachana	still water

Anthozoa

Octocorallia (soft coral)

Briareum asbestinum	4
Eunicea knighti	4
Eunicea palmeri	4
Eunicea succinea	4
Plexaurella dichotoma	4
Plexaurella fusifera	4
Pseudoplexaura flagellosa	4
Pseudoplexaura porasa	4
Pseudopterogorgia acerosa	4
Pseudopterogorgia americana	4
Pterogorgia anceps	4

Hexacorallia
Actiniarians (sea anemones)

Aiptasia pallida	3,4,5
Bartholomea annulata	2,3,4,5
Bunodosoma cavernata	2,3,4
Condylactis gigantea	2,3,4,5
Stoichactis helianthus	3,5

Madreporarians (stony coral)

Cladocora arbuscula	3
Favia fragum	3
Oculina diffusa	3
Manicina areolata	3,4,5
Porites asteroides	3
Porites furcata	4,5
Porites porites	5
Siderastrea radians	3,4
Siderastrea siderea	1,2,3,4

Zoanthideans

Palythoa mamillata	2,3
Zoanthus sociatus	2,3

Annelida

Ammatrypane fimbriata	5
Amphitrite ornatus	3,4
Armandia agilis	1,5
Cirratulus sp.	1
Eurythoe complanata	3,4,5
Hermodice carunculata	3,4,5
Hesione picta	2,3,4,5
Eunice longicirrata	3,4,5
Lumbrinereis maculata	1
Lysidice sp.	3,4,5
Nereis antillensis	3,4,5
Nereis dumerilii	3,4,5
Onuphis magna	5
Polynoe polytricha	3,4
Sthenelais sp.	5
Terebellides stroemi	5

Sipunculoidea

Aspidosiphon brocki 1
Golfingia sp. 1
Phascolion caupo 5
Phascolion cryptus 1,5
Phascolosoma antillarum 1
Phascolosoma perlucens (=P. dentiferum) 1
Phascolosoma varians 1,2,3
Themiste alutaceum 1

Mollusca

Amphineura
Acanthopleura granulata 1
Calloplax janeirensis 3
Ischnochiton limaciformis 3
Ischnochiton floridanus 3
Ischnochiton papillosus 1,3

Gastropoda
Acmaea pustulata 1,3,4
Aplysis dactylomela 3,4,5
Aspella paupercula 3
Astrea tecta americana 3,5
Astrae phoebia 5
Batillaria minima 1
Bulla occidentalis 4
Busycon contrarium 3
Caecum floridanum 5
Caecum nebulosum 5
Caecum pulchellum 5
Cerithium eburneum 3,5
Cerithium literatum 5
Cerithium muscarum 5
Cerithium variable 1,5
Cittarium pica 1
Columbella mercatoria 2,3,5
Conus mus 3
Crepidula fornicata 3,5
Cymatium nicobaricum 3
Cymatium femorale 5
Cypraea zebra 3,4
Cypraea cervus 3,4
Diadora listeri 3
Dolabrifera dolabrifera 3
Echininus nodulosus 1
Eupleura caudata 3
Fasciolaria tulipa 2,3,4,5
Favartia cellulosa 1
Leucozonia nassa 1,2,3
Leucozonia ocellata 1,2
Littorina angulifera 1
Littorina mespillum 1
Littorina lineata 1
Littorina lineolata 1
Littorina ziczac 1
Lucapina sowerbyi 3
Lucapina suffusa 3
Modulus modulus 5
Morula nodulosa 2,3
Nerita fulgurans 1

Nerita peloronta 1
Nerita tessellata 1
Nerita versicolor 1
Neritina virginea 1
Nitidella ocellata 1
Nodilittorina tuberculata 1
Pisania tincta 1,2,3
Planaxis lineatus 1
Pleuroploca gigantea 5
Pedipes mirabilis 1
Prunum apicinum 3,4,5
Purperita pupa 1
Rissoina bryerea 5
Siphonaria alternata 1
Spiroglyphus annulatus 1
Strombus gigas 3,5
Strombus pugilis 5
Strombus raninus 5
Tectarius muricatus 1
Tegula fasciata 3,5
Thais deltoidea 1
Thais haemastoma floridana 1
Thais rustica 1
Tridachia crispata 3,5
Truncatella scalaris 1
Vasum muricatum 5

Pelecypoda
Americardia media 5
Anadara notabilis 5
Antigona listeri 3,5
Arca imbricata 3
Arca zebra 3,4
Acropsis adamsi 1,3
Atrina rigida 5
Barbatia cancellaria 3,4
Brachidontes exustus 1
Chione cancellata 5
Chlamys sentis 4
Codakia orbicularis 5
Isognomon alatus 1
Isognomon bicolor 1
Isognomon radiatus 1
Laevicardium laevigatum 5
Lima scabra 3
Lithophaga antillarum 3
Lithophaga nigra 3
Lucina pensylvanica 5
Modiolus americanus 3,4
Ostrea equestris 3
Petricola lapicida 3
Pinctada imbricata 1
Pteria colymbus
Sanguinolaria sanguinolenta 5

Cephalopoda
Octopus vulgaris 3

Arthropoda

Cirripedia (Barnacles)
Balanus declivis 4
Balanus trigonus 3
Chthamalus stellatus 1
Lepas anserifera (on floating objects)

Lithotrya dorsalis 1
Tetraclita squamosa 1

Isopoda
Ligia baudiniana 1

Natantia (Shrimp)
Alpheus formosus 3,4
Latreutus fucorum 5
Leander tenuicornis 4
Palaemonetes intermedius 5
Penaeus brasiliensis 5
Penaeus duorarum 5
Periclimenes americanus 5
Synalpheus brevicarpus 4
Tozeuma carolinensis 5

Reptantia (Lobster etc)
Panulirus argus 3,4,5

Anomura (Hermit crabs)
Calcinus tibicen 3,4
Clibanarius antellensis 1,5
Clibanarius tricolor 1
Clibanarius vittatus 5
Dardanus venosus 3,4,5
Paguristes grayi 3,4,5
Paguristes puncticeps 3,4,5
Paguristes tortugae 2,3,4,5
Petrochirus diogenes 4,5

Brachyura (Crabs)
Calappa flammea 5
Callinectes ornatus 3
Callinectes sapidus 3
Cyclograpsus integer 1
Dromidia antillensis 3,4
Glyptoxanthus erosus 5
Libinia emarginata 4,5
Macrocoeloma trisponsum 2,3,4,5
Microphrys bicornutus 2,3,5
Mithrax hispidus 3,4
Mithrax spinosissimus 3,4,5
Pachygrapsus transversus 1
Panopeus herbsti 4
Pilumnus gemmatus 2,3,4,5
Pinnotheres maculatus 3,4
Pithos aculeata 3,4,5
Portunus spinimanus 3,5
Porcellana sayana 3,5
Sesarma cinereum 1
Stenorynchus seticornis 3

Stomatopoda (Mantis shrimp)
Gonodactylus oerstedii 3,4
Pseudosquilla ciliata 3,4,5

Echinodermata

Asteroidea (Starfish)
Echinaster sentus
Linckia guildingii
Oreaster retaculata

Echinoidea (Sea urchins)
Arabacia punctulata
Diadema antillarum
Echinometra lucunter
Eucidaris tribuloides
Lytechinus variegatus
Tripneustes ventricosus

Holothuroidea (Sea cucumbers)
Actinopyga agassizi
Astichopus multifidus
Fossothuria cubana
Holothuria floridana
Isostichopus badionotus
Leptosynapta sp.
Thyone briareus

Ophiuroidea (Brittlestars)
Amphiura palmeri
Astrophyton muricatum
Ophiactis savignyi
Ophicoma echinata
Ophioderma brevispinum
Ophioderma rubicundum
Ophiolepis paucispina   5
Ophiomyxa flaccida   3,5
Ophionereis reticulata   3
Ophionereis squamulosa   5
Ophiophragmus septus   5
Ophiostigma isacanthum   4,5
Ophiothrix oerstedii   3,4

Chordata

Tunicata
Polycarpa obteca   5

Fishes
Family Atherinidae:

Atherinomorus stripes
(Hardhead Silverside) 5

Family Blenniidae:
Blennius cristatus
(Molly Miller)   3
Entomacrodus nigricans
(Pearl Blenny)   5
Hypleurochilus springeri
(Orange Spotted
Blenny)   5

Family Clinidae:
Paraclinus fasciatus
(Banded Blenny)   5
Paraclinus marmoratus
(Marbled Blenny)   5
Starksia lepicoelia
(Blackcheek Blenny)   5

Family Cyprinodontidae:
Cyprinodon variegatus
(Sheepshead Minnow)   5

Family Gerreidae:
Ulaema lefroyi
(Mottled Mojarra)   4

Family Gobiidae:
Bathygobius curacao
(Notched Tongue
Goby)   1
Bathygobius soporator
(Frillfin Goby)   5

Family Labridae:
Halichoeres bivittatus
(Slippery Dick)   3

Family Lutjanidae:
Lutjanus apodus

(Schoolmaster)   3

Family Pomacentridae:
Abudefduf saxatilis
(Sargaent Major)   3
Eupomacentrus dorsopunicans
(Dusky Damselfish)   2,3

Family Pomadasyidae:
Haemulon aurolineatum
(Tomtate)   3
Haemulon flavoleatum
(French Grunt)   3

Family Scaridae:
Sparisoma rubripinne
(Yellowtail Parrotfish) 5
Sparisoma viride
(Stoplight Parrotfish) 5

Family Syngnathidae:
Corythoichthys brachycephalus
(Crested Pipefish)   5
Micrognathus crinigerus
(Fringed Pipefish)   5
Syngnathus elucens
(Shortfin Pipefish)   5
Syngnathus floridae
(Dusky Pipefish)   5

Family Gobiesocidae:
Acytrops beryllina
(Emerald Clingfish)   5

Family Clupeidae:
Harengula pensacolae
(Scaled Sardine)   4

Family Brotulidae:
Ogilbia sp.   5

CARBONATE SAND BEACH ENVIRONMENTS

BAHIA HONDA and LONG KEYS (Florida Keys): Found along U.S. Route 1, on the southwest and northwest ends of Bahia Honda Key (index map 1) and the southwest end of Long Key (index map 3).

Loose, carbonate beach sand is not a very common feature of the Florida Keys, but quantities of it may be found, particularly on the seaward-facing shores of many Keys such as those named above. Some common characteristics of these beaches are (1) poorly developed, inclined (seaward) bedding, (2) high aragonite/high magnesium calcite composition, and (3) general decrease (seaward) in size-grade composition. Well-developed ripple patterns are common (figs. 2.28 and 2.29). Interpretation of such ripples is given by Imbrie and Buchanan (1965). The sand on beaches of the Keys is often found as a thin veneer over a seaward sloping bedrock surface

(fig. 2.31). Benham *et al.,* (1970) describe a sand beach; Huffman *et al.,* (1970), an offshore bar along the northwest shore of Bahia Honda Key.

Longshore current drifting along the seaward side of the Keys is retarded by mangrove development and irregular shorelines as well as frequent tidal channels normal to the trend of the Keys. The large percentages of quartz found in the sand of the Miami area (see figures 2.30 and 2.6) is effectively stopped to the south by the embayment and tidal action of Biscayne Bay. Such southerly change from a calcareous quartz sand to a calcarenite is described and mapped as a bottom sediment facies change by Earley and Goodell (1968) at the south end of Biscayne Bay (Card Sound).

NORTH BIMINI: The exposed sand beach along the western shore of north Bimini commonly shows steep (18°) seaward dipping stratification and peculiar sorting due, probably, to (a) contamination from poorly indurated cliffs in back of the

Figure 2.28. Calcareous sand beach, southwest end of Long Key, Florida Keys.

Figure 2.29. Rippled sand at low tide off point of beach shown in figure 2.28.

Figure 2.30. Quartz-rich calcarenite sand beach overlying mangrove reef rock, Bear Cut, Key Biscayne, Miami.

Figure 2.31. Thin veneer of carbonate sand overlying seaward sloping bedrock floor at Sands Key, northern Florida Keys.

beach, and (b) frequent storm action. Submerged bars in 2-3 meter depths offshore show excellent sorting. The highly polished shells found in the beach sand are, with the exception of some open ocean forms, similar to shells of the Bimini lagoonal fauna. Incipient beachrock formation is found along this beach and is discussed under the appropriate heading below.

Imbrie and Buchanan (1965) present a comprehensive study of sand structures in the Bahamas. These authors describe the various types of ripples found along beaches and describe coastal beaches as characterized by a dominance of sheet deposits. The abstract of their paper is quoted below (see figure 2.32 for their classification of primary structures).

SEDIMENTARY STRUCTURES IN MODERN CARBONATE SANDS OF THE BAHAMAS
by
J. Imbrie and H. Buchanan

Sedimentary structures in carbonate sands of portions of the Great Bahama Bank are described and interpreted using data from air photographs, visual inspection of the bottom, echo-sounding traverses, and 50 box-cores impregnated with epoxy resin. Surface forms described include: avalanche ripples, with angle-of-repose lee slopes; accretion ripples, with lee slopes less than the angle of repose; flat, swashsurfaces; large-scale current lineations on the order of 1,000 feet long; bars; and irregular shoals. Strata are classed as avalanche deposits if they are cross-strata dipping at the angle of repose with coarser particles concentrated down their dip; accretion deposits if they are cross-strata less than the angle of repose without down-dip assortment; and sheet deposits if they are flat or gently dipping strata approximately parallel to underlying truncation surfaces. From hydrodynamic theory and field observations the conclusion is reached that avalanche deposits form where the current separates from the bed on the lee side of avalanche ripples and other migrating embankments; that accretion deposits form where a traction carpet is moving down the lee side of an accretion ripple or other embankment with a significant component of tangential fluid flow; and that sheet deposits are formed in the upper flow regime on beaches, and on level-bottom areas by rare, high velocity currents. Meta-ripples and cross-ripples are described. Biogenic structures, including biolaminites and bioturbites, dominate deposits in intertidal and sheltered subtidal areas of the shelf-lagoon where bottom traction is small compared to the rate of organic activity.

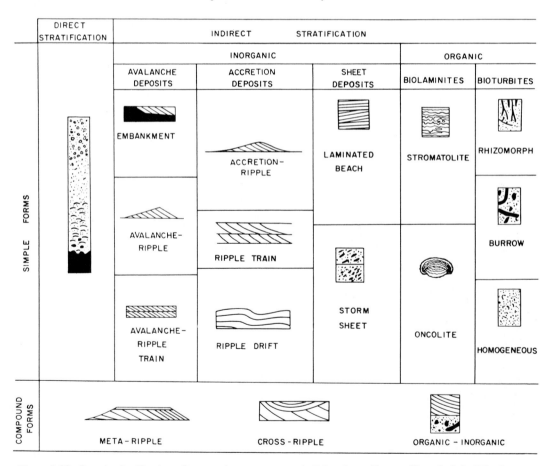

Figure 2.32. Genetic classification of some primary structures in Bahamian sediments. From Imbrie & Buchanan (1965).

BEACHROCK ENVIRONMENTS

NORTH BIMINI: In the littoral zone of the west shore of north Bimini, one can find areas showing various stages of recent cementation (fig. 2.33). T. P. Scoffin (personal communication) describes these areas as cemented beachrock composed of three components: (1) boulders of rock from the cliff bordering the beach; (2) conch shells, coral and Coca Cola bottle glass debris, all of which appear to have been at one time wedged between the boulders; and (3) fine sand. See figures 2.33, 2.34, and Scoffin (1970b).

DRY TORTUGAS: Beachrock occurs along the northwestern shore of Loggerhead Key, Dry Tortugas, Florida, as a series of seaward dipping (5-15°) beds parallel to the shore along the littoral sandy

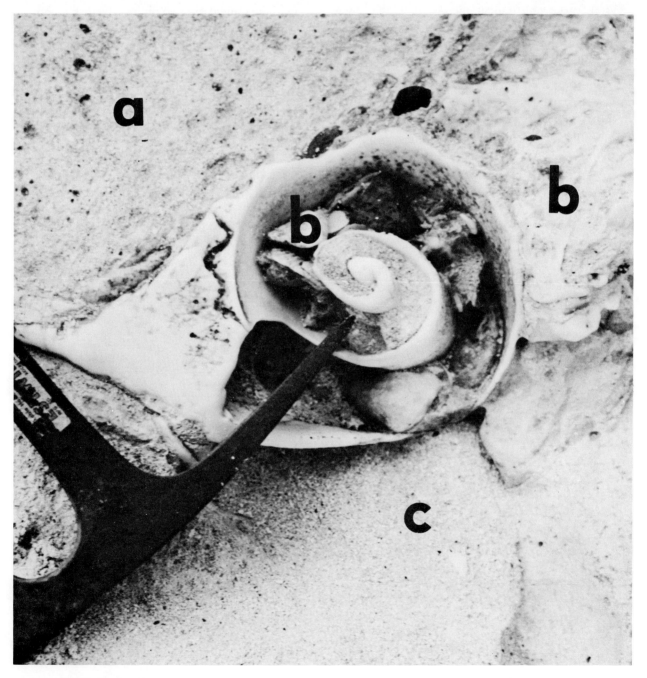

Figure 2.33. Cemented mass of cliff boulder (a), mollusk, coral and Coca Cola glass debris in fine sand matrix (b) and fine sand (c) found at low tide zone along the west shore of north Bimini.

beach (see figure 2.35). It has been discussed in some detail by Ginsburg (1953a). Multer (1971d) briefly described the beachrock cementation. A 6 m wide area of these sloping beds, displaying a pitted and jointed surface, is exposed at low tide. The rock consists mostly of mollusk, *Halimeda,* coral, and incrusting algal debris showing no recrystallization or solution effects. Certain portions of this sand are cemented together by various amounts of acicular aragonite with long axes. Individual aragonite crystals vary from 10 to 50 microns along their long axis perpendicular to the skeletal grain surfaces. The amount of cementation varies, yielding a friable to hard indurated beachrock. Usually the finger grained phases are best cemented. Cement occurs as coatings around individual grains and as fillings between groups of grains (see figure 2.36).

Origin of specific beachrock, an excellent indicator of the littoral zone for paleoenvironmental inter-pretation, is often controversial. At Loggerhead Key, present evidence indicates that cementation is due to alternate wet and dry saltwater spray conditions with skeletal grains providing nuclei for precipitation from a supersaturated calcium carbonate solution. In other words, evaporation of films of salt water on the exposed beachrock provides the mechanism for cementation during low tide intervals. The limited ground water conditions on Loggerhead Key and lack of evidence of grain solution for providing aragonite cement are two indirect factors supporting the above cited evaporation origin for the cement in this rock.

See Raphael (1975) for a description of beachrock exposed only along eroding coastlines of Great Abaco, Bahamas. For a general discussion of beachrock, mechanisms for their formation and bibliography related to this subject see Russell (1962) and Moore (1973).

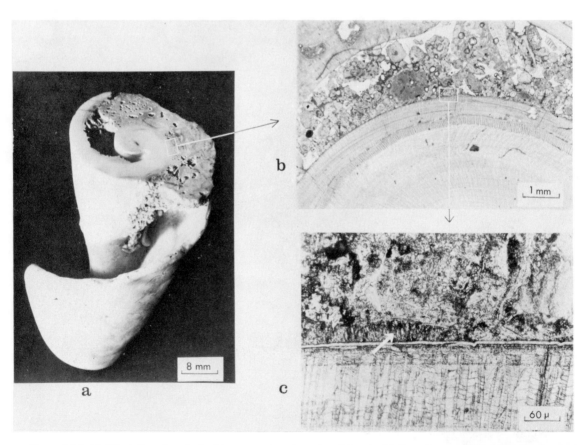

Figure 2.34. (a) Sliced *Strombus* shell with cemented sand coating, (b) peel of indicated area and (c) detail showing 5-50 micron aragonite crystals with long axes perpendicular to grain borders. Thickness of aragonite fringe varies. From Scoffin (1970b) who also notes lack of evidence for local solution and hence he believes aragonite is from precipitation during evaporation of seawater.

Figure 2.35. Northwest ocean-facing shore of Loggerhead Key, Dry Tortugas showing beachrock dipping seaward at 5° to 15°. Strip of sand beach occurs to the left.

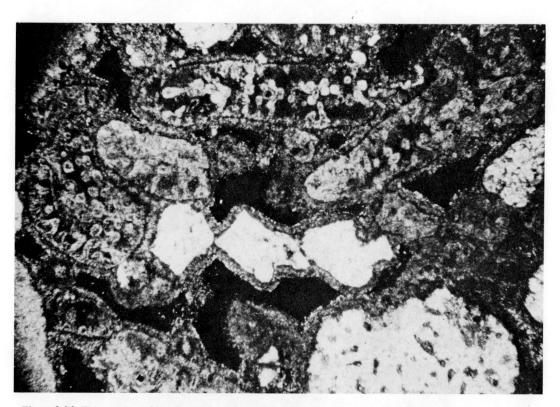

Figure 2.36. Thin section of beachrock from Loggerhead Key, Dry Tortugas, Florida. Aragonite cement occurs as coatings around individual grains and as fillings between grains. Individual aragonite crystals vary from 10 to 50 microns along their long axes perpendicular to the skeletal grain surfaces (X20, polarized light).

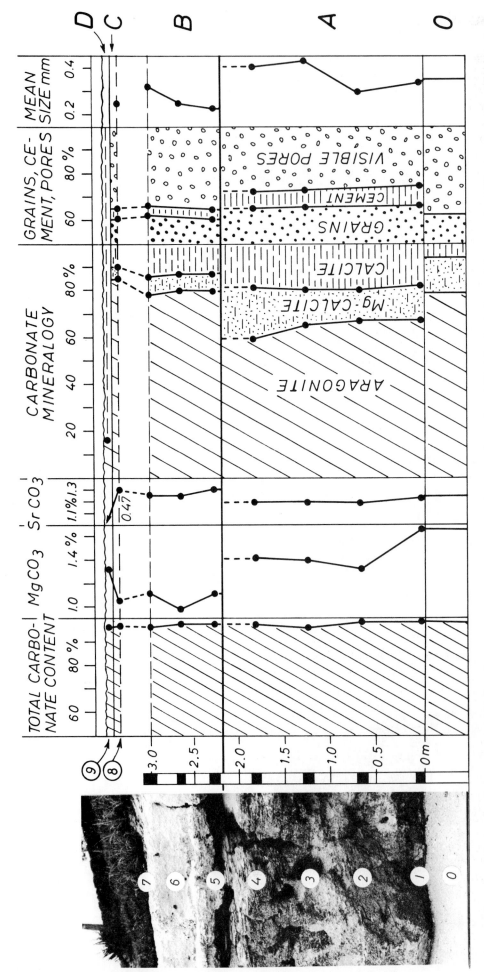

Figure 2.37. Analysis of bedrock samples by Dr. German Müller (1970, in press) from outcrop along west shore of north Bimini near the Lerner Laboratory site. Müller notes meniscus cement in the A (biopelsparite), B (biopelsparite) and C (oosparite) zones similar to the type found from meteoric cementation in the vadose zone. Data emphasizes break between A and B zone.

TIDAL AND SUPRATIDAL MUD FLAT ENVIRONMENTS

GENERAL STATEMENT: A variety of examples of tidal and supratidal mud flat environments can be examined in the Florida-Bahama region. Diagnostic organic and physical sedimentary structures associated with Holocene tidal and supratidal environments have been described by many investigators.

For a discussion of the formation of modern algal mats and biscuits and comparison with ancient analogs, see Ginsburg (1960 and 1967). Ginsburg *et al.,* (1954), using quartz grains as reference horizons on living algal mats, note that laminae are not necessarily annual but depend upon environmental controls. Neuman *et al.,* (1969) describe the composition, structure and erodability of subtidal mats in Abaco, Bahamas (see page 93 of this Guidebook).

Monty (1965), studying one type of blue-green algae (*Schizothrix calcicola*) along the eastern coast of Andros, describes some parameters dealing with growth rates, origin of laminations and depth range which could be of interest in paleoenvironmental studies. The interested reader is urged to read the complete paper, portions of which are noted below.

1. The lamination is noctidiurnal and caused by a growth cycle of the algal colony.
2. A doublet composed of an organic lamina overlain by a corpuscular lamina is formed during an interval including one day and one night.
   a. . . .
   b. The organic lamina is deposited during the first part of the day, while the overlying corpuscular lamina is formed at the end of the day and (or) during the night.
   c. Experiments conducted in the field also showed that the colonies grew very rapidly; the daily growth recorded by the lamination doublet was locally as much as 600 microns. . .
3. . . .
   a. Intertidal and infralittoral colonies produce similar laminations of similar origin.
   b. Accordingly, the lamination does not result from alternating periods of emersion and immersion, nor from a mere alternation of episodes of growth and episodes of sedimentation.
   c. The lamination originates from two processes, algal growth and sedimentation, which operate continuously or nearly so, but vary in intensity. . . .
   d. Thus, a stromatolitic lamination does not necessarily depend solely on a rhythmic variation in the physical environment (sedimentation), it may be formed by a biological cyclicity in the algal metabolism. . . .

Types of sedimentary structures produced by burrowers found in the supratidal and intertidal sediments of the Florida Keys-Bahama area are described by Shinn (1968a), who also suggests that such may be responsible for some of the fossil *Stromatactis* structures. The abstract of this paper and a map (fig. 2.38) showing the widespread distribution of burrowers are given below.

BURROWING IN RECENT LIME SEDIMENTS OF FLORIDA AND THE BAHAMAS
by
Eugene A. Shinn

Abstract

Five organisms, viz. four crustaceans (*Alpheus* Weber, 1795, *Callianassa* Leach, 1814, *Curdisoma* Latreille, 1828, and *Uca* Leach, 1814) and one coelenterate (*Phyllactis* Milne-Edwards & Haime, 1851), produce extensive and characteristic subsurface burrow complexes. *Alpheus, Callianassa,* and *Phyllactis* burrow in marine and intertidal sediments; *Cardisoma* and *Uca* burrow in supratidal sediments. Supratidal, intertidal, and subtidal crustacean burrowers produce extensive 3-foot deep open networks in muddy marine sediments which can remain open beneath overburdens exceeding 8 feet. Knowledge of depth of burrowing and relative rate of sea level rise suggests that many inactive burrows have remained open for 1,000 years or more. Burrows commonly become infilled, but compaction of either filled or open burrows has not been observed.

Striking resemblances of Recent burrow networks to ancient reef-associated *Stromatactis* Dupont, 1881 and "reef-tufa" structures suggest burrows as the origin for some of these much debated structures.

Burrowing structures can be useful in interpreting ancient rocks. This study shows that "nested-cone" structures in cross-bedded oolite of Pleistocene age were probably produced by an anemone like the Recent form *Phyllactis,* which makes similar structures in Recent cross-bedded marine oolites. Such structures can be used to differentiate between marine- and eolian-deposited sands. Recycling and mixing of sediment by burrowers can produce sediments having grain sizes coarser than the originally deposited sediments. Grain size analyses of extensively burrowed sediments may give little information regarding the energy level of the depositional environment.

Polyester plastic casts of burrow networks and other new techniques described in the text have aided greatly in this study.

Ancient analogs of Florida-Bahama tidal and supratidal mud flat environments have been described in the literature. Criteria for their recognition in ancient rocks is an important goal for the petroleum industry. A new approach to the interpretation of ancient tidal flats based on a Bahama

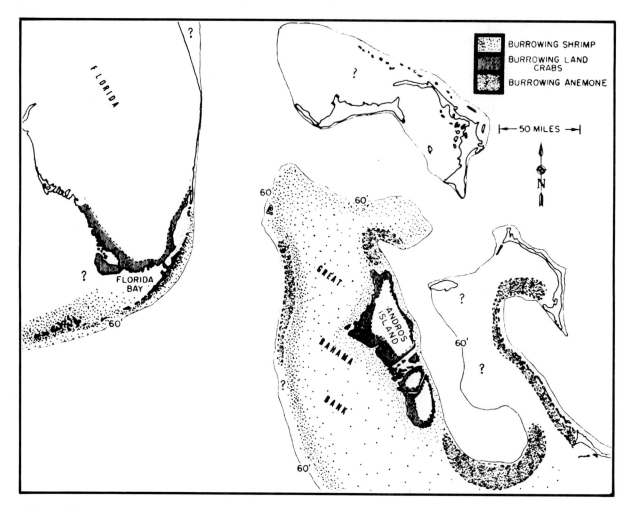

Figure 2.38. Generalized distribution of burrowers described in this report. Areas marked by question mark have not been sufficiently studied to be included. Burrowing shrimp are most abundant near the margins of banks and near shoreline. From Shinn (1968c).

tidal flat study is described and illustrated by Ginsburg *et al.,* (1970) below:

EXPOSURE INDEX AND SEDIMENTARY STRUCTURES OF A BAHAMA TIDAL FLAT

by

Robert N. Ginsburg, Owen P. Bricker,
Harold R. Wanless, Peter Garret

Sedimentary structures, both physical and organic, are the criteria for recognizing and interpreting ancient tidal flats deposits. These deposits, a major element of shelf and platform sequences from Pre-Cambrian to Tertiary, are up to 1,000 m thick and cover thousands of km²; some have major accumulations of petroleum and metallic ores, others contain the first evidence of life on earth.

Interpretations of sedimentary structures are based on their callibration in recent environments where individual structures are restricted to tidal zones; subtidal, intertidal, supratidal. Varying definitions of the three zones prevents accurate comparison of structures from different tidal flats; lack of precision in positioning structures within zones has produced apparent differences of interpretation. To provide the much-needed precise callibration of diagnostic structures, a new approach is suggested here: Measuring the amount of subaerial exposure of each structure in modern examples. This "exposure index" of a sedimentary structure is an unambiguous statement of its position in the range of tides, it makes possible direct comparison of structures from different tidal flats, and it opens up the possibility of highly refined interpretation of "tides" in ancient seas.

Three Creeks Study Area. Within the belt of recent tidal flats, 20 × 160 km, west of Andros Island, Bahamas, where quiet-water conditions are produced by the sheltering of Andros and the broad, shallow Great Bahama Bank.

Physiography. The mirror image of a deltaic plain with landwardbranching tidal channel systems; natural levees along adjacent channel systems form semi-isolated "intertidal'' ponds; landward of the channeled belt, 4 km wide, is a broad marsh covered with mats of blue-green algae; the same mats fringe the ponds.

Climate. Subtropical, humid, wet; average annual rainfall 114 cm, mostly in short squalls May to October; prevailing easterly winds, gentle in spring and summer, frequently fresh in fall and winter; high pressure areas (cold fronts) bring strong winds from the north and west quadrants an average of 40 times yearly.

Sediments. Pelleted lime mud with generally minor amount of skeletal debris (gastropod and Foraminifera); up to at least 2.7 m of Holocene sediments underlain by ''freshwater'' peat with C14 age of 4890 ± 200 years BP.

Exposure Index. Total period of exposure at closely spaced intervals throughout the tidal range determined from almost two years of continuous tide records at two stations; results for one station, levee-pond area, shown below:

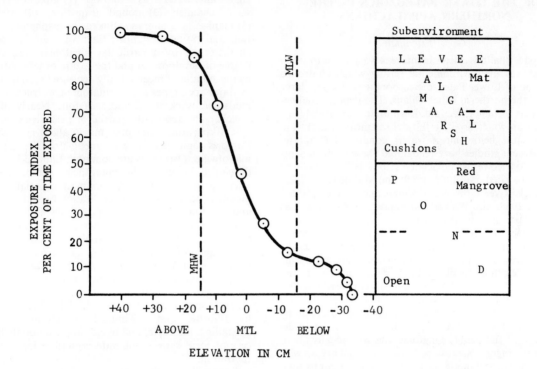

Exposure Indices of Sedimentary Structures   Elevations determined with a
Dumpy Level to a precision of 1.5 cm.

EXPOSURE INDEX

| 0 | 10 | 20 | 30 | 40 | 50 | 60 | 70 | 80 | 90 | 100 |

Bath-tile mud cracks (polygons 1-3 cm)

Fenestral pores (birdseyes)

Blistered and curled algal mat
Millimeter laminations

Knobby algal stromatolites, SH (2-5 cm)

Algal stromatolites, with pallisade
   structure, LLH (Scytonema sp.)

Finger burrows (Fiddler crabs)

Wide, shallow mud cracks
   (polygons 5-15 cm)

Deep prism mud cracks
   (polygons 20-30 cm)

Antler burrows (Alpheus sp.,
   Callianassa major)

Spaghetti burrows (1-3 mm)
   (Polycheate worms)

Friedman (1969b) lists illustrative papers and gives an excellent summary of characteristics to look for in ancient counterparts of various tidal flat environments. The abstract of his paper follows.

RECOGNIZING TIDAL ENVIRONMENTS IN CARBONATE ROCKS WITH PARTICULAR REFERENCE TO THOSE OF THE LOWER PALEOZOICS IN THE NORTHERN APPALACHIANS
by
G. M. Friedman

A broad, shallow epicontinental sea very close to mean sea level was the environmental setting which controlled deposition of Lower Paleozoic carbonate rocks over large areas of the northern Appalachians. The depositional environments include those laid down under supratidal (just above mean water level), intertidal (at mean water), and subtidal (just below mean water) conditions (Laporte, 1967). Recent studies have recognized these environments in rocks of Lower Ordovician (Tribes Hill Formation) (Braun and Friedman, 1968, 1969) and Middle Ordovician age (Black River Group) (Friedman and Sanders, 1967; Textoris, 1968; and Walker and Laporte, 1968) as well as those of Devonian age (Laporte). The supratidal facies are characterized by finely laminated dolostones or dolomitic limestones, stromatolitic (undulating) structures, "birdseye" textures, mudcracks, bituminous material, authigenic feldspar, dearth of fossils, but abundant trace fossils and burrow mottles. Intertidal environments can be divided into tidal flat and tidal channel facies. Tidal flat environments are characterized by mudcracks and lumpy structures, flaser structures, scour and fill, sporadic "birdseyes," flat-pebble conglomerates with locally abundant fossil fragments, cross-bedding, erosional breaks with shale stringers, and mottling. The tidal flats are cut by tidal channels which have pronounced truncations at their base. The channels contain abundant fossil fragments and may include foundered blocks of tidal flat lithology. The subtidal sediments are well-bedded and highly fossiliferous with round pebble conglomerates, and locally may be oolitic.

For a review of modern and ancient tidal flats see Ginsburg and Hardie (1975). Two pertinent studies of Precambrian stromatolites are given by Hoffman (1967) and Knight (1968). A paper by Donaldson (1962) includes descriptions of six basic types of Proterozoic stromatolites, a comparison of these stromatolites with similar (often confusing) structures and an extensive bibliography. Ancient counterparts of Bahaman channels and natural levee deposits are described by Kahle (1968) whose abstract is quoted below. See also figure 2.60.

PETROLOGY AND STRUCTURE OF A SALINA (SILURIAN) DOLOMITIZED ALGAL STROMATOLITE COMPLEX, NORTHWESTERN OHIO
by
C. F. Kahle

A unique algal stromatolite complex is located southwest of Maumee, Ohio, on the Niagaran reef-bank system that contributed to the isolation of the Michigan basin during the Cayugan Epoch. Overlying strata have been eroded so that the individual features of the complex may be studied in essentially their original paleogeographic setting. Major features in the study area include tidal channels, natural carbonate levees, algal mounds, scour channels, laminations, ripple marks, desiccation cracks, and gypsum and anhydrite molds.

The dolomite rocks in the area may be subdivided into eight microfacies as follows: (1) mound; (2) mound scour channel; (3) mound transition; (4) intermound; (5) blanket; (6) laterally linked hemispheroid; (7) mud flat; and (8) tidal channel. Each of these microfacies is characterized by a particular algal mat frequency, the degree of development and the nature of laminations, and by the amount of mechanically deposited sediment.

Major rock types are stromatolite-constructed dolomite mudstone, wackestone, and packstone. Nearly all samples are characterized by the partial or pervasive development of a filamentous (and presumably algal) microstructure. Burrows and pellets are uncommon in all of the microfacies. Ostracods are common. Laminations are well developed in all of the microfacies except the mound microfacies. Breccias are common especially in algal mounds and are due to mineralization by sulfate minerals, and dedolomitization, desiccation, and solution collapse.

The general sequence of mineralization was: (1) formation of microcrystalline xenotopic dolomite, probably by replacement of an initial $CaCo_3$ mud, accompanied by the formation of gypsum and anhydrite; (2) dedolomitization, primarily in the mound microfacies; (3) partial replacement of calcite and dolomite by celestite; (4) replacement of some calcite, dolomite, and celestite by fluorite; and (5) limited formation of medium crystalline, idiotopic, and hypidiotopic dolomite by a process of local source dolomite crystallization and recrystallization.

Comparison with several Recent carbonate analogs in the Persian Gulf and Bahamas suggests that most of the eight microfacies were formed in a supratidal environment.

CRANE KEY (Florida Bay): Crane Key and adjacent local keys are typical supratidal mud flat islands within a modern lagoonal environment. The larger of the Crane Keys is roughly triangular shaped (see index map 3 and figure 2.39) and readily accessible at the north end with a shallow draft boat. **Note:** Crane and adjacent keys are within Everglades National Park and thus permission to visit them should be obtained from the Park Supt. Office—see page 378 of this Guidebook. For details of the Florida Bay environment and the origin of contained islands, the reader is referred to the next chapter of this Guidebook dealing with lagoons.

Common features of supratidal environments are laminated algal mats of blue-green filamentous and unicellular algae and their desiccated by-products. These algae are found in both marine and freshwater environments and are important because of their (1) stabilizing effect, (2) accreting ability due to their mucilaginous sheaths which entrap and bind fine grain material, and (3) paleoenvironmental

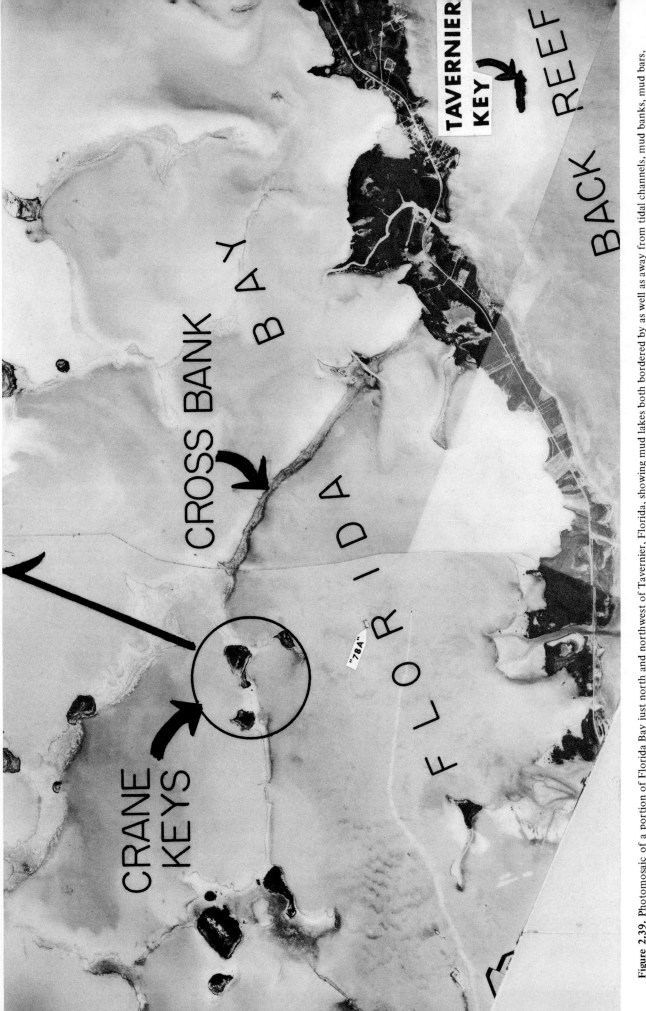

Figure 2.39. Photomosaic of a portion of Florida Bay just north and northwest of Tavernier, Florida, showing mud lakes both bordered by as well as away from tidal channels, mud banks, mud bars, submergent and emergent mud mounds. Core taken by Muller and Muller (see page 71) was located approximately at point of arrow showing Cross Bank. An isolated submergent "mud" mound is located just above the "R" in the word "FLORIDA," at the C.G.S. marker "78A."

significance. Crane Key and similar Florida Bay Keys are an excellent place to observe the formation of such laminated algal mats.

Wave action and mangrove baffle action have developed a more or less continuous natural levee or rim of sediment around the perimeter of Crane Key, allowing the center of the Key to be several inches above or below mean Bay level. This border rim prevents flooding of the interior of the island except for periods during exceptionally high wind and severe storms (Perkins and Enos, 1968, report 5 centimeters of lime mud was deposited above the level of normal high tide on Crane Key during Hurricane Donna in 1960). When the latter conditions do occur, a layer of sediment floods over the interior covering the algal layer (figs. 2.40 and 2.41). Subsequent drainage and evaporation provide time for the algae to grow up through the light-colored detrital sediment layer and present another dark-colored mucilaginous surface ready for the next storm deposit.

Descriptions of the specific blue-green algae and details of the mechanics of this layering process are described by Ginsburg *et al.,* (1954). Measurements by these authors indicate (p. 24) a range of from 0 to 3 mm thickness of fine grain sediment was deposited over Crane Key during one 10-month interval at various measuring stations producing a maximum of two new sediment lamina. Such rates of deposition, common on mud flat islands, are extremely low in comparison with depositional rates of similar algal and sediment mat layers measured by Ginsburg *et al.* (1954) on the exposed rocky platforms of West Summerland and Big Pine Keys which are more frequently subjected to tidal flooding. The speed by which the effective binding action of the filamentous blue-green algae can take place is surprising. Ginsburg *et al.,* (1954, p. 27) note that laboratory experiments suggest that "periods of sedimentation separated by as little as 48 hours could be recorded in the sediments as distinct events." See Monty (1965) and page 37 for additional data. For illustrations of Crane Key mats see figures 2.42, 2.43, and 2.44.

The work of Black (1933) and Ginsburg *et al.,* (1954) indicates that there is no clear evidence of precipitated calcium carbonate in the soft, laminated, algal mats. The layering is essentially a sequence of mechanically entrapped sediments. For subsequent induration the possibility of physiochemical or bacterial precipitation within the mats is suggested by the above workers. Others, such as Drs. S. Golubic and L. J. Greenfield (see Greenfield, 1963, 1964, 1965, 1967), are currently studying such possibilities. Shinn *et al.,* (1969) note that algal filament tubes are often cemented with tiny carbonate crystals (fig. 2.59).

Desiccation of these supratidal laminated sediments (fig. 2.44) produces distinctive mudcracks which break up into curled chips (fig. 2.43) and undulating surfaces displaying bubble-type structures.

Birdseye structures are common in the muds of Crane Key (figs. 2.44 and 2.40). Such small, open planar or bubblelike voids in Holocene carbonate sediments (calcite or anhydrite filled in ancient carbonates) have been associated with shallow water environments by many workers (i.e., Folk 1959, Illing 1959, Perkins 1963, Shinn *et al.,* 1965, Wolf 1965 and La Porte 1967). Shinn (1968b) gives an excellent summary of the birdseye structure problem and concludes that the structures are never subtidal, sometimes intertidal, but most commonly they are supratidal phenomenon. In the latter environment Shinn (1968b, p. 218) notes from field experiments that "gas bubbles make most spherical vugs and that alternate internal shrinkage and swelling, due to alternate wetting and drying, produce the planar vugs." He further notes that such birdseye structures are particularly important as supratidal indicators in ancient carbonates containing no other diagnostic structures.

On Crane Key, according to Rubin and Suess (1955), a radiocarbon date of 3300 ± 240 years was given a "core sample from peat layer 26 in. thick below 75 in. of pure calcilutaceous sediment." This sample was taken on the largest of Crane Keys, 66 m east of the western shore near the middle of the island by R. N. Ginsburg.

SUGARLOAF KEY (Florida): The Sugarloaf Key supratidal area, about 20 kilometers northeast of Key West (fig. 2.45), is of particular significance in that it (1) illustrates supratidal Holocene sediments containing up to 25 percent dolomite, and (2) provides a clue for interpreting planar nodules similar to those described in ancient carbonate rocks.

Shinn (1968a) describes this area in detail. The abstract and several illustrations from his paper are given below.

SELECTIVE DOLOMITIZATION OF RECENT SEDIMENTARY STRUCTURES
by
E. A. Shinn

Relatively nondolomitic limestone nodules which float in lithified dolomitic sediment are forming on a supratidal mud flat in the lower Florida Keys. Storm tides periodically deposit layers of lime mud above normal high tide level which dry and crack to form typical mudcrack polygons. These polygons erode into flattened nodules that subsequently become buried in relatively more porous and permeable sediment. Magnesium-enriched brines concentrated through evaporation are more readily transmitted

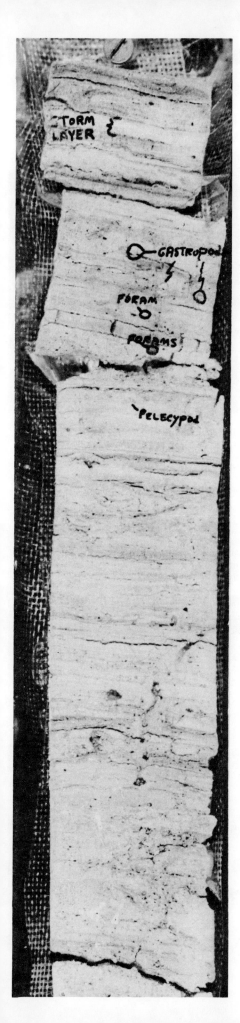
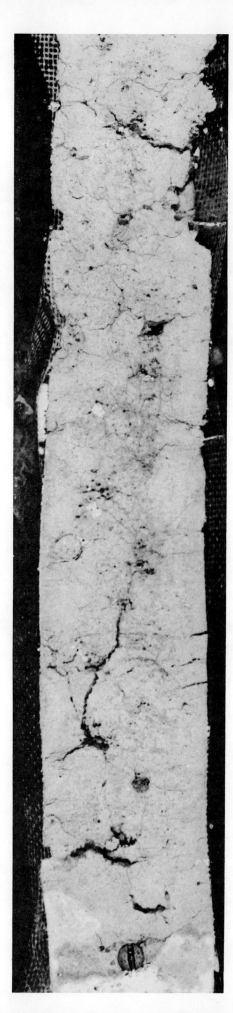

STORM
LAYER

GASTROPOD

FORAM

FORAMS

PELECYPOD

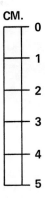

CM.

0
1
2
3
4
5

Figure 2.40. Impregnated core taken from the interior flat of the larger of the Crane Keys. Total length of core from top (upper left) to base (lower right) is 60 cm. Thick light "storm layer" is 5 mm thick. Note how laminations disappear with depth and the presence of birdseye structures.

Figure 2.41. Freshly dug hole reveals laminated blue-green algal mat structure in interior of Crane Key, Florida Bay. Note pencil for scale.

Figure 2.42. Ready to pull core taken on Bay side of natural levee or rim of Crane Key. Note pneumatophores (foreground) of black mangrove (background).

Figure 2.43. Close-up of curled desiccated algal mats on interior flat of Crane Key, Florida Bay. Note pencil for scale.

Figure 2.44. Box core of Crane Key supratidal flat deposits with finger pointing to blue-green algal mat. Note birdseye structures (mostly planar with some spherical vugs).

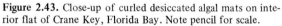

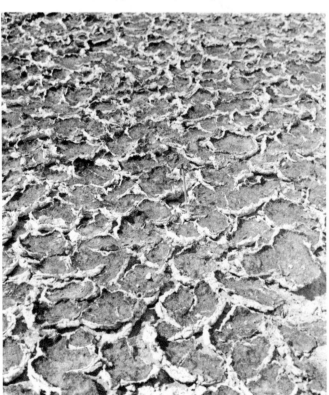

through the more permeable sediment that are the less permeable nodules. Selective dolomitization of the more permeable sediment produces features bearing a strong similarity to sedimentary structures commonly found in ancient dolomitic rocks.

Age determinations for the 2.5 to 10 cm thick upper crust layer of supratidal sediments which contains mudcracks, algal mats and birdseye structures is noted by Shinn (1968a) as follows: upper .6 cm—300 years; 10 cm below surface—900 years.

The dolomite is reported as calcium rich (60 mole percent Ca, 40 mole percent Mg), poorly ordered 1 to 3 micron crystals. The maximum dolomite content of the more impermeable nodules is 10 percent whereas the intervening more permeable layers contain up to 25 percent dolomite (fig. 2.46). As noted by Shinn (p. 615) "If the sediments described above remain in the supratidal environment, it is likely that they will eventually become completely dolomitized. However, if relatively rapid subsidence or sea level rise results in a change to a less Mg-enriched intertidal or marine environment, then these sedimentary structures may be preserved undolomitized."

Relatively nondolomitic planar nodules in ancient rocks have been described by Van Tuyl (1914) and by Matter (1967). Completely dolomitized counterparts are described by Sander (1936), Friedman and Sanders (1967), La Porte (1967), and Matter (1967).

Recent investigations of the tidal flat environment of Sugarloaf Key by Atwood and Bubb (1968) delimit occurrences of dolomite at the Sugarloaf Key site described by Shinn (1964) and suggest alternative mechanisms for origin of the supratidal dolomite.

WESTERN ANDROS (Bahamas): Western Andros (figs. 2.47 and 2.1) represents a tidal and supratidal area of significant size and complexity. It has been the object of considerable study. Comprehensive regional papers which can serve as introductory guides to the area are found in Newall *et al.,* (1959), Purdy and Imbrie (1964), Shinn *et al.,* (1969), Gebelein (1974) and Ginsburg and Hardie (1975). For field trip itineraries see Purdy and Imbrie (1964), Friedman (1970—reproduced in the appendix of this Guidebook) and Gebelein (1974—includes illustrations of common organisms and extensive bibliography).

The Andros area is best seen by chartered amphibious plane out of Miami or Naussau. Effective arrangements can sometimes be made linking these flights with small boat use for longer specific studies of such areas as Joulters Cay or the windward reef tract. Detailed surveys of the west coast of Andros usually require establishing camps accessible to channels or beach ridges along which amphibious planes can land. Common areas for field trips are:

1. Traverse from channel to palm hammock described by Shinn *et al.,* (1965) just east of Williams Island (see index map 2, arrow in figure 2.48, and figures 2.51 through 2.55).
2. Williams Island described by Purdy and Imbrie (1964) and pages 62 and 63 of this Guidebook.
3. Three Creeks area of northwestern Andros described by Ginsburg *et al.,* (1970).
4. The prograding tidal flats southeast of Williams Island, figures 2.61 through 2.63 described by Gebelein (1974).

The low, flat, western third of Andros contains a wedge of fine grain sand and pelletoid mud lying in the lee of the higher, exposed bedrock surface to the east (see figure 2.47). Shinn *et al.,* (1969) present an excellent study of the anatomy of this western third of Andros. They subdivide the area into three vertical tidal zones which they call the (1) subtidal, (2) intertidal, and (3) the supratidal. Only the latter zone will be discussed here.

The supratidal portion of western Andros, according to Shinn *et al.,* (1969) can be divided into three geomorphic subdivisions called (a) beach ridges, (b) levees, and (c) marshes, each having an ordered, spatial distribution. These various subdivisions are flooded only by storm or semimonthly high (Spring) tides. At any one time the above writers' note that the maximum vertical range of supratidal sedimentation is about 1.2 meters.

Beach ridges, .3 to 1.5 m high (fig. 2.56) form a boundary between the marine areas and the tidal and supratidal area of the channel belt (fig. 2.1). They have a geographic position similar to the beach ridges described on Crane Key in Florida Bay on page 42. The beach ridges of western Andros display spurlike projections and steepest slopes on the seaward side. They show 1°-5° slopes on the landward side. Shinn *et al.* (1969) also report that low beach ridges occur on headlands and are composed of laminated pellets and fine, sand-sized, skeletal fragments, whereas high ridges occur in embayed areas and are made up of festoon cross-bedded gastropod sand. Individual laminations show graded bedding, birdseye vugs and lateral discontinuity. Laminated, dried, mud chips are also reported as common.

Back from the open marine environment *levees* border tidal channels (fig. 2.56) and perform a role similar to that of the beach ridges. Levees are composed of laminated (less than 3 mm thick) pelletoid, sand-size sediments which build up to about 30 cm above the high tide mark along both sides of straight channels and along the outer banks of meandering chan.nels. Shinn *et al.* (1969) note that the laminated levee deposits are graded, contain birdseye vugs and between 5 and 10 percent dolomite.

Figure 2.45. Oblique aerial view of a portion of western Sugarloaf Key showing various subdivisions. Recent selective dolomitization can be observed in the supratidal area noted. From Shinn (1968a, p. 613).

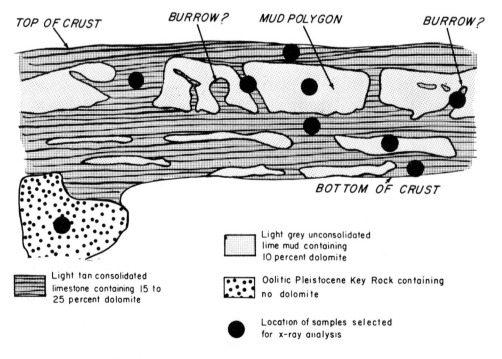

Figure 2.46. Sketch of slabbed face of supratidal crust showing relatively nondolomitic limestone nodules floating in dolomitic sediment. Sugarloaf Key, Florida. From Shinn (1968a, p. 614).

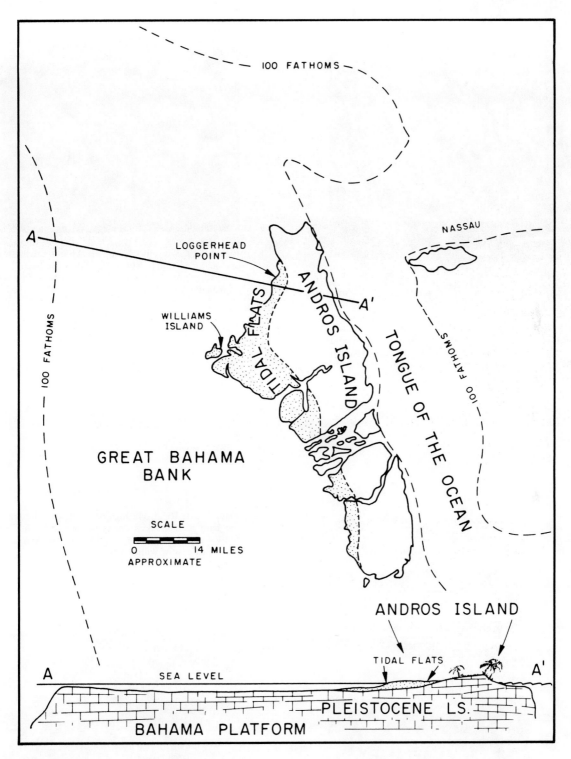

Figure 2.47. Index map after Shinn *et al.* (1969) showing location of tidal flats on western Andros, Bahamas.

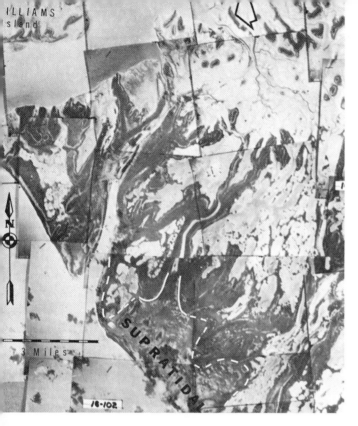

Figure 2.49. Western edge of the Andros tidal flats showing several major channels which may "flow landward" during storm depositing mud over bordering levees. Note sinkhole. Area east of Williams Is.

Figure 2.48. Aerial photomosaic of western tip of Andros taken from Shinn *et al.* (1969) who states . . . "notice straight southwest shoreline. Supratidal flats extend more than 2 miles inland from the beach ridge. The one-time mouth of an active channel is shown filled by supratidal sediments more than 2 miles from present shoreline.

Figure 2.50. Close-up of sinkhole area shown in figure 2.49. Dark dolomitic crust band (D) rings higher hammock area.

Figure 2.51. Relatively high percentage of dolomite is found in the hard mud cracked crusts (foreground) on the back side of the levees just above the high tide level (background). Palm hammock just east of Williams Is.

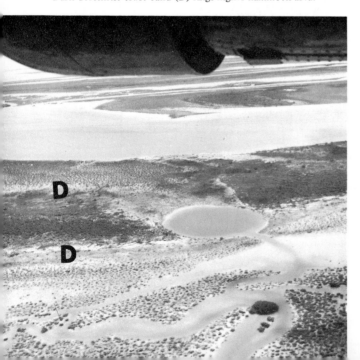

Figure 2.52. Harder, less dolomitic crusts found about 45 cm above high tide mark. Same locality as shown in figure 2.51.

Figure 2.53. Supratidal marsh core from Shinn *et al.* (1969). Shows alternation of light storm-deposited layers with dark organic-rich exposure layers which generally contain algal filaments as shown in figure 2.59.

Figure 2.54. Common tidal flats organisms. Pellet-making *Batillaria* (top); land snails *Cerion* (center); foraminifer *Peneroplis* (bottom). From Shinn *et al.* (1969).

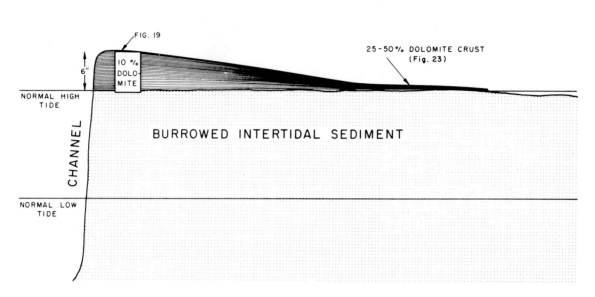

Figure 2.55. Schematic cross section of levee illustrating more concentrated dolomite on low flanks just above normal high-tide level similar to mud-cracked area shown in figure 2.51. Higher part of levee shows less dolomitization. Cross section from Shinn *et al.* (1969).

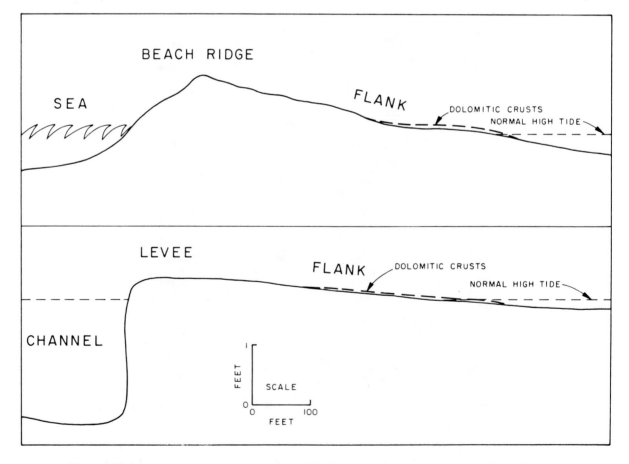

Figure 2.56. Schematic cross-sectional comparison of beach ridges and levees from Shinn *et al.* (1969).

In both supratidal levees and beach ridges the living fauna consists of the gastropod *Batillaria* (fig. 2.54) and the burrowing fiddler crab *Uca*. Flora consists of the black mangrove *Avicennia*. Higher portions of the beach ridges display cabbage palms and the Australian pine tree, *Casurinia*. The blue-green algae *Scytonema* is found on the landward beach ridge and level slopes where bordered by low, more burrowed and nonlaminated intertidal flats.

The third subdivision of the supratidal environment of western Andros is the supratidal *marsh* which forms a broad, dark (due to algal mats) belt landward from the channel belt shown in figure 2.1. According to Shinn *et al.,* (1969) most of this area is above the normal high tide and contains low areas often filled with fresh to brackish water. The pelleted silt and mud of the marsh contains more organic material, sedimentary laminations and particularly more blue-green algal mats than the other supratidal subdivisions (fig. 2.57). The algae is composed mostly, according to Shinn *et al.,* (1969), of filaments of *Scytonema*. Root structures (from mangrove and marsh grass) and birdseye structures are common. Living fauna is noted by the above authors as consisting of the large land crab *Cardisoma* and land snail *Cerion*. Peneropolid and miliolid forams as well as some gastropods found in the sediment of the marsh have been washed in from the west by storm action.

The percentage of dolomite in supratidal sediments of western Andros varies from 0 to 80 percent increasing with the amount of induration of surface crusts. These crusts are best developed only a few cm above the high tide mark (fig. 2.51). In beach ridges and levees this maximum development of dolomitic crusts occurs on the landward slopes farthest from the open marine or channel environment (fig. 2.50). The seaward part of the marshes contains the harder, more dolomitized (maximum of 20percent dolomite) crusts according to Shinn *et al.,* (1965). Regarding the origin of this dolomite, Shinn *et al.,* (1965) state:

It is believed that much of the dolomite is a penecontemporaneous replacement of calcium carbonate. The dolomite forms where tidal flooding and storm sedimentation is followed by many days of subaerial exposure. Surface evaporation during these periods of exposure increases the concentration of dissolved salts near the surface but no evaporites are preserved. Magnesium calcium ratios as high as 40 to 1 are present in the concentrated interstitial waters where dolomite is forming.

The supratidal sediments of western Andros are allochthonous in that they represent deposits brought in and built up, by successive storm wave action. Such transported lime mud from the adjacent marine environment is described by Purdy and Imbrie (1964) as "welding together" the initial emergent islands of Andros and building up a broad belt of tidal flat sediments displaying abandoned (fossil) channels and small residual lakes. A major factor in this mud accumulation is the channel system (fig. 2.49) which Shinn *et al.,* (1969) describe as the pathways of sediment-laden water during storms. The above writers compare such tidal-flat deposition to that of a river delta turned landward so that the sea is the river and the "channels, levees, flats and ponds are analogous to those of fluvial deltas." As with fluvial deltas, deposition occurs mainly during flood periods which in this case are tropical storms.

Shinn *et al.,* (1969) further note that the northwestern edge of Andros (Loggerhead Point area) is undergoing transgression while the southwestern coast (south of Williams Island, figure 2.48) may well be in a regressive phase. They note that rates of sediment supply and exposure to wave action may explain this divergence. If this is true, caution should be exerted before interpreting ancient transgressive/regressive sequences as due to sea level changes or land fluctuations.

The results of a study by Gebelein (1974) describe the sedimentology and stratigraphy of the actively accreting tidal flat and shallow marine sediment wedge along the southwest coast of Andros Island. Data were gathered from probe and alidade transects and analysis of over 200 unipod piston cores. Results of this study have indicated "recognition criteria" for identification of ancient analogs. Initial results were presented by Gebelein (1973) in abstract form as noted below:

SEDIMENTOLOGY AND STRATIGRAPHY OF RECENT SHALLOW-MARINE AND TIDAL-FLAT SEDIMENTS. SOUTHWEST ANDROS ISLAND, BAHAMAS

by
Conrad D. Gebelein

Shallow-marine and tidal-flat carbonate sediments form a seaward-prograding wedge (maximum 6 m thick) on the southwest coast of Andros Island. Sediments have been deposited in a large (85 × 25 km), arcuate embayment in Pleistocene limestone. Incursion of the sea over freshwater peat reached the inner, central margin of the embayment 5,000-7,000 years ago. Subsequent lateral progradation (maximum 30 km perpendicular to strike) of marine and tidal-flat sediments has occurred in several phases. 1. Continuous lateral progradation of a very shallow (less than 1 m), wide (3-5 km) subtidal platform adjacent to the shoreline. Sediments on the platform are poorly sorted, massive, fossiliferous, white-pellet muds with prominent filled burrows. 2. Development of 5 major shorelines, roughly parallel and 2-5 km apart, during the past 5,000 years. Each new shoreline developed, presumably by storm action, onto the subtidal platform, and isolated a shallow linear lagoon. Abandoned shorelines form parallel bands of discrete to coalesced, symmetrical. V-shaped hummocks. These vegetated ridges (up to 1 m high) are positions of former beach ridges and tidal-channel levees. Sediments in the hummocks are thin-bedded (3-15 cm), unlithified pellet sands with pronounced fenestral fabrics. Sloping flanks of the hummocks are composed of lithified crusts (up to 3 cm thick), commonly separated by unlithified sediment layers. 3. Initial infilling of the lagoons in the form of closely spaced, circular to ellipsoidal Carolina bays. Bay margins form by spit accretion of well-sorted pellet and skeletal sands. The bays themselves fill with poorly sorted, muddy pellet sands. 4. Bay sedimentation generates a mosaic of isolated small sand bodies with a muddy, pelletal, massive sediment. Vertical infilling of the lagoons in the form of laterally continuous, alternate thin beds of blue-green algae and pellet mud, (5) Capping of the sequence by laterally continuous, thin crusts of aragonite-dolomite.

Gebelein (1974) also indicates that in contrast to the southwest coast described above, shorelines of northwest Andros display surface morphology and contain a sediment wedge indicative of conditions where reworking exceeds or is equivalent to sediment supply. He feels that such a morphology and sediment wedge are probably only rarely preserved in the geologic record.

Current interstitial water chemistry investigations in the tidal flats of southwestern Andros by Gebelein (1976—personal communication) indicate "mixing zone dolomitization" as being an important mechanism for generating dolomitized sediments, where the dolomite is *not* confined to surface crust types of facies.

Figure 2.58. Curled edge of thick filamentous blue-green algal mat between intertidal and supratidal zones. From Shinn *et al.* (1969).

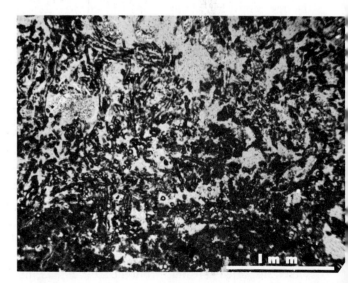

Figure 2.59. Thin section of algal filaments preserved in marsh sediment core. From Shinn *et al.* (1969) who report that tubes are often cemented with tiny carbonate crystals.

Figure 2.57. Supratidal marsh core from Shinn *et al.* (1969). Core taken from spot several centimeters higher than core shown in figure 2.53. Sediment more desiccated (note mud cracks at arrow; spots show paraffin which filled voids during impregnation). Preserved algal filaments (fig. 2.59) are abundant.

Figure 2.60. View of exposure of Greenfield (Silurian) algal stromatolite complex near Maumee, Ohio. Key: MF = mud flat microfacies; TC = tidal channel microfacies; CL = natural carbonate levee microfacies. The mud flat microfacies consist of desiccated, flat laminated algal stromatolites. The upper portion of the natural carbonate levees are made up of laterally linked algal stromatolites (note surface of levee in background). The tidal channels and natural carbonate levees appear to be nearly identical to certain modern tidal channels and levees which have been described in parts of the Bahama Bank area (Purdy and Imbrie, 1964). Photo and data from C. F. Kahle.

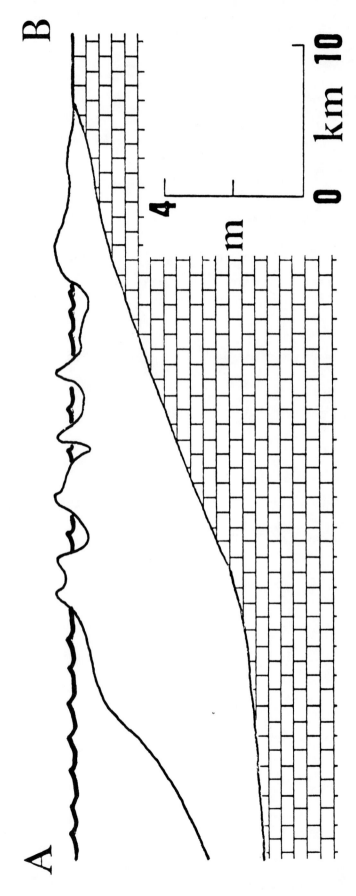

Figure 2.61. Wedge of marine and tidal-flat sediments which are prograding onto a shallow epieric platform, southwest Andros. See figure 2.62 for location. Figure courtesy of C. D. Gebelein.

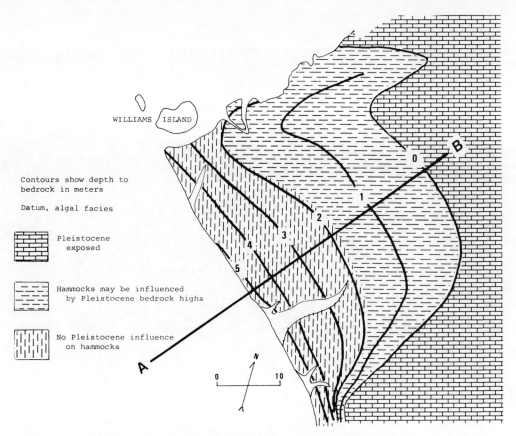

Figure 2.62. Tidal-flat sediments deposited in arcuate-shaped embayment on Pleistocene bedrock, southwest Andros. Figure courtesy of C. D. Gebelein.

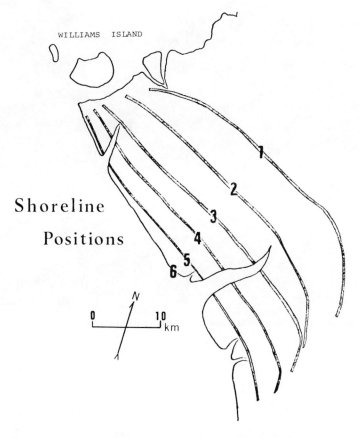

Figure 2.63. Shoreline positions which have developed during the last 3-4,000 years along the southwest Andros coast. Figure courtesy of C. D. Gebelein.

Figure 2.64. Roots of the mangrove *Avicennia nitida* provide an effective stabilizing and entrapping mechanism in the supratidal area. Bear Cut, Miami.

FIELD QUESTIONS
Intertidal Environments

BEAR CUT (Miami)

1. A series of platforms of mangrove reef rock, each of decreasing elevation, extends out into Bear Cut. What is the implication of such platforms relative to sea level fluctuations during the last several thousands of years?

2. Red mangroves are usually found seaward of black mangroves. What is the implication if the exposed fossilized mangroves of this area are only black mangroves?

3. Are there other possible mechanics (besides the ones suggested by Hoffmeister and Multer, 1967) for producing these indurated rods?

4. Does the combination of quartz and calcareous sand, necessary for the above hypothesis, occur very often as seen here along the shore of Bear Cut?

5. What are the characteristics which differentiate these rods from fossil worm tubes and other burrow and root structures.

6. What is destroying and what is protecting the present mangrove reef rock?

7. Is any mangrove root material now in the area presently undergoing fossilization? Where would you look for such and why?

8. Does the worm reef need a stable substrate for attachment?

9. What limits the spread of worm reef rock? Who is winning the war between the encrusting worm reef and the barnacles?

10. What is the difference between size, composition and sorting of the beach sand and of the same composing adjacent worm tubes?

11. What conditions would be necessary for the preservation of the worm reef rock?

   Notice the common pockets and random shallow ponds filled with fecal pellets on the innertidal and supratidal zones along Bear Cut. Such pellets are derived from the common marine crustaceans and worms that inhabit the area. The majority of pellets probably are destroyed by mechanical or scavenger action, or their pellitory shape is lost under compaction. It is interesting to note that Ginsburg (1957, p. 81) indicates that abundant pellets in ancient rock may not indicate the most intense scavenger action, but rather specific bathymetric and geographical environments where induration preserves soft pellets.

ROCKY SHORE (Florida Keys and Bimini)

12. Estimate the percentage of total destruction to the bedrock done by (a) organisms, (b) chemical activity, (c) wave action.

13. Can you observe any *constructional* features in the intertidal—supratidal zones?

14. Specifically what causes the different color zones? Will evidence of such zones be preserved in the fossil record?

Sample No.   **Photo No.**

FIELD QUESTIONS
Intertidal Environments

SAND BEACH

BAHIA HONDA—Long Key (Florida)

1. How many types of ripples are present and what are their significance?

2. Cut a two feet deep trench (or take a series of box cores) normal to the shore and describe the stratification. Are there evidences of storm or organic activity? Does size-grade decrease seaward?

3. Compare the grain polish from these beaches with surfaces found on similar constituents in the lagoon and back-reef areas.

4. Compare the constituent particle composition of sand beaches on the lagoon side of Bahia Honda with that found on the seaward side.

5. What are the differences in size grade and constituent composition between the beach sand and grass—held shallow submarine sand bodies (offshore from these same beaches)? What are the causes for changes found?

6. Take three 100 cc samples for size grading from (a) 20 feet above high tide, (b) midtide mark, and (c) 20 feet seaward of the low tide mark. Record data on chart in appendix.

NORTH BIMINI (Bahamas)

7. Compare the variety and type of fauna found in the beach sand on the west shore with that found in Bimini lagoon.

8. What is the relationship between the steep seaward slope and stratification of this sand beach and the size grade of its constituents? Take three 100 cc samples for size grading from (a) 20 feet above high tide, (b) midtide mark, and (c) 20 feet seaward of the low tide mark. Record data on chart in appendix.

   Newell *et al.,* (1959, p. 193-194) describe the lithified beach ridges and dunes such as seen at Bimini to be a product of a late or regressive phase of the Sangamon interglacial interval. A date of 13,200±400 years BP is given (p. 194) for the oolite dune rock about 8 feet above high tide level, Till Hotel, Bimini, although the authors indicate that the rock may be actually much older.

9. Examine the lithified dunes along the west shore of North Bimini.

10. How many criteria can you find which support an aeolian origin:
    a. **Root casts?** (distinct from burrows which do not decrease in diameter as they branch out.)
    b. **Large scale fine grain x-bedded sand laminae?** (hard to form underwater)
    c. **Very well-sorted grains?** (how well sorted and what size should aeolian carbonate sand be as distinct from marine sand?)
    d. **Other criteria?**

11. What is the prevailing dip direction of the cross-beds? Throughout the Bahamas what should the prevailing dip direction of cross-beds indicate about former winds and sand source?

12. Compare the amount of polish on sand grains from the layered beach on the west side of North Bimini with sand in Bimini lagoon. Can differences in polish be due to a relative difference in turbulence?

BEACHROCK

NORTH BIMINI (Bahamas)

1. Observe all stages in the development of the indurated "beachrock" in the littoral zone along the west beach.

    a. What is the composition and source of the constituents? (can include coke bottle fragments)
    b. What is the nature and origin of the cement?

LOGGERHEAD KEY (Dry Tortugas, Florida):

1. Compare and contrast the composition and size grade of constituents found in the (a) supratidal beach sand, (b) lithified beachrock, and (c) loose offshore sand.

2. What is the nature and origin of the cement and of the bedding?

3. What evidence is there for change in shape and size of the north and south end of this Key?

Note: The foundations of the former Carnegie Laboratory can be seen toward the north end of this Key.

Sample No.

Photo No.

FIELD QUESTIONS
Supratidal Environments

CRANE KEY (Florida Bay):

1. Examine the algal mats.
   a. How widespread are they and what controls their distribution on the island?
   b. How many individual layers can you observe? Do these layers persist with depth? How deep are the lowest observable algal bands? What are the limiting factors in their preservation?
   c. Could these laminations have formed without the aid of the mucilagenous algae?
   d. Are these mats strictly supratidal in origin? Could such layering have been accomplished while under continual submersion? Is there evidence for such submarine algal mats forming below low tide level in Florida today?
   e. Is there any evidence of lithification of these soft mats?
   f. How does the structure of these soft mats compare with ancient lithified stromatolite equivalents?

2. Observe the various types of desiccation features.
   a. List the various types of desiccation features present.
   b. What would happen if sea level covered this desiccation surface for a short period?
   c. Is there any evidence for such an event from your cores?

3. Description: Cut out a smooth square sample section down through the mats and draw a detailed sketch of what you see below.

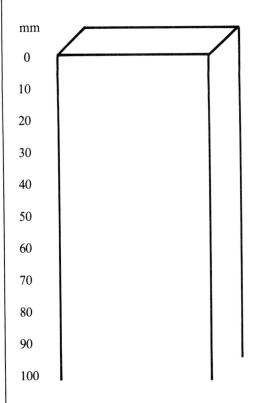

mm

0

10

20

30

40

50

60

70

80

90

100

Note the role of mangrove trees in originating, maintaining and enlarging emergent mud islands. In many cases they form the expanding edge of a saucer-shaped carbonate mound island with a low, central area. Vaughan (1910, p. 464) believed that mangroves . . . "are among the most important constructional geologic agents of southern Florida," and he estimated that one-third to one-half of the total area of the Florida Keys is occupied by them. Davis (1940) did a classic study on the ecology and geologic role of the mangroves in Florida and indicated that about 1,500 acres of new "land" had been formed in the past 30 to 40 years in the Florida Bay region because of the entrapment mechanism of mangroves.

Pioneer red mangrove communities start when floating seedlings come to rest in quiet, shallow water. Their developing root system tends to drop velocity and hence the sediment load of currents, while acting as a protective home for invertebrate animals and a catch area for drifting material. After an island starts to be established, a beach ridge or natural levee forms, back of which the brackish marsh black mangrove thrives. The red mangrove (*Rhizophora mangle*) with its prop roots can easily be distinguished from the black mangrove (*Avicennia nitida*) with its numerous vertical pneumatophores, which sprout up from a radiating root system. See Multer and Multer (1966) for a short summary of various Florida Key mangrove communities and their evolution.

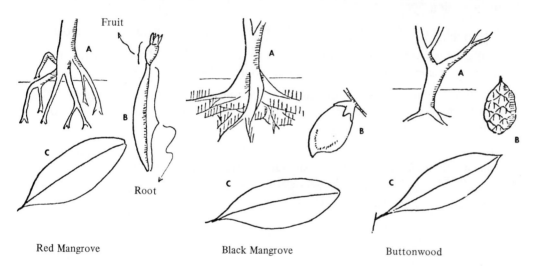

Red Mangrove   Black Mangrove   Buttonwood

KEY: A–Tree Base, B–Fruit, C–Leaf.

4. Examine the role of the mangroves wherever you see them. At what stage of development in forming new land are the mangroves observed at different locations?

5. How many different species of mangroves can you find? Sketch their root system.

Ornithologists Note: In addition to its geological fame, Florida Bay and the adjacent Everglades are well known for their birds, including bald eagles, spoonbills, frigate birds, great white herons, reddish egrets, and other wading and shore birds.

Sample No.

Photo No.

FIELD QUESTIONS
Supratidal Environments

WILLIAMS ISLAND

Comparison of aligned topographic features and composition indicates that Williams Island probably represents an eroded remnant of the Andros tidal flat belt. It is essentially a supratidal mud island with bordering laminated beach ridges and low interior intertidal-flats. Man-made cliffs and cuts near the abandoned sponge fisherman shed contain abundant shells of the pelecypod *Pseudocyrena colorata* and of the gastropod *Cerithidea costata*—both described by Purdy and Imbrie (1964) as apparently limited to the marine poikilohaline habitat adjacent to or in proximity to Andros Island. The terrestrial gastropod *Cerion* can be found living off vegetation on the island.

1. Using the cross section given on the next page, make a traverse across the various environments and encircle on the section features observed. Add other features not included on the section.

2. What kind of mangroves grow on the island? What is their role in the island's evolution?

3. Compare the fauna found offshore of Williams Island with that found in Bimini Lagoon as to number of genera and abundance. What do your results indicate?

4. Note how the mangrove pneumatophores break up the hardened algal mat. Will these algal mats be preserved? Do the algal laminations disappear with depth? What is the distribution pattern of the Recent dolomite crusts?

5. Are the offshore sediments laminated? Do they contain burrows or roots of any kind? Compare these results with those of similar investigations on the Island.

DEPOSITIONAL ENVIRONMENTS

WILLIAMS ISLAND

FACIES	SUPRATIDAL	INTERTIDAL	SUBTIDAL
SEDIMENTARY STRUCTURES	Laminated Ripples	Burrows, (common) Rootholes	Burrows, (common)
COLOR	Tan	Tan	Gray W/H_2S
TEXTURE	Laminated Mud- Pellet & Skeletal	Mud-Pellet	Mud-Pellet
FAUNA & FLORA	Algal Matt Shell Hash	Gastropods-(Abun.) Battalaria Cerithium Pelecypods Alga-mats	Gastropods Battalaria Cerithium Pelecypods Forams Algae Thalassia

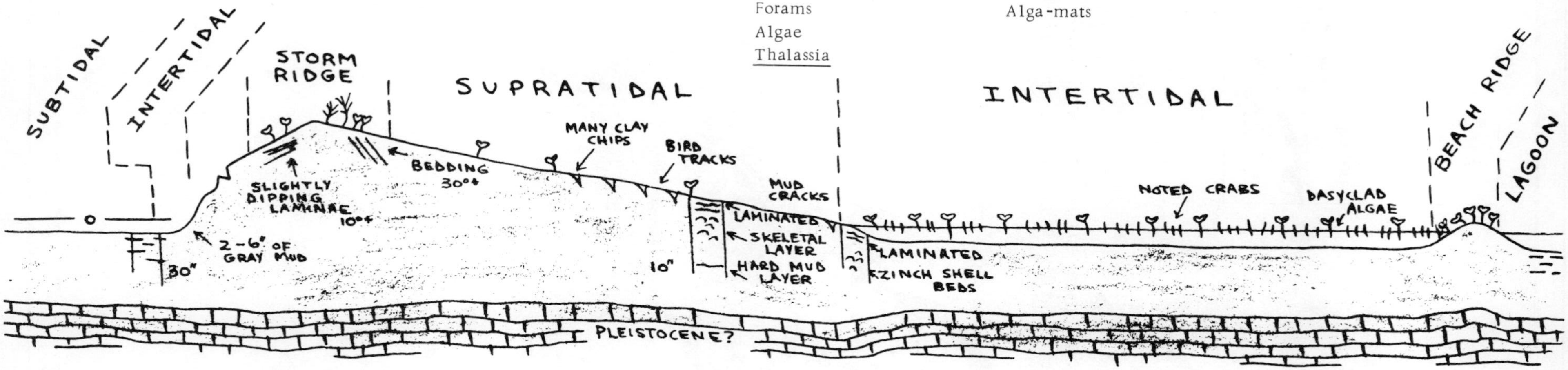

STORM RIDGE	SUPRATIDAL	INTERTIDAL SWAMP	BEACH RIDGE	LAGOON
Dipping Laminae Cross bedding Mud Cracks & Chips Birdseye rugs	Laminated Algal Matt Mud Cracks & Chips Rootholes	Burrows? Rootholes		Burrowed
Light Gray	Gray	Gray W/H_2S	Gray	Gray
Mostly Skeletal w/Mud-Pellet layer at surface	Mud-Pellet	Mud-Pellet w/Skeletal Layers	Gunky Mud	Gunky Mud w/Skeletons
Abundant Shells Mostly Pelecypods & Gastropods Palmetto	Algal Matt Mangrove Shell Hash	Algal Thalassia	Mangroves Grass Leaves	Gastropod & Pelecypods Common Thalassia, common Leaves

Diagram labels:

LAGOON — BEACH RIDGE — SUPRATIDAL — INTERTIDAL — SUBTIDAL

DASYCLAD ALGAE
NOTED CRABS
LAMINATED FZINCH SHELL BEDS
MUD CRACKS
BIRD TRACKS
MANY CLAY CHIPS
LAMINATED
SKELETAL LAYER
HARD MUD LAYER
STORM RIDGE
BEDDING 30°+
SLIGHTLY DIPPING LAMINAE 10°+
2-6" OF GRAY MUD
10"
30"
PLEISTOCENE?

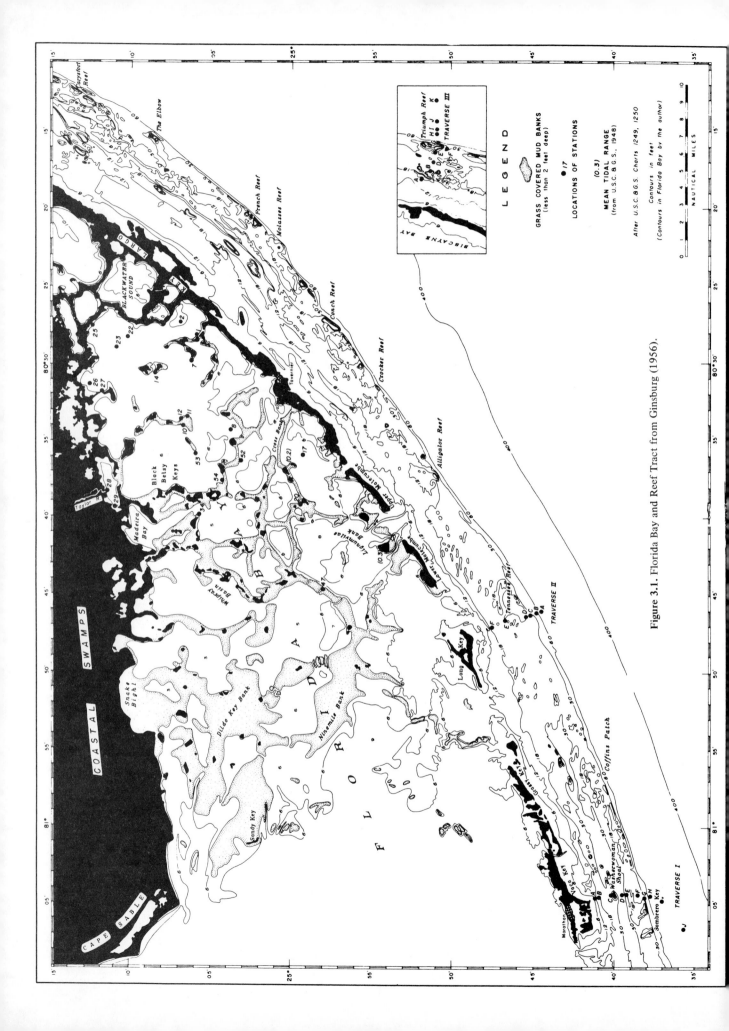

Figure 3.1. Florida Bay and Reef Tract from Ginsburg (1956).

LAGOONAL AREAS

Many examples of what may be called "lagoonal areas" are found in the Florida Keys-Bahama region. These areas all contain marine to brackish, usually low energy, sites for carbonate deposition and are often protected from the open sea by some type of barrier or platform edge. Examples of four types of such areas are noted below.

1. **Large, Complex, Subdivided Lagoonal Environments Bordering Mainland:** *Florida Bay*.(approximately 62 × 24 km) is characterized by a lacework of individual "lakes" or mud bank-bordered small basins and contains many "lagoonal"-type environments. The majority of these different types, found around the borders of the Bay, owe their differences to seasonal runoff from the bordering mainland on the north, to the open ocean conditions on the west and to the tidal influence from beneath porous, emergent Keys and through several open sea channels on the south. The interior region of the Bay with less than 6 cm tidal range and excessive seasonal salinity and temperature ranges, provides still another lagoonal-type environment.

2. **Small, Complex Lagoonal Environments Bordering Mainland:** *Card Sound* is a small yet complex lagoonal environment with sediment displaying facies change from calcareous quartz sand to calcarenite.

3. **Large, Sea Island Lagoonal Environments:** . The *Bight of Abaco* represents a large semirestricted marine lagoon far from the mainland. It is bordered by low-lying carbonate bedrock and separated from deep adjacent basin by a shallow rock and shoal "lip" or ridge.

4. **Small, Sea Island Lagoonal Environments:** *Coupon Bight* is a small tidal lagoon isolated from the mainland and surrounded by low-lying carbonate sediment and bedrock. *Bimini Lagoon* represents a small sea island lagoon lacking the usual mud component. It is surrounded by carbonate bedrock and

sediment stabilized by mangrove and *Casuarina* trees.

Each of the above types of lagoonal areas will be discussed and/or illustrated on the following pages.

LARGE, COMPLEX, SUBDIVIDED, LAGOONAL ENVIRONMENTS BORDERING MAINLAND

FLORIDA BAY: Florida Bay is a roughly triangular-shaped, rock-floored lagoon lying like a wedge between the southern mainland of Florida and the Florida Keys. It is bounded by the Gulf of Mexico on the west, by the mainland on the north and by the Florida Keys to the south. It has an area of approximately 62 × 24 km. It is readily accessible on eastern and southern boundaries by U.S. Route 1 and on the north from Flamingo in the Everglades National Park.

The pioneer geological studies in Florida Bay were done by Ginsburg who describe among other things the hydrography, size-grade, constituent particles (see figure 3.2), early diagenesis and baffle mechanisms found in Florida Bay. See Ginsburg *et al.,*. (1954), Ginsburg (1956), Ginsburg (1957) and Ginsburg *et al.,* (1958).

One of the most striking features of Florida Bay is its lacework of individual "mud lakes" and pattern of mud islands (figs. 2.39 and 3.1). The question immediately arises relative to the origin of such features within an essentially nontidal lagoon. Origins by storm action, prevailing wind and currents, mangrove and grass baffle growth and differential algal growth have been proposed along with other theories. Most of these theories are reviewed by Price (1967, p. 388-398) who himself considers that the basin-in-basin honeycomb of Florida Bay is due to "drowning and embayment of oriented lakes formed in drowned marsh rills." He indicates that continued marine embayment of the oval-oriented marsh lakes bordering the north and east edges of Florida Bay today (see figure 3.1) will produce more of this honeycomb subdivision. See also Craighead (1964) and White (1970) who give discussions of the distribution of mud banks and mangrove islands in Florida Bay from the effect of current action.

An excellent recent summary of the Florida Bay environments and problems by Scholl (1966) is given below:

FLORIDA BAY: A MODERN SITE OF LIMESTONE FORMATION*
by
David W. Scholl

Introduction

Florida Bay, located at the southern tip of Florida, has been of special interest to the geological scientist since the late nineteenth century when it became generally known that "limy" sediments were forming in this shallow-water tropical embayment. Interest arose because detailed studies of the neritic calcareous deposits (otherwise rather rare today) would afford a better understanding of the depositional environment of their lithified counterpart, limestone. Limestone is substantially composed of calcium carbonate in the form of the mineral calcite and is a much-studied fossiliferous rock type, of particular interest to oil geologists, that probably forms slightly more than 20% of the sedimentary rocks (a volume amounting to about 30×10^7 km^3 or 7×10^7 cubic miles) of the lithosphere.

Important papers on the calcareous sediments of Florida Bay have been written by Thorp (1936, 1939), Ginsburg (1956, 1957), Stehli and Hower (1961), Fleece (1962), Gorsline (1963), and Taft and Harbaugh (1964).

Physical Setting

Florida Bay is a triangular-shaped embayment lying immediately south of the Florida peninsula (fig. 3.1). A curved archipelago of elongate islands, the Florida Keys, on which are exposed Late Pleistocene (last interglacial) coral reefs of the Key Largo Limestone, forms the eastern and southeastern boundaries of the bay. To the west, the bay is open and faces the Gulf of Mexico; this side is conveniently limited by longitude 81° 05′W. The average depth over the central part is between 4 and 5 feet; the bay as defined above occupies an area of about 841 square miles (table 3.1).

The most characteristic feature of Florida Bay is an anastomosing array of shallow mud banks composed of shelly calcareous silts that cordon the bay into a lacework of interconnected shallow basins referred to as "lakes" (fig. 3.1). The lakes are typically 5-6 feet deep, 3-4 miles long, and 2-3 miles wide, and except for a thin veneer of sandy calcareous sediment, they overlie a nearly flat limestone surface composed of the Miami Oolite of Late Pleistocene age. The tops of the banks are partially ex-posed at times of low water; however, portions of many of them rise about 0.5 feet above the average bay level to form mangrove-crested or fringed keys (islands). The inner or eastern banks are sinuous and chain-like; they are narrow, about 0.5 mile across, and more keys rise from them than from the much broader (1-3 miles) amoeba-shaped banks along the western or open side of the bay.

Based on radiocarbon dates, calcareous sediments began to accumulate over 4,000 years ago when the Flandrian transgression (rise in sea level that accompanied the melting of the last continental ice sheets) flooded the area of Florida Bay. Bank formation probably began early in the history of the bay, presumably in areas where slack water was produced by converging currents. As the bay evolved, the banks grew upward and outward to their present configuration; colonization of the banks by marine grasses and mangroves must also have contributed to their stability. The keys that crown the banks may not have developed in their present form until about 3,000 years ago when the rate of filling decreased considerably. The keys are evidently built up above the average level of the bay during storm surges, when the water in the bay may deepen by 3-4 feet.

Since the bay was formed, approximately 1.5×10^9 cubic meters of calcareous sediment have accumulated. Over the banks and their slopes, the average sedimentation rate has been about 6 g/cm^2/100 yr. Because most of the banks are about 6 feet thick, the average maximum sedimentation rate over them can also be expressed as 0.15 ft/100 yr (.04 mm/yr), if a slight amount of compaction is neglected.

Environmental Factors

Florida Bay can be divided into two subenvironments, the interior and the marginal (Ginsburg, 1956). The interior subenvironment is characterized by a low tidal range, less than 0.2 foot, weak currents and a large seasonal variation in water salinity and temperature. The surrounding marginal subenvironment is one of more uniform water characteristics and stronger water motion because of good tidal exchange with the Gulf of Mexico and the Atlantic Ocean. A dashed line on figure 3.1 approximately delineates the two subenvironments.

A tropical wet-and-dry climate prevails over southern Florida and Florida Bay. This climate is characterized by a long dry season, lasting from midfall to late spring, and by an intense rainy period during the summer and early fall months. Precipitation is approximately 40 in./yr with about 70% of this falling during the rainy season. Evaporation probably exceeds rainfall by a factor of 1.5 (table 3.2).

During the rainy season, water salinity in the interior subenvironment may fall as low as 10-15‰ (parts per thousand, i.e., weight of dissolved salts in grams per kilogram of bay water) owing primarily to the influx of runoff from the extensive freshwater swamps (the Everglades) which

*Reproduced from Encyclopedia of Oceanography, edited by Rhodes W. Fairbridge, by permission of Reinhold Book Corp., a subsidary of Chapman-Reinhold, Inc., N.Y., 1966.

Average Grain Size and Constituent Composition of Sediments from Florida Bay and Reef Tract

	Florida Bay (17 samples[1]) Average %	Range %	Reef Tract (25 samples) Average %	Range %
Grain size				
Weight percentage less than $\frac{1}{8}$ mm.	49	10–85	17	0–68
Constituent composition of fraction greater than $\frac{1}{8}$ mm.				
Algae	$\frac{1}{2}$	0–1	42	7–61
Mollusk	76	58–95	14	4–33
Coral	—	—	12	2–26
Foraminifera	11	1–32	9	3–32
Non-skeletal	3	0–3	12	3–24
Miscellaneous	$\frac{1}{2}$	0–4	9	2–23
Unknown	1	0–3	8	4–15
Ostracods	2	1–6	—	—
Quartz	6	0–20	—	—

[1] Only surface scoop samples and the upper parts of cores are included in this average (Table X).

Grain Size and Constituent Particle Compositions of Florida Bay Sediments

Sample Number[1]	Depth Below Top in Cores (cm.)	Grain Size Weight Percentage: Greater than 1 mm.	Greater than $\frac{1}{2}$ mm.	Greater than $\frac{1}{4}$ mm.	Greater than $\frac{1}{8}$ mm.	Less than $\frac{1}{8}$ mm.	Less than $\frac{1}{16}$ mm.	Constituent Particle Compositions[2] Percentage of Fraction Greater than $\frac{1}{8}$ mm.: Mollusk	Foraminifera	Aggregates	Ostracods	Quartz	Halimeda	Miscellaneous	Unknown
1C-a*	0–6	18.7	11.5	9.4	6.6	53.8	50.0	95	2	—	1	1	—	—	1
7C-a*	0–8	0	9.5	25.6	16.1	48.8	43.1	76	14	5	2	2	—	—	1
7C-B	16–22	5.8	4.7	4.5	8.6	76.4	67.8	72	17	2	8	—	—	—	1
8S*	—	9.9	17.0	17.2	21.3	34.6	29.4	79	4	—	3	13	—	—	1
10C-a*	0–5	14.3	7.7	7.1	5.8	65.1	52.7	69	7	18	5	—	—	1	—
11S-2*	—	1.4	8.0	31.0	30.1	29.5	27.8	58	23	2	5	7	1	1	3
12C-b	11–18	2.1	1.4	0.9	1.1	92.7	91.6	79	14	—	7	—	—	—	—
14S*	—	3.6	3.1	2.5	3.0	87.8	85.1	72	11	—	2	7	—	4	4
17D*	—	2.9	7.7	7.6	12.2	69.6	52.4	82	7	—	—	5	1	2	3
20S*	—	14.3	14.5	19.9	11.6	27.4	25.0	86	6	2	1	4	1	—	—
22C-a*	0–6.5	4.1	6.1	18.2	29.6	42.0	35.3	71	3	1	2	20	1	—	2
22C-b	35–42	4.3	4.8	9.5	13.0	68.4	61.5	82	3	—	3	11	1	—	—
22C-c	47–60	6.7	2.9	2.7	3.9	83.8	80.0	92	1	—	4	2	—	1	—
22C-d	72–76	12.6	7.4	6.3	8.3	65.4	60.0	91	1	—	2	3	1	—	2
22C1-a*	0–6	6.4	5.5	4.7	12.3	71.1	57.8	82	6	4	3	4	—	—	1
23S*	—	6.3	7.6	7.6	12.9	65.6	59.3	72	6	1	—	19	1	—	1
25C-a*	0–4	11.1	6.8	5.5	6.6	70.0	66.5	83	6	3	3	4	—	—	1
26C-a*	1–10	9.7	5.5	8.8	11.2	64.7	54.1	64	31	—	3	2	—	—	—
27C-a*	0–4	9.5	7.8	19.3	14.9	48.6	41.8	62	32	—	1	5	—	—	—
28C-a*	7–12	8.8	6.4	8.1	8.4	68.5	63.0	92	3	1	4	—	—	—	—
28C-b	20–28	8.9	5.6	5.4	4.4	75.8	72.8	76	16	2	5	1	—	—	—
28C-c	32–40	4.2	2.6	2.7	2.7	87.8	85.5	77	16	—	6	—	—	—	1
29C-a*	10–18	10.3	5.9	4.6	3.9	75.3	73.3	74	18	4	2	—	—	—	2
29C-b	35–45	5.4	4.2	3.5	2.9	84.1	81.6	79	13	2	6	—	—	—	—
34S-2*	—	68.4	11.9	5.2	3.4	11.2	9.6	86	11	2	—	—	—	—	1

* These samples were used for the average values and ranges given in Tables II and IV.
[1] The locations of the samples are shown in Figure 5. The capital letter indicates the type of sample (C = core, S = surface scoop, D = dredge) and the small letter indicates the divisions of the cores based on examination.
[2] From counts of about 300 grains in each size fraction.

Figure 3.2. Florida Bay data from Ginsburg (1956). For location of numbered samples see figure 3.1.

Molluscan Fauna from Banks in Florida Bay
(From Allen, 1942, p. 111)

Water Depth	Character of Area	Average Number[1] of: Live Mollusks	Genera	Dominant
(Inches)				
$\frac{1}{2}$–5	Broad marl flat; fairly hard bottom; scattered grass	1,220	6	*Cerithium minimum*
1–2$\frac{1}{2}$	Narrow marl flats with soft bottom; no grass	68	6	*Modulus modulus*
2–2$\frac{1}{2}$	Broad marly flat; soft bottom; no grass	258	4	*Cerithium minimum*
2$\frac{1}{2}$–6	Broad marl flat; fairly hard bottom; scattered grass	150	9	*Modulus modulus*
3–5	Semi-enclosed slough; soft bottom	1,072	5	*Anomalocardia cuneimeris*
4$\frac{1}{2}$–5$\frac{1}{2}$	Broad marl flat; fairly hard bottom; no grass	129	8	*Cerithium minimum*
4$\frac{1}{2}$–5$\frac{1}{2}$	Broad marl flat; fairly hard bottom; sparse grass	88	6	*Cerithium muscarum*
5–7	Broad marl flat; fairly hard bottom; scattered grass	227	5	*Pinctada radiata*
(Feet)				
2	Edge of shoal; heavy grass on bottom	83	7	*Pinctada radiata*
3	Edge of shoal; heavy grass on bottom	76	1	*Pinctada radiata*
6	Open bay; marly bottom with sparse grass	75	1	*Pinctada radiata*
6$\frac{1}{2}$	Open bay, marly bottom; no grass	0	0	

[1] The samples were standardized as to the size of area dredged.

TABLE 3.1
Major Physiographic Elements of Florida Bay

	Area		% of Total Area	Number in Bay
	(sq miles)	(sq km)		
Key (islands)	26	67	3	approx. 105
Mud banks (including keys)[a]	192	496	23	
Intra-bay lakes[b]	406	1050	48	approx. 17
Florida Bay[c]	598	1546	74	
Florida Bay[d]	841	2179	100	

[a] Areas of bay having depth less than 2 ft.
[b] Areas of bay ~-7 ft deep and essentially surrounded by mud banks.
[c] Area if western boundary is taken along outer edge of westernmost banks.
[d] Area if western boundary of bay is taken at longitude 81° 05′W.

TABLE 3.2
Environmental Parameters of Florida Bay

	Yearly Average	Range
Rainfall[a]	41 in.	21–83 in.
Evaporation (estimated)[a]	60–70 in.	
Air temperature[a]	75°F; 24°C	34–94°F; 1–34°C
Water temperature[b]		59–104°F; 15–40°C
Water salinity		
Northwest corner of bay[a]	39‰	20–70‰
Bay in general[b,c]		10–70‰
Current velocities[b,c]		
In passes between Florida Keys		3–4 knots setting NW or SE
Within intra-bay lakes	Few hundredths of a knot, counterclockwise rotation	
Turbidity (after two days of 10–20 knot winds)[b]		
Surface water		11.0–17.1 mg/liter
1.5 ft above bottom (depth about 4 ft)		9.4–17.6 mg/liter
Tidal range [b,d]		
Vicinity of Cape Sable		2.9 ft
Along inner (bayside) margin of Florida Keys		0.3 ft
Interior sub-environment of bay		< 0.2 ft

[a] Tabb, Durbrow, and Manning, 1962.
[b] Ginsburg, 1956.
[c] Gorsline, 1963.
[d] U.S. Coast and Geodetic tide tables.

overlie southern Florida. Water salinity during the dry season is typically 35-40‰; however, over the shallowly submerged mud banks, excessive evaporation during spring months may raise salinity values to 70‰, twice that of typical seawater for this region (table 3.2). During years of normal rainfall, water salinity throughout the outer parts of the marginal sub-environment stays near 35-36‰ owing to good tidal exchange with adjacent oceanic and gulf water masses.

Water temperature within the interior subenvironment changes markedly with the seasons and ranges from 15-40°C. Sudden temperature changes accompany strong winds that thoroughly mix and muddy the water of the shallow intrabay lakes (table 3.2).

Owing to the baffling effect of the mud banks, water circulation in Florida Bay is very complex. Tidal and wind-stress currents are weak owing to this baffling and also to the shallowness of the bay, which appreciably dampens the height of the tidal wave as it sweeps northwestwardly across the bay. Based on hydrographic data, Gorsline (1963) computed a weak counterclockwise rotation (table 3.2) for the water mass within several of the lakes. Strong tidal currents of several knots, however, carry water into the bay from the Atlantic through between-island passes separating the Florida Keys.

Fauna and Flora

The bottom fauna in Florida Bay is dominated by mollusks (at least 100 genera are known) and to a far lesser extent by Foraminifera, especially large tropical forms (table 3.3). The flora is predominatnly turtle grass, *Thalassia testudium,* a true marine phanerogam or seed producing plant. Turtle grass thrives on the shallow mud banks, and its root system may be instrumental in their stabilization. Calcified or partially calcified green algae, principally species of *Halimeda* and *Penicillus,* are common in current-swept areas near the Florida Keys. Small corals also thrive in these areas.

Unconsolidated Calcareous Sediments

Texture. Unconsolidated sediments in Florida Bay are typically more than 90% composed of calcium carbonate; the remainder primarily comprises quartz, opal (as tests of Radiolaria and as sponge spicules), and disseminated and fibrous organic matter. The grain size of the calcareous matter ranges from large mollusk or coral fragments, measuring several centimeters or more in length, to ultrafine particles less than 1μ in greatest dimension. The mud banks are approximately 30% (by weight) composed of sand-size (larger than 0.062 mm) carbonate particles, 48% silt- and clay-size constituents (0.062-0.001 mm) and 22% subclay-size particles (less than 0.001 mm or 1μ). The average median grain is close to 0.028 mm, which means the average bank sediment is a medium silt (table 3.4). In lithified form (as limestone), these sediments would be classified as fossiliferous calcilutites or as biomicrites.

Ginsburg (1956) determined that for particles larger than 0.125 mm, 76% are derived from mollusks and 11% from Foraminifera. Under the microscope, virtually all of the calcareous matter down to a size of 0.020 mm can also be recognized as skeletal in origin, i.e., derived from calcified parts of plants and animals (Taft and Harbaugh, 1964). Fragmentation of coarse skeletal parts to particles as small as 20μ may to some extent be accomplished by physical abrasion, but doubtless sediment-ingesting and sediment-burrowing organisms account for much of the comminution (Ginsburg, 1957). Interestingly, some sediment-ingesting organisms also increase the grain size of the calcareous muds by aggregating fine-grained matter, during the digestive process, into firm ellipsoidal fecal pellets. The pellets are up to several millimeters in length and may constitute as much as 50% of the sediment. The origin of calcareous grains smaller than 0.020 mm is much debated; this problem is largely centered around the precipitation of minute needles of aragonite, which form a substantial proportion of particles less than 0.010 mm in length.

Mineralogy. Four minerals form the calcareous fraction of silty muds in Florida Bay: aragonite, high-magnesian calcite, low-magnesian calcite, and dolomite, listed here in their order of abundance (table 3.4).

Aragonite is composed of calcium carbonate and crystallizes in the orthorhombic system; this mineral is an unstable polymorph of calcium carbonate and converts to calcite (perhaps even dolomite) during some phase of limestone formation. In Florida Bay, aragonite forms about 65% of the sediment and is primarily contributed by mollusks (tables 3.3 and 3.4). Most of the aragonite occurs as aggregates of light-brown crystals, but a small amount consists of acicular needles generally less than 0.010 mm in length (Taft and Harbaugh, 1964). Aragonite needles are found in the sediments of other shallow tropical seas as well, especially in those of the neighboring Bahaman platform. Initially, denitrifying bacteria living in the bottom mud were thought to bring about precipitation of the needles. Subsequently it was thought that loss of carbon dioxide, either through high photosynthetic activity of marine plants or by evasion of the gas produced by evaporation or solar heating, prompts precipitation of aragonite from supersaturated water. Most recently aragonitic skeletal rods found in partially calcified green algae, especially species of *Penicillus,* have been considered the source of the aragonite needles. A thorough discussion of these ideas and of the physical chemistry involved in precipitating calcium carbonate in the metastable aragonite form is given by Cloud (1962).

High-magnesian calcite is calcite with more than 4%, and up to about 19%, randomly substituted $MgCO_3$ in the lattice of this hexagonal mineral. High-magnesian calcite constitutes 20-30% of the mud banks and occurs as microcrystalline aggregates of light-brown crystals similar to the aragonite aggregates (Taft and Harbaugh, 1964). Like aragonite, high-magnesian calcite is metastable and during some phase of limestone formation converts to low-magnesian calcite or dolomite. Calcite containing large amounts of magnesium is undoubtedly derived mostly from Foraminifera, especially the species *Peneroplis proteus* and *Archais angulatus.* Other organisms, both plant and animal, supply secondary amounts of high-magnesian calcite to the sediments of Florida Bay.

Only about 15% of the deposits of Florida Bay is composed of low-magnesian calcite (essentially common calcite), the stable hexagonal polymorph of calcium carbonate. The bulk of this mineral is derived from mollusks. However, some of the low-magnesian calcite occurs as colorless transparent hexagonal prisms less than 0.062 mm in length; the source of these prisms is not fully known. Prisms of low-magnesian calcite are typically pitted and etched, and the impression is gained that they are undergoing solution (Taft and Harbaugh, 1964). Crystals of aragonite and high-magnesian calcite do not show this effect. This is paradoxical because at ambient earth temperature and pressures, thermodynamic calculations predict and experiments show that low-magnesian calcite is the least soluble of the three.

Dolomite is a stable double-carbonate mineral bearing equal atomic amounts of calcium and magnesium; this mineral crystallizes in the hexagonal system typically as unit rhombohedrons. Dolomite accounts for a maximum of about 5% of most sediments in Florida Bay. This mineral, a subject of intense research, was not found in the bay until 1961 (Taft and Harbaugh, 1964). Within the bay, dolomite occurs as vitreous rhomboid crystals measuring less than 0.062 mm across; they commonly have a dark central area.

Implicit in the foregoing discussion is the fact that the calcareous silts of Florida Bay are 80-85% composed of metastable carbonate minerals (aragonite and high-magnesian calcite) that will convert to calcite or dolomite during some phase of limestone formation. The obvious question to ask, therefore, is how soon and under what conditions will the transformation take place? Stehli and Hower (1961) have shown that exposures of the young (about 100,000 year-old) Miami Oolite and Key Largo Formations, which immediately underlie the sediments of the bay, contain little to no high-magnesian calcite and moderate to low concentrations of aragonite (typically less than 17%). Because both limestone units contain fossils that were originally substantially composed of these minerals, it is obvious that relatively early postdepositional (diagenetic) changes have induced conversion of these minerals to low-magnesian calcite. Two studies of unconsolidated bank sediments in Florida Bay have shown that little if any diagenetic mineral transformation has occurred since calcareous sediments began to accumulate in the bay approximately 4,000 years ago (Ginsburg, 1957; Taft and Harbaugh, 1964). However, Fleece (1962) believes he has detected a significant loss of high-magnesian calcite and some aragonite in sediments about 4,000 years old which immediately overlie bedrock at the base of mud banks in eastern Florida Bay. The general consensus, nonetheless, is that Recent aragonite and high-magnesian calcite are "stable" within the environmental framework of Florida Bay. Conversion of these minerals to low-magnesian calcite and/or dolomite must probably await a pronounced change in this environment—for ex-

TABLE 3.3

Important Animal and Plant Genera That
Contribute Calcium Carbonate to Sediments in Florida

Animals			Plants
Mollusks	Corals[b]	Foraminifera	Green Algae[b]
Anomalocardia	*Porites*	*Archaias*	*Halimeda*
Bulla	*Siderastrea*	*Peneroplis*	*Penicillus*
Brachidontes		*Quinqueloculina*	*Udotea*
Cerithium			
Chione			
Modulus			
Pinctada			
Tellina			

[a]Data are from Thorp (1936 and 1939) and Ginsburg (1956).
[b]Common only in vicinity of Florida Keys and in tidal channels.

TABLE 3.4

Physical, Mineralogical and Chemical Characteristics of
Sediments in Florida Bay

	Western Florida Bay		Eastern Florida Bay	
Mass properties	Average[a]		Average[b]	
Water content (dry weight basis), %	71.5			
Porosity, %	65.9			
Grain density, g/cc	2.71			
Bulk density, g/cc	1.58			
Grain size distribution				
Median diameter, mm	0.028		0.025	
Trask sorting coefficient	6.81			
	Average[c]	Range[c]	Average[d]	Range[d]
Wt % > 0.125 mm (sand)			51	15–90
Wt % > 0.062 mm (sand)	30	1–52		
Wt % < 0.062–0.001 mm (silt-clay)	48	9–70		
Wt % < 0.001 mm (subclay)	22	5–50		
Carbonate mineralogy	Subsurface and surface sediments[c]		Surface sediment[c]	
	Average	Range	Average	Range
% aragonite	59	35–77	46	20–78
% high-Mg calcite	27	3–54	37	0–51
% low-Mg calcite	14	1–29	17	4–80
	Surface sediment[c]		Surface sediment[c]	
	Average	Range	Average	Range
% aragonite	59	41–70	67	40–100
% high-Mg calcite	26	10–47	19	0–39
% low-Mg calcite	15	10–20	16	0–44
General sediment chemistry	Average[c]	Range[c]	Average[e]	Range[e]
Ca/Mg	25	6–41		
Sr/Ca × 10³	8.7	6.3–11.7		
% Sr			0.42	0.20–0.64
% Mg			1.4	0.06–4.1
% Mn			0.006	0.0005–0.04
% Ba			0.002	0.001–0.004
	Average[d]	Range[d]	Average[b]	
% calcareous minerals	87	81–90		
% non calcareous minerals	9	8–13		
% organic matter	4	2–6	6.2	
% organic carbon	2.1	1.3–3.7	3.5	
% organic nitrogen	0.15	0.29–0.09	0.1	
Organic carbon/organic nitrogen	17	13–26	27.5	

[a]Previously unpublished data (from author's files) based on entire thickness, up to 5 ft, of bank sediments.
[b]Fleece (1962).
[c]Taft and Harbaugh (1964).
[d]Ginsburgh (1956).
[e]Stehli and Hower (1961).

ample, the marked change in interstitial water chemistry that would accompany subaerial exposure of the sediments of Florida Bay.

Value of Shallow-water Carbonate Studies

The results of sedimentological research on calcareous deposits in Florida Bay and over the nearby Bahaman platform have in many ways enhanced our ability to reconstruct the amalgam of physical, chemical and biological processes that created limestones in the past. A number of papers incorporating some of these results are collected in a publication edited by Bass and Sharps (1963). Considering the number of symposia on carbonate sediments that have been held within the last several years, there is little doubt that scientific interest in limestone formation and diagenesis will continue in the future. Florida Bay will continue to be a focal point of this interest because it is one of our most readily accessible natural laboratories for the investigation of shallow-water carbonate sedimentation and carbonate mineralization.

References

Bass, R. O., and Sharps, S. L., 1963 (editors), Symposium on Shelf Carbonates of the Paradox Basin, Four Corners Geological Society, Denver, Colo., 273pp.

*Cloud, P. E., 1962, "Environment of calcium carbonate deposition west of Andros Island, Bahamas, *U.S. Geol. Survey Profess. Papers*, **350**, 138pp.

*Fleece, J. B., 1962, "The Carbonate Geochemistry and Sedimentology of the Keys of Florida Bay, Florida," Unpublished M. S. Thesis in Geology, Florida State University, Tallahassee, Florida, 112pp. (available as report to U.S. Office of Naval Research, Sedimentological Res. Lab., Dept. Geol.).

*Ginsburg, R. N., 1956, "Environmental relationships of grain size and constituent particles in some south Florida carbonate sediments," *Bull. Am. Assoc. Petrol. Geologists,* **40**; 2384-2427.

Ginsburg, R. N., 1957, "Early diagenesis and lithification of shallow-water carbonate sediments in south Florida," in (Le Blanc, R. J., and Breeding, J. G., editors) "Regional aspects of carbonate deposition," *Soc. Econ. Paleontologists Mineralogists Spec. Publ.,* **5**, 80-100.

Gorsline, D. S., 1963, "Environments of Carbonate Deposition Florida Bay and the Florida Straits," in (Bass, R. O., and Sharps, S. L., editors) Symposium on Shelf Carbonates of the Pardox Basin, pp. 130-143. Four Corners Geological Society, Denver, Colo., 273pp.

Stehli, F. G., and Hower, J., 1961, "Mineralogy and early diagenesis of carbonate sediments," *J. Sediment. Petrol.,* **31**, 358-371.

Tabb, D. C., Dubrow, D. L., and Manning, R. B., 1962, "The ecology of northern Florida Bay and adjacent estuaries," *State of Florida Board of Conservation Technical Series,* **39**, 81pp. (Tallahassee, Florida).

*Taft, W. H., and Harbaugh, J. W., 1964, "Modern carbonate sediments of southern Florida, Bahamas and Espiritu Santo Island, Baja, California; a comparison of their mineralogy and chemistry," *Stanford Univ. Publ. Univ. Serv. Geol. Sci.,* **8**, No. 2

*Thorp, E. M., 1936, "Calcareous shallow-water marine deposits of Florida and the Bahamas," *Carnegie Inst. Wash. Publ.,* **452**, *Papers Tortugas Lab.,* **29**, 37-119.

———, 1937, "Florida and Bahama Marine Calcareous Deposits," in (Trask, P. D., editor) "Recent Marine Sediments," pp. 283-297, Amer. Assoc. Petroleum Geologists, Tulsa, Oklahoma, 736pp.

Algal Mats

For a discussion of the blue-green algal mats so common in Florida Bay see chapter 2, pages 37 to 45 of this Guidebook. See also Ginsburg *et al.,* (1954), and Ginsburg (1960) and Eutsler (1970). For details of algal mats in other areas see Monty (1965) and Neumann *et al.,* (1970—abstract on page 93 of this Guidebook). See also figures 3.5 through 3.8.

Sediments

In many respects Florida Bay can be considered an independent mud producing factory. Stockman, Ginsburg, and Shinn (1967) demonstrated that the green algae *Penicillus* is a major fine aragonite mud (less than 15 microns) contributor of Florida Bay. They further note that the present mud production rate by *Penicillus* (.3 cm/1,000 years) could account for 1/3 of the present mud in northeastern Florida Bay today. Production of skeletal silt (15 to 62 microns) and mud by the biological and mechanical breakdown of resistant skeletons, mollusks, algae, corals, etc., is also noted as effective sediment producers in Florida Bay. See Matthews (1966) for such lime mud production by physical breakage and abrasion in lagoons of southern British Honduras. Lowenstam and Epstein (1957) describe aragonite needle mud from organic sources. See Barron (1976) for a discussion of source and dynamics of suspended sediments in the nearshore regions of southwestern Florida.

The abstract and graphic synopsis of a detailed study of a single core (see figure 2.39 for core location) taken on Cross Bank by Müller and Müller (1967) is given below. See also figures 3.3 and 3.4.

MINERALOGISCH-SEDIMENTPETROGRAPHISCHE UND CHEMISCHE UNTERSUCHUNGEN AN EINEM BANK-SEDIMENT (CROSS-BANK) DER FLORIDA BAY, U.S.A.

by

German Müller and Jens Müller

Summary: The sediments of a core of 1.55 m length taken on the windward side of the Cross Bank, Florida Bay, are clearly subdivided into two portions, as shown by grain size analysis: silt-sized particles predominate in the relatively homogeneous lower two-thirds of the core. This is succeeded abruptly by a thin layer of sand, containing fragments of *Halimeda*. They indicate a catastrophic event in the Florida Bay region, because *Halimeda* does not grow within Florida Bay.

Above this layer, the amount of sand decreases at first and then continuously increases right to the present sediment-water-interface. The median and skewness in-

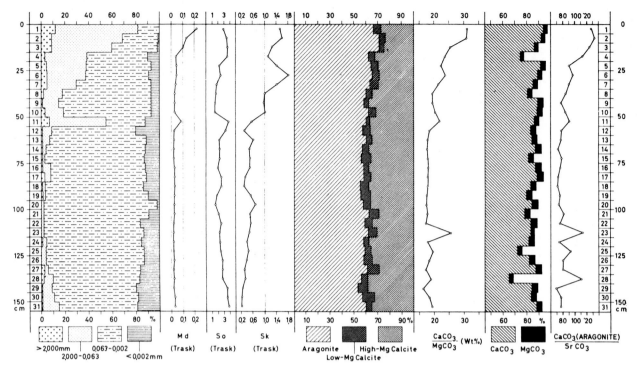

·Synopsis of grain size analysis, mineralogy, and chemical composition of sediments.

crease simultaneously with the increase in the sand and granule portion. We assume that the changing grain size distribution was determined chiefly by the density of the marine flora: during the deposition of the lower two-thirds of the core a dense grass cover acted as a sediment catcher for the fine-grained detritus washed out of the shallow basins of the Florida Bay, and simultaneously prohibited renewed reworking. Similar processes go on today on the surface of most mud banks of Florida Bay.

The catastrophic event indicated by the sand layer probably changed the morphology of the bank to such an extent that the sampling point was shifted more to the windward side of the bank. This side is characterized by less dense plant growth. Therefore, less detritus could be caught and the material deposited could be reworked. The pronounced increase in skewness in the upper third of the core certainly indicates a strong washing out of the smaller-sized particles.

The sediments are predominantly made up of carbonates, averaging 88.14 percent. The average $CaCO_3$-content is 83.87 percent and the average $MgCO_3$-content amounts to 4.27 percent. The chief carbonate mineral is aragonite making up 60.1 percent of the carbonate portion in the average, followed by high-magnesian calcite (33.8 percent) and calcite (6.1 percent).

With increasing grain size the aragonite clearly increases at the cost of high-magnesian calcite in the upper third of the core. Chemically, this is shown by an increase of the $CaCO_3$: $MgCO_3$-ratio. This increase is mainly caused by the more common occurrence of aragonite fragments of mollusks in the coarse grain fractions. The bulk of the carbonates is made up of mollusks, foraminifera, ostracods, and—to a much lesser extent—of corals, worm-tubes, coccolithophorids, and calcareous algae, as shown by

microscopic investigations. The total amount of the carbonate in the sediments is biogenic detritus with the possible exception of a very small amount of aragonite needles in the clay and fine silt fraction.

The individual carbonate components of the gravel and sand fraction can be relatively easy identified as members of a paticular animal or plant group. This becomes very difficult in the silt and clay fraction. Brownish aggregates are very common in the coarse and medium silt fraction. It was not always possible to clarify their origin (biogenic detritus, faecal pellets or carbonate particles cemented by carbonates or organic slime, etc.).

Organic matter (plant fragments, rootlets), quartz, opal (siliceous sponge needles), and feldspar also occur in the sediments, besides carbonates.

The lowermost part of the core has an age of 1365 ± 90 years, as shown by C^{14}-analysis.

A review of investigations in Florida Bay from 1851 to 1962 is given by Fleece (1962) who presents the results of his geochemical and sedimentological study of cores from Florida Bay. Such studies indicate (1) loss of high-magnesium calcite at the base of cores, (2) presence of strontium enriched sediment, and (3) evidence that the overall depositional environments of Florida Bay have remained fairly constant over the last 4,000 years. Manker and Griffin (1971) describe both source and mixing of chlorite and smectite in Florida Bay. They describe x-ray diffraction techniques and illustrate typical x-ray patterns. Their abstract is noted below.

Figure 3.3. Undignified walking prevails in the grass-held mud banks of Florida Bay. Photo was taken at low tide.

Figure 3.4. Box core of mud taken on top of Cross Banks showing stabilizing carpet of turtle grass (*Thalassia testudinum*) and its deep network of roots and rhizones; some mollusk shells are also present.

Figure 3.5. Lacework of shallow individual "mud lakes" made possible by grass-held mud banks connecting with commonly algal mat-covered supratidal mud islands. Florida Bay. U.S. Army Map Service photograph #16113 VV AJ M110 AMS 11 Jan. 51.

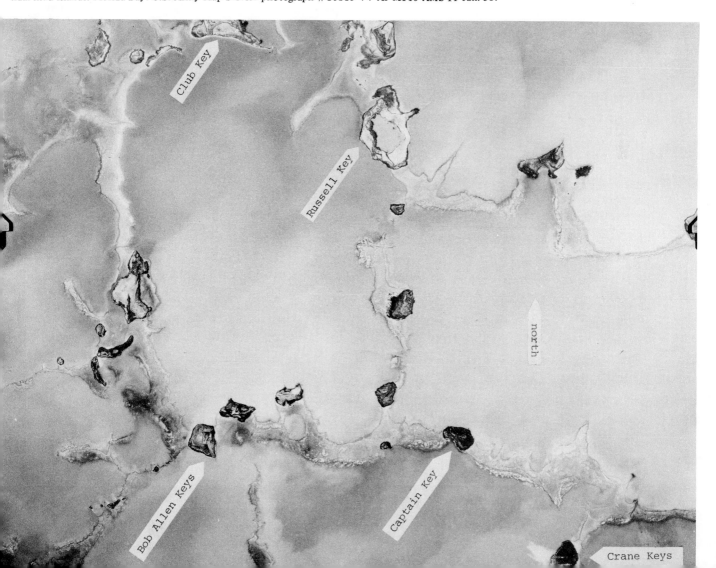

Figure 3.6. Example of supratidal island (Cluet Key) in Florida Bay. Beach ridge seen here prevents low interior of island from being flooded except during storm periods.

Figure 3.7. Supratidal blue-green algal mat showing varieties of desiccation cover the low interior of Cluet Key.

Figure 3.8. Close-up of desiccated curled mats, Cluet Key.

SOURCE AND MIXING OF INSOLUBLE CLAY MINERALS IN A SHALLOW WATER CARBONATE ENVIRONMENT— FLORIDA BAY
by
J. P. Manker and George M. Griffin

Chlorite and smectite dominate the clay-size insoluble residue of Recent carbonate sediments of Florida Bay. Illite and kaolinite also occur in very small quantities.

Chlorite is derived from the Atlantic Coast and eastern Everglades provinces and is introduced by streams and by tidal channels through the northern Florida Keys. Smectite is derived from the Gulf of Mexico province to the west. In the northern part of Florida Bay, water flow is greatly impeded by a complex bank and basin system, and the clay mineral suites remain relatively segregated near their respective sources. However, in the southern part of the bay, banks are less frequent, water flow is less impeded, and the clay mineral suites mix gradually across the area.

Ginsburg (1956) provides the "classic study" of the sediments of Florida Bay (and adjacent reef tract). For the abstract of his paper see page 185 of this Guidebook. For an illustrated discussion of the Florida Bay area including petrography and implication of sediments see Bathurst (1971, p. 154-161). Metastable aragonite and high-magnesium calcite are the predominant minerals found in Florida Bay today. Stehli and Hower (1961) and Taft and Harbaugh (1964) present the results of mineralogical and geochemical studies of these unstable minerals.

A sequence of cores taken across a series of emergent mud islands and adjacent lakes by Multer (1967—unpublished) suggests that some islands of Florida Bay migrate laterally in time. Discussions of shapes and distribution of these islands with long-term local residents of the area substantiate this idea of shifting environments. The vertical sequence and evolution of Florida Bay sediments is discussed by Enos and Perkins (1976) whose abstract is given below.

EVOLUTION OF FLORIDA BAY INTERPRETED FROM ISLAND STRATIGRAPHY
by
P. Enos and R. D. Perkins

The sedimentary record of most Florida Bay islands is an asymmetric cycle consisting of a transgressive sequence followed by a regressive sequence formed during a continuous Holocene rise in sea level. The principal sedimentary environments of Florida Bay and the south Florida mainland are represented in an upward succession of: (1) freshwater pond; (2) coastal mangrove swamp; (3) shallow bay ("lake"); (4) mud bank; and (5) island. Some parts of the cycle may be missing, but the sequence is always the same. Supratidal carbonate sedimentation on islands may develop from coastal mangrove swamps or by mangrove colonization of mud banks. Islands have developed from mud banks at many different times during the rise of sea level into Florida Bay, indicating that mud banks must have existed throughout most of the bay's history. Present trends of island formation and growth suggest that Florida Bay will evolve into a coastal carbonate plain with inland mangrove swamps and freshwater ponds, very similar to the present southwest Florida mainland.

The effect of sinkhole topography on lagoonal sediment thickness and vegetation patterns is described by Dodd and Siemers (1971) and figure 3.9 below. See also Kotila *et al.,* (1975) for study of carbonate sediments off Long Key overlying karst topography in the Key Largo limestone.

EFFECT OF LATE PLEISTOCENE KARST TOPOGRAPHY ON HOLOCENE SEDIMENTATION AND BIOTA, LOWER FLORIDA KEYS
by
J. R. Dodd and C. T. Siemers

Detailed mapping of bedrock topography on Bahia Honda and Big Pine Keys has revealed a buried karst topography not previously documented in the lower Florida Keys. This topography, developed during lowered sea level of the Pleistocene strongly controls Holocene sediment thickness and present biotic distribution. Circular to oval sinkholes, which are up to 75 m or more in diameter and over 4 m deep, are usually completely filled with peat and carbonate sediment. Sinkholes are well developed on both the Miami limestone (oolitic facies) and the Key Largo limestone (both late Pleistocene in age). Thick sediment in buried sinkholes in more than a few inches of water favors the growth of thick patches of turtle grass (*Thalassia testudinum*). Shallower water and supratidally located sinkholes (that is, those partly or wholly surrounded by subaerially exposed bedrock) are generally marked by thick growths of either red or black mangroves (*Rhizophora mangle* and *Avicennia nitida*). These distinct, nearly circular vegetation patterns are extremely abundant in the study area, as shown by aerial photographs which suggest that Bahia Honda and Big Pine Keys are "riddled" with sinkholes.

Coalescing depressions on top of bedrock are found beneath a small enclosed lagoon on the northeast side of Big Pine Key by Fowler *et al.,* (1970) who also give a description of sediment and organisms.

In the northeast corner of Florida Bay there are a series of semi-isolated bodies of water which have been studied by various workers. Buttonwood Sound (see figure 6.1 aerial photo) has been described by Lynts (1966a) whose abstract is quoted below. See also figure 3.10. Lynts also describes Foraminifera of Buttonwood Sound—see page 164 of this Guidebook.

Figure 3.9. Vertical aerial photo of Bahia Honda Key showing the abundance of sinkholes as indicated by dark vegetation patches. Width of photo is approximately 3.2 kilometers. From Dodd and Siemers (1971, p. 216).

RELATIONSHIP OF SEDIMENT-SIZE DISTRIBUTION TO ECOLOGIC FACTORS IN BUTTONWOOD SOUND, FLORIDA BAY
by
G. W. Lynts

Seventy-four sediment samples were collected from 19 stations located in Buttonwood Sound, Florida. Stations were occupied on August 14th, 17th, 20th, 1962, and February 9th, 1963. At each occupation of stations, the following environmental parameters of sediment-water interface were measured: depth, temperature, salinity, pH and Eh. Techniques outlined for measuring pH and Eh must be strictly adhered to in order to obtain valid measurements of *in situ* environment. Factor-vector analysis of numerical data indicate that ecologic factors were not linearly related to sediment-size and were not linearly interrelated amongst themselves. Sediment-size distribution is closely correlated to turtle grass occurrence, which acts as an effective sediment-stabilizer and trapper. Temperature and salinity are related to climate, while pH and Eh are related to organic activity. Carbonate sediments of Buttonwood Sound are probably almost wholly organically derived.

Barnes Sound lies just northeast of U.S. Route 1 as it crosses from the mainland to the Keys. Some investigations by Multer (unpublished) show a sequence of storm and current deposited shell debris in-terbedded with algal mats which lie from 45-50 cm below present sea level (figs. 3.11 and 3.12).

For many years one of the most fascinating areas to the writer has been the molluscan hash mound in the center of a mud "lake" of Florida Bay. Examination of this locality has been of high interest to geologists both from universities and oil companies. This mound is approximately 3 km north of Windley Key adjacent to the red intercoastal waterway marker "78A" (see figure 2.39 and page 125 of this Guidebook). The X-section sketch drawn by the writer illustrates the main characteristics of this mound. In a survey by G. W. Gray (1974-unpublished senior thesis), the peat beneath this mound at .75 m depth below mean low water was dated at 1900 ± 185 years from radio carbon analysis. A bedrock map made from data gathered by probing and coring indicated three depressions up to 2 m depth which were cited as responsible for initial trapping of mud development by a *Thalassia* baffle mechanism which built the mound. Using Scholl's 1964 data and conclusions, Gray (1974) indicates probably freshwater origin for the sediment beneath the peat (calcite more abundant) and marine origin for sediment above the peat (aragonite more abundant). Figures 3.13 and 3.14 are from Gray's 1974 study.

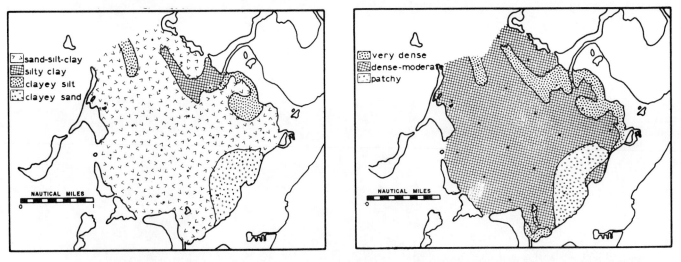

Figure 3.10. Sediment-size (left) and *Thalassia* (right) distribution in Buttonwood Sound from Lynts (1966).

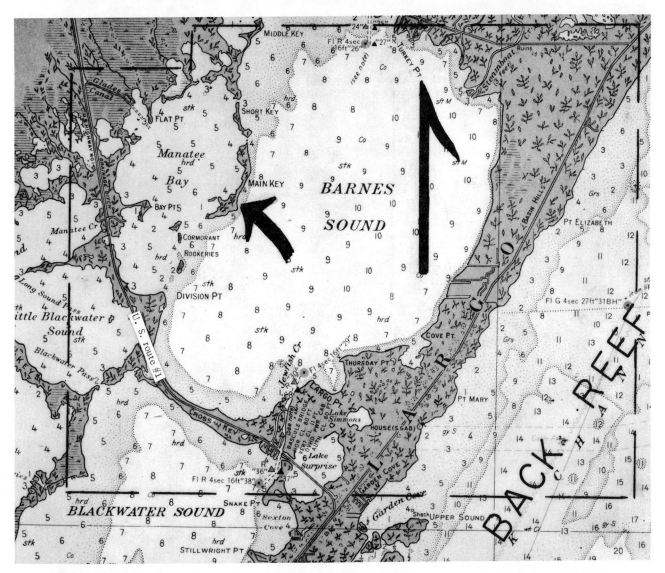

Figure 3.11. Index map of Barnes Sound with arrow showing location of core pictured in figure 3.12. Map from C. G. S. chart #1249.

Figure 3.12. Core taken 4 m offshore, south side of Main Key. Top of core (upper left) is 10 cm below low tidal level. Note graded bedding down to algal mat followed by current bedding underlain by contoured mat; basal shell unit lies on thick sequence of mud containing mangrove roots and partially dissolved shells.

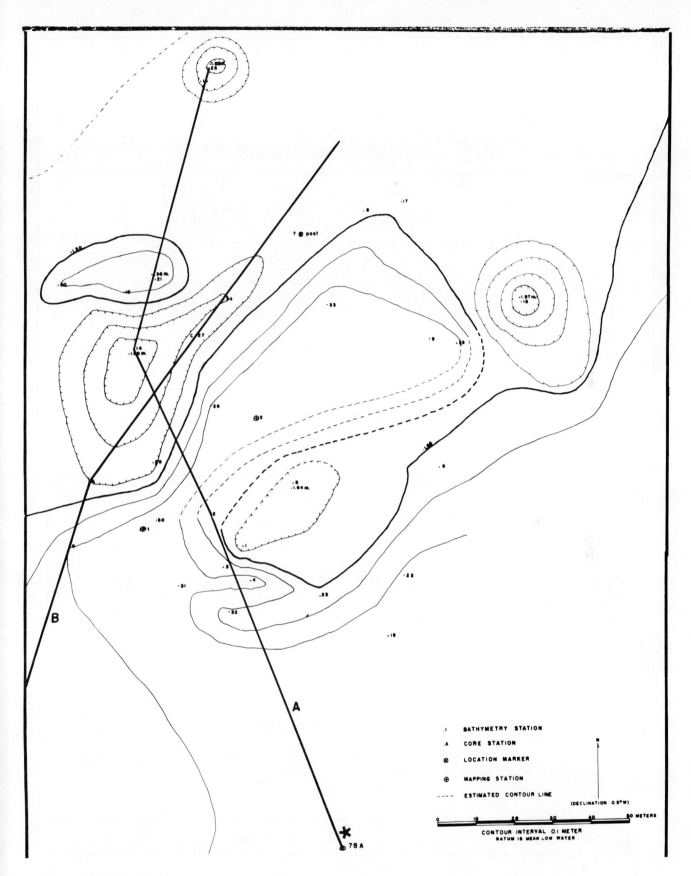

Figure 3.13. Bedrock surface contour map beneath mound just north of C. G. S. marker "78A" (see *), Florida Bay. From Gray (1974).

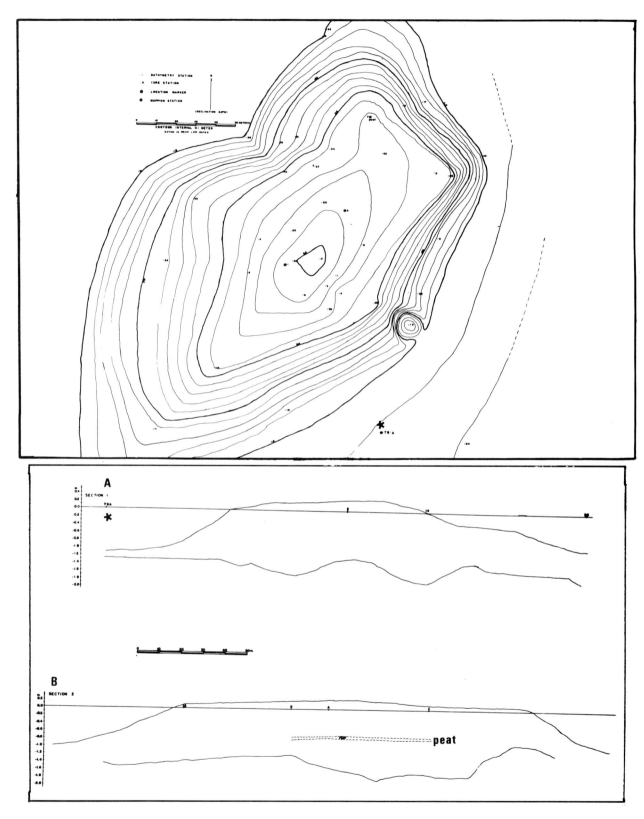

Figure 3.14. (top) Sediment surface contour map of mound just north of C. G. S. marker "78A" (see *), Florida Bay, Contour interval .1 m datum mean low water. (bottom) Cross sections showing surface of emergent mound and underlying depressions in bedrock. See figure 3.13 for location of sections. From Gray (1974).

Hurricane Effects

Hurricanes have a variety of effects on lagoonal sedimentation. Large variations in supratidal mud sedimentation (layers from less than .6 cm to 10 cm) due to hurricanes passing over Florida Bay are reported by Perkins and Enos (1968). See also Ball, Shinn and Stockman (1967), Thomas (1974), and Warzeski (1976).

An abstract by Pray (1966) and two of his illustrations concerning the Hurricane Betsy influence on Florida carbonate sediments are given below.

HURRICANE BETSY (1965) AND NEARSHORE CARBONATE SEDIMENTS OF THE FLORIDA KEYS
by
L. C. Pray

The center of Hurricane Betsy (1965) passed westward over the northern Florida Keys with 120-140 mph winds and "tides" 3-5 feet high. The geologic record of Betsy well illustrates the complex interplay of normal and catastrophic processes for various nonreef carbonate shoals near Tavernier.

Hurricane erosion and deposition was negligible or minor (<1/2 inch) on most subtidal and intertidal areas, such as the Tavernier Bank (ocean-facing "back-reef" shelf), tidal deltas, and sinuous mud mounds of Florida Bays. The strikingly minor changes reflect the turbulence resistance of these biota-protected carbonate shoals, and the bottom "insulation" created by coincidence of high tide, storm surge, and hurricane winds.

Supratidal deposits formed locally on the low-relief (1 foot) interiors of Florida Bay islands. Storm sediment consists of carbonate clay, silt, and minor sand—variably admixed with *Thallassia* and mangrove debris. Sediment thickness rarely exceeds 1-2 inches and diminishes toward island interiors. Large areas that were inundated received only a sediment film. Deposits form marginal inward-sloping sheets (levees) 20-100 feet wide and patchy irregular layers extending tens to several hundreds of feet into interior algal-mat swamps. Most sediment occurs directly inland of those island sides subjected to maximum wave attack; apparently most sediment was transported by large waves during relatively normal water levels. The nature, thickness, and geologic preservation of the interior supratidal deposits is strongly influenced by microtopography and shrub distribution. Hummocks or swells of only a few inches relief permit localized destructive bioturbation or peat-induced dissolution of entire carbonate layers.

Foraminifera

For a summary paper on species definition and the taxonomy of benthic Foraminifera of Florida Bay and adjacent waters see Bock (1971) who also summarizes previous work in the south Florida area. In this paper Bock identified 235 species belonging to 99 genera and correlates faunal groups with areal changes in the physical environment. He distinguishes (p. 2) five faunal groups as described below.

Straits fauna. Although this faunal assemblage was taken from a single station it is included to emphasize the faunal differences between the various habitats. The major fa.ctor controlling faunal distribution within this group is depth. The most abundant species are: *Cibicides cicatricosus, Hoegelundina elegans* and *Pyrgo comata.* Lesser elements include *Cassidulina subglobosa, Ehrenbergina pacifica, Pyrgo fornasinii, Pyrulina cylindroides, Rectobolivina advena, Robertinoides bradyi, Spiroloculina soldanii, Technitella legumen,* and *Trifarina bradyi.*

Back-reef fauna. This group has specimens in common with the bay fauna, but is sufficiently different to be recognizable. The major distinguishing species are: *Discorbis rosea, Homotrema rubrum* and *Pyrgo murrhina.* Lesser elements include: *Fissurina wiesneri, Ammodiscus anguillae, A. tenuis. Laticarenina holophora, Pyrgo elongata, Schlumbergerina alveoliniformis* var. *occidentalis* and *Spiroloculina rotunda.* It has two characteristic species in common with the Gulf fauna: *Textularia agglutinans* and *Uvigerina flintii.*

Brackish-water fauna. This group is characteristic of the small bays immediately adjacent to the Florida mainland. These bays are subject to pronounced runoff of fresh water during the wet season and evaporation during the dry season, and therefore the fauna in them is controlled predominantly by salinity. There are only two species characteristic of this group: *Ammonia beccarii* var. *parkinsoniana* and *Elphidium discoidale.* Although these species are present in other groups they never occur in large percentages other than in the brackish-water areas, where they may be the only species present.

Gulf fauna. This group is characterized by the following species: *Bigenerina irregularis, B. nodosaria, B. textularoidea, Eponides repandus, Marginulina planata, Lenticulina calcar, L. iota, Quinqueloculina bicostata, Spiroplectammina floridana, Textularia agglutinans, T. candeiana, Textulariella barrettii, Uvigerina flintii, U. peregrina* and *Textularia mayor.* There are many minor species found in this group. The distribution of individual species is discussed in the section on Systematic Paleontology.

Bay fauna. This group contains a very large number of species of which only the most abundant are listed here. It includes: *Articulina lineata, A. mayori, A. mexicana, A. mucronata, A. multilocularis, A. pacifica, A. Sagra, Bolivina lanceolata, B. lowmani, B. paula, B. pulchella* var. *primitiva, B. striatula, Cyclogyra involvens, C. planorbis, Rosalina floridana, Cymbaloporetta squammosa, Neoconorbina orbicularis, Elphidium advenum, E. sagrum, Cribroelphidium poeyanum, Eponides antillarum, Fursenkoina complanata, F. compressa, F. pontoni, Hauerina bradyi, Miliolinella circularis, M. fichteliana, M. labiosa, M. suborbicularis, Nonion depressulum* var. *magordanum, N. grateloupi, Peneroplis pertusus, Broeckina orbitolitoides, Pyrgo denticulata, P. subsphaerica, Quinqueloculina bidentata, Q. bosciana, Q.*

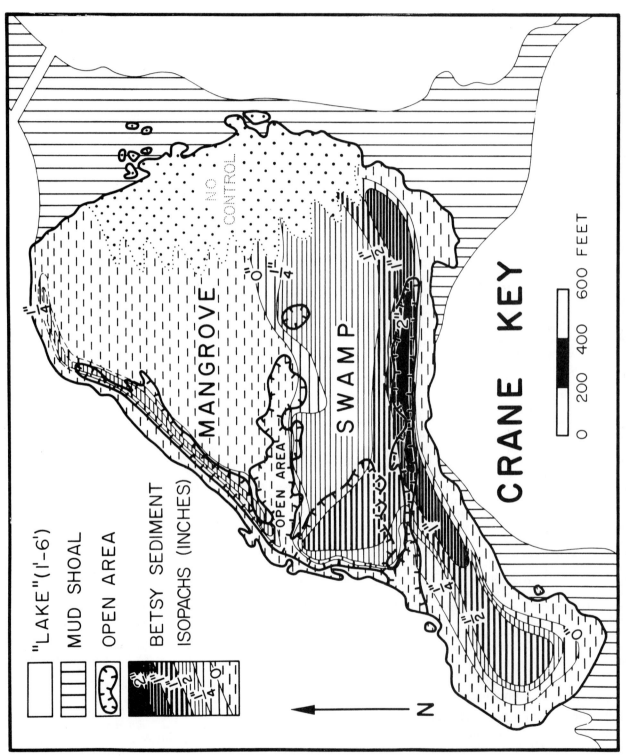

CRANE KEY

MANGROVE SWAMP

NO CONTROL

OPEN AREA

N

0   200   400   600 FEET

"LAKE"(1'-6')

MUD SHOAL

OPEN AREA

BETSY SEDIMENT
ISOPACHS (INCHES)

Figure 3.15. Crane Key Betsy Sediment isopach map. See Pray (1966). Map courtesy of L. C. Pray.

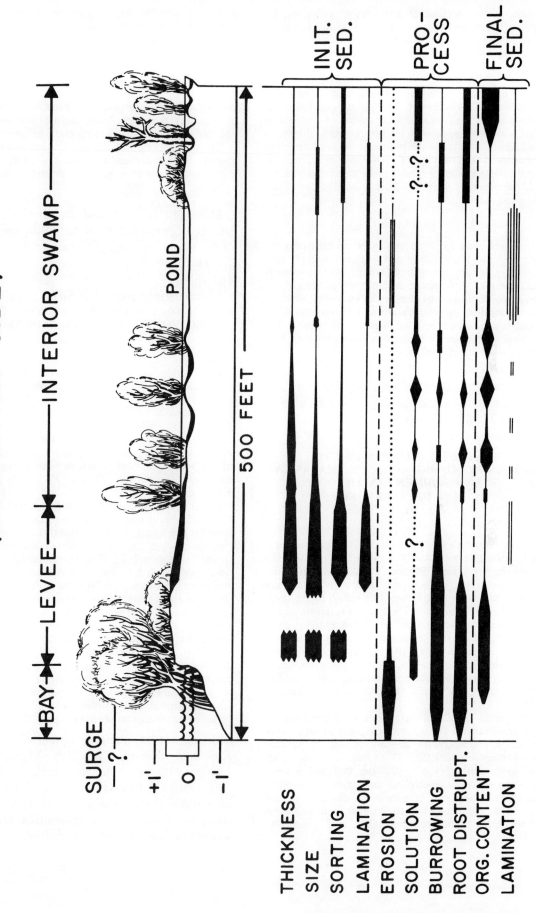

Figure 3.16. "Typical" Island Profile. See Pray (1966). Profile courtesy of L. C. Pray.

laevigata, Q. lamarckiana, Q. poeyana, Q. polygona, Q. tenagos, Q. sabulosa, Q. seminula, Q. subpoeyana, Q. tricarinata, Q. wiesneri, Sorites marginalis, Spirillina vivipara, Spiroloculina antillarum, S. arenata, Spirolina acicularis, S. arietinus, Triloculina bassensis, T. bermudezi, T. bicarinata, T. carinata, T. fitterei var. *meningoi, T. linneiana, T. oblonga, T. placiana, T. rotunda, T. sidebottomi, T. tricarinata, T. trigonula* and *Valvulina oviedoiana.* All the above species appear to be able to tolerate major changes in both temperature and salinity.

An additional factor in the bay fauna which affects the distribution of dead tests is sorting by wave and current action resulting in the larger species being associated with the coarsest sediment while the smaller are seemingly confined to the finest sediment.

Conclusions

Ninety-nine genera and 235 species are represented in the sediments from Florida Bay and adjacent waters. Five faunal groups are recognized which are apparently characteristic of different environments: the Straits and Gulf faunas controlled by depth; the back-reef and bay faunas controlled by temperature and salinity; and the brackish-water fauna controlled primarily by salinity. The bay fauna's distribution appears to be secondarily controlled by wave and current action.

An abstract by Bock (1967) of a detailed study concerning Foraminifera in the Big Pine Key area of Florida Bay is quoted below.

MONTHLY VARIATION IN THE FORAMINIFERAL BIOFACIES ON THALASSIA AND SEDIMENT IN THE BIG PINE BAY AREA, FLA.
by
W. D. Bock

A fauna consisting of 38 genera and 81 species of benthonic Foraminifera inhabits the waters between Big Pine Key and the Torch Keys, Florida. The distribution of this fauna was controlled by the type of substrate which in turn appeared to be controlled by the distribution of the turtle grass, *Thalassia testudinum.* The fauna was dominated by 9 species living on the grass and by 11 species living on or in the sediment. A few species were common to both environments. Eighteen species lived primarily on the grass and 10 species were restricted to the sediment. Four species showed preference for coarse sediment and two for fine sediment. One new genus and two new species are described.

Ten species exhibited temperature dependence in population changes. The other species showed no correlation between temperature or salinity and population changes except at one station. This station was located adjacent to deeper, more oceanic water and had a lower salinity range. Most species at this station showed a temperature dependence in population changes. Several species of the fauna were limited to this station, which was a transition zone between polyhaline and stenohaline waters.

The distribution and abundance of the Foraminifera were also controlled by competition with other organisms, living in the same environment as well as by interspecific competition. Dead blades of *Thalassia,* subject to transportation by currents and wave action, provided a means of dispersal for foraminiferal species living on them.

The coiling direction of *Rosalina floridana* was temperature dependent, with sinistrally-coiled specimens increasing in percentage as temperature decreased.

Trace element analyses of tests of *Archaias angulatus* and *Sorites marginalis* showed no monthly variation in mineralogic composition from site to site.

Lynts (1971) conducted distribution and model studies on Foraminifera living in Buttonwood Sound of the Florida Bay area. The abstract of his paper is given below.

DISTRIBUTION AND MODEL STUDIES ON FORAMINIFERA LIVING IN BUTTONWOOD SOUND, FLORIDA BAY
by
G. W. Lynts

Distribution of foraminiferal standing crop in Buttonwood Sound, Florida Bay, was investigated to define ecological factors controlling or influencing species distributions. A total of 74 samples were collected from a grid system consisting of 19 stations. Stations were occupied three times during August 1962, and once during February 1963. The following environmental parameters of the water-sediment interface were measured: depth, temperature, salinity, pH and Eh. Each sample was analyzed for foraminiferal standing crop and ratio of sand, silt and clay. Q-modal factor-vector analysis divided the standing crop into 17 faunal assemblages, consisting of three to five assemblages in a single collection. Distribution of most of these assemblages appeared to be controlled by an interaction of ecological parameters. A fauna which was related to sediment size and one which was related to bathymetry persisted throughout all collections. Distribution of foraminiferal species and environmental parameters indicated some species were influenced by the measured environmental parameters, but most species showed no such relationships. The faunal information indicated there was no simple linear relationship between distribution of Foraminifera and environmental parameters. Distribution of Foraminifera living in Buttonwood Sound was controlled by a complex interplay of physicochemical and biologic factors, only partially reflected by the measured parameters.

Steinker (1974) discusses variation in cultures of a *Rosalinid* foraminifer of Florida that may be correlative with environmental factors. Habitats of foraminifers in Florida (and Bahamas) are discussed by Steinker (1976). See Smith (1971) for a study of the present benthonic Foraminifera fauna of the western portion of lower Florida Bay.

Mollusks

For a comprehensive study of the molluscan fauna in Florida Bay the reader is referred to the paper by Turney and Perkins (1972), the abstract of which is noted below.

MOLLUSCAN DISTRIBUTION IN FLORIDA BAY
by
W. J. Turney and R. F. Perkins

Within Florida Bay, four subenvironments can be recognized by the physical characteristics of salinity and variability of salinity, water circulation, and wind. The *Northern Subenvironment* is characterized by low and variable salinities due to freshwater runoff from the Florida mainland. The *Interior Subenvironment* is characterized by restricted circulation and is relatively unaffected by tidal exchange with either the Gulf of Mexico or the Atlantic Ocean. It is subject to large salinity variations due to seasonal or annual climatic variations. The *Atlantic Subenvironment* has near-normal marine salinity, with mixing of waters with the Atlantic Ocean through tidal passes in the Florida Keys. The *Gulf Subenvironment* has near-normal marine salinity, but its position in a wind and current "shadow" causes its waters to be more stagnant than those in the Atlantic Subenvironment.

The fauna of Florida Bay is dominantly molluscan, principally gastropods and bivalves which are represented by approximately 100 genera and 140 recognized species. A few "index species" and several "consistently common species" define four molluscan suites whose distributions appear to be controlled by the environmental influences characterizing the four subenvironments.

Molluscan debris comprises 58 to 95 percent of the sediment particles greater than 1/8 mm. It is believed that the disintegration process is almost entirely organic and effected by crabs, boring sponges, perforating algae, holothurians, worms, and *Thallasia* roots. Thin-shelled bivalves tend to breakdown more rapidly than thick-shelled bivalves and gastropods.

Variations in the oxygen and carbon isotope ratios of Florida Bay mollusks and their environmental significance is discussed by Lloyd (1964).

SMALL, COMPLEX LAGOONAL ENVIRONMENTS BORDERING MAINLAND

CARD SOUND: Card Sound is a small (approximately 5 × 16 kilometers) lagoon bordered on the north by the Florida mainland and on the south by the Florida Keys (Key Largo). The Sound opens to the north into Biscayne Bay, to the east by tidal passes to the ocean and to the south into Barnes Sound and Florida Bay. See index map 8 in Appendix.

Eardley and Goodell (1968) conducted a detailed examination of Card Sound. The abstract of their report is noted below. See also figure 3.17.

THE SEDIMENTS OF CARD SOUND, FLORIDA
by
C. F. Eardley and H. G. Goodell

The surface sediments of Card Sound, Florida, and of the adjacent continental shelf differ considerably. The Sound is almost completely separated from the shelf by the northern end of the Florida Keys. The shelf sediments are entirely carbonate, composed primarily of aragonite, high-magnesium calcite, and lesser amounts of low-magnesium calcite. Grain size generally decreases seaward but increases with proximity to the series of coral knolls which parallel the shore landward of the present living reef. In the trough between the knolls and the Keys, sediments are better sorted and higher in calcite. High-magnesium calcite increases toward the coral knolls.

In contrast, the presence of detrital quartz in Card Sound creates a dual-component sediment, with low-magnesium calcite being the predominant carbonate mineral, although some aragonite and high-magnesium calcite also occur. The sediments are poorly sorted, and mean grain size and sorting decrease both down the long axis of the Sound and toward the Keys. The noncarbonate fraction is mainly well-sorted quartz. Both the amount of quartz and the size of the individual grains decrease toward the southwest. The distribution of quartz in several size fractions shows that sediment transport is primarily by waves and wave-generated currents whose direction and magnitude are seasonal and, secondarily, by tidal currents. The facies change from a calcareous quartz sand to calcarenite is transitional over a 10-mile interval.

LARGE SEA ISLAND LAGOONAL ENVIRONMENT

BIGHT OF ABACO: The shallow Bight of Abaco represents a large (80 × 35 km) sea island lagoonal environment completely isolated from mainland influence. Perched on the eastern end of the Little Bahama platform, this average 7 m deep lagoon is mostly contained within the borders of a low relief emergent carbonate bedrock and a shallow rock and shoal "lip" or "rim" which marks the edge of the drop-off into the adjacent deep basin (see figure 3.18 and figure 8.1). Such a restrictive perimeter distinguishes large sea island lagoons from the more open and physically unrestricted large Interior Platform Environments such as can be seen on Great Bahama Bank (see chapter 4). For a description of coastal morphology of southwest Great Abaco Island see Raphael (1975).

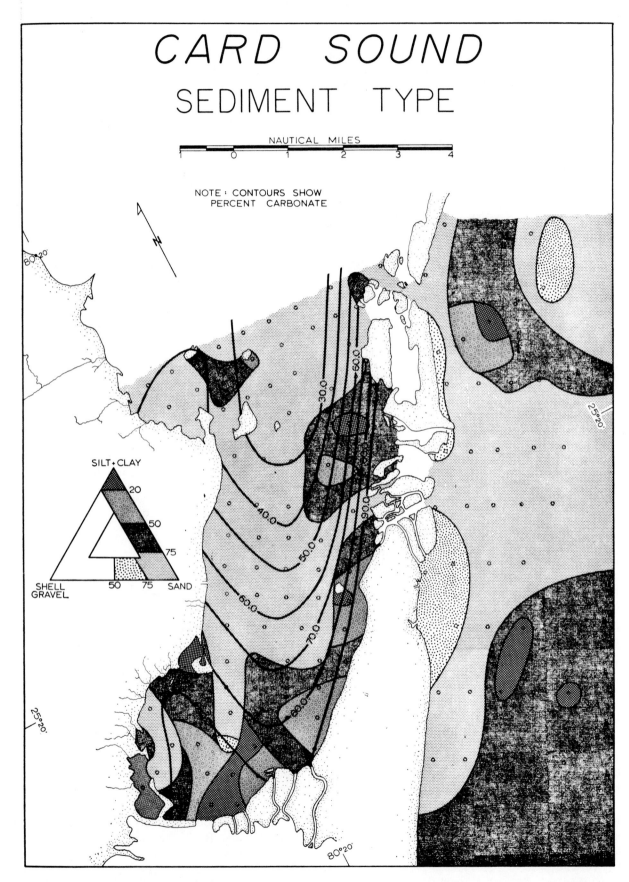

Figure 3.17. Sediment type in Card Sound. Contours give the percentage of CaCO$_3$ in the Sound; the shelf is 100% CaCO$_3$. From Eardley & Goodell (1968, p. 997).

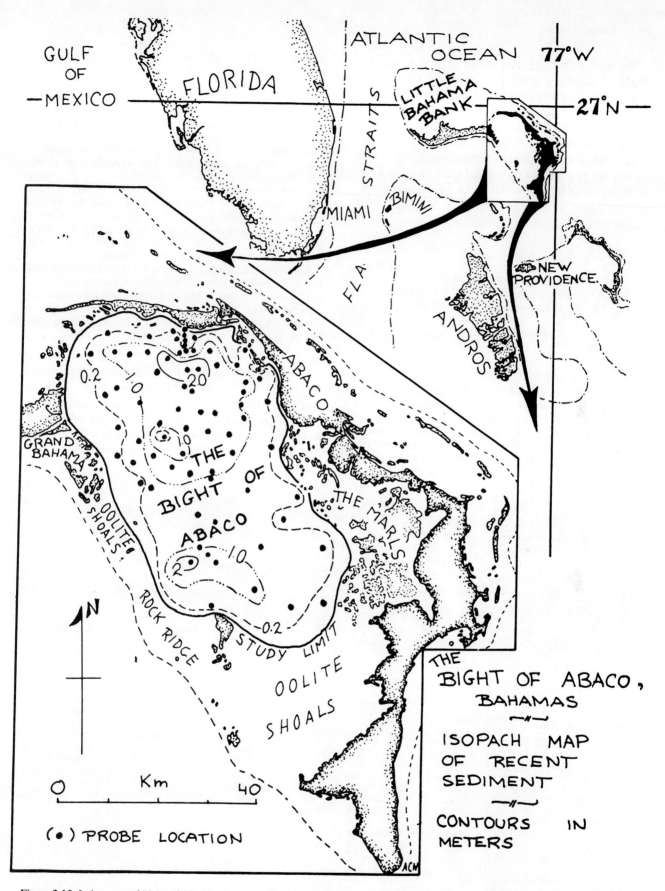

Figure 3.18. Index map of Bight of Abaco area on southeastern end of Little Bahama Bank. Contours show thickness of unconsolidated carbonate sediment over limestone bedrock in meters. Probe data control is shown by black circles. Map and description from Neumann and Land (1975).

As with other large lagoonal environments, the Bight of Abaco is a prolific sediment (mud) producer. In a comprehensive paper by Neumann and Land (1975) the Bight of Abaco is described in terms of a model for carbonate algal mud production (or overproduction) and dispersal. Data on sediments include composition, volume, density, mass, size range as well as dispersal as suspended load into the adjacent basin (see figure 3.19). Algal populations, growth rates and breakdown are described. The abstract of this paper is given below.

LIME MUD DEPOSITION AND CALCAREOUS ALGAE IN THE BIGHT OF ABACO, BAHAMAS: A BUDGET
by
A. C. Neumann and L. S. Land

Abstract

The 7 m deep Bight of Abaco, 80 × 35 km in size, is a semirestricted marine lagoon on Little Bahama Bank. A community of *Thalassia* and calcareous green algae grows on and contributes to a 2 m thick sediment carpet of calcareous muddy sand, which is accumulating at a net rate of 120 mm/10^3 yrs. Measurements of suspended sediment concentrations yield an estimate of a total off-bank sediment loss of 172 × 10^{10} kg. A budget is prepared which compares total algal carbonate production with sediments of possible algal derivation now in the basin. Sediment thickness was determined by probing. Density, mineralogy, grain composition and texture were determined from grab samples and cores. Laboratory experiments on the breakdown of calcareous algae reveal that products include sand from *Halimeda* plus material less than 62μ m for all algae. The standing crop of the mud producers *Penicillus, Rhipocephalus,* and *Halimeda* averages 22 plants m^2 and is highest in the winter. Calcification ranges from 45 to 64 percent dry weight depending upon the species. Work by others indicates a growth rate for *Penicillus* of 6 to 12 crops per year. Radiocarbon ages indicate that marine conditions similar to the present have prevailed for 5,500 years.

Results reveal that calcareous green algae have produced 1.5 to 3 times the mass of aragonite mud and *Halimeda* sand now in the basin. Thus it is concluded that the study area, a typical interior Bahama Bank lagoon, produces much more sediment by calcareous green algae alone than can be accommodated on the bank tops.

A detailed study of the Holocene depositional history of the Bight of Abaco area based on core analysis, subbottom profiles, isopach studies and surface sediment analysis has been done by Boardman (1976). An abstract of the results of his study and selected illustrations are given below (see figures 3.20 to 3.23).

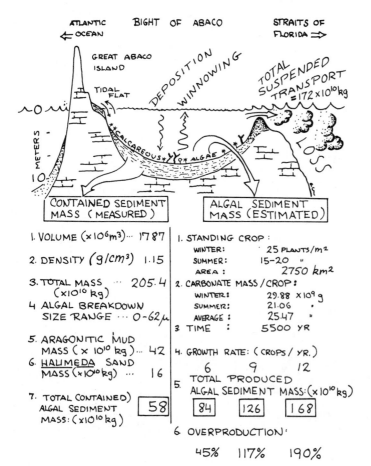

Figure 3.19. Diagrammatic presentation of the results of the budgetary considerations of lime mud production by calcareous green algae in the Bight of Abaco, Bahamas. On the left are listed the parameters from which is calculated the mass of all sediment in the basin that could have been derived from calcareous green algae, *i.e., Halimeda* sand plus aragonitic mud. On the right are listed the parameters from which is calculated the mass of sediment that could be produced by a population of calcareous green algae like the present one over the Holocene marine lifetime of the basin at rates of growth known or assumed. Comparison of the resultant figures of sediment mass yields an overproduction of sediment by the algae relative to that contained in the basin. Diagram and description from Neumann and Land (1975).

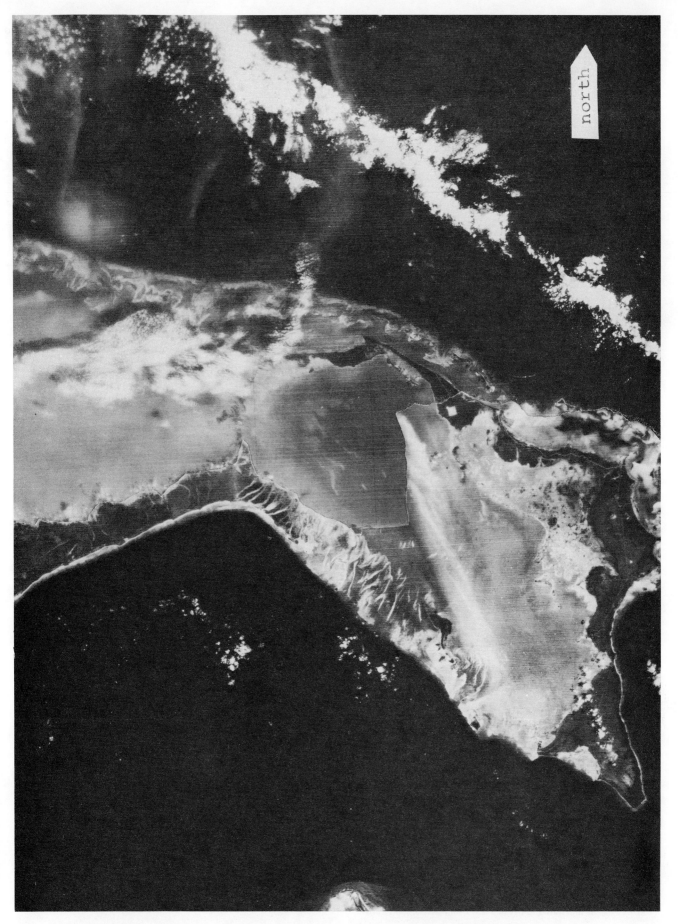

Figure 3.20. Photomosaic of satellite view of Bight of Abaco. Photos from NASA and courtesy of M. Boardman.

Area	Sample #	Hali-meda %	Bi-valves %	Sori-tids %	Lumps %	Mud %	Area	Sample #	Hali-meda %	Bi-valves %	Sori-tids %	Lumps %	Mud %
	37d	8	16	20	15	42		6d	11	8	45	4	?
	4d	4	23	23	15	27		7d	7	8	38	20	17
A	27a	7	8	26	19	38	D	32a	19	9	35	17	16
	Cliff	8	15	18	10	84		Rim	24	15	15	12	7
	average	6.8	15.5	21.8	14.8	47.8		average	15.3	10.0	33.3	13.3	13.3
	std. dev.	1.89	6.14	3.50	3.69	24.98		std. dev.	7.68	3.37	12.87	6.99	5.51
	30b	13	15	25	3	47		34a	10	9	14	41	11
	10d	26	9	25	6	29		35a	1	4	4	83	4
	18d	21	18	20	3	75		36a	1	5	3	77	1
B	21d	20	11	26	9	48		2c	10	3	7	74	3
	34d	16	15	30	9	37		4c	1	10	40	9	3
	Deep	22	17	25	1	75		15d	8	7	19	21	3
	average	19.7	14.2	25.2	5.2	51.8		44d	2	8	35	24	7
	std. dev.	4.59	3.49	3.19	3.37	19.25		48d	1	2	22	20	1
	16d	14	13	24	11	38		average	4.3	6.0	18.0	43.6	4.1
	17d	11	11	22	8	43		std. dev.	4.27	2.93	13.89	29.88	3.36
C	30d	19	4	31	9	26	More's Is.	1c	25	3	5	24	44
	41d	20	10	29	6	28							
	average	16.0	9.5	26.5	8.5	33.8	Marls	47d	14	20	37	6	29
	std. dev.	4.24	3.87	4.20	2.08	8.10							

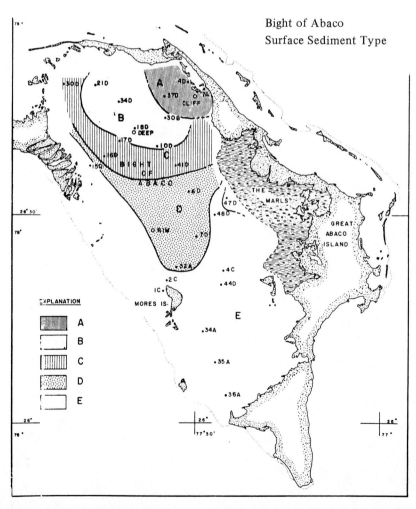

Bight of Abaco
Surface Sediment Type

Figure 3.21. Surface sediment types and their respective constituents. This map also provides location sites for three cores (DEEP, RIM & CLIFF) as illustrated in figure 3.23. From Boardman (1976).

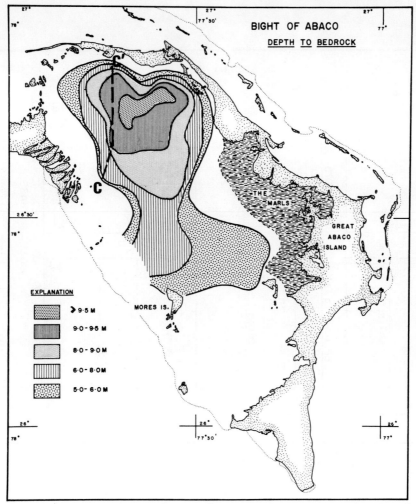

BIGHT OF ABACO
DEPTH TO BEDROCK

THE MARLS

GREAT
ABACO
ISLAND

MORES IS.

EXPLANATION

>9·5 M

9·0 - 9·5 M

8·0 - 9·0 M

6·0 - 8·0 M

5·0 - 6·0 M

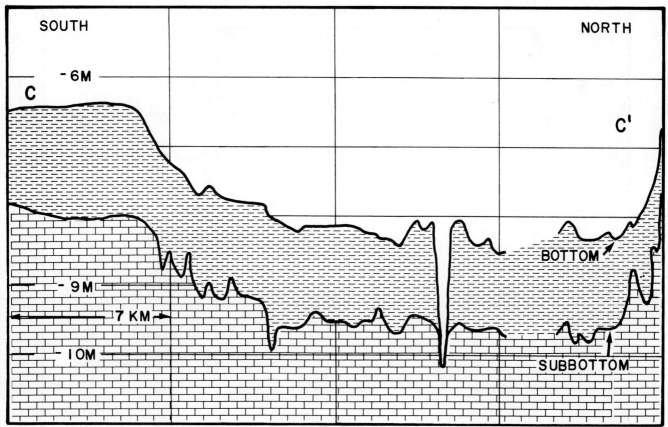

SOUTH

NORTH

- 6M

C

C'

BOTTOM

- 9M

7 KM

- 10M

SUBBOTTOM

Figure 3.22. Map showing depth to bedrock below water surface in Bight of Abaco and cross section of subbottom profile along section C-C'. Note karstlike bedrock topography. From Boardman (1976).

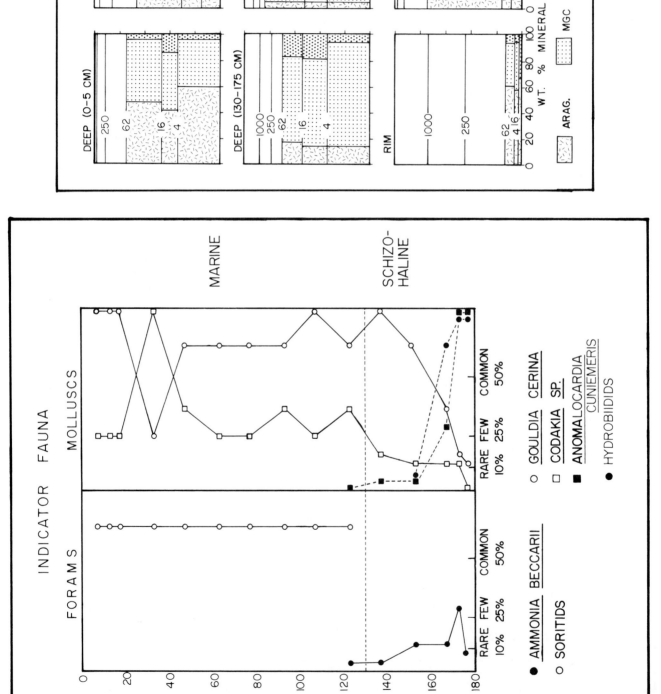

Figure 3.23. Bight of Abaco core analysis. (left) Indicator Fauna from DEEP core. (right) Mineralogy and size analysis from DEEP core at four different intervals and averaged data from the RIM and CLIFF cores. See figure 3.21 for location of core sites. Data from Boardman (1976).

HOLOCENE DEPOSITIONAL HISTORY OF THE BIGHT OF ABACO, BAHAMAS
by
M. Boardman

The Bight of Abaco is a shallow tropical silled lagoon in the northern Bahamas and is accumulating muddy carbonate sediment. Analysis of subbottom profiles, sediment probes, sediment cores (texture, mineralogy, constituent grain analysis, SEM analysis), and surface sediments provide evidence of 3 major depositional facies (environmental changes) in the lagoon caused by the Holocene rise of sea level.

A thin (< 5 cm) soil zone composed primarily of quartz and boehmite of probable aeolian origin developed on the karstlike bedrock topography until sea level rose high enough to affect the central, deepest portion of the basin by infiltrating through the porous limestone (~6100 C-14 years BP). A brackish to schizohaline (variable salinity) environment prevailed from 6100 BP to 3600 BP and resulted in 50 cm thick muddy carbonate deposit with a characteristic macrofauna and a mud characterized by a high clay content (49% of the mud or 33% of the sediment) and low aragonite content (15% of all mud fractions). The end of the schizohaline environment was ended and a marine environment began in the B/A when sea level rose above the sill depth and allowed a direct oceanic influence. This marine unit (130 cm thick) is characterized by a much lower clay content (23% of mud or 15% of the sediment) and a much higher aragonite content (aragonite > 45%) especially in the clay fraction (aragonite = 63%).

The textural and mineralogic analyses coupled with this environmental analysis indicate that the marine environment allowed winnowing and export from the basin of carbonate mud, primarily algal-derived clay. There is no evidence for diagenesis in the marine sediment, however, the early schizohaline sediments show evidence (SEM and mineralogic evidence) of high-Mg calcite precipitation accompanying pyrite formation.

Subtidal sediment-binding algal mats in the shallow parts of the Abaco area are described in detail in an excellent paper by Neuman *et al.,* (1970) who discuss the composition, structure and erodabil-

ity of these mats. The abstract and three illustrations from this paper are given below.

THE COMPOSITION, STRUCTURE AND ERODABILITY OF SUBTIDAL MATS, ABACO, BAHAMAS
by
A. C. Neumann, C. D. Gebelein and T. P. Scoffin

The composition and microstructure of widespread subtidal biological mats binding sandy carbonate sediments in the Rock Harbour Cays vicinity of Little Bahama Bank were examined in detail; these mats were subjected to *in situ* flume experiments. The mats consist of various assemblages of green algae, red algae, blue-green algae, diatoms and animal-built sand grain tubes. Green algae, red algae and/or sand grain tubes provide a rigid open network into which grains infiltrate and are trapped. The mucilaginous secretions of both blue-green algae and diatoms in association with the fine filaments of blue-green algae bind the grains to each other and to the mat network. On the basis of composition and microstructure, three basic mat types were recognized: a fibrous, rigid, *Cladophoropsis* mat; a thin, gelatinous, *Lyngbya* mat; a cohesive, aggregated, *Schizothrix* mat. The erosin by artificial currents of initially undistrubed mats was studied in the field using an underwater flume, and the complex manner in which the mats disintegrated was recorded. The surface sediment from each mat area was then treated with NaOCl to remove the organic matter and erosion tests repeated in a tank in the laboratory with the same apparatus. The natural, *in situ,* mat-bound sediment could withstand current velocities at least twice as high and, in some cases, as much as five times as high as those that eroded the treated, unbound sediment. The intact mat surface could withstand direct current velocities three to nine times as high as the maximum tidal currents (13 cm/sec) recorded in the mat environment. Each mat type eroded in a characteristic manner and sequence dependent upon the mat composition and microstructure. This breakdown process differed markedly from the erosion behavior of loose sediment. Observations indicate that grain size, sorting,

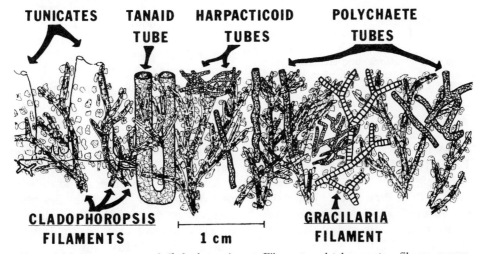

Figure 3.24. Microstructure of *Cladophoropsis* mat. Filaments and tubes create a fibrous, porous fabric within which grains are bound by diatom, blue-green and animal mucilage to form a thick mat. From Neumann *et al.* (1970).

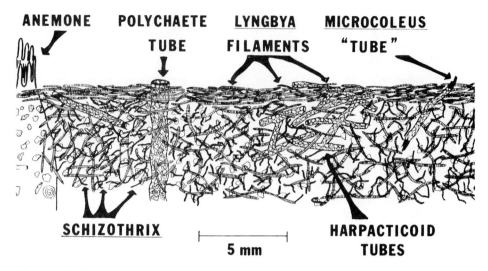

Figure 3.25. Microstructure of *Lyngbya* mat. High surface concentrations of flat-lying *Lyngbya* filaments create a thin, smooth, cohesive and gelatinous mat. From Neumannn *et al.* (1970).

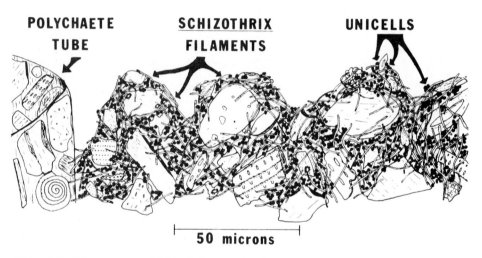

Figure 3.26. Microstructure of *Schizothrix* mat. Abundance of fine filaments and mucilage create a rough-surfaced mat lacking rigid fibrous structure but within which grains are completely enmeshed. From Neumann *et al.* (1970).

packing, structure and sediment surface morphology are influenced by mat formation. This study demonstrates the need for close consideration of interfacial biological communities when examining depositional and erosional processes at the sediment-water interface, or when making interpretations from the products of these processes in ancient rocks.

SMALL SEA ISLAND LAGOON ENVIRONMENT

COUPON BIGHT (Big Pine Key, Florida): Coupon Bight is a small (2 × 3.2 km) horseshoe-shaped lagoon located at the south end of Big Pine Key (see aerial photo figure 3.27). It is open to the Newfound Harbor channel on the west and to the open ocean through a limited number of low, intertidal, and supratidal banks between Keys forming its southwestern boundary. It is readily accessible from U.S. Route 1. Boats for visiting the Bight may be rented along Newfound Harbor channel or at the southeast end of Big Pine Key.

The close proximity of patch reefs within .8 km of the southern boundary of Coupon Bight (bordering Newfound Harbor Keys, see Dodd *et al.,*, 1973) makes this locale the most convenient of any in the Keys for seeing both lagoonal and coral patch reef environments.

An interesting feature found on the western edge of Coupon Bight Lagoon as described by Kissling (below) and mapped by Multer (unpublished) is the apparent control of topographic highs (grass-held mud banks) by bedrock lows. In other words, it appears that initial topographic lows provided the "niche" for incipient grass growth. Once started, such mounds have perpetuated themselves by their

Figure 3.27. Aerial photograph of southern end of Big Pine Key, showing Coupon Bight and Newfound Harbor Keys. Patch reefs can be found off the south shore of many of the Keys. U.S.A.F.

baffle action and deep root hold-fast network. The question that arises from this observation is—"could large-scale ancient mud mounds be associated with or detected by similar depressions as shown on structural contour maps drawn on the base of shale units?" It is interesting to note that Turmel and Swanson also described a topographic low in the rock floor beneath Rodriguez Bank mud mound on the back side of the Florida reef tract (see page 167).

Detailed investigations of Coupon Bight have been done by James P. Howard, Don L. Kissling and Jerry A. Lineback. The abstract of a summary report from Howard *et al.,* (1970) is given below.

SEDIMENTARY FACIES AND DISTRIBUTION OF BIOTA IN COUPON BIGHT, LOWER FLORIDA KEYS
by
J. F. Howard, D. L. Kissling and J. A. Lineback

Coupon Bight is a 9-sq-km bay separated from an offshore Inner Reef Tract by low-lying islands and a baymouth bar. Coupon Bight has less wave activity and less complete tidal exchange than Inner Reef Tract waters, and is no more than 250 cm deep. Temperatures and salinities in Coupon Bight are more variable and usually higher than those in waters of the Inner Reef Tract. The harsher environmental conditions of Coupon Bight are reflected in its less diverse biota, compared to that of similar areas in the Inner Reef Tract. Environemntal gradients persisting between the various parts of Coupon Bight produce more pronounced faunal variations than are found in the more homogenous Inner Reef Tract. Physical and chemical conditions, sedimentary facies, and biofacies of the bight are closely related. Several major environments can be recognized in Coupon Bight, each characterized by a particular assemblage of organisms. Open bay, nearshore, restricted bay, baymouth bank, and tidal channel environments are present.

The unconsolidated sediment of Coupon Bight is mainly autochthonous skeletal debris, with minor contributions washed in from the Inner Reef Tract and from eroded limestone bedrock. The texture of bight sediments ranges from sand to mud, with most of the mud facies found in banklike deposits associated with thick growths of the marine grass *Thalassia*. Most of the fine-grained sediment in Coupon Bight is the product of disintegration of coarse shell debris. Fine materials are winnowed out by waves and currents and either trapped on *Thalassia* covered mud banks or carried out of the bight.

The variability of and relations between environments, sedimentary facies, and biofacies that can occur in a small area such as Coupon Bight are clarified by the data collected during this study. As a result, similar variability in shallow-water marine carbonate rocks can be better interpreted.

Papers presented separately by Lineback (1968), Kissling (1968) and Howard and Faulk (1968), at the Ohio Academy of Science meeting in Toledo, Ohio (4/19/68) include details of particular interest for visiting investigators and are reprinted here (with permission of the authors).

MACROFAUNAL AND FLORAL DISTRIBUTIONS AND CONTROLS IN COUPON BIGHT, LOWER FLORIDA*

by

Jerry A. Lineback

Distributions of macrofauna and flora in Coupon Bight indicate distinct zonation of species present, of total abundance, and of diversity. The highest levels of abundance and diversity occur in portions of the Bight where there is maximum tidal interchange with offshore waters. Shallow areas of the Bight, where there is restricted interchange by tidal currents, are characterized by low abundance and diversity.

The general benthonic macrobiota can be divided into three communities reflecting differences in environmental conditions. They are:

1. The *Thalassia* community—characterized by thick accumulations of unconsolidated sediment and abundant growth of *Thalassia*. Also present

are the algae *Acetabularia, Phipocephalus,* and *Penicillus;* a bivalve, *Pinctata radiata;* and unidentified small brown encrusting and purple finger sponges.
2. The sandy bottom-sponge community—characterized by very thin sediment cover and a sparse growth of *Thalassia*. Sponges attaching themselves to the rock bottom such as *Ircinia strobolina, Ircinia campana,* and *Speciospongia vesparia* are common. A few algae including *Udotea* and an unidentified brown algae, and the corals *Siderastraea* and *Porites* are present.
3. Shallow water-high salinity community—characterized by shallow water, a cover of foraminiferal sand of intermediate thickness, elevated salinity and temperature, and restricted water circulation. This environment has a low abundance and diversity of macrobiota including a few algae, low lumpy black sponges, red mangroves, jellyfish, and Ophiuroids.

Detailed examination of the molluscan faunas indicate the existence of five major subenvironments. Small areas of no great lateral persistence are characterized by other molluscan communities. The major subenvironments delineated are:

1. Open bay—characterized by thin sand cover over bedrock and water deeper than 0.8 m, dominant constituents are bivalves *Chione cancellata, Laevicardium mortoni, Nucula proxima,* and *Pitar* cf. *fulminata.*
2. Shallow water, near shore—characterized by thin sediment cover over rock bottom and shallow water, dominant species are gastropods *Bulla occidentalis, Cerithium floridanum,* and *Nassarius ambiguus.*
3. Restricted, high salinity—characterized by high salinity and temperature, restricted circulation, intermediate sediment thickness, and low diversity but relatively high abundances. Typical mollusks are *Anomalocardia cuneimeris, Parastarte triquetra, Melongena corona,* and *Polymesoda floridana.*
4. *Thalassia* spit—characterized by shallow water, thick unconsolidated sediment cover, and abundant *Thalassia*. Dominant mollusks are *Bulla striata, Modulus modulus,* and *Tellina* cf. *mera.*
5. Tidal channel—characterized by maximum tidal exchange, rocky bottom, and water deeper than 1.6 m. Dominant bivalves are *Codakia orbiculata, Lucina nassula,* and *Tellina candeana.*

The areal variation in faunal and floral assemblages within the small confines of Coupon Bight indicates that relatively minor variations in water depth, sediment depth, temperature, salinity, water circulation, and other factors can result in considerable variation in the composition of the benthonic biota within small distances. Geologists may therefore expect to find similar small scale areal variations in the fossil content of some ancient shallow water limestones.

*See figure 3.28.

SEDIMENTARY CONTROLS AND FACIES IN COUPON BIGHT, LOWER FLORIDA KEYS*
by
Don L. Kissling

Banks of carbonate mud represent one of two principal sedimentary facies in Coupon Bight. These banks accumulate in depressions on the underlying Pleistocene limestone surface to thicknesses of 3 meters, and although relatively elevated, they are little affected by sediment winnowing. Their localization results from antecedent topographic irregularities and protective cover furnished by the seagrass, *Thalassia*. The surrounding substrate is veneered by calcarenite consisting of Recent skeletal grains and reworked Pleistocene ooliths, together possessing less constituent variability than inner reef tract sands. This facies is controlled by local intensification of wave-generated turbulence combined with tidal current transport, effecting a gradual loss of sediment from Coupon Bight. Limited sediment influx into southwestern Coupon Bight under storm conditions is indicated by local anomalous constituent content.

MICROFAUNAL DISTRIBUTION AND CONTROLS, COUPON BIGHT, SOUTHERN FLORIDA KEYS*
by
James F. Howard and Kenneth L. Faulk

Variations in microfaunal distributions within Coupon Bight allows delineation of 6 major environments on the basis of living faunal abundance, living faunal diversty, and dominance of certain given species.

a. Shallow, Restricted Bay—Salinity, 39.5—40.4°/°°; O_2, 3.9-7.4 ppm. Foraminiferal diversities (4.1-11.7) and foraminiferal numbers 21 to 167. Dominant species is *Archaias angulatus* (50-80% of total fauna). Ten genera and twenty-two species are represented.

b. Open-Rocky Shore—Salinity, 37.3-38°/°°; O_2, 5.2-6.4 ppm. Foraminiferal diversities (4.0-6.7) and foraminiferal number ranges from 34 to 73. Dominant species are *Archaias angulatus* and *Quinqueloculina bosciana* (20-48% and 28-49% of total fauna respectively). Fourteen genera and thirty-two species occur in this facies.

c. Thalassia Flat—Salinity, 37.0-38.0°/°°; O_2, 7.8-9.4 ppm. Foraminiferal diversity, 3.3-4.6 and foraminiferal numbers, 69 to 81. Dominant species are *Quinqueloculina bosciana* and *Scutuloris dilatata* (27-35% and 16-18% respectively). *Archaias* is sparse (0-15%). Twenty-two genera and 40 species are represented.

d. Open Bay—Salinity range 37.7-40.0°/°°; O_2; 6.6-7.1 ppm. Foraminiferal diversity, 1.7-3.4 and foraminiferal number, 13 to 51. No single species dominates but *Q. bosciana, Scutularis dilatata* and various species of *Elphidium* are locally abundant. *Archaias* is very rare, ranging from 0-5% of total fauna. Total genera and 40 species are represented.

e. Mangrove Bay—Represented by a single station.

Salinity—39.9°/°°; O_2-11.7 ppm. Foraminiferal diversity is 3.7 and foraminiferal number 90. Characterized by numerous species of *Quinqueloculina,* including *Q. costata* (10%) and *Q. quadrilateralis* (8%), *Scutularis dilatata* (11%) and the paucity of *Archaias*. Nine genera and 21 species are represented.

f. Channel—Salinity—39.1°/°°; O_2—6.8 ppm. Foraminiferal diversity is 3.5 and foraminiferal number is 64. Characterized by *Q. quadrilateralis* (7%), *Q. bosciana* (8%), *Archaias angulatus* (38%), *A. compressus* (5%) and *Q. reticulata* (4%). Twelve genera and 26 species are represented.

A hierarchy of controls is suggested by the distributions observed, in which turbulence is most important, followed by substrate and oxygen content. Salinity and depth appear to exert no direct control within the confines of Coupon Bight. Diversity in general decreases from the mouth towards the interior of the bay and maximum total abundance occurs in the protected lee shadow of Newfound Harbor Keys.

Steinker (1974) discusses Foraminifera as important biotic elements and as skeletal constituents in tropical shallow water carbonate environments and includes data from his study of Coupon Bight Foraminifera. The abstract of his paper is given below. See also Steinker (1976). See Grant *et al.,* (1973) for some habitats of Foraminifera in Coupon Bight.

FORAMINIFERAL STUDIES IN TROPICAL CARBONATE ENVIRONMENTS
by
D. C. Steinker

The foraminifera are important both as biotic elements and as skeletal constituents in tropical marine, shallow water, carbonate depositional environments. Although a great deal is known about the taxonomy and distribution of modern foraminifers, the general areas of ecology, physiology, life cycles, and variations have been somewhat neglected. Geologists have been interested mainly in establishing correlations that may be significant in reconstructing ancient depositional environments, commonly disregarding the potential paleoecological significance of differentiating between factors of the environment affecting the distribution of living Foraminifera and those affecting the distribution of total populations (living and dead) in the sediments.

Several studies of Foraminifera from tropical carbonate environments, either in progress or recently completed, deal with sampling methods, methods for recognition of living individuals, foraminiferal habitats and ecology, and reef assemblages, as well as culture studies pertaining to the biology of certain species and taxonomic implications. Some summary statements are presented.

Most distributional and ecological studies have been

*See figure 3.29.

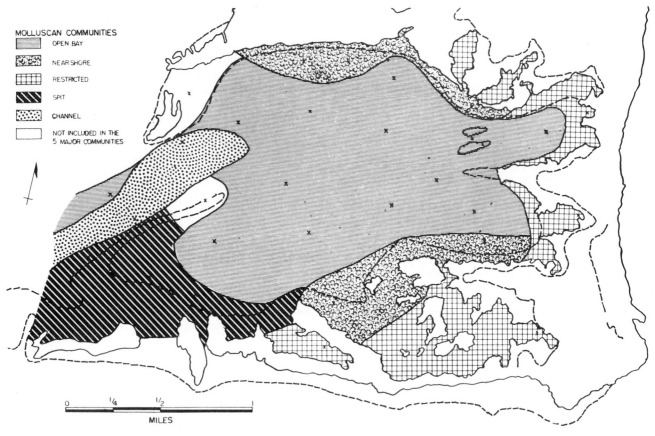

MOLLUSCAN COMMUNITIES
- OPEN BAY
- NEAR SHORE
- RESTRICTED
- SPIT
- CHANNEL
- NOT INCLUDED IN THE 5 MAJOR COMMUNITIES

0   1/4   1/2   1
MILES

Figure 3.28. Coupon Bight, south Big Pine Key. *above* Five major subenvironments of molluscan fauna. *below* The general benthonic macrobiota can be divided into three communities reflecting differences in environmental conditions. From Lineback (1968); for Abstract see preceding pages.

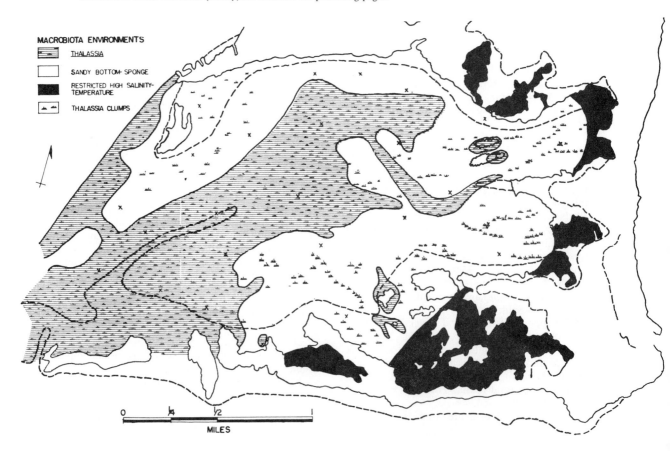

MACROBIOTA ENVIRONMENTS
- THALASSIA
- SANDY BOTTOM-SPONGE
- RESTRICTED HIGH SALINITY-TEMPERATURE
- THALASSIA CLUMPS

0   1/4   1/2   1
MILES

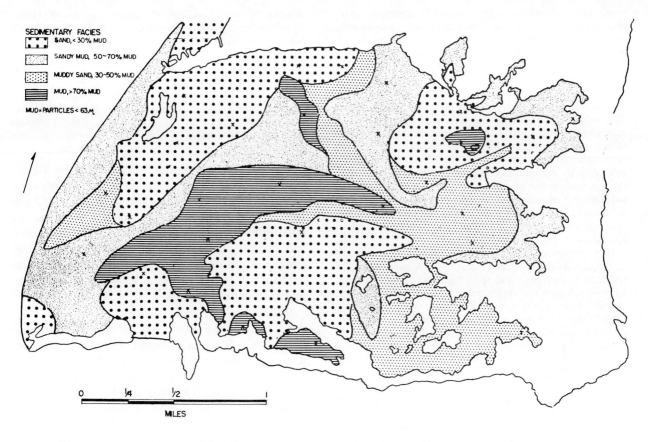

Figure 3.29. Coupon Bight, south Big Pine Key. *above* Sedimentary facies of the area by Kissling (1968). *below* Distribution of microfaunal biofacies according to Howard and Faulk (1968); for Abstracts see preceding pages.

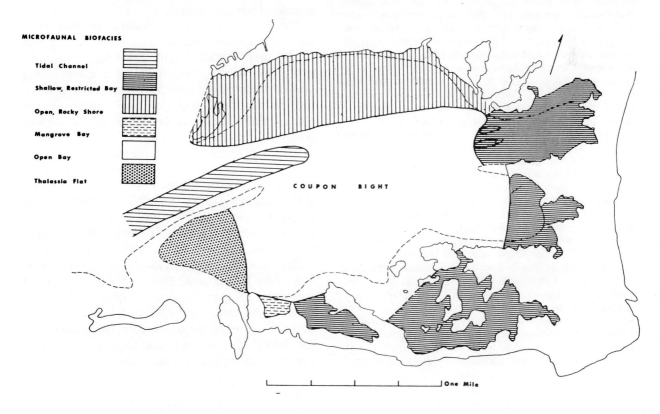

based upon sediment samples obtained by means of cores, grab samplers, or dredges, which generally seem to miss the bottom vegetation and foraminifers thereupon. In my experience in shallow water tropical environments, the foraminifers in such samples are represented mainly by empty tests, and even living specimens commonly show the effects of current sorting. Examination of foraminifers living on the benthonic vegetation may reveal a different distributional pattern than that exhibited by the sediment assemblage. Both juvenile and mature specimens are likely to be represented on the plants, whereas those tests occurring in the sediments may be size sorted. Also, individuals belonging to species that attain only a small test size may be common on the plants but rare or absent among the sediments. And species with organic tests or with loosely aglutinated tests may be well-represented on marine algae and grasses but absent from the adjacent bottom sediments.

One of the initial problems in ecologic studies is to distinguish living from dead individuals. A reevaluation of techniques for recognition of living foraminifers has lead to the conclusion that the commonly employed rose bengal staining technique is unreliable and should be used only with caution. No easy method for determining living individuals has been found. The only reliable methods involve direct observation soon after collection in an attempt to recognize various signs of life, such as distinctive protoplasmic coloration, pseudopodial activity, and protoplasmic streaming within the test.

In a study of ecology and habitats of shallow water foraminifera in southern Florida, 34 species were identified from a small area along the northwest side of Coupon Bight, Big Pine Key. Many living specimens were found on benthonic plants, whereas most specimens in the bottom sediments were dead. Important substrates for foraminifers included the plants *Thalassia testudinum, Dasycladus vermicularis, Penicillus capitatus,* and *Halimeda* spp. Living specimens were most abundant on those plants that supported diatom growths and associated organic detritus, which apparently serve as a food source. Some species living on the vegetation were not represented in the surrounding sediments. These include *Allogromia laticollaris, Iridia* sp., *Nubecularia lucifuga,* and *Planorbulina mediterranensis.* A similar situation was observed in the upper reaches of the bight where handfuls of peneroplids (almost all empty tests) could be scooped up from the bottom and the branchlets of *Dasycladus* were laden with living peneroplids. Comparable results have been obtained in the Bahamas where the living population around one small island constituted only a small part of the total population and the occurrence of living forms was correlative with the occurrence of detrital organic material (Paula J. Steinker, personal communication).

Several species of *Rosalina* have been maintained and studied in culture for periods of several years. One of these is *R. floridana,* originally described by Cushman from the Dry Tortugas. Clonal lineages of this species have been used in studying its ecology, life cycle, growth rate, reproduction rate, and variation. This seems to be a widespread species with a relatively broad temperature tolerance. No indications of sexual reproduction were observed. Chambers are added at the approximate rate of one every two days, and reproductive maturity is reached at about 30 days. Considerable variation occurs in various morphological features, some of which have previously been considered to be of diagnostic taxonomic significance.

Because of the paleoecologic and biochronologic significance of fossil Foraminifera and their common occurrence in carbonate rocks, it is suggested that more critical and discerning work on living forms, including studies of productivity, experimental ecology, and habitats, will contribute to our understanding of ancient carbonate environments.

BIMINI LAGOON (Bahamas): There are a great many examples of carbonate environments within a short distance of the Bimini Islands (index map 2 and figure 3.30). Beach sand and beachrock deposits of north Bimini are discussed on pages 30 and 33 and rocky coasts are described on page 19 of this Guidebook. The area is readily available by air from Miami to the south Bimini airport. Outboard skiffs can be rented from local sources.

Bimini Lagoon (figs. 3.32 and 3.33) is between north and south Bimini along the northwestern edge of the Great Bahama Platform. The lagoon occupies an area of approximately 21 square kilometers, rarely has depths exceeding two meters, and a tidal range of from .7 to 1 meter. Tidal channels in the southeastern portion of the lagoon have velocities greater than 2 knots. In the western portions, away from the eastern tidal swept portions of the lagoon, shallow hypersaline conditions can be found. Mangrove communities ring portions of the lagoon and are in the process of reducing its size.

The Bimini Lagoon is unique in that (1) it represents a shallow enclosed basin on the very edge of the deep Florida Straits, (2) it lacks the usual mud component of lagoons, (3) it is a low energy area in which oolitic films have been noted as forming around grains, and (4) it lacks the usual terrestrial and clastic runoff dimension of lagoons bordering mainlands. Many investigators have described different aspects of the Bimini Lagoon and surrounding area. Some of these studies are Newell and Imbrie (1955), Newell *et al.,* (1957), Turekian (1957), Squires (1958), Newell *et al.,* (1959), Voss (1960), Kornicker (1960), Purdy (1963), Hay (1967), Bathurst (1967), Kelly and Conrad (1969), Scoffin (1970a), Supko *et al.,* (1970), and Bathurst (1971).

Newell *et al.,* (1959, p. 192-193) discuss the evolution of the Bimini area. They describe flooding and deposition up over the exposed edge of the Great Bahama platform about 4,000-5,000 years ago. One of the evidences cited for this recent rise in sea level is data procured from the Bimini Lagoon. A pertinent portion of their paper is noted below.

"Just how ancient are the existing distributions and ecological situations? In 1956, during dredging operations in the Bimini Lagoon, a thin bed of mangrove peat with *in situ* roots extending into underlying weathered bedrock was encountered 9 feet beneath low-water level (fig. 3.34). The position of the peat below the unconsolidated strata was determined at a point immediately west of the channel by means of a soil auger and sampler. Radiocarbon analysis of the peat by Dr. Wallace Broecker at the Lamont Geological Observatory gave a date of 4370 ± 110 years."

Dr. Robin Bathurst submitted the following excellent summary of Bimini Lagoon for the 1969 edition of this Guidebook.

BIMINI LAGOON
by
R. G. C. Bathurst

The sedimentology of the lagoon that lies in the shelter of the Bimini Islands is worth looking at for several reasons. Collected data are relatively abundant, the lagoon is small and largely enclosed so that the source of the carbonate grains is known with some accuracy, and there is a development of oolitic films on grains in an environment markedly different from that of the mobile oolite habitat

The Bimini Islands lie at the northwestern margin of the Bank, bounded on the ocean side by the coralgal lithofacies, to the east by the oolitic and on the south by pellet-mud. The number of stations occupied by Purdy, a half dozen or so, are too few to yield any detail. The lagoon itself, shown on Purdy's map (1963, fig. 1) as mud lithofacies, is more truly described as stable sand habitat

and *Strombus costatus* community (Newell *et al.,* 1959, p. 222). Quantitative analyses have yet to be made of the sedimentary constituents, but it is clear that the dominant skeletons are *Halimeda,* Peneroplidae, or foraminiferids and molluscs as in the oolitic lithofacies; grains of coralline alga are also present. There is no mud and only at one station in the East Lagoon does a sample rank as a sandy silt. All other samples from about 80 stations occupied by Bathurst, are medium to fine sand. Faecal pellets and cryptocrystalline grains are abundant in some parts of the lagoon. Information is mainly derived from Bathurst's unpublished and published (1967) data. Mean sea level varies generally between 0.3-0.7 m.

Following Turekian (1958) the lagoon is divided into three natural regions, the North Sound, Main Lagoon and the region east of Pigeon Cay. This last has been named the East Lagoon by Bathurst (1967). Turekian's data on temperature and salinity, collected in the spring and summer of 1955, suggest that these three regions are occupied by three fairly distinct water masses (fig. 3.35). The Main Lagoon (fig. 3.36) has direct connection with the unstable sand habitat of the shelf to the southwest through a narrow tidal channel. The East Lagoon is open to the east. The entire lagoon is, nevertheless, sheltered and largely enclosed by three islands. North Bimini rises to 5 m or more about mean sea level and is clothed mainly with trees of *Casuarina equisetifolia.* East Bimini is chiefly mangrove swamp (*Rhizopora mangle* and *Avicennia nitida*) and south Bimini bears for the most part of casurinas, thorn scrub and sporadic trees of *Manilikara mastichodendon* and *Ficus* on slightly elevated limestone.

The biota in the lagoon can be regarded as typical of the *Strombus costatus* community with local variations. *Thalassia* densities are shown in figure 3.37; there is no striking difference in density between the three regions.

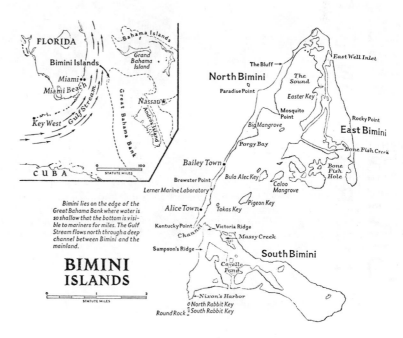

Figure 3.30. Index map to the Bimini Islands, Bahamas. Map drawn by I. E. Alleman for article by Zahl (1952); provided by courtesy of National Geographic Magazine. © National Geographic Society.

BIMINI AREA
DISTRIBUTION of BOTTOM TYPES

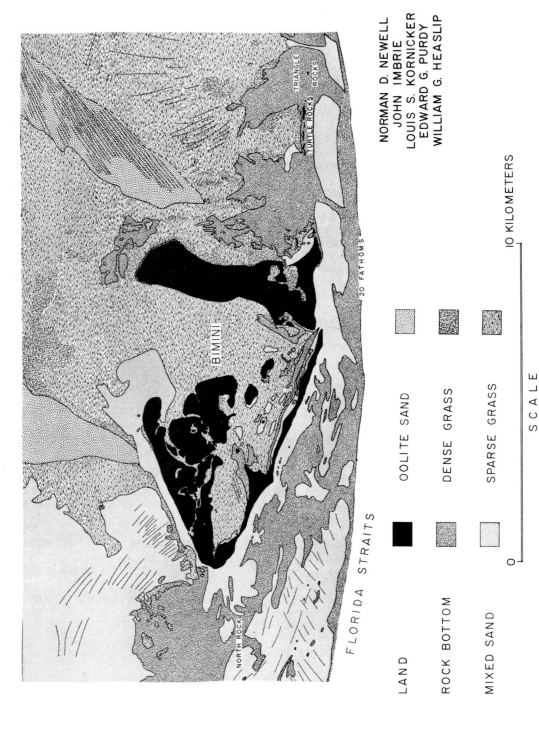

N

BIMINI

FLORIDA STRAITS

NORTH ROCK

20 FATHOMS

TRIANGLE ROCKS

TURTLE ROCKS

NORMAN D. NEWELL
JOHN IMBRIE
LOUIS S. KORNICKER
EDWARD G. PURDY
WILLIAM G. HEASLIP

LAND

ROCK BOTTOM

MIXED SAND

OOLITE SAND

DENSE GRASS

SPARSE GRASS

SCALE

0   10 KILOMETERS

Figure 3.31. Distribution of Bimini bottom types from Newell *et al.* (1959).

The North Sound. The North Sound is peculiar in that, although the salinities are at times the highest in the lagoon (fig. 3.35), owing to its shallowness and remoteness from the ocean, they are sometimes the lowest as a result of direct rainfall. Runoff from the mangrove swamps, though rich in calcium carbonate dissolved from the limestone by organic acids, further reduces the salinity (Seibold, 1962). In this part of the lagoon particularly, the day-night cycle of plant photosynthesis and respiration causes fluctuations in the dissolved CO_2 and thus in the concentration of calcium carbonate. A reconnaissance by Seibold (op. cit.) suggests a possible influence of this cycle on carbonate precipitation. Whether precipitation occurs is not yet known.

On the floor *Batophora* and *Laufencia* are specially plentiful, and, near the beaches, the 2 cm long gastropod *Batillaria minima* makes characteristic pellets (Kornicker and Purdy, 1957). The discarded shells are apt to move surprisingly swiftly but erratically over the bottom, inhabited by little hermit crabs. The underlying sediment can be cut into blocks: it is firm because of a high content of plant debris from the mangroves and other trees. The biota is a poor relative of the *S. costatus* community.

The Main Lagoon. The community of the Main Lagoon is typically a rich, sheltered *S. costatus.* There are numerous holothurians and the long-spined echinoid *Diadema:* asteroids and ophiuroids are common. Small areas of rock pavement, with a thin sand cover of a few millimeters, support a simplified pavement community of plexaurids, the coral *Porites* and the sponges *Ircinia, Speciospongia* and the red sponge *Tedania.* Algae grow in the extensive patches of *Thalassia* (fig. 3.37). *Strombus costatus* and *S. gigas* may be seen any day waddling with undignified gait over the *Thalassia* carpet. Fish, nosing the sand, are numerous over the rock pavement and small barracuda, sharks and sting rays are ubiquitous.

The East Lagoon. The East Lagoon is generally poor in variety of biota. *Thalassia* and densely arranged *Callianassa* mounds cover much of the stable sand habitat, the mounds being so close in places as to interfere. Walking over such a hummocky bottom is tedious and undignified. There is a weak tidal channel along the south Bimini shore. Much of the western part of the area enclosed by the 0.5 m line is exposed at low tide.

Oölitic films. When thin sections of the calcarenite are examined between crossed nicols and with a gypsum first order red compensator, certain grains show a very thin oolitic coat or film. These layers of oriented aragonite, in every respect like those of other Bahamian oöids, are only about 3μ thick.

Few samples of the calcarenite are without some trace of oölitic films on some grains. To justify comparison between samples of varied composition an ooid index was devised. It is based on the occurrence of an oolitic film where these occur *on peloids only.* Thus comparison is made between grains of the same shape and surface on the assumption that they are equally susceptible to the process of oolitic deposition. It is, in fact, clearly apparent that irregularly shaped grains, unlike the well-rounded peloids, rarely have oölitic films on them. All samples having films on less than half of the peloids are recorded as *scarce.* Samples having films on more than half the peloids are recorded as *common,* but those wherein 95 percent or more (commonly 100 percent) of the peloids have films are called *abundant.*

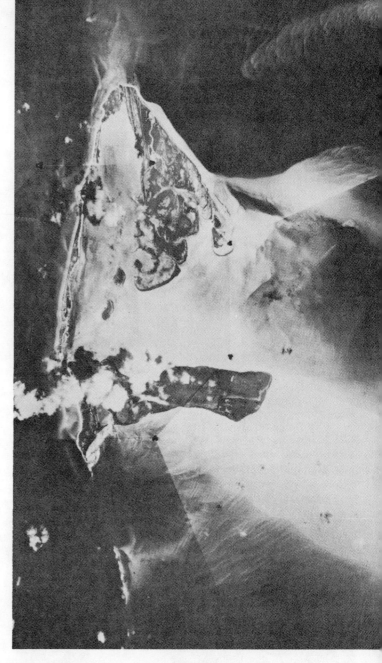

Figure 3.32. Aerial photograph of Bimini Islands. Courtesy of M. M. Ball.

The remarkable feature is the distribution of the samples with *abundant* ooids. Out of 35 *abundant* samples all but 5 lie within a region closely corresponding to that occupied by the distinct southeasterly water mass of Turekian (op. cit.): he drew the boundary just east of Alec and Pigeon Cays. Yet this region, the East Lagoon, is quite unlike the turbulent, shifting sands of the oolite shoals of Browns Cay or Joulters Cays. Twice each day the tidal water seeps shyly in and then drains quietly away. The densely packed mounds are so little disturbed that the subtidal mat commonly spreads up their flanks.

The strangeness of the East Lagoon as a region of ooid growth seems greater when one considers that the highest recorded salinities are not there but in the North Sound: however, the occasional low salinities in the North Sound may offset the effect of high values. It is also true that samples taken in the East Lagoon from a boat (vide Turekian) within the 0.5 m line were necessarily collected around *high tide*: the area is too shallow to be crossed by

Figure 3.33. Aerial view looking toward southern tip of North Bimini with South Bimini in left background. Note tidal channels and *Thalassia* flats in the Main Lagoon to the left and cliffed beach with breakers to the right. Photo from article by Zahl (1952) provided by courtesy of National Geographic Magazine. © National Geographic Society.

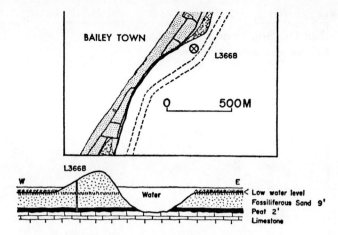

Figure 3.34. Peat layer exposed in Lyon's channel, Bimini. *Above:* Index map of a portion of North Bimini showing location of bore-hole sample L-366B just west of channel. *Below:* Cross section showing bore hole and peat layer; no vertical exaggeration. From Newell, *et al.* (1959, p. 192).

DISTRIBUTION OF SALINITY AND TEMPERATURE IN BIMINI LAGOON (SPRING AND SUMMER 1955)

(After TUREKIAN, 1957)

	Salinity (p.p.m.)	Temperature (°C)
North Sound	increases northward from 40.4–41.9, with a drop to 40.5 in the far northeast corner	increases northward from 29.3–29.8, with a rise to 31.3 in the far northeast corner
Main Lagoon	36.0; no trend except for an increase to 39.4 near Mosquito Point	27.9–29.2 no trend
East Lagoon	36.6–37.5 no trend	28.3–30.7 no trend

Figure 3.35. Distribution of salinity and temperature in Bimini Lagoon measured by Turekian (1957) *in* Bathurst (1967, p. 103).

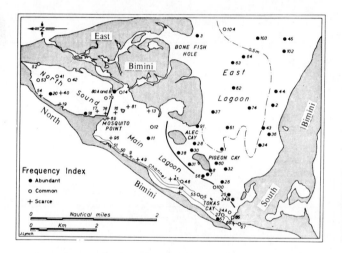

Figure 3.36. Index map showing natural subdivisions of Bimini Lagoon and station and frequency indices for coated grains. After Bathurst (167, p. 90).

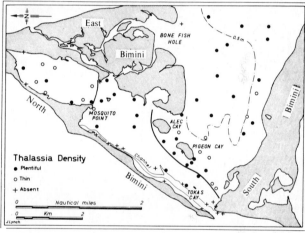

Figure 3.37. *Thalassia* density in Bimini Lagoon after Bathurst (1967, p. 93).

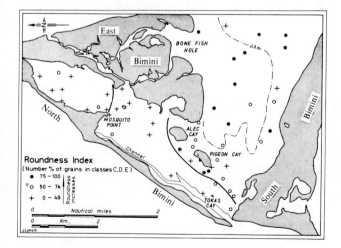

Figure 3.38. Roundness indices of sand found in Bimini Lagoon after Bathurst (1967, p. 96).

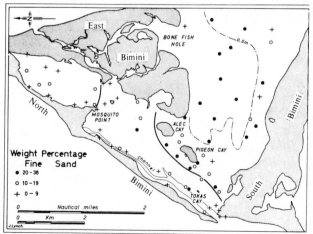

Figure 3.39. Weight percent of fine sand in Bimini Lagoon after Bathurst (1967, p. 107).

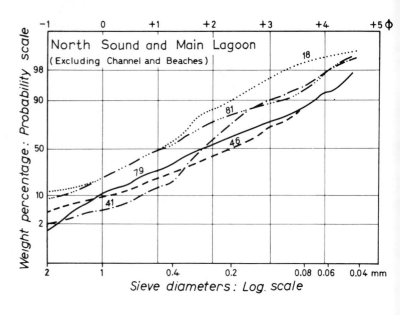

Figure 3.40. Representative sieve analyses of Bimini Lagoon sands from North Sound and Main Lagoon, excluding channel and beaches. From Bathurst (1967, p. 94).

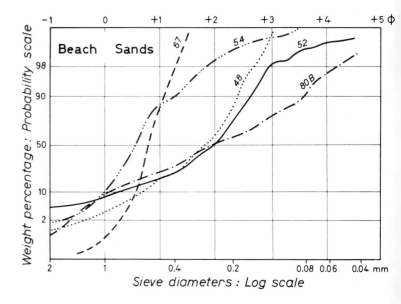

Figure 3.41. Representative sieve analyses of beach sands, Bimini Lagoon. From Bathurst (1967, p. 94).

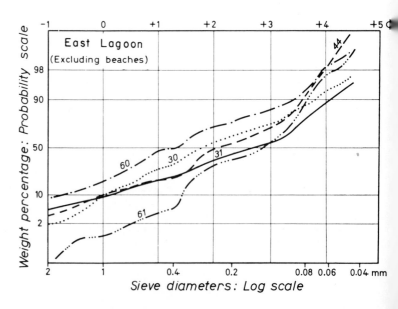

Figure 3.42. Representative Bimini Lagoon sieve analyses of sands from the East Lagoon. From Bathurst (1967, p. 95).

boat at other times. Thus it is still possible that, as in Florida Bay (Taft and Harbaugh, 1964, p. 23), extreme conditions of high salinity may exist for a short time about *low tide,* but the appropriate measurements of salinity have yet to be made. These would arise where small shallow bodies of water were temporarily isolated from the deeper, moving water: evaporation under these conditions causes the salinity of water in parts of Florida Bay to rise for a short while by as much as $7°/°°$. Still, in Florida Bay this process does not apparently produce ooids.

Fine sand and boring algae. Linked in a striking manner to the distribution of samples with abundant ooids is distribution of fine sand. Sieve analysis of the calcarenites at 0.25 intervals reveals that the samples from the East Lagoon have more grains between $150-90\mu$ than have samples from other parts of the lagoon (fig. 3.39). It has been shown that this high value has no connection with the trapping of finer grains by *Thalassia* baffles. Under the microscope the fine sand of the East Lagoon is seen to be very different from that in the rest of the lagoon. Whereas in the North Sound and Main Lagoon the fine sand consists almost entirely of recognizable skeletal debris, often broken and angular, in the East Lagoon it is composed of rounded peloids (fig. 3.38). These are variable in shape and are, in form and content, unlike faecal pellets. They are akin to the grains of cryptocrystalline carbonate of the grapestone lithofacies. Experience of the various stages of their formation in the lagoon suggests that most, if not all, are skeletal grains which have been micritized by boring algae. They are also characterized by greater roundness (Pettijohn, 1949, fig. 24) than is found in the fine sand outside the East Lagoon (fig. 3.38). This is doubtless caused by the ready abrasion of grains already weakened by boring algae (Swinchatt, 1965, p. 81-82). This same abrasion may have caused a shift of the frequency distribution of grain diameters towards the finer fractions in the East Lagoon. It is particularly noteworthy that a high degree of roundness is attained in water of low turbulence as a result, predominantly, of biological attack.

Source of the ooids. Before going any further it is necessary to consider if by any chance the ooids could have grown outside the lagoon and been brought in. Fuchtbauer has reported similar ooids a few kilometres to the south of south Bimini (unpublished). To begin with there is no sign of any tidal or wave driven system of grain transport which could carry ooids into the lagoon. If it be argued that the ooids were swept in at some earlier time under different conditions or, indeed, that they were extinguished, then the evidence from cores contradicts both these hypotheses. Two cores were taken at points east of Pigeon Cay (fig. 3.36). They were sampled at 25 cm intervals and, throughout their 1.5 m depths, they show abundant ooids in all samples. If ooids were added from without to the nonoolithic grains formed in the lagoon, then a lower content of ooids would be discovered than the nearly 100 percent value in the abundant samples. If the ooids represented an earlier oolite environment no longer active then, again, dilution with more recently formed, nonoölithic grains would have taken place. There is no sign of this. The calcarenites have probably been intensively mixed by the burrowing of *Callianassa,* so that any vertical change representing successive environments would have been blurred or destroyed. At the same time, any lengthy deposition of nonoölithic grains would be reflected in some degree of dilution of the virtually pure composition of film-covered peloids. No such dilution appears to have occurred. It seems that the oolitic films must have grown on grains inside the area of the East Lagoon.

The problem of turbulence and ooïd growth. One problem here is to reconcile the growth of aragonitic oölite with the near absence of that day-to-day turbulence of the water which, it is claimed, has an essential function in the growth process. Of course the films are unusually thin, indicating, perhaps, an environment unsuitable for the most rapid growth. But growth has taken place and is, presumably, going on now. If turbulence is essential, then we have to decide what kinds of turbulence are possible in the gentle environment of the East Lagoon. The binding effects of *Thalassia,* codiaceans and *Batophora* and, above all, the subtidal mat, ensure that the day-to-day movement of grains in the tidal currents is negligible. The mat surface of the grains confirms the lack of that abrasion which brings about the polish of ooïds on the typical Bahamian oolite shoal. The matter would be easier to discuss if the reasons for the need for turbulence were understood: but they are not known. Presumably enough movement must take place for the oölitic film to be able to grow without interruption all over the grain surface. The possible ways in which grains can be moved on, or even under, the sediment surface vary between the extremes of the rare hurricane, perhaps every twenty years, and the gentle but persistent stirring and mixing of fish, holothurians and mound-makers. It is impossible, however, with the information at present available, to reach any conclusion on how much motion the ooid must undergo or how this is caused.

Algae and ooïd growth. The extraordinary correlation of oolitic growth and the activity of boring algae harks back to the insistence by Nesteroff (1956) on the role played by boring algae in the growth of ooïds. Nesteroff's suspicions are encouraged by the work of Shearman and others on Persian Gulf ooïds. The Bahamian ooïds of all types, like those of the Persian Gulf, are extensively permeated with mucilaginous material. The effect of this on both the precipitation and stability of aragonite is at present an open question about which very little is known.

For a well illustrated and detailed summary of the Great Bahama Bank including Bimini Lagoon, the interested reader is urged to see Bathurst (1971, p. 93-143).

See Scoffin (1970) for a detailed account of the marine vegetation found in Bimini Lagoon including maps of sediment thickness, water depth, tidal currents, algal distribution and cross sections of cays showing bedrock foundation. The abstract of Scoffin's 1970 paper is given below.

THE TRAPPING AND BINDING OF SUBTIDAL CARBONATE SEDIMENTS BY MARINE VEGETATION IN BIMINI LAGOON, BAHAMAS
by
T. P. Scoffin

In the shallow water lagoon of Bimini, Bahamas, the following plants are sufficiently abundant to influence sedimentation locally—mangroves (*Rhizophora mangle*), marine grass (*Thalassia testudinum*), macroscopic green algae (*Penicillus, Batophora, Halimeda, Rhipocephalus and Udotea*) and microscopic red, green and blue-green algae forming surface mats of intertwining filaments (*Laurencia, Enteromorpha, Lyngbya* and (?) *Schizothrix*). Plants were observed under conditions of natural tidal currents and artificial unidirectional currents produced in an underwater flume and measurements were made of the abilities of the plants to trap and bind the carbonate sediment. The density of plant growth is crucial in the reduction of current strength at the sediment-water interface. The most effective baffles are *Rhizophora* roots exposed above the sediment, dense *Thalassia* blades and *Thalassia* blades with dense epiphytic algae, *Laurencia intricata* and *Polysiphonia havanensis*. All three types can reduce the velocity of water from a speed sufficiently high to transport loose sand grains along the bottom in clear areas (30 cm/sec) to zero at the sediment-water interface in the vegetated areas. The strongest binders of sediment are the roots of *Rhizophora* and *Thalassia*. These two hardy plants trap and bind sediment for a sufficient time to produce an accumulation higher than in nearby areas without dense mangroves or grass. Macroscopic green algae growth is not sufficiently dense and the holdfasts too weak to appreciably affect the accumulation of sediment although they provide a degree of stabilization to the substrate. Algal mats trap sediment chiefly by adhesion of grains to the sticky filaments. Their ability to resist erosion by unidirectional currents varies considerably depending on mat type, smoothness of surface and continuity of the cover. The intact areas of dense *Enteromorpha* mat can withstand currents five times stronger than those that erode loose unbound sand grains. Premature erosion of mats by currents occurs at breaks in the mat surface caused by the burrowing or browsing action of animals. Algal mats were found to be ephemeral features and consequently do not build up thick accumulations of sediment as do dense grass and mangroves. The thickest accumulation of sediment in the lagoon correlate with deepest bedrock surfaces. The distribution of many plants in the lagoon is directly or indirectly controlled by the depth to bedrock; for example, mangroves on bedrock highs, marine grass in sediment-filled depressions.

Till (1970) describes the results of a geochemical study conducted in Bimini Lagoon. His abstract is noted below:

THE RELATIONSHIP BETWEEN ENVIRONMENT AND SEDIMENT COMPOSITION (GEOCHEMISTRY AND PETROLOGY) IN THE BIMINI LAGOONS, BAHAMAS
by
R. Till

Surface sediments from the Bimini Lagoon were analyzed for their major element content (CaO, SrO and MgO), minor element content (Si, Al, Fe, Ba, Cr and Mn), mineralogy (aragonite, high and low magnesium calcite) and petrographic constituents (grain size, intraclasts, pellets, oolites, skeletal material and mud content). The overlying waters were also analyzed and environmental data was collected. Methods of analysis are briefly described. The data was analyzed statistically to discover controls on overall geochemical variation.

Linear product moment correlation coefficients were calculated for all variable pairs. A correlation matrix was generated and used to identify covarying groups of variables. The results of this test allowed a simple interpretation to be made. The physical environment in terms of water quietness (rather than the chemical environment) is shown to govern the organic and inorganic phase constituents of the sediments; these in turn govern the bulk chemistry. Correlation analysis was repeated for the sample population from each of the three water masses in the lagoon (as recognized by Turekian, 1957, in terms of salinity). The same basic control of water quietness is shown to be operative in each area, though in certain localized environments its government of the bulk chemistry and mineralogy is reflected in a different way.

A simple analysis of variance for the sample groups from each water mass showed that the sediments from these three areas of different salinity in the lagoon are also significantly different in bulk constitution. But the salinity (and hence chemistry) of the waters is not simply related to the sediment composition; for the three masses are also distinct in terms of physical environment and it is this which is reflected in the mean sediment composition.

Alexandersson (1972) describes micritization of carbonate particles and accompanying precipitation and dissolution including (p. 215) an SEM photo of aragonite needles on a mollusk shell from Bimini Lagoon. Kelly and Conrad (1969) describe results of photo studies of bottom communities along an area extending from 37 kilometers south of Bimini along the edge of the Bahama Banks.

Supko, Marszalek and Bock (1970) have compiled a guidebook entitled, "Sedimentary Environements and Carbonate Rocks of Bimini, Bahamas." In this guidebook, Supko (1970) describes and illustrates ten bedrock exposures found in the Bimini area and summarizes information known concerning the "New Bimini" and "Old Bimini" bedrock units. One of the outstanding papers in the above mentioned guidebook describes the results of a comprehensive study of modern organism communities in the Bimini Lagoon area by Hay, Wiedenmayer and Marszalek (1970). This article is quoted below.

MODERN ORGANISM COMMUNITIES OF BIMINI LAGOON AND THEIR RELATION TO THE SEDIMENTS
by
W. W. Hay, F. Wiedenmayer and D. S. Marszalek

The considerable variations in water depth, salinity, temperature, and the degree of exchange with open ocean water in Bimini Lagoon promotes a great diversity of shallow water benthic marine communities. Almost all of the community types of the northern Bahama Banks are represented within or near Bimini Lagoon, the major exception being the true reefs, which are found only on the eastern margins of the Banks.

The lagoon consists of two regions: (1) an outer lagoon, which communicates with the open ocean and Gulf Stream through a narrow, but deep channel between north and south Bimini and with the banks to the east by a broad, but very shallow tidal flat area between east and south Bimini, and (2) an inner lagoon which communicates only with the outer lagoon. Salinities in the inner lagoon are subject to more abnormal variation than those in the outer lagoon, and faunal and floral diversity of the communities is correspondingly less than that encountered in the outer lagoon.

Sediment accumulation varies from nil to about 6 feet, and is most closely related to the amount of plant cover. The marine grasses *Thalassia* and *Cymodocea* act as sedi-ment stabilizers by slowing currents and wave motion near the bottom, and by binding the sediment with their extensive root systems. The density of *Thalassia,* in particular, bears a special relation to the thickness and nature of the accumulating sediment. In areas with tidal current velocities below 1 knot, the *Thalassia* plants tend to be widely spaced, and wave action can winnow the bottom to produce a well-sorted sand. Above a certain threshold value, probably about 0.8 knot, the *Thalassia* grows more luxuriantly and densely. The denser growth serves as a more effective sediment trap, producing a shallow bank. The lessened depth of water increases the velocity of tidal currents, until an upper value of approximately 1.8 knots is reached. Higher current velocities will quickly erode the bank once the *Thalassia* cover is breached, but the dense growth protects the sediment from winnowing or movement below 1.8 knots. The sediment of the medium *Thalassia* areas is less well sorted than that of the sparse *Thalassia* areas, but the dense *Thalassia* areas have an essentially unsorted sediment which is commonly rich in organic matter. The *Thalassia* "makes its own sediment" to a certain extent by serving as substrate for a wide variety of microorganisms, particularly diatoms, foraminifera, bryozoa, and a number of species of encrusting algae. The dense system of *Thalassia* roots in medium and dense grass areas serves as an effective barrier to the movement of potential infauna, with the result that the infauna of the grassy areas is restricted to worms with vertical burrows and the pelecypod *Pinna carnea.*

The areas with thin sediment or bare exposed rock bottom commonly lie at a lower level than the surrounding grass flats, but support communities rich in sponges and sometimes corals, so that the geologist would tend to class them as biostromes. It is interesting to note that these biostromes are not the sites of sediment accumulation, but are sites of the production of biogenic carbonate and silica particles transported by the currents to become trapped in the grass flats.

The accompanying diagrams illustrate some of the communities of the lagoon which may be visited on the field trip. Each diagram represents 100 m² of the bottom, and all surface dwelling animals with a cross-sectional area greater than 10 cm² are indicated by an appropriate symbol. The size of the symbol is proportional to the area of the bottom occupied by the individual. The locations of individual animals were not surveyed, and are represented only schematically on these diagrams. The plants have been omitted from the diagrams for one very simple reason: in regions with sparse grass, the surface area of the blades on plants in 1 m² is 1 m²; in medium grass areas, the surface area of the blades of grass in 1 m² is 4 m²; in dense grass area, 1 m² of bottom may support 8 m² of grass blades. *Thalassia,* if indicated, would then obscure almost all diagrams, except those of rock bottoms.

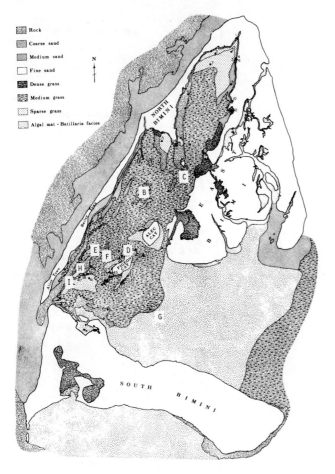

Figure 3.43. Index map to localities in Bimini Lagoon from Hay *et al.* (1970).

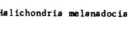

SPONGES

 Halichondria melanadocia

 Haliclona permollis

 Haliclona molitba

 Epipolasis angulospiculata

 Dysidea crawshayi

 Verongia fistularis

 Anthosigmella sp. nov.

 Ircinia fasciculata

 Ircinia strobilina

 Cinachyra cavernosa

Dysidea fragilis

Haliclona viridis

 Verongia aurea

 Dysidea etheria

 Sphericospongia vesparia

 Anthosigmella varians

 Spongia obliqua

 Siphonochallina siphona

 Spongia sterea sp. nov.

 Xytopsene sigmatum

 Tedania ignis

Chondrilla nucula

 Cryptotethya crypta

Callyspongia vaginalis

Organism Key 1

110

HYDROZOAN
Millepora alcicornis

Porites asteroides

ALCYONARIANS
Pseudopterogorgia acerosa

Porites porites
var. furcata

Porites porites divaricata

Pterogorgia citrina

Siderastrea siderea

Siderastrea radians

Eunicea tourmefuti

Flavia fragum

Manicina areolata

Pterogorgia anceps

Dichocoenia stokesi

Eunicea mammosa

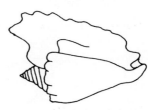

MOLLUSCS
GASTROPODS

Strombus raninus

Muricea atlantica

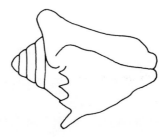

Strombus gigas

ANEMONES
Chondylactis gigantea

Aiptasia annulata

Strombus costatus

Organism Key 2

GASTROPODS

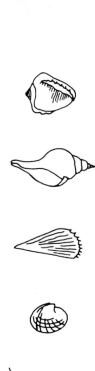

Cassis flammea

Fasciolaria tulipa

MOLLUSCS
PELECYPODS

Pinna carnea

Codakia orbicularis

ARTHROPODS

Petrochirus diogenes
in Strombus gigas

Callinectes exasperatus

ECHINODERMS

Oreaster reticulatus

Tripneustes variegatus

Clypeaster rosaceus

Holothuria mexicana

Actinopyga agassizi

Stichopus badionotus

TUNICATES

Black encrusting tunicate

Didemnum

Perophora viridis

Organism Key 3

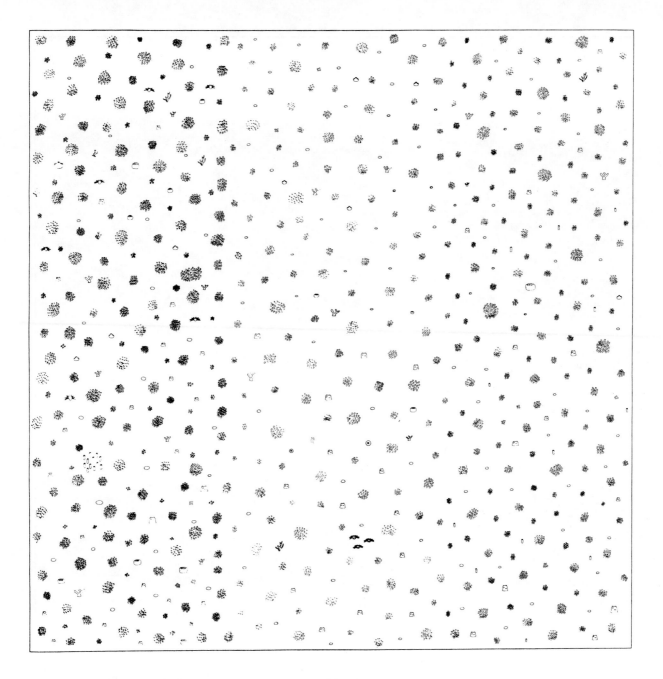

LOCALITY A: The bottom in this area is covered by a thin layer of sediment; two different strata are usually present, the upper layer is light colored and coarse grained, the lower layer is dark and fine grained. Neither of these layers exceeds one inch in thickness. This thin layer of sediment may be discontinuous, and the bare rock substrate is exposed in places.

The dominant organism at this site is a green tunicate, *Perophora viridis.* Individuals of this species are up to 1 cm in length, and they tend to occur in clusters which may be as large as 500 cm². The coral *Siderastrea radians* is common, along with the sponge *Anthosigmella varians.* Rarer animals include the sponges *Ircinia fasciculata, Dysidea etheria, Chondrilla nucula,* and *Haliclona molitba,* the anemones *Condylactis gigantea* and *Aiptasia annulata,* the coral *Porites porites furcatas,* and the crab *Callinectes exasperatus.*

In the geologic record, this sort of community would probably be classed as a poorly developed coral biostrome, and because of the lack of species diversity among the fossils, the paleontologist might correctly infer that it grew in restricted waters.

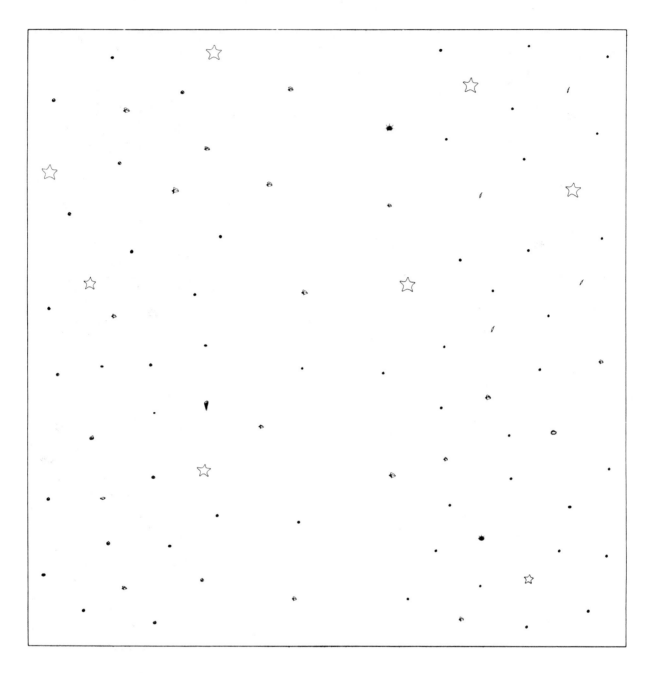

LOCALITY B: Sediment in this area is several feet thick. The dominant organism is *Arenicola marina,* the lugworm which produces the obvious mounds and depressions in the area. It is not shown on the diagram which illustrates only the surficial fauna, (the infaunal diagrams are not completed yet). *Arenicola* mounds in three areas in this part of the lagoon numbered 64, 77, and 83 per 100 m². The dominant organisms remaining on the surface of the sediment are the shells of *Codakia orbicularis,* a clam which lives at a depth of several inches in the sediment. Live animals are relatively rare, but include the sediment and detritus feeders *Strombus gigas* and *Strombus raninus, Oreaster reticulatus,* and the coral *Manicini areolata.*

The infauna in this area has reworked the sediment thoroughly. Digging in this area, we found a 7-Up bottle of recent vintage buried to a depth of more than 1 foot. Incidentally, *Arenicola marina* produces the depressions by feeding, and the mounds are composed of material which has been passed through the lugworm. The clean appearance of the sediment on top of the mounds is testimony to the efficiency of the lugworm's digestive systems in removing organic debris from the sediment. Crab holes may also be found in this area; they are open, and not accompanied by the large mound and depression made by *Arenicola.*

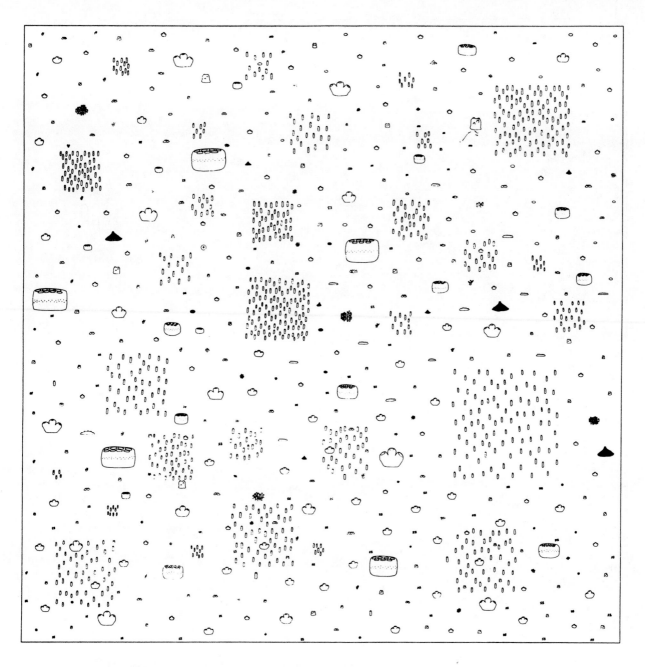

LOCALITY C: This is a typical sponge patch in the inner part of the Lagoon; it is floored by rock bottom with practically no sediment accumulation.

The dominant organism here is *Spheciospongia vesparia*, with the other sponges *Ircinia fasciculata, Anthosigmella varians, Tedania ignis,* and *Epipolasis angulospiculata* also common. The sponges *Verongia aurea, Cinachyra cavernosa, Xytopsene sigmatum, Chondrilla nucula, Myriatra kallitella, Haliclona permollis, Haliclona viridia,* and *Dysidea etheria* are rare. The corals are more diverse than in the northern part of the lagoon: *Manicina areolata, Siderastrea radians, Porites porites furcatus* are all present. The tunicate *Prasopora viridis* occurs as isolated individuals or small clusters.

LOCALITY D: This represents part of another sponge patch; the bottom is rocky with some sediment accumulation in depressions in the rough rock surface.

Six sponges dominate the assemblage: *Anthosigmella varians, Ircinia fasciculata, Chondrilla nucula, Verongia aurea, Epipolasis angulospiculata,* and *Spongia barbara dura.* There are a few other sponges present in lesser quantity: *Ircinia strobilina, Dysidea etheria, Spheciospongia vesparia,* and *Spongia obliqua.* The coral assemblage is still more diverse than those found further to the north in the lagoon: *Siderastrea radians, Porites porites furcata, Porites astreoides,* and *Manicina areolata.* The little sediment found in this area passes through the gut of the common holothurian found here, *Actinopyga agassizi.*

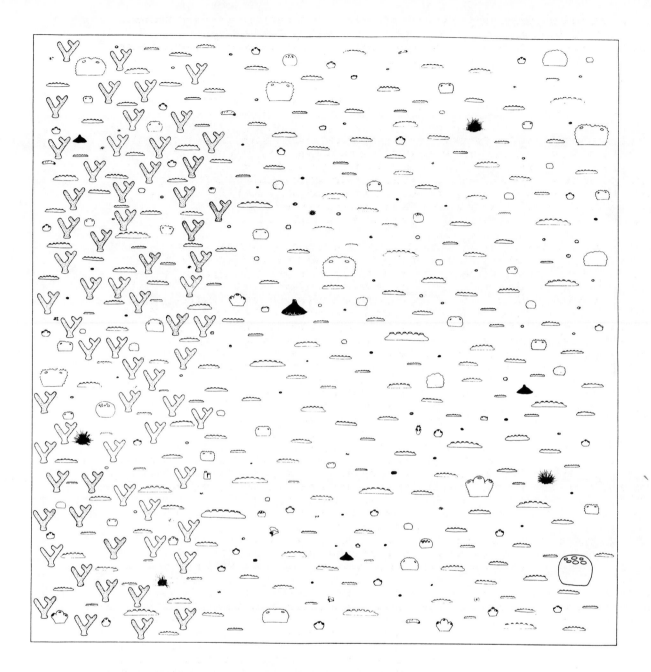

LOCALITY E: This is a sponge and coral community, growing on a rock bottom with thin sediment distributed in patches. Parts of the area are dominated by the coral *Porites porites furcata,* elsewhere the dominant animal is the sponge *Haliclona viridis.* A diverse assemblage of other sponges occurs: *Ircinia strobilina, Ircinia fasciculata, Spongia obliqua, Chondrilla nucula, Epipolasis angulospiculata, Siphonochalina siphona, Dysidea crawshayi* are common to rare. The coral assemblage is the most diverse found in the lagoon, due to the supply of oceanic water from the nearby channel: *Porites porites furcata, Porites porites clavaria, Porites astreoides, Siderastrea radians, Manicina areolata,* are common, and isolated individuals of *Diploria clivosa* and other corals characteristic of Bahamian reefs occur. The long spined sea urchin *Diadema antillarum* is present, along with the holothurian *Actinopyga agassizi.* The fish are reefal types, although only the smaller species of the reef assemblage are present here.

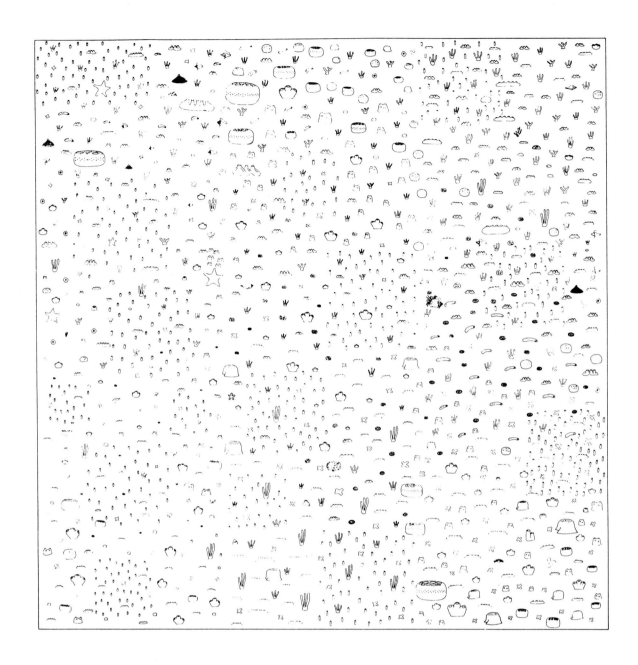

LOCALITY F: This is another sponge and coral community, exhibiting great faunal diversity. The dominant sponges are *Verongia aurea* and *Haliclona viridis*, with *Ircinia fasciculata*, *Ircinia strobilina*, *Chondrilla nucula*, *Tedania ignis*, *Spheciospongia vesparia*, *Anthosigmella varians*, and *Dysidea crawshayi* common. A number of other sponges are rare. The corals are *Porites porites furcata*, *Porites astreoides*, *Manicina areolata*, and *Siderastrea radians*. The holothurian *Stichopus badinotus* is rare, but *Actinopyga agassizi* is relatively widespread. This is one of the few areas in the lagoon where *Strombus gigas* may still be found; probably the confusing nature of the bottom makes its fishing difficult.

LOCALITY G: The sediment is several feet thick in this area, and tends to be poorly sorted. In many ways, the tidal flats between east Bimini and south Bimini resemble the area west of the Andros Island. In both areas, a substantial part of the sediment is composed of fine aragonite needles.

The dominant organism here is the white colonial tunicate *Didemnum*. This animal produces large quantities of small (20-30 micron) spicules of spherical shape. The spicules are easily recognized by their construction; they appear to be made of many calcite scalenohedral calcite crystals. The mounds and depressions in this area are probably caused by either *Arenicola marina,* or the shrimp *Calianassa.* There are 50-60 mounds per 100 m².

LOCALITY H: This area with medium dense grass supports a community that is very widespread in the outer lagoon. The dominant animal is *Tedania ignis,* the fire sponge. Few other large invertebrates are present. The blades of grass support a rich epifauna of foraminifers, ostracods, encrusting algae, etc., and the moderately sorted sediment reflects this microfauna.

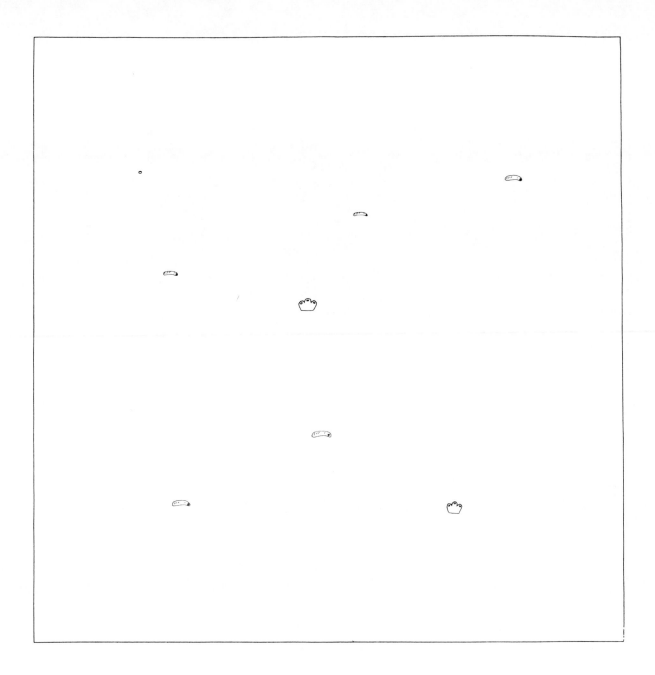

LOCALITY I: This is a dense *Thalassia* area. The current velocities over this area are the highest in the lagoon, and there is an interesting relation between the currents, the growth of the *Thalassia,* and sediment accumulation. The stronger the current (up to about 2 knots), the more dense and luxuriant the growth of the *Thalassia*; the more dense the growth of *Thalassia,* the more sediment trapped; the more sediment trapped, the shallower the water becomes; the shallower the water, the higher the current velocity during flow and ebb tide.

The only large invertebrate found in this area is the holothurian *Actinopyga agassizi.* Crabs, particularly *Callinectes exasperatus,* and the anemones *Condylactus gigantea* and *Aiptasia annulata* also occur in the area. The sponge *Ircinia fasciculata* may also be present.

In contrast to the paucity of larger invertebrates, the blades of grass, which in this area have a surface area almost eight times that of the bottom they cover, support a rich diverse microfauna. The microfauna and flora is largely responsible for the sediment accumulation here.

The sediment is unsorted, and contains large amounts of organic debris. No infauna exists here, however, because the dense root system of the *Thalassia* plants excludes burrowing organisms.

LOCALITIES J, K, L: These are all on the bank, to the east of Turtle Rocks, to the south of Bimini. They represent medium grass communities of a type no longer found in the lagoon. They are dominated by the edible conch, *Strombus gigas,* which forms a significant part of the Bahamian diet. The lagoon has been fished so heavily for these conchs that only a few individuals are still encountered there. However, at a distance of 4-5 miles from the islands, the populations are again normal. J and L represent normal *Strombus* populations, K represents an area that was probably recently fished.

K

L

FIELD QUESTIONS
Lagoonal Environments—Florida

MUD LAKES

1. What is the origin of the mud banks or bars which define individual lakes?

2. Are these really broad banks or narrow bars, or both? Do they always extend from an emergent mound (island) like a sand spit or do they appear to sometimes evolve independently?

3. How does the composition and texture of the tops of the banks or bars differ from that found in adjacent lake bottoms? Is there any difference in the floral or faunal assemblage found on one (windward, seaward?) side of your bar than that found on the other? Plot your findings on the following cross section. Use probe and core to check vertical persistency.

scale

What will be preserved in the fossil record to identify the above environment?

4. Would an indurated equivalent of such a long, narrow bar or bank ever yield a "stringer sand" or zone of relatively high porosity/permeability? Would the available mud choke it up?

5. Examine the bottom grasses—pull up some and examine its root system and blades.

   a. Are there any forams, bryozoa or other "critters" attached to the blades? If so—what are they?
   b. What role does the root system play—how extensive is it?
   c. What is the result of the baffling effect of large patches of such grass? Be specific.

6. What is the source of the mud? Be specific. Was it formed locally or brought in?

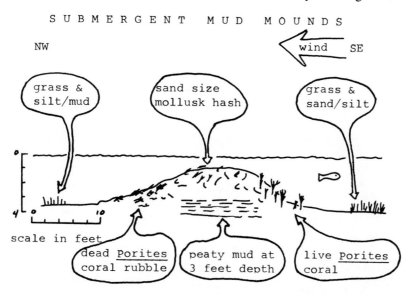

SUBMERGENT   MUD   MOUNDS

NW

wind   SE

grass & silt/mud

sand size mollusk hash

grass & sand/silt

scale in feet

dead Porites coral rubble

peaty mud at 3 feet depth

live Porites coral

The above submerged "mud" mound is an example of those found in mud lakes bordering tidal channels of Florida Bay. For location of one such isolated "mud" mound see figure 2.39.

7. What is the origin of isolated round "mud" mounds found within mud lakes? What are the characteristics of such mounds that might be preserved in the geologic record? The following space is for your own sketch to record information you found at such mud mounds. How much actual mud is present in comparison with shell hash?

8. What does a core indicate as to persistency of present conditions?

9. What would happen if world sea level dropped 1 foot? 5 feet? 50 feet?

EMERGENT MUD ISLANDS

See field questions under Supratidal Environments, pages 60 to 62.

FIELD QUESTIONS
Bimini Lagoon

1. Examine the bedrock-floored channels covered in places with sand.

   a. What organisms are living on the bottom?

   b. What forms are living *under* the surface?

   c. Which live forms which you can observe are also present as dead skeletal particulate material?

   d. What is the size grade of the sand? Is there any mud mixed with this sand?

   e. Observe what the animals are doing (browsing, burrowing, predatory forms?). Estimate the velocity of the current and temperature of the water.

2. Examine the restricted sand and grass flats.

   a. Take a core to bedrock. Depth? What does this indicate as to rate of sedimentation during last five thousand years?

   b. List the types of organisms found in the grass and in flat areas.

   c. What causes the burrow mounds?

   d. Compare temperature and current activity to the bedrock—floored channel area.

   e. Are the mangroves encroaching on this part of the lagoon?

3. What is the minimum thickness of sand necessary to support a flourishing area of *Thalassia?* How does the sand trapped in this effective grass baffle differ in size and composition from that found on (a) the exposed beach, and (b) the bare rock channel bottom?

Bathurst (1967) points out that unlike the broken and angular skeletal debris found in other areas of the Bimini Lagoon the sand in the east lagoon portion is of finer grain, more round and composed of former skeletal grains now micritized by boring algae. Note these features and describe how such a sand in such a low turbulence area attains such a high degree of roundness and sorting. This is also the site of the most abundant distribution of oölitic (3μ thick) films, coating grains. How does this occur under such a low turbulence condition?

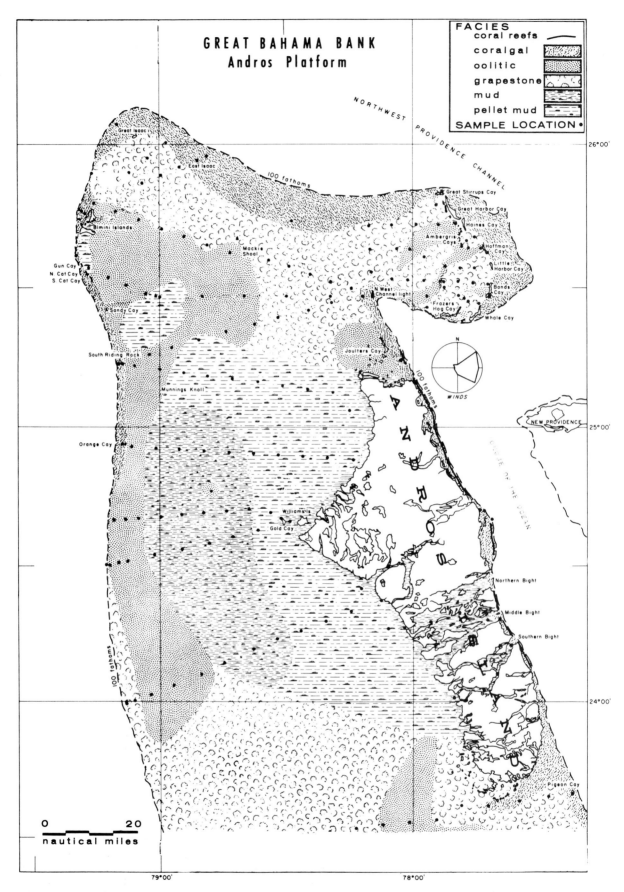

Figure 4.1. Index map to some platform interior areas of the Great Bahama Bank. Map and facies distribution from E. G. Purdy, "Recent Calcium Carbonate Facies of the Great Bahama Bank. . . ." JG 71 (1963). Copyright © 1963 by the University of Chicago Press.

4

PLATFORM INTERIOR AREAS

The term "Platform Interior Areas" as used in this Guidebook refers to the inner areas of open and unrestricted submerged platforms. They contain extensive environments (usually exceeding 600 square km in area) for the deposition of blanket deposits of certain carbonate facies. Such facies include:

1. grapestone
2. mud
3. pellet mud.

These deposits are particularly significant for study as they may represent modern analogs of many widespread carbonate facies deposited on ancient continental shelves.

Examples of these platform interior environments can be found from north to south along the interior parts of the Great Bahama shelf west of Andros (see figure 4.1). Some studies of these environments are those of Purdy (1963a and 1963b), Purdy and Imbrie (1964), Traverse and Ginsburg (1966), Gebelein (1974) and Winland and Matthews (1974). See Busby and Dick (1964) for platform interior temperature-salinity data of eastern Great Bahama Bank.

The following material is taken (with permission) from a portion of the excellent "Modern Bahaman Platform Environments" Field Guide by Gebelein (1974) in which he describes the grapestone and various mud deposits of the platform interior.

PLATFORM INTERIOR ENVIRONMENTS

At a distance of ten km or more from the platform margin, the character of the Bank deposits changes remarkably. This change is related to the decreased competency of the waters to move sand, silt and clay-sized materials. Decrease in competency, in turn, can be related straightforwardly to: (a) increased distance from the margin, and virtual absence of the effect of oceanic waves and swell; (b) increased depth of water, platformward of the raised marginal "lip."

Water depth over most of the platform interior is quite uniform. However, important gradients in water turbulence, current strength and water chemistry do occur. These changes can be related to the character of the bottom sediments: grain type, grain size and sorting, cementation. We will be able to observe the more important of these trends during the field trip.

Stop No. 10. Platform Interior Grapestone Deposits

PURPOSE: To examine the sediments and organisms of portions of the platform interior characterized by good water circulation, but by relatively high bottom sediment stability.

DISTRIBUTION OF GRAPESTONES: Grapestone deposits are characteristic of platform environments where wave and current strength are sufficient to remove fine silt and clay-sized materials but insufficient to move sand and coarse silt-sized materials. Such areas possess relatively strong "cross bank currents." Water depths are generally greater than 3 meters, although the maximum depth for grapestones depends on the wave turbulence: grapestones occur in 10 to 15 meters of water along the north central portion of the platform, where strong wave swell from Northwest Providence Channel crosses the deep platform "ramp." (Areas on Little Bahama Bank with similar depths are well protected from wave swell, and contain pellet mud deposits.)

Effective winnowing of fines occurs over most of the broad shallow (2 to 4 meters) platform south of Andros Island, and also characterizes most of the eastern lobe of the Great Bahama Bank. These areas are blanketed with grapestone deposits.

SEDIMENTS: Grapestone grains are aggregate grains with a typical size range from about 0.5 to 2.5 millimeters. The grains have a lumpy, irregular surface, hence their same. As demonstrated by Purdy (1963) and Winland (1974), there is no real genetic distinction between grains variously called organic aggregates, grapestones, encrusted lumps and irregular well-cemented lumps. In thin section, these aggregate grains may be seen to be composed of a number of well-rounded cryptocrystalline grains, which are cemented together mostly by cryptocrystalline aragonite.

As demonstrated by Winland (1974) and discussed below, the composition of grapestone deposits is identical to the oolite deposit, *if* we assume that all cryptocrystalline grains are derived from ooids. Both facies are characterized by a very low concentration of silt and clay-sized material (less than 5%) and by low concentrations of both fecal pellets (about 10%) and skeletal particles (about 10%). The following discussion of the origin of these grains is taken from Winland (1974).

Grapestone grains consist for the most part of cemented recrystallized ooids. The initial cementation of grains into composites is accomplished by growth of forams and algae in the substrate: additional organic growth and precipitation of cement in interior voids of the composite gradually produces a solidly infilled aggregate (fig. 4.2). The presence of ooids in grapestone indicates that these composite grains are not produced in the same depositional environment as the constituent individual grains: the oolitic environment is characterized by bottom mobility; the grapestone environment by bottom stability. Carbon-14 dates suggest that the altered ooids making up grapestone were formed during transgression of the sea onto the Bahaman platform at the end of the last glacial low-stand. With continued deepening of water over the original sites of oolite formation, the depositional environment has changed from an oolitic mode to a grapestone mode.

BIOTA: The bottom is generally covered with a sparse to rarely thick growth of marine grasses. Molluscs are the dominant megafaunal elements, and occur in much greater abundance and diversity than in the pellet mud areas. *Strombus costata* is a common member of the epifauna.

Calcareous algae (*Halimeda, Penicillus, Rhipocephalus,* and *Udotea*) occur in greater abundance than in the mud areas. Large sponges are common.

In areas where cemented grapestone is exposed, high concentrations of sea whips, sponges, and small corals are found. These areas are identical to the "biostromes" described from grapestone areas on the eastern lobe of the Bank near Yellow Bank (Taft *et al.*, 1968).

SEDIMENTARY STRUCTURES: Reworking of the surface sediments by grazers and burrowers prevents the preservation of most bedding. Discontinuous layers of more highly cemented grapestone occur in some areas.

CEMENTATION: The degree of cementation of grapestone grains into lithified chips, flakes and beds appears to be a function of rate of water movement over the surface sediments. On the Great Bahama Bank, clear-cut gradients in cementation can be observed. Highest cementation of grapestone into well-lithified layers (often 10 centimeters thick) occurs in areas of shallow water with strong cross bank currents. Thus, greatest cementation of grapestone, north of the Joulters area, occurs close to the contact with the oolitic facies, and in water depths less than 4 meters. As distance from the platform margin and water depth increase westward, the amount of intergranular cementation decreases rapidly.

South of Andros Island, the broad shallow grapestone sands show the same margin to interior gradient in intergranular cementation. In this area, water depth is relatively constant. However, currents are slowed during their passage across the wide bank. Hence, rate of water movement is decreased toward the interior, and cementation decreases.

The narrow eastern lobe of the Great Bahama Bank is characterized by depths less than 4 meters, and by strong cross bank currents over most portions of the platform: development of islands along the eastern platform margin is much less than on the western lobe, and cross bank currents are ubiquitous. On this platform, wide areas of grapestone sand have been cemented into well-lithified layers (Taft *et al.*, 1968).

Significance of the Grapestone Deposits

Two general principles can be derived from the development of grapestone grains on the Bahama Bank. The formation of well-sorted aggregate grains sands (which happen to be composed of ooids in the Bahamas) occurs within a well-defined environment. Intergranular cementation which forms the aggregate requires an absence of fine-grained material, and a high rate of water flow over the surficial sediments. These two requirements appear to go hand-in-hand. At the high energy end of this spectrum, sand grain movement is too high for initial grain cementation (i.e., ooids form and accumulate). At the low energy end of the spectrum, fine-grained particles fill the intergranular pores, and prevent (or retard) cementation (i.e., pellet muds or muds accumulate).

Within the environment of grapestone formation, gradients in cementation provide additional information about the setting. Highest rates of cementation will occur wherever water flow is highest—again with the constraint that individual grains themselves are not moved by the currents.

The other principle concerns the environments for ooid formation. Quite obviously, the distribution of environments suitable for ooid formation was different during the early stages of sea level rise over the Bahama platform. The formation of ooids occurred over large areas of the platform where cross bank currents carried supersaturated water into the platform interior. Thus, it is clear that degree of supersaturation of the water is not the sole limiting factor on distribution of oolite deposits at the present time. The key factor is turbulence. During the early stages of sea level rise, shallow, turbulent supersaturated waters covered the portions of the platform now designated "grapestone." These early oolite deposits formed a transgressive sheet, or blanket, across the platform. At some point, rate of sea level rise outstripped rate of vertical ooid accumulation, and water depths began to increase across the platform interior ooid deposits. Increase in water depth continued until rate of flow along the bottom was no longer sufficient to keep the ooids in constant motion. At that time (about 1500 BP?) ooid formation became confined to the only remaining platform areas with sufficient energy to keep grains in motion, and hence to generate ooids: the platform margin.

One important exception to this scheme remains. In the area of Elbow Bank and Mackie Shoal, a linear Pleistocene ridge reaches to within two meters of present sea level. On this elevated ridge, oolite deposits are forming at the present time (at a slow rate), even though this region is in the center of the platform. Water saturation is sufficient for precipitation, and the shallow depths allow sufficient turbulence for this precipitation to take the form of tangential aragonite needles around a nucleus.

The recognition of the older Holocene blanket oolite provides a different, and important, geometry for oolite sand deposits. We might expect to find such broad, continuous ooid deposits on any low sloping platform where

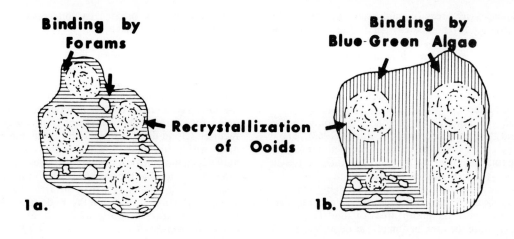

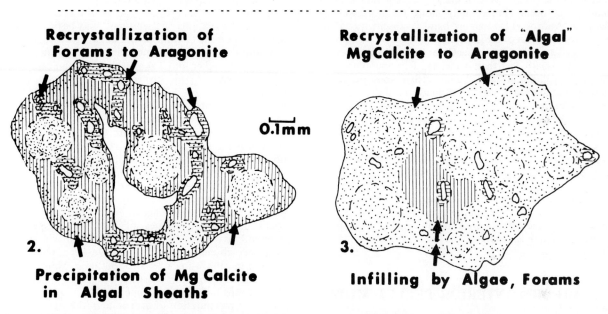

Figure 4.2. Stages in the development of grapestone on the Great Bahama Bank. From Gebelein (1974).

rate of sea level rise (or subsidence) is slow enough to provide a continuous locus for high rates of water movement and relatively high turbulence. Many such deposits are known from the geologic record.

The abstract of the paper by Winland and Matthews (1974), which is discussed above by Gebelein, is given below.

ORIGIN AND SIGNIFICANCE OF GRAPESTONE BAHAMA ISLANDS
by
H. D. Winland and R. K. Matthews

Bahaman grapestone is produced by cementation of recrystallized ooids in the marine environment. Initial binding of grains is accomplished by growth of algae and encrusting forams in the substrate. Later cementation and infilling of the composite may be due to continued growth of these organisms in the interior of the aggregate and to chemical or biochemical precipitation of cement. Data indicate the recrystallized ooids making up grapestone were formed at an earlier stage in sedimentation as the sea transgressed onto the Bahaman Platform at the end of the last glacial lowstand. Thus, grapestone is produced through a sequence of depositional conditions rather than a single set of conditions. Given a supply of firm grains, uneven water turbulence, high water circulation rates, and very low sedimentation rates favor the formation of grapestone.

A description of platform muds and pelletoid muds from the Field Guide by Gebelein (1974) continues below.

PLATFORM INTERIOR MUDS

In areas with little current or wave generated turbulence, fine-grained carbonate sediments are accumulating on the Great Bahama Bank (Cloud, 1962; Purdy, 1963). As previously discussed, areas of mud accumulation on the Bank correspond to those areas, in the lee of Andros Island, where cross bank currents are absent and where winnowing by the easterly winds is minimal. Gradients in water residence time and salinity tend toward maxima along the central western Andros coast. These gradients create a more and more restricted and stressed environment for the biota.

The areas of mud accumulation correspond to the "Mud" and "Pellet Mud" facies of Purdy (1963). These facies have an average silt-clay component of about 62 percent of the total sediment (by weight). The facies differ in the amount of this fine-grained material which is pelleted.

The mud facies accumulates at depths ranging from intertidal to about 6 meters. There is a general correspondence between the major areas of mud facies and deeper water (note the area of "mud" at depths of 3 to 6 meters in the central western bank). A secondary area of mud facies occurs immediately along the northwestern coast of Andros. This narrow band is absent along the southwestern coast of the island.

The pellet mud facies on the Great Bahama Bank occurs in depths usually less than 4 meters. The boundaries between the pellet muds and the adjacent grapestone facies are quite gradational, and correspond to a decrease in intergranular cementation, and an increase in the percent fines. Pellets are well lithified in this transitional zone.

Our field trip will take us to the pellet muds off the northwestern coast of Andros. While we can see the general character of the facies, no detailed work on the sedimentology or ecology of these muds has been done. For that reason, we have supplemented the guidebook with a summary of the trends from the nearshore "pellet muds" to the "muds" off the southwestern Andros coast. We consider that these trends are applicable in analogous positions throughout the Banks.

PLATFORM INTERIOR PELLET MUDS

Stop No. 11

PURPOSE: To observe the sediments and organisms which characterize the protected interior of the Bahama Bank, i.e., areas with long water residence times, high salinities, high temperatures and sluggish currents.

SEDIMENTS: The sediments in this area are composed of 50 percent or more silt and fine sand-sized pellets (Purdy, 1963). Depending on local hydrologic conditions, these pellets may have or may lack a surrounding matrix of silt and clay-sized carbonate. Major skeletal components are molluscs and forams.

The vast majority of the preserved pellets are produced by one or perhaps two organisms. Both are polychaete worms which live in the sediment in low concentration. These pellets typically possess some degree of internal lithification by cryptocrystalline aragonite. Other organisms (gastropods, crustaceans) may produce pellets, but these are destroyed by mechanical action prior to lithification (see below).

Organic concentration in the pellet muds is very low, averaging less than one percent. Along the southwestern coast, the average organic carbon concentration is less than 0.5 percent (by weight).

The aragonite needles which make up the vast majority of the initial sedimentary particles are produced by calcareous algae, and, possibly, by inorganic precipitation. Data from the southwestern coast confirm Cloud's (1962) finding that the standing crop of calcareous algae, *within* the mud facies area, cannot account for the volume of sediments present. This statement is strengthened by the fact that Cloud did not include in his volume estimate the muds within the tidal flats. These more than double the mud which must be produced. Nonetheless, it is important to keep in mind that the area over which the fines are produced is much, much larger than the area of accumulation. Thus, production rates within the mud areas may not be of great significance in trying to deduce an origin for the aragonite needles.

While a complete discussion of the origin of the aragonite needles is beyond the scope of this Guidebook, a few field observations on "whitings" may be pertinent. No data exists on the Great Bahama Bank which conclusively demonstrates the inorganic precipitation of aragonite in whitings. Broecker and Takahashi (1966) indicate that at least 85 percent of the suspended needles in the whiting which they monitored had characteristics identical to the bottom sediments, and different from that expected for equilibrium precipitation from the water. We have made less elegant observations on whitings on the southwestern coast of Andros for the last four years. Pertinent observations are: (1) whitings may develop following and during periods of many days of absolutely calm water conditions—mechanical stirring of the bottom under these conditions is impossible; (2) long-term biologic observations indicate a total absence of bottom feeding fish which could stir large quantities of mud into suspension—observations before, during and after several whitings confirm this absence; (3) whitings have been observed to develop in calm water over rock bottoms, eliminating the possibility of sediment stirring altogether. These latter observations indicate that some whitings may, in fact, be related to precipitation of aragonite in the water column. Geochemical observations are now being made along this coast.

BIOTA: The following description will apply throughout the pellet mud areas. A more detailed description of the southwestern Andros coast follows.

The bottom is covered with a sparse growth of marine grasses. Scattered runners of the noncalcareous green alga *Caulerpa* are interspersed with sparse stands of calcareous green algae (*Penicillus, Udotea,* and *Rhipocephalus*). The bottom is everywhere covered by a surficial organic mat. The mat is broken only by the scattered *Callianassa* mounds.

Living molluscs are the most important faunal elements. Both bivalves and gastropods occur alive with a very low abundance and species diversity.

Further offshore (15 to 20 km) an epifaunal filter-feeding community becomes evident. Dominant organisms are the tunicate, *Didenum,* and various sponges.

SEDIMENTARY STRUCTURES: None are preserved. However, filled *Callianassa* burrows are common. Cores reveal the presence of possible "layers" marked by surfaces with incipient intergranular cement.

Trends in the Pellet Mud Facies, SW Andros

Onshore to Offshore (Shore to 20 km Offshore)

Macrofaunal Feeding Types	Total Diversity (# spp./1,000 individuals)	Relative Diversity	Relative Abundance
Predators	Increase (7 to 23 spp)	Increase	Increase
Filter Feeders	Increase (14 to 23)	Decrease	Increase
Herbivores	Increase (8 to 10)	Decrease	Same
Deposit Feeders	Sl. Increase (3 to 4)	Sl. Dec.	Decrease

Sediments	Mean Grain Size	Sorting	Kurtosis	Skewness	Pellets	Cement
Sediments	Decrease	Decrease	Decrease	Coarse to Normal	Decrease	Decrease

Several important general features can be derived from this data (J. M. Queen, pers. comm.). As expected, the total faunal diversity increases offshore. This certainly corresponds to an amelioration of physical and chemical conditions (salinity, temperature, fluctuations in salinity and temperature, periodic bottom turbulence) offshore. However, the total preserved diversity, even in the most restricted environments nearshore, is much higher than previously estimated (32 to over 100 species, depending on location). This data can be directly correlated to the very low standing crop of organisms. High diversity in this environment is a reflection of the total lack of biological competition. Individuals are maintained at a non-competitive abundance by the extremely low concentration of available food. This conclusion is substantiated by size frequency data for the various feeding types. Predators and herbivores, which are mobile and which can seek out the isolated high concentrations of available food (another animal, or a patch of grass or algae, respectively) have a normal size frequency distribution for their species. However, filter feeders, which are restricted to the ubiquitous, very low organic concentrations immediately around their fixed position, are always very small, and die at an early age: implying an extreme food restriction.

Sediment parameters reflect primarily a decrease in the cementation of pellets offshore. The rationale for this is not completely clear, but the data suggest the following. Only a few organisms produce pellets at all in the mud environment: a few annelids, several gastropods, and several crustaceans. Of these organisms, only the annelids deposit their pellets *within* the upper layers of the sediment. The other organisms excrete their pellets onto the seafloor, where they are rapidly broken up and dispersed by periodic wave energy. Thus, only the annelid pellets survive for a sufficient period for cementation to occur. The increase in cementation onshore apparently is related to the amount of water pumped through the surficial sediments. Clearly, the further offshore one goes, the slower the rate of water movement directly over the bottom, and the lower the amount of water pumping through these sediments. Note that as water depths decrease again toward the margins of the mud facies, the degree of pelleting (as preserved in the sediments) and cementation both increase. The relative decrease in abundance of pellets along the shoreline of the northwestern Andros coast likely is related to the frequent high energy events along that coast (see below). All types of pellets, including lightly cemented ones, would tend to be disaggregated under these conditions.

Applicability of the "Mud" Facies Trends to the Record

On the scale of the Great Bahama Bank, it is not difficult to distinguish and interpret the various subtidal platform interior deposits. The transitions from oolitic sand to grapestone, grapestone to hardened pellets, pellet muds to "muds," can be recognized on the basis of changes in grain type and lithology, and to a lesser extent on the basis of faunal changes. Regardless of the geometry or three-dimensional relationships of these deposits, they can be individually related to particular regimes for turbulence, water movement, and water chemistry. To date, only the muddy environments have been subdivided on the basis of water depth, water chemistry, turbulence and food resources. For this particular environment, trends in organism diversity and abundance (related only to feeding types so that, hopefully, the trends may be applied to the record) can be recognized and interpreted in terms of the physical environment. While sedimentologic trends may also be recognized within this mud complex, the most sensitive indicators of environmental parameters are the fauna.

See Traverse and Ginsburg (1966) for a description of pine pollen in the water and sediment of the Bahaman platforms. They note that the pollen distribution is explicable in terms of sedimentation (low turbulence) and not source. These authors also give graphic summaries of wind, wave motion, currents, surface salinity and sediment data for the Bahaman platforms.

The phenomenon of "whitings," mentioned by Gebelein on page 132 of this Guidebook, is further discussed by Harris (1976) whose abstract follows.

THE DISTRIBUTION OF WHITINGS, BAHAMA BANKS
by
F. W. Harris

Whitings are patches of suspended sediment of unknown origin, commonly observed in shallow marine waters overlying carbonate banks. Mapping the distribution of whitings from satellite imagery (ERTS) shows that they occur only on Great Bahama Bank, leeward of Andros Island, and Little Bahama Bank, leeward of Grand Bahama Island. Whitings are numerous and widely distributed in the spring, but fewer and more localized in the fall; they show no relationship to projected isohalines or nearshore areas of high residence times; they occur near bank margins in areas of open circulation. Most whitings were observed over muddy sediments, although some were found in areas deficient in mud. Observed patterns are more consistent with resuspension of bottom sediment (possibly by fish) than with a chemically precipitated origin.

For an illustration of a "whiting" see figure 4.3 taken from the air over the Great Bahama Bank west of Andros.

Figure 4.3. Aerial view of a "whiting" on the Great Bahama Bank west of Andros Island. Photo courtesy of A. Conrad Neumann.

FIELD QUESTIONS
Platform Interior

GREAT BAHAMA BANK

1. Realizing all the various known mechanisms for the production of carbonate mud:

   a. what can you look for in the field to isolate the origin of a specific mud sample? List mud producing organisms or processes observed.

   b. what could you design in the way of a controlled field or lab experiment to evaluate the origin of a specific mud sample?

Sample No.

Photo No.

2. Is there a chance that some (or most) carbonate mud observed today is the product of a variety of processes (physical, biological and chemical)?? List all such processes under appropriate headings and cite examples of each from personal field observations or from the literature (if not personally observable).

Physical   **Examples**

Biologic

Chemical

Combination of Above

3. List the physical and biologic criteria necessary for mud to be pelleted. List the organisms (and note their estimated number per square meter) that you observe in the field that are responsible for mud pellets of various sizes. Is there a natural sequence? (i.e., does mud start out pelleted or nonpelleted?).

4. What conditions are necessary for the preservation of pelletoid mud?

5. Watch for "whitings" as you fly over or sail over the Bahama Banks. If possible investigate a "whiting" area in person and observe and sample it comparing conditions with adjacent non-whiting situation. Can you observe (with visibility zero . . . good luck!) any fish or other special bottom activity?

6. What are the physical conditions necessary for the formation of grapestone? How does the depositional environment necessary for grapestone formation differ from that necessary for formation of oolite sand bars?

7. What environmental interpretation can be made when oolitic sand is found overlying grapestone deposits? Explain.

8. Observe the thickness of cemented layers of grapestone at various places. What is the relationship between water depth, current velocity and amount of cementation observed? Explain.

9. Observe the boundary between grapestone and mud facies. Is one "transgressing" over the other? Cite your evidence. How does the mud influence grapestone development and/or cementation?

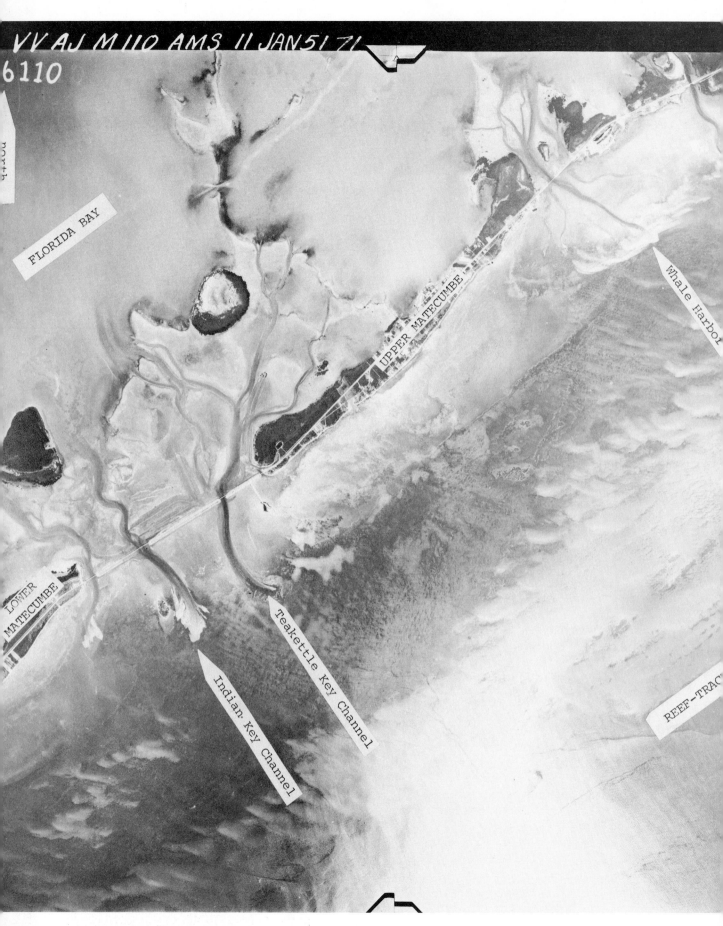

Figure 5.1. Tidal channels and tidal banks represent examples of transitional areas between Florida Bay and the Florida reef tract. Upper Matecumbe area. Photo from U.S. Army Map Service.

5

TRANSITIONAL AREAS

Transitional areas as defined in this Guidebook are those areas which lie as stabilized or migrating channels, banks or belts between two other distinctly different environmental areas, are subject to tidal action and are generally limited in lateral extent. Their unique physical characteristics are valuable in application to paleoenvironmental studies.

Three such transitional type environments which will be dealt with in this Guidebook are:

1. tidal channel areas and tidal banks, such as are present between Florida Bay and the Florida reef tract or Gulf of Mexico and the lower Florida Keys.
2. migrating patchlike accumulations of transgressive sediments of Biscayne Bay.
3. the oolitic sand shoals and flats which lie between deep ocean basins and platform interiors of the Great Bahama Bank.

Examples of each of these sites which can be visited by the reader will be discussed and illustrated below.

TIDAL CHANNEL
AND TIDAL BANK ENVIRONMENTS

SNAKE CREEK AND WHALE HARBOR CHANNELS (Windley Key, Fla.): Snake Creek and Whale Harbor Channels cut through the emergent Florida Keys and link Florida Bay with the reef tract (see figure 5.1). Examination by Multer (1967-unpublished) of the channel bottoms and channel banks bordering flats on the north and south ends of these two tide-swept cuts show transitional variations in organisms, sediment and size-grade. In other words, on the north (Florida Bay) side of these tidal channels (figs. 5.2 and 5.3) a small percentage of reef tract fauna such as finger corals (*Porites divaricata*) and rose corals (*Manicina areolata*) may be found in the grass-held mud banks and on the rock-floored channel bottoms, respectively. On the southern (reef

tract) side of these channels (fig. 5.4), grass-held mud banks with typical Florida Bay mollusks give way to coarser *Halimeda*-rich sand banks and increasing amounts and variety of corals.

BAHIA HONDA CHANNELS (Between Bahia Honda and Big Pine Keys): These channels represent a major tidal passage between Florida Bay and the reef tract (see figure 5.5). Hastings (1970) found that these channels permit some reef tract characteristics (constituent composition, percentage of fines, median phi and standard deviation) to penetrate up to 3.5 meters into Florida Bay. Ancient analogs of similar situations would certainly make proper environmental interpretations challenging.

MATECUMBE KEYS TIDAL BANK (Between Upper and Lower Matecumbe Keys): A comprehensive study of the approximately 3 × 5 km size carbonate tidal bank which lies between upper and lower Matecumbe Keys (see figure 5.1) has been made by Ebanks and Bubb (1975). Their study includes a contour map on top of the Key Largo bedrock for the area, core results in terms of geometry of units encountered as well as sediment descriptions. Facies and the history of bank development are described. A summary of diagnostic characteristics for south Florida carbonate banks is also given (see figures 5.6 and 5.7).

CAESARS CREEK BANK (East of the Northern End of Old Rhodes Key—Northern Florida Keys): Banks of relatively restricted outlet channels in transitional areas can preserve unique records of past history as they often are the recipient of products from storm tides as well as areas of active seagrass entrenchment (the latter providing both a stabilizing and entrapment mechanism). An abstract and two illustrations from a master's thesis study by Warzeski (1976a) describing one such channel bank at the south end of Biscayne Bay is given below. See also figure 5.8.

Figure 5.2. Florida Bay side of Snake Creek tidal channel. Man is walking near edge of grass-held mud bank (left) and deep channel (right).

Figure 5.3. Box core of top of grass-held mud bank seen in figure 5.2. Note extensive *Thalassia* root system which stabilizes the bank.

Figure 5.4. Reef tract side of Snake Creek Channel, Florida Keys. Edge of grass-held skeletal mud bank extends from left margin of photo to just under bow of boats. Channel along right margin of photo goes to approximately 1.5 m depth.

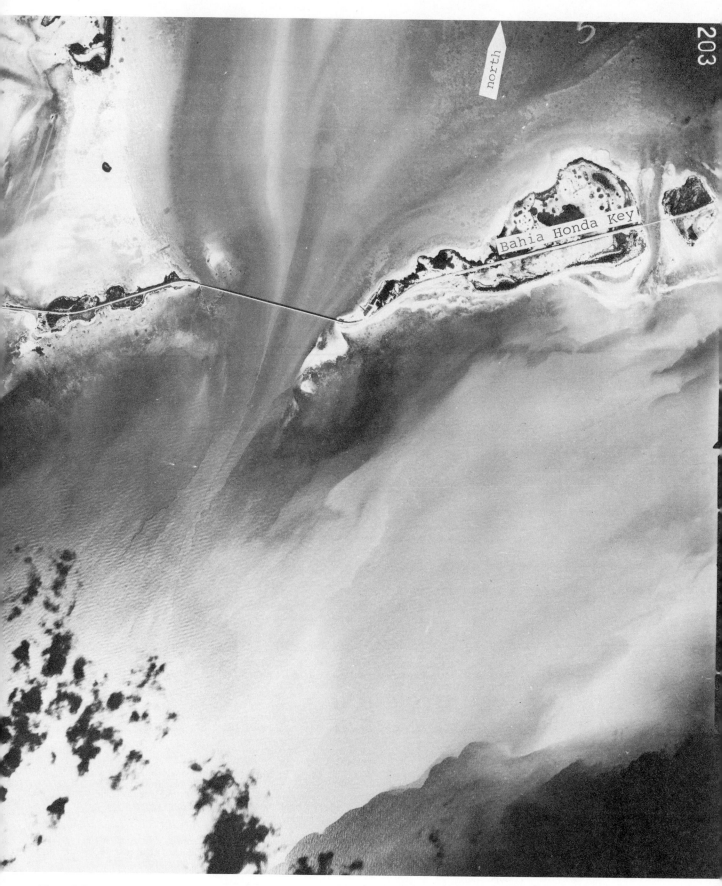

Figure 5.5. A major tidal channel between Florida Bay to the north and the Florida reef tract to the south. West of Bahia Honda Key. U.S. Air Force photograph, Dec. 9, 1959.

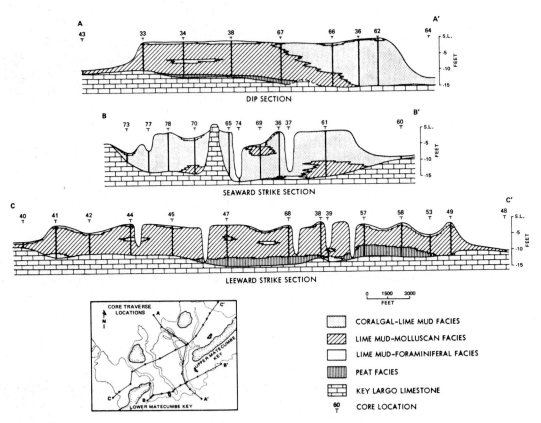

Figure 5.6. Holocene sediment facies, Matecumbe Keys tidal bank. All cores completely penetrated the Holocene sediment to the underlying Pleistocene limestone surface. Figure and description from Ebanks and Bubb (1975).

Characteristics	Skeletal bank	Tidal bank	Mud bank
GEOMETRY	Small and oval shaped; three times thicker than adjacent sediment; surface area about 1 square mile.	Fairly large to small; four times thicker than adjacent sediment; wider on leeward side; surface area 1½ to 10 square miles.	Wide range of sizes and shapes; surface area ¼ to 15 square miles; long and narrow to amoeboid; two to three times thicker than adjacent sediment.
MAIN CONSTITUENTS	Green algae, mollusks, small corals, branching red algae, foraminifers, marine grass, lime mud in some parts.	Green algae, mollusks, foraminifers, marine grass, lime mud.	Mollusks, foraminifers, lime mud, marine grass.
SURFACE ZONATION AND FACIES	Conspicuous faunal zonation and facies asymmetry reflect directionality of environmental factors; supratidal cap facies limited in area.	Weak zonation of fauna into seaward and leeward assemblages; more active grain production on seaward side; greater current-shadow deposition on leeward side; supratidal facies within and on top of bank may be areally important.	None; no strongly directional factors in environment; supratidal facies on crest may be important.
LOCATION AND TREND	Occur in narrow belt on shallow shelf near break in slope; commonly localized above slight depression in underlying surface.	Form at break in hydrographic barrier well leeward of shelf margin; localized at tidal passes opposite restricted embayment.	Occur widely in restricted, shallow embayment; control on individual bank location not apparent.
POROUS ZONES (ORIGINAL)	Coarse skeletal sediment on seaward margin and capping island facies of bank top.	Coarser skeletal sediment in seaward bank and on channel margins.	Only spotty, coarse, skeletal sediment; algal-mat deposits on capping island facies.

Figure 5.7. Summary of characteristics, south Florida Holocene banks. From Ebanks and Bubb (1975).

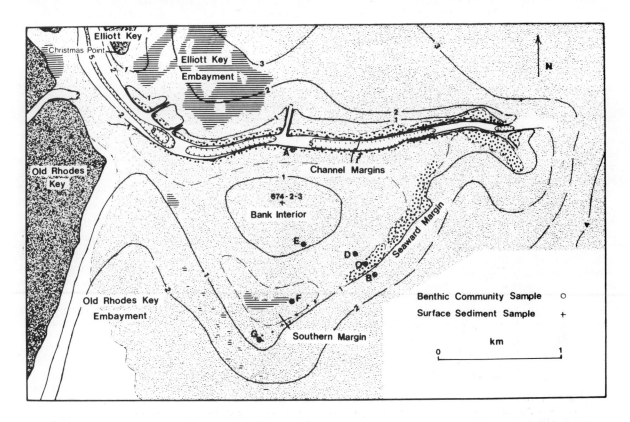

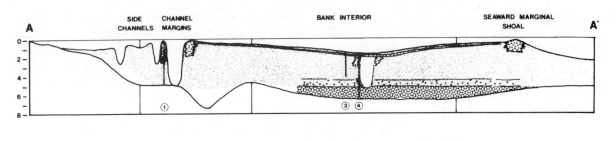

Figure 5.8. *(above)* Bathymetry and bottom character, Caesar Creek Bank. *Light stipple* = seagrass covered muddy bottom; *horizontal lines* = bare mud bottom; *solid small triangular* pattern = shoal fringe or *porites porites* community. *(below)* Cross sections showing bank stratigraphy. *Circles* = molluscan packstone; *light stipple* = *Halimeda* wackestone; *solid triangles* = coralgal packstone; *dark stiple* = calcisphere mudstone; continuous layer beneath bank interior is a storm silt layer. From Warzeski (1976a).

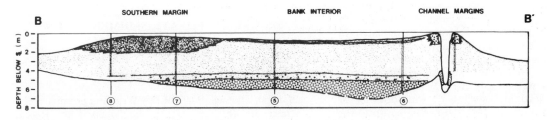

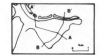

143

GROWTH HISTORY AND SEDIMEN-
TARY DYNAMICS
OF CAESARS CREEK BANK
by
E. R. Warzeski

Caesars Creek Bank, a modern biogenic carbonate bank on the southeast coast of Florida, was studied in detail to determine (a) its depositional history, and (b) relative roles of physical and biological sedimentary processes in bank formation. The bank is distributed assymetrically to the south of the 3.5 km-long seaward channel of Caesars Creek, a 4 to 6 meter deep tidal pass connecting southern Biscayne Bay with the inner shelf margin. Coralgal channel levees, a coralgal seaward marginal shoal and a broad, muddy southern bank margin are nearly exposed at MLW. They form an elevated bank rim surrounding a deeper bank interior. Coralgal levees and shoal support thriving shoal fringe or *Porites* communities, while the rest of the bank is covered by medium to dense growth of *Thalassia*. Periodic blooms of the alga *Acetabularia* occur on the southern bank margin.

The bank contains 4 to 6 meters of mudstone, wackestone and muddy packstone, resting within a broad, shallow bedrock embayment. Bank sediments consist of four basic units. (1) A basal molluscan packstone up to 2 meters in thickness overlies Pleistocene bedrock within the bedrock embayment. (2) *Halimeda* wackestone 2 to 3.5 meters in thickness forms the bulk of the bank, and is depositing at present within the bank interior and beyond the shallow bank rim, in nearby off-bank environments. (3) Coralgal packstone 0.3 to 1.7 meters in thickness forms elevated channel levees and the seaward marginal shoal. (4) Calcisphere mudstone up to 2 meters in thickness forms the broad, muddy southern bank margins. In addition, firm, sometimes cross laminated mudstone layers 0.5 to 25 cm in thickness, characteristically overlying erosional surfaces, are found (a) just below the present sediment surface in the bank interior, and (b) in a zone 0.7 to 1.2 meters above bedrock in two cores directly seaward from the tidal pass entrance.

The depositional history of Caesars Creek Bank consists of four overlapping phases, each characterized by one or more depositional themes. (1) Initial flooding phase lasted from 6500 to 5500 B.P., and was marked by deposition only within the deep bedrock channel of Caesars Creek. (2) Prebank phase lasted from 5500 to 2500 B.P., and comprised two depositional themes. A small, seagrass covered creek-mouth buildup formed in 1 to 1.5 meters of water prior to initiation of tidal exchange with Biscayne Bay. Beginning at the same time, a basal wedge of sparsely seagrass covered molluscan packstone accumulated in deeper water (1.5 to 4 meters) within the bedrock embayment. Molluscan packstone accumulation was displaced seaward by progressive seagrass stabilization, but persisted within the study area until 2500 B.P. (3) Tidal bank phase began between 4600 and 4000 B.P. with opening of Caesars Creek to Biscayne Bay, and has lasted up to the present. It consisted of growth of tidal bank and associated channel system. Tidal bank growth consisted primarily of vertical accretion during rapid sea level rise prior to 3000 B.P., but changed to rapid seaward progradation when the rate of sea level rise slowed around 3000 B.P. (4) Biogenic buildup phase encompasses major organismal modifications of the tidal bank, and comprises three depositional themes: (a) buildup of coralgal channel levees since

at least 2500 B.P.; (b) construction of the seaward marginal shoal within the past 500 to 1,000 years; and (c) buildup of the southern bank margin within the past 200 years.

Primary depositional controls—hydrographic setting, preexisting bedrock terrain, the climate of south Florida, and the Holocene sea level rise—have, in part, determined the location, size, shape and pattern of growth of Caesars Creek Bank. The most important aspect of its climatic-hydrographic setting is the location of Caesars Creek at the southern end of Biscayne Bay, making it an outlet for storm tides of sporadic northerly winter storms and rare major storms. The Holocene rise of sea level, by gradually submerging bedrock energy barriers, has varied the effects of other primary depositional controls and thus controlled the timing and pattern of bank growth.

The two most important sedimentary processes in growth of Caesars Creek Bank have been stabilization of the substrate by the seagrass *Thalassia testudinum* and focused deposition of fine detrital sediment by storm discharge. Also important, but not necessary for tidal bank growth, were sediment trapping by seagrass, and local sediment production.

Biogenic buildup phase deposits were deposited as a result of intense local biogenic production of sediment. The shoal fringe or *Porites* community produced large amounts of coarse skeletal debris, constructing coralgal levees and the seaward marginal shoal; the alga *Acetabularia* produced extremely large amount of aragonitic mud, causing buildup of the southern bank margin at a rate of 1 centimeter/year.

Intense storm tidal discharge during rare major hurricanes destroyed grass cover and eroded sediment (a) seaward of the creek mouth between 4600 and 3700 B.P., and (b) in the bank interior in 1926 A.D. Waning currents of these storms deposited preservable mudstone storm layers.

BLUEFISH CHANNEL (North of Key West): Channels in the lower Florida Keys such as Bluefish Channel (6.4 kilometers north of Key West) have been studied by Jindrich (1969) and found to contain characteristic size-grade patterns, bottom sand structures and evidence of submarine cementation. A summary of Jindrich's results is quoted below. See also figure 5.9.

RECENT CARBONATE SEDIMENTATION
BY TIDAL CHANNELS
IN THE LOWER FLORIDA KEYS
by
Vladimir Jindrich

Bluefish Channel, north of Key West, is a prominent ebb channel (3.5 km long, 100 m wide) incised into Recent carbonate banks up to 3.5 meters thick that overlie Pleistocene limestones. The sediments of the channel and associated environments (fig. 5.9) consists of four major constituents: *Halimeda*, mollusks, foraminifera, and rock fragments of the Pleistocene limestone. Texturally, the sediments can be divided into three major groups (fig. 5.9): (1) *Halimeda*—rich calcarenites with considerable grain-size dispersion of polymodal character. This rather poorly-

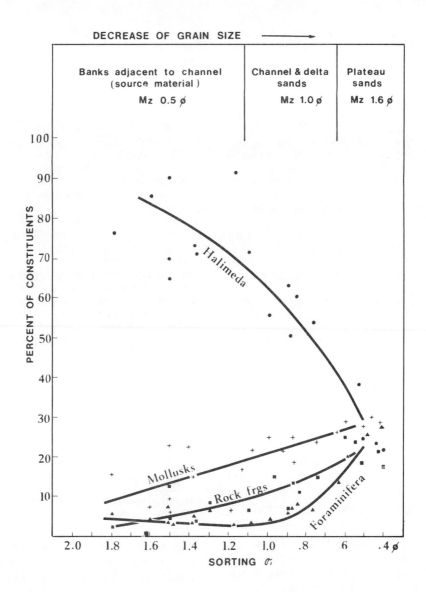

Figure 5.9. *(above)* Diagram showing compositional changes of the sediments during the process of sorting and mechanical breakdown. *(below)* Geological profile along the axis of Bluefish Channel and environmental subdivisions of the area. Bluefish Channel area north of Key West. From Jindrich (1969).

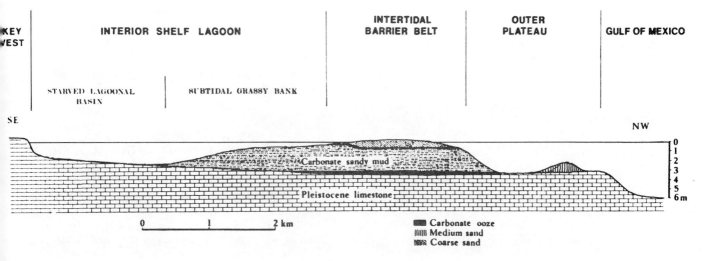

sorted material constitutes the present, *in situ* forming substrate of the intertidal barrier and subtidal grassy bank bordering on the channel. (2) Coarse to fine, submature calcarenites of the channel and its tidal delta, characterized by textural inversions (abnormal size to roundness relation) and strong negative skewness as a result of periodic sediment influx along the course of the channel. (3) Mature, medium calcarenites of the plateau with major constituents represented equally, of unimodal, leptocurtic grain-size distribution. The latter, well-sorted calcarenites are here interpreted as a product of submerged beach environment. Such materials are commonly held to be basically single-constituent sands while diverse skeletal mixtures are interpreted as poorly-sorted products of low-energy environments.

Some calcarenites deposited below the low tide mark exhibit Recent intergranular aragonite cementation. The process appears to be qualitatively similar to that of forming the initial lumps of Bahamian grapestone as described by Illing (1954).

Hydrodynamically, the channel represents the upper-flow regime, the channel mouth and tidal delta display bed forms of the lower-flow regime. Megaripples generated by ebb currents are the dominant structures forming the rapidly-growing tidal delta. The delta is a large, flat-crested channel-mouth bar of cross-bedded megaripple drift sets, grading seaward into delta foreset beds. The entire delta (0.5 km in length, 0.5 km in width) is laterally and vertically differentiated both in constituent composition and texture.

At the channel mouth, submarine erosion of cohesive carbonate ooze produces intraformational breccias, flat pebbles and armored carbonate mud balls. The flat pebbles are very similar to those produced in supratidal environments.

BARRACUDA KEYS BANKS (Northeast of Key West): Basan (1973) in discussing carbonate sedimentation on banks of the Barracuda Keys indicates that daily tidal activity (particularly flood tide currents) may be more important than periodic storms in considering the overall construction of banks. The abstract of his paper documenting the evolution of local carbonate banks is given below.

ASPECTS OF SEDIMENTATION AND DEVELOPMENT OF A CARBONATE BANK IN THE BARRACUDA KEYS, SOUTH FLORIDA
by
P. B. Bassan

An extensive carbonate bank in the Barracuda Keys, Florida, was studied to ascertain those factors influencing its growth and present configuration. Five hydrodynamically or biologically controlled sedimentary subenvironments were distinguished: tidal channels, unstable banks, stable banks (including bare-sand, *Thalassia,* and mangrove island) and silty lagoons.

The bank is a closed system wherein local biological production of sediment is in equilibrium with physical dispersal of sediment. Small amounts of fine-grained sediment are derived from the Gulf of Mexico, but this material is insignificant relative to continued bank development. Sedi-

ment is generally of uniform size, and responds to current flow more as unit "sheets" than as individual particles, thereby permitting a maximum amount of sediment transport. The major constructional process is the flood tide current, which transports sediment by traction, saltation, and to a lesser extent, suspension and flotation. Steady southeasterly wind-waves cause cross-bank transportation but are subordinate to tides as an agent to bank construction.

The basin-shaped Pleistocene bedrock surface exerted principal control on localization of the overlying bank. A resistant limestone ridge on the northern margin of the study area is a barrier to the dispersal of sediment by waves.

Development of this bank may be summarized as follows: preferential accumulation of fine sediment in sinkholes, forming coalescing silty banks; contemporaneous colonization of these banks by calcareous algae and marine grasses; entrapment and accumulation of coarse sediment by these marine plants, forming a single, contiguous sand bank; and continued growth by accretion of sediment over avalanche slopes.

The bank is probably extending itself into the adjoining lagoon by a process of differential growth. This process is dependent upon stabilization of one part of the bank, while growth continues in another.

ENVIRONMENTS OF MIGRATING PATCHLIKE ACCUMULATIONS OF TRANSGRESSIVE SEDIMENTS

BISCAYNE BAY: A second example of transitional areas in the Florida Keys is well illustrated by Wanless (1969) in the Biscayne Bay area. He describes in a detailed study the distribution and depositional history of six major sediment regimes lying in close proximity and undergoing dynamic change (erosion and redistribution) during the last 6,000 years of postglacial Holocene rise of sea level. A study of the Biscayne Bay bedrock basin (see figure 5.10) and accompanying cross sections (figures 5.11 through 5.13) helps provide a better understanding of the controls which permit transitional boundaries to develop and migrate with time. Wanless (1974) also describes "fining-upwards" sequences generated by seagrass beds.

The abstract of the report by Wanless (1969) on Biscayne Bay is given below.

SEDIMENTS OF BISCAYNE BAY DISTRIBUTION AND DEPOSITIONAL HISTORY
by
H. R. Wanless

Three shallow elongate bays, Biscayne Bay, Card Sound and Barnes Sound trend south from Miami along the southeast Florida coast. Hand probing and coring through Recent sedimentary sequences within and underlying bedrock topography, has revealed the general features of

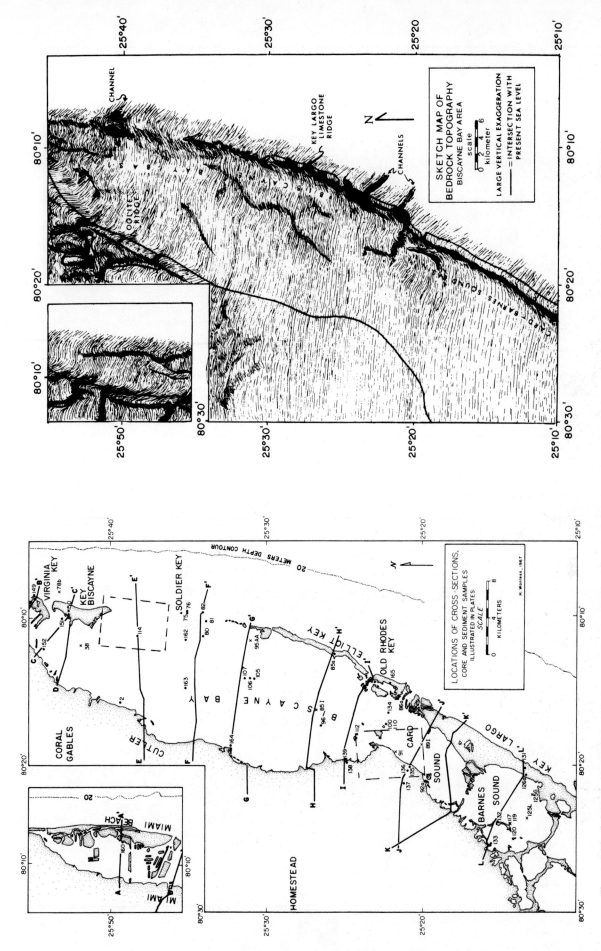

Figure 5.10. *(left)* Index map of Biscayne Bay area showing location of cross sections and aerial photos. Dots represent core locations. *(right)* Sketch map of bedrock topography in Biscayne Bay. Heavy contour is intersection of present sea level with the bedrock surface. From Wanless (1969).

147

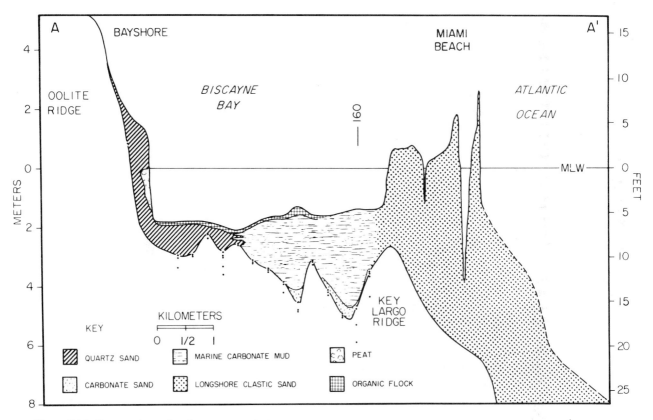

Figure 5.11. Cross sections of sediment accumulations, northern Biscayne Bay including Miami Beach (Section A-A′ *above*) and Key Biscayne (Section C-C′ *below*). Bottom line is preexisting bedrock surface. Variations in bedrock depth at each probing station are shown by vertical dots. Numbers refer to core localities. See data relative to mean low water. See figure 5.10 for locations. Vertical exaggeration x500. From Wanless (1969).

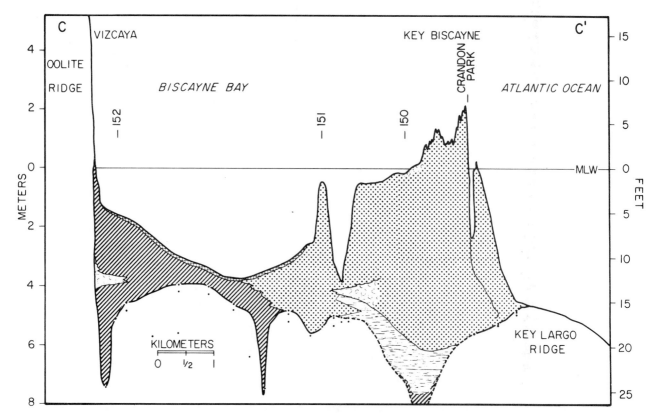

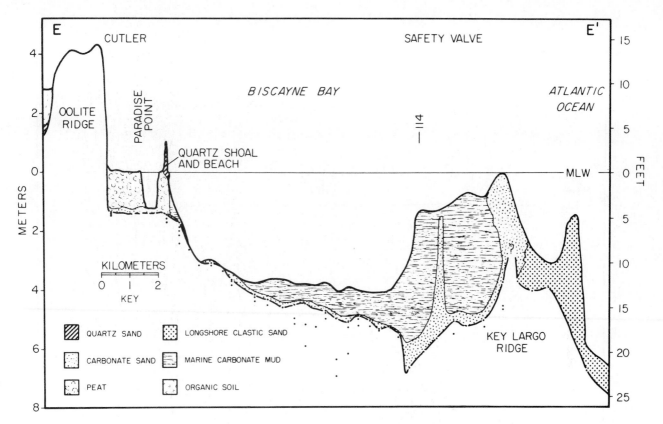

Figure 5.12. Cross sections of sediment accumulations, including central Biscayne Bay (Section E-E′ *above*) and Card Sound (Section J-J′ *below*). Bottom line is preexisting bedrock surface. Variations in bedrock depth at each probing station are shown by vertical dots. Numbers refer to core localities. All data relative to mean sea low water. See figure 5.10 for locations. Vertical exaggeration x500. From Wanless (1969).

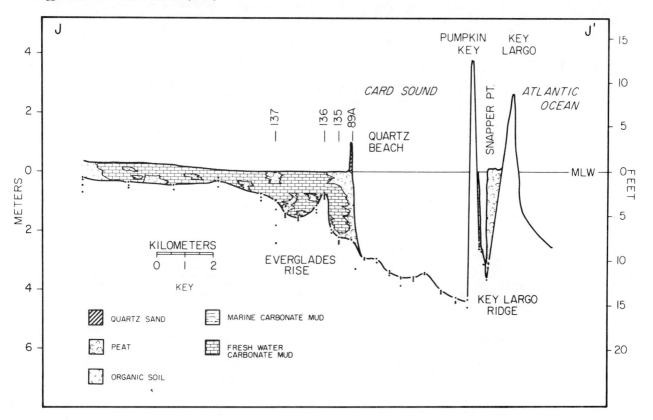

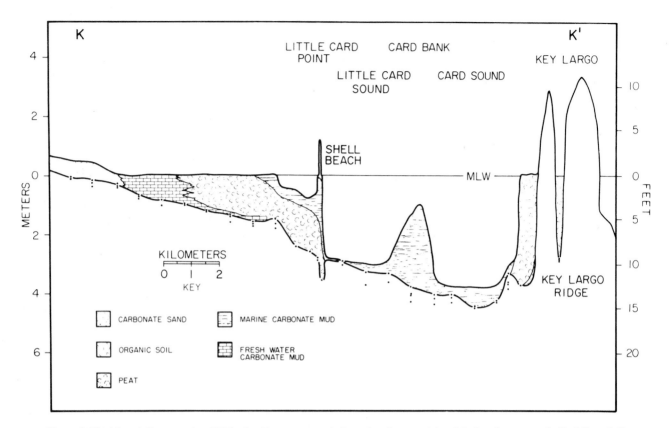

Figure 5.13. *(above)* Cross section K-K′ of sediment accumulations showing transitional facies change, south Card Sound. See figure 5.10 for location. *(below)* Radiocarbon dates together with sample descriptions, positions in centimeters and ages in years B.P. For core locations see figure 5.10. All data from Wanless (1969).

Marine Lab. Sample No.	Core No.	Location	Material Dated	Midpoint of dated interval below sediment surface*	**Sediment** surface elevation above (+) or depth below (–) mean sea level	Midpoint of dated interval below mean sea level	Estimated position of mean sea level relative to elevation of midpoint when deposited	Depth of former sea level relative to present stand when dated interval midpoint was deposited	Age of interval midpoint in C-14 years B.P. (1950)
ML-462	114	Safety Valve, tidal bar belt, west side	Halimeda plates >250μ	365 ± 8	-147 ± 10	512 ± 18	Above	?	910 ± 85
ML-463	114	Safety Valve, tidal bar belt, west side	Carbonate mud <62μ	365 ± 8	-147 ± 10	512 ± 18	Above	?	2,300 ± 90
ML-469	96	Pelican Bank, southwest Biscayne Bay	Fibrous mangrove peat	165 ± 8	- 35 ± 10	200 ± 18	0 ± 30	-200 ± 48	3,540 ± 100
ML-472	152	Spoil island, 0.5 km. west of Vizcaya, northwest Biscayne Bay	Fibrous mangrove peat	360 ± 5	- 20 ± 10	380 ± 15	0 ± 30	-380 ± 45	4,270 ± 100
ML-481	149	North end of Virginia Key, beneath oldest visible beach ridge	Shells	495 ± 8	+ 5 ± 5	490 ± 13	150 ± 40	-340 ± 53	4,200 ± 100

*Correction from mean low water according to tidal ranges given in Tide Tables (1966).

the sediment bodies and the spatial relations of the sediment types and has given insight into the developmental history of the Recent sediment accumulations.

Biscayne Bay, Card Sound, and Barnes Sound are underlain by a shallow, north-south trending late Pleistocene bedrock basin 2 to 6 meters in depth. It is bordered to the east by a ridge formed by the Key Largo limestone, to the northwest by a ridge of the oolite member of the Miami limestone, and to the southwest by a low platform of the Everglades. The basin was first invaded by the sea about 6,000 years ago during the postglacial Holocene rise of sea level. Sedimentation that has taken place during the subsequent period of slowly rising sea level has been controlled by sediment supply and bedrock topography through its influence on wave energy, tidal currents and wind-driven circulation.

Six major Recent sediment regimes are recognized on the basis of sediment type, sediment body geometry and depositional controls:

(a) The influx of a quartz carbonate longshore sediment supply from the north during the past 3,500 years has been the dominant influence in the formation of the sedimentary barrier islands of Miami Beach, Virginia Key, and Key Biscayne and the associated lagoon and offshore shoals.

(b) Intrabay quartz sand accumulations fill depressions and channels in northwestern Biscayne Bay and form beaches and shoals along the mainland shoreline of Biscayne Bay and Card Sound and on bedrock rises within the bay. The quartz is derived from the late Pleistocene Pamlico formation adjacent to the northern end of the bay.

(c) Mud and sand carbonate tidal bars are present where tidal currents are intensified and directed by shallow thresholds of the bedrock topography. Where the bedrock threshold is entirely submerged the tidal bar belt parallels the trend of the bedrock restriction (Safety Valve and Cutter Bank). Where currents are restricted to channels through the bedrock rise the tidal bars are transverse to the bedrock restriction (Featherbed Bank and Caesar Creek Shoal).

(d) Paralic peat and freshwater peat and calcitic mud swamp deposits have developed along the transgressing shorelines since marine waters first entered the bays. Except along the more protected shorelines in Card and Barnes Sounds, these deposits have largely been eroded.

(e) The open bay contains two distinct sediment types. In areas where bedrock is less than 3 to 3.5 meters below sea level a winnowed quartz and carbonate sand forms a veneer (less than 15 cm) over bedrock. In deeper bedrock areas lime mud has accumulated in association with turtle grass, *Thalassia,* which may increase bottom stability.

(f) Non-tidal mud banks in Barnes Sound appear to be actively migrating shoals which have developed in response to wind-induced circulation and wave energy in the adjacent bay in a manner comparable to that of classical cuspate spits.

The dominant feature of the Holocene transgression in Biscayne Bay has not been the preservation of the transgressive history in successive sheet-like deposits, but rather the erosion and redistribution of products of either present or previous deposition in patch-like accumulations.

Transitional tidal areas can be observed for a distance of about 6.5 km south of Key Biscayne (see outline of bars seen just south of Cape Florida Light on index map 1 and figure 5.14). This area, called Safety Valves, is a shallow grass-held bank cut by a series of deep, parallel, tidal channels having currents of approximately 1 to 2 knots. Winnowing action in the channels results in coarse, sorted, bottom sediment which stands in sharp contrast to the adjacent *Thalassia*-held mud and quartz-silt banks containing common *Halimeda* segments and some mollusk debris. This tidal bar belt, approximately 2.5 km wide, is discussed by Ball (1967a, p. 580) who compares it with other tidal bar belts in the Bahamas where an increase in velocity in the tidal channel results in a reversal of sediment types—namely, muddy pelletoid sands in the channels and current-cleaned sorted sands on the bars.

OOLITIC SAND
SHOAL ENVIRONMENTS

Environments of Holocene ooid sand development are found marginal to shelf edges in many Bahaman areas, i.e., Cat Cay, Joulters Cay, Berry Islands, east of Eleuthera (Schooner Cay area), southern rim of the Tongue of the Ocean and between some Exuma Islands (see Newell and Rigby, 1957, plates 11-13 for aerial views of these sites). Amphibious planes from Miami or Naussau are recommended for reconnaissance surveys; chartered boats can be used for more extended investigations.

Ooid sand development can be found as:

1. *active ooid shoals* marginal to open sea and representing the major site of ooid formation, and
2. *stabilized oolite sand flats* usually behind or landward from any active oolite shoals and often forming large blanket deposits extending onto the platform interior (see index map 2) where they intergrade with other platform interior sediment types.

Ooid environments described in this Guidebook are from two areas both of which display active ooid shoals and stabilized oolite sand flats. These are the Cat Cay and Joulters areas.

CAT CAY TO BROWN'S CAY AREA (South of Bimini, Bahamas): A discontinuous belt of oolitic, cross-bedded lenticular sand can be found today in the process of formation along the shallow western edge of the Bahama platform (index map 2). The area is one of active ooid shoals involving alternate tidal action of cool oceanic waters of the deep Florida Straits and warm shallow waters of the Grand Bahama Banks. See figure 5.15 for an aerial view of this oolite belt.

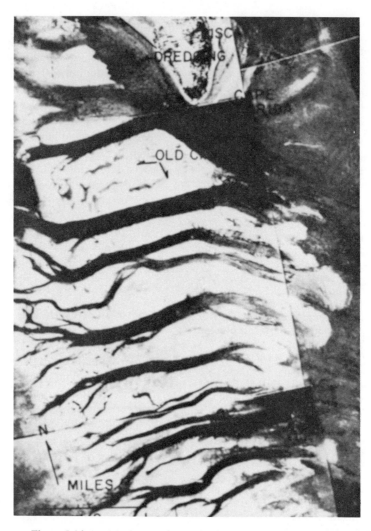

Figure 5.14. Aerial photo (1960) of tidal bar belt (Safety Valve) development to the south of Key Biscayne. Tidal channels are cut to bedrock across axis of shoal. See figure 5.10 for location and figure 5.12 for cross section view. From Wanless (1969).

Figure 5.15. Nine-lens aerial photo at Cat Cay oolitic sand belt from south Cat Cay at top of photo to Sandy Cay in south—distance about 14 km. Transverse tidal channels with deltaic ends show clearly. USC & G Survey, Aerial Photo 48386.

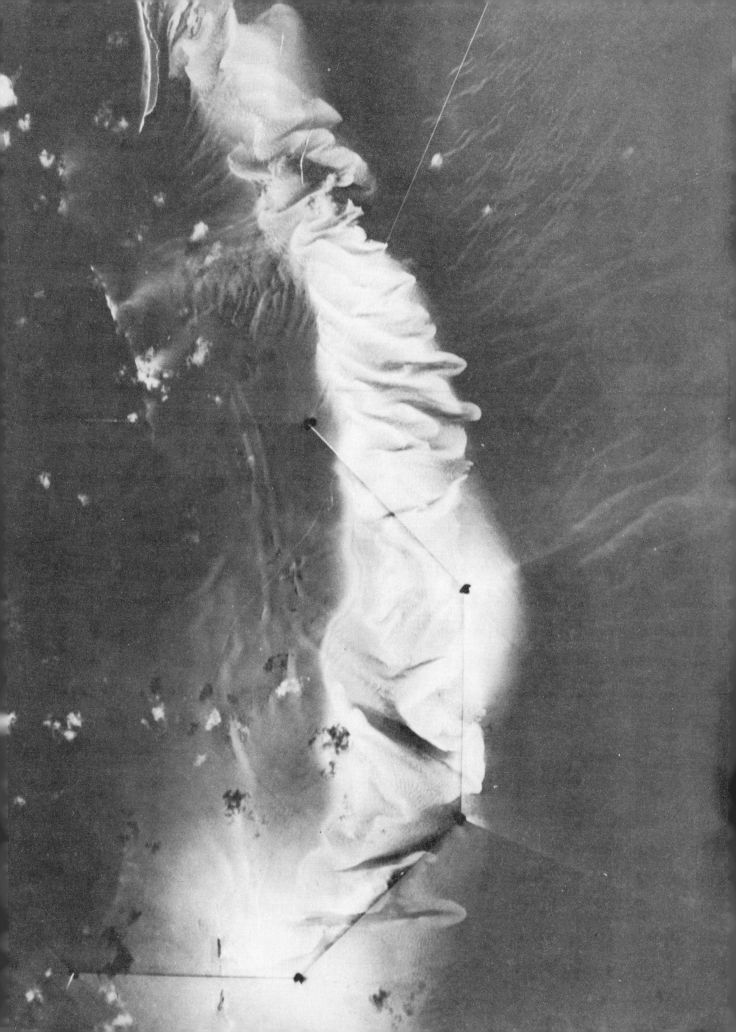

Illing (1954) and Newell et al., (1960) give detailed accounts of previous investigations as well as distribution, chemical composition, structure and origin of the ooids found in this area. Purdy and Imbrie (1964, p. 22) note the following relative to the origin of the oolitic facies:

The oolitic facies is typified by an abundance of oolite and an extremely low skeletal content. Constituent particle analyses of samples from the south Cat Cay-Brown's Cay area reveal an inverse correlation between depth and the abundance of oolitically coated grains. However, depth alone is not sufficient to explain the distribution of oolite, for along the east and west coasts of Andros and along the west coast of Bimini, there are extensive shoal water bottom deposits that contain few or no ooids. Evidently then, oolite formation is related to factors correlated with shoalness in the south Cat Cay-Brown's Cay but not correlated with shoalness elsewhere. One such factor is bottom agitation. In the south Cat Cay-Brown's Cay area, the decrease in depth from the marginal escarpment to the barrier rim of cays and shoals causes an increase in tidal current velocity in the same direction and consequently an increase in bottom agitation. Along the west coast of Andros decreasing depth is not accompanied by increasing bottom agitation, for the large island shelters the shoal areas from vigorous tidal flow and wind waves.

The evidence available strongly supports a physicochemical origin for Bahamian ooids. As previously noted, the progressive increase in temperature and salinity of oceanic waters moving bankward across the outer platform causes a progressive decrease in calcium carbonate solubility. The relatively rapid decrease in depth from the marginal bank escarpment to the barrier rim insures an increase in tidal current velocity in the same direction, a consequence of which is a progressive increase in turbulence. The increased current agitation drives off any excess carbon dioxide which happens to be present and thus further reduces the solubility of calcium carbonate. The warming, evaporation, and agitation of such supersaturated solutions as they move across the oolite shoals represents ideal conditions for the precipitation of calcium carbonate.

Purdy (1963, p. 476) separates the oolitic facies of the Cat Cay-Brown's Cay area into two divisions, namely (a) the oolite facies, and (b) the mixed oolite facies. The chief difference between the two is that the oolite facies has a higher content of oolite, lower percentage of skeletal grains, and lower percentage of particles smaller than 1/8 mm. In light of the above paragraph it is interesting to note that the distribution of the *oolite* facies is along the shoals (shallower crest) of the edge of the bank, thus at a position for optimum conditions of oolite formation (agitation) as described by Purdy and Imbrie. The distribution of this better sorted and essentially nonskeletal oolite is shown as an unstable oolite belt on figure 5.15. See figure 5.16 for a table describing the mean constituent composition and grain size of oolite and mixed oolite facies from the south Cat Cay-Brown's Cay area.

Concerning the structure of individual ooids Purdy and Imbrie (1964, p. 16) note the following (see also figures 5.17 and 5.18).

MEAN CONSTITUENT COMPOSITION AND GRAIN SIZE OF OÖLITE AND MIXED OÖLITE FACIES FROM THE SOUTH CAT CAY–BROWNS CAY AREA*

	OÖLITE FACIES			MIXED OÖLITE FACIES		
	Mean	Observed Range	Standard Deviation	Mean	Observed Range	Standard Deviation
Coralline algae	0	0 – 0.3	0.1	0.3	0 – 1.1	0.4
Halimeda	0.4	0 – 3.7	0.8	2.5	0.7– 6.5	1.8
Peneroplidae	0.1	0 – 0.5	0.1	1.5	0 – 8.3	2.3
Other Foraminifera	0.1	0 – 0.5	0.1	0.9	0.2– 2.0	0.6
Corals	0	0 – 0.4	0.1	0.3	0 – 2.0	0.6
Mollusks	0.4	0 – 2.8	0.2	2.4	0.6– 7.3	1.8
Miscellaneous	0.1	0 – 0.8	0.2	0.3	0.1– 1.0	0.3
Unknown	0.7	0.1– 1.5	0.4	2.4	1.2– 3.4	0.7
Total skeletal	1.7	0.1– 6.8	1.5	10.5	3.8–25.9	6.1
Fecal pellets	0.1	0 – 1.9	0.2	1.4	0 – 6.9	2.0
Mud aggregates	0.2	0 – 1.7	0.4	2.9	0.4– 8.9	2.4
Grapestone	1.0	0 – 6.9	1.5	8.1	1.6–18.4	6.0
Organic aggregates	0	0 – 0	0	0.2	0 – 0.8	0.3
Oölite	92.6	80.2–98.0	1.9	60.1	42.3–73.7	12.4
Cryptocrystalline grains	2.6	0.5– 6.9	1.7	9.1	3.5–18.8	4.9
Cay rock	0	0 – 0.4	0.1	0.7	0 – 3.3	1.3
Calcilutite	0	0 – 0	0	0	0 – 0	0
Weight percentage less than ⅛ mm	1.9	0.8– 5.5	1.2	7.1	1.7–25.0	7.6

* No. of samples from oölite facies = 25; no. of samples from mixed oölite facies = 11.

MEAN CONSTITUENT COMPOSITION AND GRAIN SIZE OF OÖLITIC FACIES*

	Mean	Observed Range	Standard Deviation
Coralline algae	0.1	0– 1.1	0.2
Halimeda	1.8	0–13.1	2.7
Peneroplidae	0.5	0– 8.3	1.2
Other Foraminifera	0.5	0– 3.4	0.7
Corals	0.1	0– 2.0	0.3
Mollusks	1.4	0– 4.8	1.4
Miscellaneous	0.3	0– 1.6	0.4
Unknown	1.6	0– 6.9	1.3
Total skeletal	6.2	0.1–25.9	6.3
Fecal pellets	7.2	0–28.5	9.1
Mud aggregates	2.8	0–15.1	3.2
Grapestone	4.5	0–18.4	5.3
Organic aggregates	0.2	0– 1.9	0.4
Oölite	66.6	8.4–98.0	25.0
Cryptocrystalline grains	7.4	0.2–34.5	6.8
Cay rock	0.1	0– 3.3	0.5
Calcilutite	0.0	0.0– 0.4	0.0
Weight Percentage less than ⅛ mm	5.0	0.8–25.0	5.0

* No. of samples from oölitic facies = 71. The oölite and mixed oölite samples from the South Cat Cay–Browns Cay area are included in this facies

Figure 5.16. Constituent data on the oolite and mixed oolite facies of the south Cat Cay-Brown's Cay area. From E. G. Purdy, "Recent Calcium Carbonate Facies of the Great Bahama Bank..." JG 71 (1963). Copyright © 1963 by the University of Chicago Press.

In thin section Bahamian ooids can be seen to consist of a series of concentric laminations surrounding a nucleus. In the larger ooids the nucleus typically consists of a grapestone grain or a large cryptocrystalline grain. In the smaller ooids cryptocrystalline grains constitute more than 90 percent of the nuclei. The elliptical outline of many of these nuclei suggests that they consist largely of recrystallized fecal pellets. Between crossed nicols the ooids typically exhibit a pseudo-uniaxial cross which is optically positive in sign. X-ray analyses demonstrate that the oolitic laminae are composed of aragonite. Optical observations by Sorby (1879, p. 74) and Illing (1954, p. 36, 38) indicate that the c-axis of the aragonite crystals is parallel to the concentric layers but randomly disposed in the laminae surfaces. In some instances recrystallization of the oolitic laminae appears to begin in the outer laminations and to progress toward the nucleus; in others, no definite direction of recrystallization is apparent. In both cases, however, the end product of laminae recrystallization is unoriented cryptocrystalline aragonite which cannot be distinguished petrographically from the usual cryptocrystalline carbonate of the nucleus.

Penecontemporaneous, submarine cementation is noted by Ball (1967a, p. 561) as common in the burrowed oolitic sands. He notes that cemented cobble-size rocks can be found on the surface and 2-3 square meter areas of cemented sand underlie only a few inches of unconsolidated sand. The cement (see figure 5.17) is fibrous aragonite.

The structure and origin of the sand belt itself is the subject of an excellent paper by Ball (1967a, p. 561) who describes the oolitic sand belt as follows (see also figure 5.19):

The sand belt is up to 12 feet thick and is composed of cross-bedded, current-sorted, medium sand size grains, many of which are thickly coated. The cross-bedded sand overlies a burrowed sand which contains some oolitically coated grains and a high percentage of fine sand size pelletoidal grains. The basal contact of the belt sand is sharp beneath the platformward margin of the sand belt and is gradational elsewhere. The lack of distinctness of this basal contact and the admixture of thickly coated ooliths in the underlying sand is probably due to burrowing across the contact. Grains in the sediments beneath the sand belt with well-defined long axes arranged at all angles to the horizontal show that these sediments have been extensively churned by burrowers.

The percentage of ooliths with coats thicker than 0.1 mm is always at least 40 percent in the cross-bedded sand belt but does not exceed 20 percent in any of the adjacent sediments. In this study, oolitic grains with coats thinner than 0.1 mm were designated "superficial ooliths" and those with coats thicker than 0.1 mm were regarded as "well-developed ooliths," in order to see if the degree of oolitization varied in the several subenvironments mentioned, in some respect other than the percentage of coated grains. Skeletal admixture is greatest in the sands in the slightly deeper water just seaward of the cross-bedded sand belt. Fine sand and silt size particles are commonest in the sediments beneath and platformward from the cross-bedded oolitic sand belt.

Hoffmeister *et al.,* (1967 and chapter 9 of this Guidebook) give a general discussion of the geology of the Cat Cay unstable oolite belt and an example of an ancient (Pleistocene) analog in Florida.

JOULTERS CAYS AREA (Northwest Coast of Andros Island, Bahamas): The Joulters Cays area contains a wide variety of carbonate environments. Although a frequent field trip stop for visiting geologists (see page 388 of this Guidebook), who are usually impressed with the oolite sand shoals, oncolites and shrimp mounds, there has been surprisingly little detailed work published on this fascinating area.

Purdy and Imbrie (1964) give an illustrated (including a colored map of depositional environments) field trip discussion of important aspects of the Joulters Cays area. They include a description of various sedimentary structures and of the transgression of oolite shoals over lime muds of the shelf.

Gebelein (1974) provides a description of tidal channels and associated oolite shoals as found along the general northeast coast of Andros. He describes associated sediments, biota, sedimentary structures, morphology and dynamics of this area and notes the interesting existence of algal-bryozoan reefs within tidal channels rising up 2 to 4 meters above the channel floor. He notes that up to 100 oncolites per square meter are found in shoal areas and flooring some tidal channels. Gebelein (1974) also provides an aerial photomosaic and facies distribution map of the Joulters Cays area.

It is interesting to note that one of the special features of Joulters Cays in contrast to other areas such as Cat Cay is the considerable distance (3 to 4 km) between the active oolite shoals and the platform edge (which itself displays an intermittent rim of coral reefs). Both the distance and presence of reefs reduce the effective wave energy and as noted by Friedman (page 388 of this Guidebook) may be a factor in limiting size of the individual ooids as well as sedimentary structures.

See Halley *et al.,* (1976) for microborings attributed to endolithic algae of some Joulters Cays ooids and a comparison with casts of similar type microborings from Miami limestone ooids.

Work by P. M. Harris (1976 and research-in-progress) in the Joulters Cays area has included coring and description of carbonate facies as well as the study of their geometry and sequence. It is interesting to note that Harris (personal communication) believes that perhaps the mere fringe of active ooid shoals, compared to the large blanket of stabilized sand flats, demonstrates an advance stage of ooid

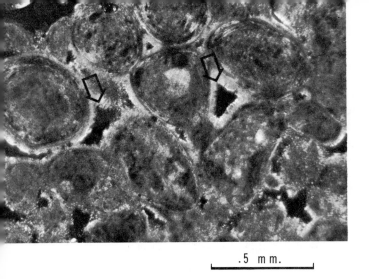

.5 mm.

Figure 5.17. Thin section from Cat Cay sand belt showing fibrous aragonite which grows radially out from grain boundaries cementing ooids. Note some concentric laminations surrounding some nuclei. From Ball (1967a, p. 566).

Figure 5.18. Recent ooids showing spherical to ellipsoidal shapes. Photo courtesy of R. J. Dunham, Shell Development Co.

Figure 5.19. Schematic block diagram of the Cat Cay oolitic sand belt from Ball (1967a, p. 563).

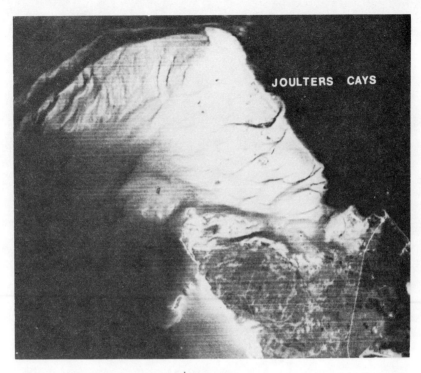

JOULTERS CAYS

Figure 5.20. Landsat I (Band 5) scene of northern Andros-Joulters Cays area, Great Bahama Bank. NASA.

sand complex development for the Joulters Cay area compared to other areas.

The following unpublished data was furnished by P. M. Harris for the 1977 edition of this Guidebook. It briefly describes essential elements of the ooid sand complex of the Joulters Cays area together with pertinent adjacent marginal areas. Figures 5.20 through 5.23, also provided by Harris, illustrate his remarks.

DEPOSITIONAL ENVIRONMENTS OF JOULTERS CAYS AREA
by
P. M. Harris

Sediments on the windward platform margin of Great Bahama Bank, north of Andros Island, appear to be unique among sand accumulations in the Bahamas. Accumulation nearly to sea level has occurred over a large portion (greater than 375 km²) of the Joulters Cays area.

Several distinct major depositional environments are recognized in the Joulters Cays area.

Environments within the ooid sand complex are:

1. the marginal sand shoal and tidal sandbar
2. the stabilized sand flat
3. tidal channels, and
4. islands.

Environments marginal to the ooid sand complex are:

1. the platform margin shelf
2. the open platform interior, and
3. the restricted platform interior.

Marginal Sand Shoal and Tidal Sandbar

The marginal sand shoal trends NW-SE and is continuous for approximately 24 km parallel to the shelf break in slope. It has a maximum width of little more than 1 km, but in most places is only hundreds of meters wide. To the north, tidal sandbars and channels trend E-W to NE-SW and extend greater than 30 km to the west onto the shallow platform. The shoal and bars make up the windward margin of the ooid sand complex—that part most affected by easterly winds and wave-dominated currents.

Water depths are shallow along the length of the shoal; the shallowest places are commonly exposed at low tide. Only on the shallow parts of the shoal are currents sufficient to keep ooids in near-constant motion.

Only those organisms adapted to living in a moving substratum are found on the shoal—a few infaunal annelid worms, burrowing mollusks, and echinoderms.

Sedimentary structure of this mobile sand environment are large-scale transverse ridges, or megaripples, with complex patterns of small-scale ripples superimposed upon them. Submarine cemented layers are present within the sediment of the shoal. Sand shoal and sandbar sediment is texturally a very well-sorted medium sand, with less than 3

Figure 5.21. Aerial photograph view looking northwest, Joulters Cays. (A) The Inner Platform Margin Shelf, where ooid-skeletal sand supports a *Thalassia*-stalked green algae covering. (B) The Marginal Sand Shoal, intertidal rippled ooid sand. (C) One of the three islands making up the Joulters Cays. (D) The Stabilized Sand Flat, where ooid-pellet-skeletal sand is stabilized by algae and grass. Photo and description courtesy of P. M. Harris.

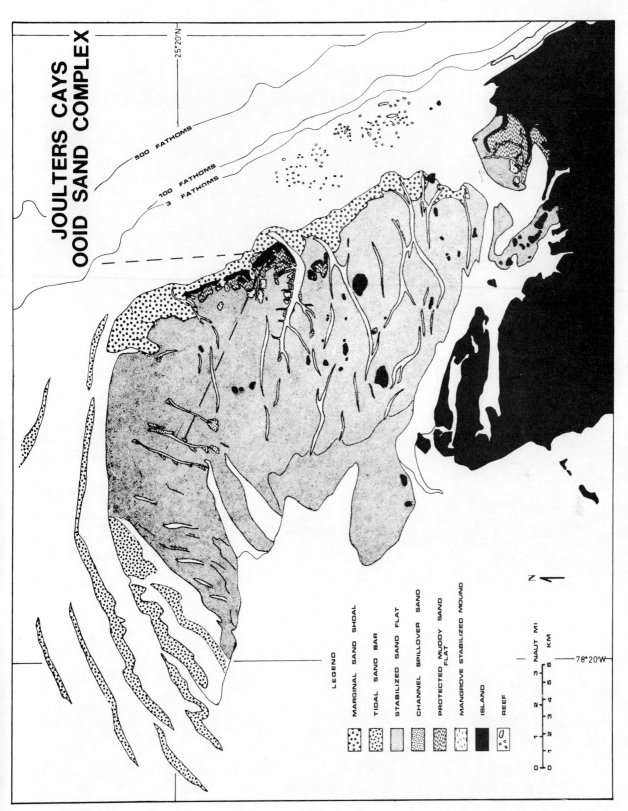

Figure 5.22. Map showing depositional environments recognized within the sand complex. Map and description courtesy of P. M. Harris (1976 – research-in-progress). Note: Dashed "v" indicates approximate view of aerial photo showing in figure 5.21.

Figure 5.23. Rippled ooid sand of marginal sand shoal at low tide, Joulters Cays. Photograph courtesy of P. M. Harris.

percent mud. Compositionally the sediment is greater than 85 percent ooids, with the ooids having a well-defined oriented cortex.

Stabilized Sand Flat

The stabilized sand flat covers the greatest area (300 km²) of any environment in the ooid sand complex. The sand flat lies bankward of the windward marginal sand shoal and tidal sandbar and extends westward for approximately 15 km where it gradually changes into platform interior environments. The sand flat is protected from easterly wave and storm turbulence by the marginal sand shoal and bars.

Water depths are shallow subtidal (generally 1 m or less) to intertidal. The sand flat is "stabilized" by the seagrass (*Thalassia*) and stalked green algae (*Penicillus, Rhipocephalus, Udotea,* and *Halimeda*) and by a covering of mucilagenous blue-green algal scum or mat. Both epifauna and infauna are abundant.

The sediment surface is generally smooth and flat with scattered burrow mounds. Burrowing by shrimp, crabs, and annelid worms results in recognizable organic sedimentary structures.

Compositionally the sediment is greater than 50 percent ooids. Other grain types are Foraminifera, pellets, and aggregate grains. The ooids are generally micritized, that is, much of the ooid cortex has been disrupted by endolithic algae. Texturally the sediment is a moderately sorted fine to coarse sand with approximately 20 percent mud.

Tidal Channels

The channels extend part of the way through the sand shoal and through the sand flat. They shallow at both ends with the bankward end often terminating in a lobate sand fan. The channels are up to 6 m deep at their seaward end, where rock is often exposed or covered by a thin sediment veneer. A depth of 3 m is common through most of their length. The channels are hundreds of meters wide at their widest points to tens of meters wide toward their bankward ends.

Both bottom and sediment type are quite variable in the tidal channels. Generally, they have a grass-covered poorly-sorted muddy sand bottom. The amount of grass covering and percentage mud increase toward the bankward end where up to 40 percent mud is often found. Sand waves are common in the channel bottom closer to the sand shoal, and a rock bottom is common toward the seaward end.

Channel spillovers or levees are found along the channel banks. The sand waves, channel sand spillovers, and terminal lobate sand fans indicate the channels are pathways by which ooids are being transported bankward by strong tidal currents.

Islands

Twenty-nine islands are scattered throughout the ooid sand complex. An additional fifteen or so are present in the areas marginal to the complex. The islands are of various sizes, shapes, and origins. The largest islands and probably the most important in the development of the sand complex are the Joulters Cays themselves.

The Joulters Cays are three islands lying on the NW-SE trend of the marginal sand shoal for a length of over 6 km. South Joulters Cay is the largest of the three islands. Barely a km wide at the widest point, the island is an impressive series of sand ridges and interdune lows. The sedimentary structures and geometry of the ridges indicate a beach-dune origin. The dominant grain type in the island sediment and rock is ooids. The sediment is a medium to coarse grained well-sorted ooid sand.

Based on the condition that the sand grains are in (that is, no major neomorphism has taken place), the limited amount of cementation that has occurred, and preliminary C¹⁴ dating, the Joulters Cays are young Holocene in age. For this reason, they must figure prominently in the development of the sand complex during the Holocene.

Immediately bankward and adjacent to the Joulters Cays and a few of the other islands is a protected muddy sand flat. The area is intertidal to supratidal; ponding of both salt and fresh water occurs. The muddy sand flat is an accumulation of mud and ooid sand that is extensively burrowed by shrimp and crabs. In the higher portions, blue-green algal mats cover the surface.

Smaller mangrove-stabilized islands, or sand and mud mounds, are found bankward of the Joulters Cays.

Platform Margin Shelf

Of the environments marginal to the ooid sand complex, the most seaward is the platform margin shelf. The shelf seaward of the sand complex lies to the north of the Andros Barrier Reef but numerous luxuriant patch reefs are present.

The shelf slopes gently seaward from depths of 2 to 3 m at the seaward end of the sand shoal for a distance of 5 to 6 km to depths in excess of 6 m where the slope increases greatly. To the north, the platform margin shelf grades into the platform interior. Much of the oceanic wave and swell action are dampened across the platform margin shelf.

The outer platform shelf is a deeper open area with a thin veneer of skeletal sand over a rocky bottom. Gorgonians, sponges, and small boulder-type corals are common. A thicker accumulation of skeletal sand (but still less than 1 m thick) is found on the inner platform shelf. *Thalassia* and stalked green algae are abundant.

Patch reefs, built nearly to sea level, are scattered throughout the southern half of the shelf. Sediments accumulated near the reefs are rich in coral, coralline algae, and mollusc debris.

Sediment typical of the platform margin shelf away from a reef is a micritized skeletal-aggregate sand. Less than 5 percent mud is present.

Open Platform Interior

Gradational between the platform margin shelf and more restricted platform interior is an open platform margin.

Tidal sandbars from 1 to 3 m deep and intervening channels up to 6 m deep are present. The sand is primarily an ooid-aggregate-pellet sand that is mobile on the shallow bars and stabilized in lower areas by *Thalassia* or by cemented layers.

Restricted Platform Interior

The restricted platform interior is the most bankward environment. It changes gradually toward the east into the stabilized sand flat and to the north to the open platform interior sandy deposits. Water depths range from two to three meters in the east to about 5 meters toward the west. The bottom is burrowed and covered by a brownish organic scum and scattered *Thalassia* and stalked green algae. The distance from the platform margin and increased water depth result in decreased competency of the bank waters and accumulation of mud.

The sediment ranges from a peloid-foraminifera muddy sand at the western boundary of the stabilized sand flat to a pelleted sand and mud toward the west.

Conclusion

The lateral distribution of different depositional environments and the sediments that are accumulating there reflect physical, chemical, and biological controls on sedimentation.

Physical controls are tidal currents, wave-induced currents, and preexisting bedrock topography. The chemical control is cementation by both submarine and meteoric waters. Biological controls are algae and grasses that stabilize the sediment by their binding ability and burrowing organisms that change sediment texture and produce organic sedimentary structures.

FIELD QUESTIONS
Transitional Areas

SNAKE CREEK— WHALE HARBOR CHANNEL (Windley Key, Florida)

1. Examine the shallow banks bordering the channel exiting on both the bay side and the ocean side. List the characteristic sediment and living forms found below. Note forms living on blades of grass.

Bay Side of Channel   **Sea Side of Channel**

Sediment

Living Organisms

Take 100 cc sample from both areas and run size-grade. Record results and tabulate constituents on chart in appendix.

2. Dive to the bottom of the channel at the above two places and observe the sorting and size-grade found in this relatively high energy environment. Are any sedimentary structures present? Take 100 cc sample and run size-grade; record on chart in appendix.

SOUTH CAT CAY (Bahamas)

3. Make east-west traverse across the unstable oolite sand belt area. Observe the spillover lobes on both edges, the several sizes of ripples and paucity of life. Sample across the traverse taking 100 cc samples, run size-grade and plot data and constituent analysis on table in appendix.

4. While underwater, take off a swim fin and create ripples yourself by using it as an underwater wave promoter. How easy is it to produce ripples? What kind do you produce? How long do they remain?

5. Note lithified pieces of rock found on the bottom or core into sand and recover some lithified pieces. What is the nature and origin of the cement? How widespread is this cementing phenomenon?

6. Compare the size-grade and constituent analysis of this loose sand with that lithified along the sea cliff on the west shore of north Bimini.

7. Although Squires (1958) lists 25 species of corals in the vicinity of Bimini they are mostly restricted to the rocky bottom west of the islands (Bimini) along the edge of the shelf. What are the various factors responsible for this growth pattern?

JOULTERS CAYS (Bahamas)

8. Compare the median size, sorting and shape of ooids found at Joulters Cays with ooids seen at Cat Cay. If there is a difference describe why.

9. Compare sedimentary structures found in the active oolite shoal with those found on the stabilized sand flats. Make scale drawings of each. Compare with similar features seen at Cat Cay and note reasons for differences in size (if any).

10. How would you *prove* that the sand facies at Joulters Cays are transgressing onto the platform? Be specific.

11. What are the observable criteria that allow the formation of oncolites?

Figure 6.1. Aerial photograph of the southern portion of Key Largo, Florida showing, from the northwest to the southeast, lagoon, inner shelf and outer shelf areas, respectively Molasses outer reef is just off the photo. Shoal areas (from the shore to approximately 2 m) are shown by vertical line pattern. For enlarged aerial view of the east end of Rodriguez Key, outlined above, see figure 6.8. Photo from U.S. Army Map Service.

6

INNER SHELF AREAS

The so-called "Inner Shelf" areas as used in this Guidebook include those areas from the intertidal seaward to the beginning of the outer reef zone. This distance in the Florida Keys area is approximately 6 1/2 km but varies from less than 2 to 6 km along the eastern edge of Andros Island in the Bahamas. Inner shelf areas are therefore restricted by nature to rather elongate zones parallel to the trend of the platform edge and stand in sharp contrast with the unrestricted wide Platform Interior Areas (subject of chapter 4 of this Guidebook).

Within the Inner Shelf area belt can be found many types of bottom environments. For discussion purposes and for ease in interpreting ancient analogs these various environments may be divided into four major types as noted below.

1. Shallow nearshore (shoal) environments.
2. Patch reef environments.
3. Blanket sediment environments.
4. Subtidal hardground environments.

These four types are often found as elongate northeast-southwest belts parallel to the bordering trend of the outer reef and the emergent Keys in the Florida area. See Gebelein (1974, p. 29-37) for a discussion of back-reef areas along eastern Andros including a comparison with the Florida reef tract.

SHALLOW NEARSHORE (SHOAL) ENVIRONMENTS

SHOAL AND MUD MOUNDS (Eastern Shore Key Largo, Florida): This environment represents shoal waters extending from the sublittoral shore zone to offshore depths of approximately 2 m. An aerial view of this environment bordering a portion of the Florida Keys is outlined by vertical lines on the aerial photos shown in figures 6.1, 6.2, and 6.17. This shoal environment is the one farthest from the submarine platform edge and is normally, therefore,

the most protected from high wave energy conditions. Such a geographic position allows the settling out of mud where abundant bottom baffles provided by grass, *Halimeda* or *Porites* coral growth are present. Shallow water conditions in this environment are also subject to the maximum effect of seasonal temperature changes. For example, maximum/minimum monthly subtidal temperatures taken by Multer (unpublished) throughout a one-year period at Harry Harris Park (figs. 6.3 and 6.4) showed variations ranging from 10 to 35 degrees C at a station in 1 m of water 35 m from shore. Measurements of the amount of suspended material that would settle out (into a plastic graduate with a round 2.5 cm diameter opening exposed 25 cm above the seafloor) at the same bottom station indicated monthly variations of from 12 to 35 cc of mostly clay size sediment.

In addition to the normal excessive mud and temperature conditions mentioned above, certain portions of this shoal environment (those areas adjacent to channels connecting the reef tract with Florida Bay) are subject to daily variable salinity, temperature and sediment conditions. Such channel effects can be observed at Indian Key and Teatable Channels (fig. 6.2).

Bottom topography in the shoal areas tends to be irregular and patchy as illustrated in figures 6.1, 6.2, and 6.17. Such irregularity is due to relief on the bedrock floor and to constructional effects of organic baffles such as *Thalassia* which build relief features on the bottom. It is interesting to note that these topographically high relief features can (as noted on page 94) reflect bedrock lows. All stages in the evolution of grass-held mud and sand banks from small submergent patches to the well-developed, mangrove-populated Rodriguez Key can be seen from examination of the shoal area.

The shoal areas can be prolific producers of skeletal carbonate grains. For example, six different species of *Halimeda* (see Appendix, page 348) have

Figure 6.2. Aerial photograph showing Indian Key and Teatable Key tidal channels connecting Florida Bay (northwest corner) with the reef tract. Vertical pattern lines indicate shoal (subtidal to approximately 2 m depth) area of reef tract bordering land. Large areas of both grass-held sediment and loose sand are seen out to the edge of the platform (Alligator reef). Photo from U.S. Army Map Service.

been identified growing in the Harry Harris area (fig. 6.3). The most common of these six species are the prostrate *H. opuntia* and the erect *H. incrassata*. *H. opuntia* attaches to hard rock bottom or to debris such as dead coral (fig. 6.4) by one or more short holdfasts. They form soft "inverted bushel baskets" or "cushionlike clumps" up to 40 cm high in shallow shoal water and broader, lower mounds in deeper water. Fossil examples of these "clumps" and mounds can be observed in the Pleistocene (100,000 years old) Key Largo limestone such as exposed in the Windley Key Quarry (chapter 10). The other most common species is *H. incrassata* (fig. 6.6) which grows on soft bottom. It is attached by a single, extensive, bulbous holdfast system which extends into the sand or mud to a maximum of 9 cm, stabilizing the local sediment. The effectiveness of this growth form as a stabilizer can often be seen by the topographic high which large numbers of them form in contrast to lower adjacent areas without erect *Halimeda* forms.

See Cuffey and Fonda (1976) for living Schizoporella bryozoans found in sheetlike "giant" colonies encrusting rock and concrete surfaces nearshore in the inner shelf regions of south Florida. See also pages 352 and 353 in Appendix 3.

A particularly significant feature of the shoal areas of the northern Florida Keys are mud mounds showing windward zonation of sediment-producing animals and plants. Such "relief-building communities of marine organisms" as described by Ginsburg and James (1974) can be found in many areas of the Florida Keys. One of these communities noted by Ginsburg and James (1974) is the Algal Bank Community made up of loose branched finger corals (i.e., *Porites porites*) and branched and segmented coralline algae (i.e., *Goniolithon strictum*). See figure 6.5 for location of the principal algal banks in the Florida Keys.

RODRIGUEZ KEY (Northeast of Tavernier, Florida Keys): One of the most "famous" and frequently visited of the "mud mounds" of the Florida Keys is the mangrove-stabilized and windward-zoned Rodriguez Key near Tavernier, Florida (see figures 6.1 and 6.8-6.14). The abstract of a detailed study by Turmel and Swanson (publication pending) concerning Rodriguez is quoted below.

THE DEVELOPMENT OF RODRIGUEZ BANK, A RECENT CARBONATE MOUND
by
R. J. Turmel and R. G. Swanson

Rodriguez Bank is a Recent mound of lime mud sediments deposited during a relative rise in sea level in the absence of vigorous wave action. This buildup has no rigid organic framework and plants are directly and indirectly responsible for the accumulation. An embayment in the Pleistocene rock floor localized the initial deposition of sediment that provided the nucleus for bank development. Although the hydrography and biotic assemblages changed, the bank maintained itself as a topographic feature. The rate of sediment accumulation is greater in the very shallow water on the bank than in the deeper water surrounding the bank. Sediments of varying textures, when produced by sedentary organisms in very shallow water accumulate at about the same rate and keep pace with the relative rise in sea level.

These same authors describe details of the physical and biologic parameters of Rodriguez Key *in* Ginsburg (1964). Their discussion is given below. (Figures cited refer to the present Guidebook; nonaerial photography by Multer.)

Topography, Tides, Winds, and Temperature

Rodriguez Bank is an elongate, flat-topped shoal surrounded by deeper water about one mile east of Key Largo, (figs. 6.1 and 6.9). Over the flat top of the bank the depths range from 1 foot above mean low water to 3 feet below this datum. The deeper water surrounding the bank ranges from 5 to 13 feet deep. There are two low topographic features on the bank: the mangrove-covered island Rodriguez Key a few inches to 1 foot above mean low water, and a narrow ridge along the eastern or windward margin of the bank that is about a foot higher than the rest of the surface (fig. 6.9).

The tidal ranges reported from nearby Mosquito Bank are mean 2.2 feet, spring 2.6 feet. During "normal" low water, neap tides, the depth of water over the bank is less than 2 feet and at low water spring tides almost the entire bank is exposed.

In the Rodriguez Bank area ocean swells are absent and only wind-generated waves are present. Winds are easterly as shown by the rose on figure 6.9. During the summer months they are predominantly from the southeast and prevail from the northeast during the winter months. These prevailing easterly winds produce almost continual wave action on the east facing margins. Because of these wind directions, only the windward bank margin receives wave action. Rodriguez Bank is well protected by Key Largo from north and northwest winds.

Temperature and salinity variations for the area are expected to be generally similar to the reef tract. The temperature range for the reef tract is 59° to 91° F while the range in salinity is 32-38°/oo. The maximum temperature recorded on Rodriguez Bank during 1959 by K. W. Stockman and E. A. Shinn was 100° F, and the minimum for the same period 52° F. Off the bank, the maximum was 94° F and the minimum 57° F for the same period.

Zonation of Plants and Animals

The surface of Rodriguez Bank and the adjacent areas have large populations of marine plants and animals. Many of these have calcareous skeletons whose whole and fragmented remains produce most of the bottom sediments. Two uncalcified plants—mangroves and turtle grass—contribute organic matter to the sediments and

Figure 6.3. Aerial view of Harry Harris Park, Tavernier. Whale-shaped shoal area shown in dotted pattern contains abundant *Thalassia*, *Halimeda*, *Penicillus*, and *Udotea* and is bordered on the seaward edge with *Porites*. Photo taken in 1965; subsequent breakwater and channel construction has altered bottom topography and biota. Spots in the upper left corner are grass and *Halimeda opuntia* patches.

Figure 6.4. Looking down on a small mound of *H. opuntia* attached to *Porites* coral in a hard rock bottom shoal area 25 m offshore of Harry Harris Park, Tavernier, Fla. Depth of water is 1 m.

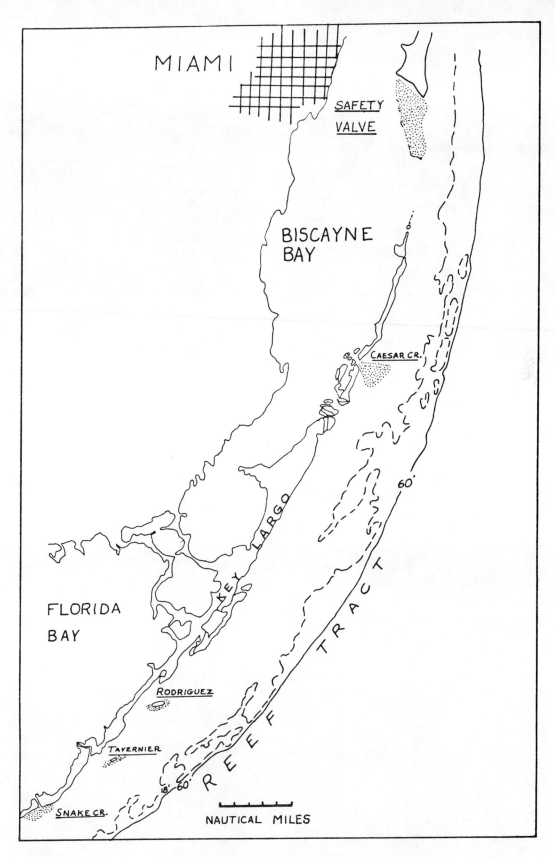

Figure 6.5. Map of a part of the Florida Reef Tract showing locations (stippled) of the principal algal banks. Contours in feet below mean sea level. From Ginsburg and James (1974b).

Figure 6.6. Thriving *Halimeda incrassata* in 1 m, of water 44 m from shore is found carpeting a 7 cm layer of muddy skeletal sand over bedrock at Harry Harris Park, Tavernier, Fla. Other green algae present are Penicillis (P) and Udotea (U).

Figure 6.7. *Halimeda simulans* (S), *Porites divaricata* (P) and Udotea (U) on the seaward side of grassy patch 45 m from shore in 1 m of water at Harry Harris Park, Tavernier, Fla. Both the flat-bladed *Thalassia* and cylindrial-bladed Manatee grass (*Syringodium*) can also be observed in this photograph.

Figure 6.8. Aerial view of the eastern end of Rodriguez Key near Tavernier, Florida. Zonation of sediment-producing animals and plants can be seen. For map of area see figures 6.9 and 6.10: for photograph of core 1 and 2 (below) see figure 6.14.

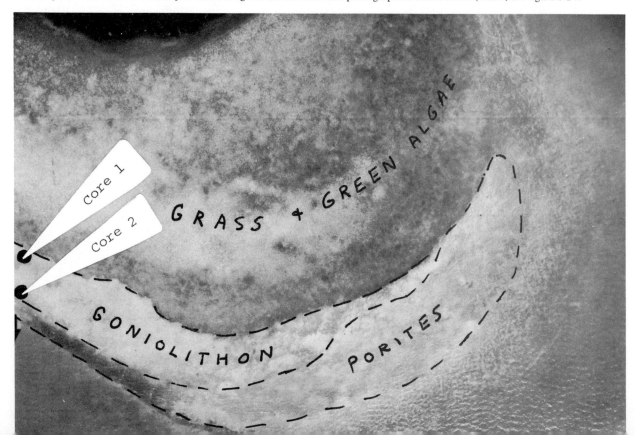

their growth influences the deposition of calcareous sediments. The bank and immediately adjacent areas can be divided into four zones based on the dominant predominant plants or animals, figure 6.9.

Goniolithon Zone. Goniolithon strictum Foslie is a branching calcareous red algae (Corallinaceae), figure 6.13. Its brittle branches 1 to 5 mm in diameter consist of high— magnesium calcite. The branching growths are not attached to the bottom but the intergrowth of adjacent plants creates a wave-resistant forestlike growth that completely covers the bottom. The brittle branches are rather easily broken down to rods that are mostly granule size, greater than 2 mm.

Goniolithon forms an almost complete bottom cover in a narrow zone along the windward margins of the bank, figure 6.9. This algal ridge, that recalls the familiar *Lithothamnium* ridge of Pacific atolls, is barely awash or exposed at normal or neap tides and completely exposed at low water spring tides.

A variety of other plants and animals occur within the *Goniolithon* forest. Sponges, tunicates, laminated algal biscuits (oncolites), and green algae are intergrown with this red alga, and many invertebrates—brittle stars, holothurians, mollusks, and crustaceans—find protection and food beneath the canopy of intergrown branches.

Porites Zone. The small branching finger coral, *Porites divaricata* (fig. 6.12) carpets the bottom immediately seaward of the *Goniolithon* zone, figure 6.9. The branched colonies of *Porites,* like *Goniolithon,* are not attached to the bottom but in the same way the intergrowth of adjacent colonies and associated plants and animals forms an interlocking wave-resistant hedge. The *Porites* branches, 5 to 15 mm, consist of porous aragonite, they are rather easily broken to coarse sand, granule, and gravel-sized fragments.

Grass and Green Algae Zone. Over most of Rodriguez Bank and the surrounding deeper water the prevailing bottom cover is turtle grass, *Thalassia testudinum,* and calcareous green algae of the family Codiaceae, figure 6.9. Turtle grass is not an algae but a true grass with leaves, roots, rhizomes, and seeds; it is not calcified but because of its morphology and abundance it has important influence on deposition and the variety and abundance of organisms. The long leaves coated with sticky microorganic slime trap fine sediment from suspension and provide surfaces for attachment, and food-gathering of other plants and animals. Locally the grass is so abundant as to form a carpet that is both a baffle promoting the deposition of suspended sediment and a protective skin of intertwined rhizomes that prevents erosion of the sediment in the same way that terrestrial grasses stabilize sand dunes.

The two most common green algae are *Halimeda* sp. and *Penicillus* sp. *Halimeda* sp. consists of articulated calcified segments or joints of aragonite. *H. opuntia* has a cushion-shaped form and is frequently intergrown with other plants and animals (fig. 6.12). *H. tridens** and related forms are erect with stems, figure 6.6, and their basal filaments penetrate the bottom sediment to form a holdfast. *Halimeda* sp. is the major sediment producer in the Rodriguez Bank area as well as in the adjacent reef tract. Individual segments separate after the death of the plant and may occur whole in the sediment or be broken down largely by organic activity to sand- and silt-sized particles.

Penicillus sp. and its near relatives *Rhipocephalus* sp. and *Udotea* sp. are lightly calcified green algae, (see Appendix, page 350). All, like *H. tridens** above, are erect forms with a holdfast, a central stem or stalk, and an upper head or crown of branched or bladed filaments. The stem and the filaments or blades of the head are encased in a delicate sheath of aragonite crystals that disintegrates postmortem to the individual silt- and clay-sized crystals, Lowenstam (1955).

Cushion-shaped growths of *H. opuntia* are most abundant in the deeper water surrounding Rodriguez Bank and especially immediately seaward of the *Porites* zone, figure 6.9. The erect *H. tridens** and *H. monile* predominate on the mixed lime mud and lime sand bottoms of the bank. *Penicillus* sp. is present all over the area, *P. capitatus* in the shallow water and *P. dumetosus* in the deeper water off the bank.

The plants and animals that occur within the grass and green algal zone are so varied and numerous that it is not possible to give a complete account of them. Among the more common ones that are seen in almost any transect across the bank are the branching coral *Cladocora,* that is common in the deeper water beyond the *Porites* zone; the sediment-ingesting holothurians that are apparent over the top of the bank; and the conical mounds of subsurface sediment brought up by the industrious tunneling shrimp *Callianassa* sp. Foraminifer are common, brittle stars will be found in sponges and around the bases of *Goniolithon* and *Porites,* and small crustaceans are everywhere.

Mangrove Zone. Rodriguez Key is covered with mangroves, figure 6.9. The red mangrove, *Rhizophora mangle,* with its characteristic prop roots reaching into the water is most abundant around the margins of the key and the black mangrove, *Avicennia nitida,* with its pneumatophores or breathing roots predominates in the interior of the key. The almost impenetrable mangrove jungle baffles the current during storm flooding to catch fine sediment and prevent erosion. The debris from both mangroves gives a peaty surface sediment on the key that is exposed at low tide and except for the center is flooded at high water. The only inhabitants of Rodriguez Key are small snails and maneating crustaceans.

Interpretation. The main feature of the zonation on Rodriguez Bank are the concentric bands of *Goniolithon* and *Porites,* figure 6.9, on the windward margins. These same organisms are almost completely absent on the leeward or west-facing sides of the bank. This preferential distribution reflects the predominance of wave action produced by the prevailing trade winds, see the wind rose, figure 6.9. Wind-driven sea and tidal exchange produce the agitation and flushing that sensitive corals require. The concentration of *Goniolithon* in the shallowest zone along the windward margin that is exposed twice daily can probably be related to a coincidence of intense light, wave agitation at high water that keeps sand and silt in suspension, and wave splash at low water that prevents complete desiccation.

A summary diagram of the resulting zonation pattern and estimated percentage of various organisms at Rodriguez Key is given in figure 6.10. Typical underwater photographs of two of these zones are shown in figures 6.11, 6.12, and 6.13. Two impregnated slab cores are shown in figure 6.14.

---

*Author's Note: *Halimeda tridens* has more recently been reclassified as *H. incrassata.*

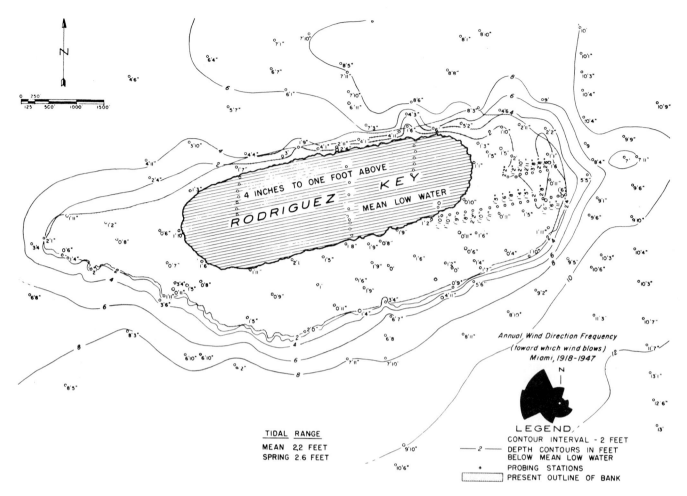

Figure 6.9. Bathymetry (above) and zonation of dominant plants and animals on the surface (below) of Rodriguez Key, northeast of Tavernier, Florida. From Turmel & Swanson *in* Ginsburg (1964). See also figure 6.10.

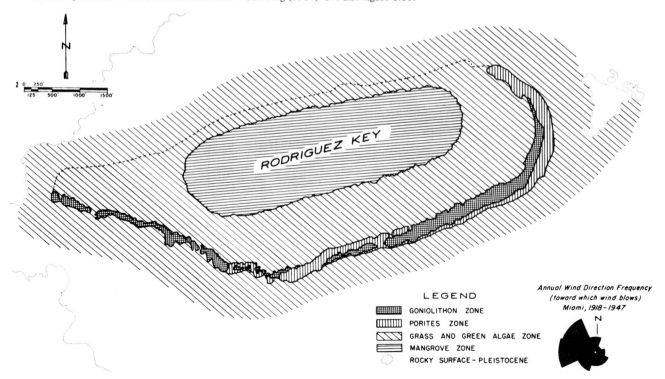

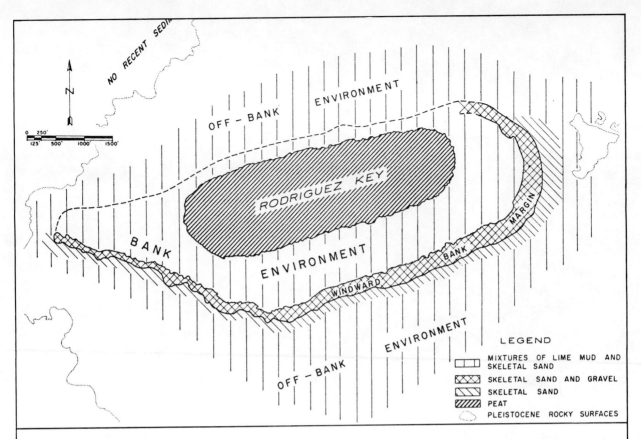

Figure 6.10. *(above)* Generalized variations of the major constituents in surface sediments. *(below)* Surface sediments and environments, Rodriguez Key, northeast of Tavernier. From Turmel & Swanson (publication pending). See also figure 6.9.

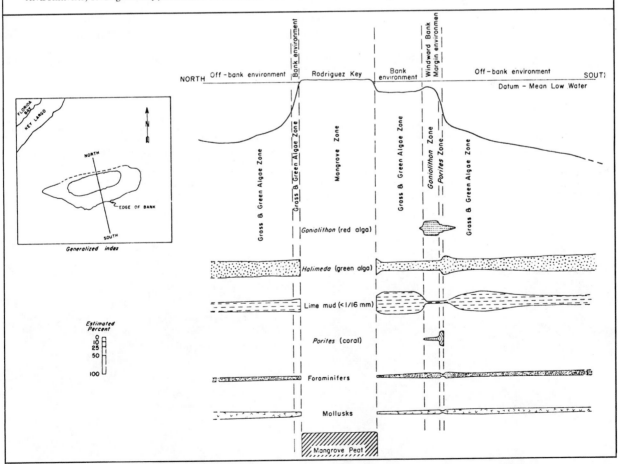

Figure 6.11. Mounds built up by burrowing shrimp *Callianassa* sp. on the flat, shallow *Thalassia* grass zone (see figure 6.9) of Rodriguez Key. Depth of water is 15 cm.

Figure 6.12. *Porites divaricata* zone (see figure 6.9) with *Halimeda opuntia* under the knife blade. A partially disintegrated *Penicillus* (P) and common *Udotea* (U) are also seen among the *Thalassia* grass.

Figure 6.13. *Goniolithon strictum* from the goniolithon zone (see figure 6.9) of Rodriguez Key. This brittle branching form readily breaks down into granule size rods of high-Mg calcite. Upper specimen encrusts a *Porites* fragment. Depth is 8 cm.

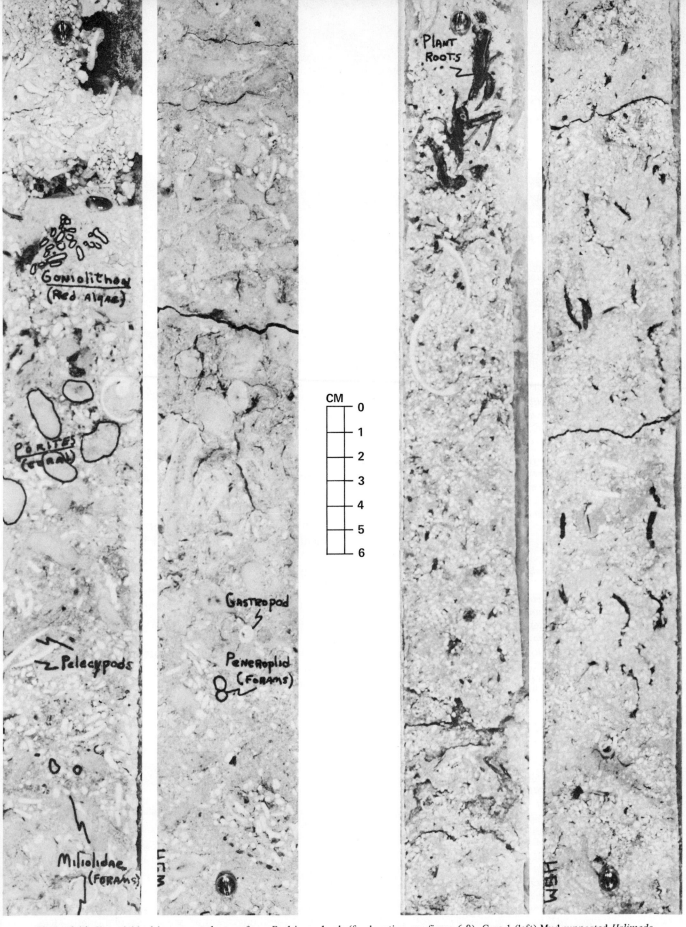

Figure 6.14. Two slabbed impregnated cores from Rodriguez bank (for location see figure 6.8). Core 1 (left) Mud-supported *Halimeda* skeletal sand with forams, mollusk and *Goniolithon* grains; common plant roots. Core 2 (right) Grain-supported *Porites* and *Goniolithon* skeletal sand with common mollusk fragments in muddy foraminiferal and *Halimeda*-rich matrix.

A partial list of organisms that inhabit the surfaces and margins of Rodriguez Bank has been compiled by D. L. Kissling (unpublished) and is given below:

Algae

Acetabularia sp.
Batophora sp.
Goniolithon strictum
Halimeda opuntia
Halimeda tridens
Laurencia sp.
Penicillus capitatus
Penicillus dumetosus
Porolithon sp.
Rhipocephalus sp.
Udotea sp.

Angiosperms

Avicennia nitida
Rhizophora mangle
Diplanthera wrightii
Thalassia testudinum
Cymadocea manatorum

Sponges

Aplysilla sulfurea
Lissodendoryx isodictyalis
Ircinia strobolina
Spheciospongia vesparia
Halichondria melanacdocia
Callyspongia vaginalis

Corals, etc.,

Cladocora arbuscula
Diploria
Manicina areolata
Montastrea annularis
Porites porites
Siderastrea radians
Solenastrea hyades
sea anemones
Millepora alcicornis
Eunicea sp.
Plexaurella dichotoma
Pseudopterogorgia acerosa
Pterogorgia anceps

Molluscs

Ishnochiton sp.
Arca umbonata
Atrina rigida
Barbatia cancellaria
Chlamys sentis
Codakia orbicularis
Pinctata radiata
Tellina lineata
Astraea americana
Astraea longispina
Batillaria minima
Fasciolaria tulipa
Littorina angulifera
Pleuroploca gigantea
Strombus raninus
Octopus vulgarus

Arthropods

Callinassa sp. burrows
Limulus sp.
Panulirus argus
small green crabs
large land crabs

Echinoderms

Clypeaster rosaceus
Echinaster sentis
Echinometra lucunter
Holothuria floridana
Tripneustes esculentis
serpent stars

Also

Several small fishes,
numerous foraminifers,
bryozoans and ostracodes.

PATCH REEF ENVIRONMENTS

Patch reefs as found in inner shelf areas of the Florida/Bahama regions usually represent unzoned, slightly topographically elevated, solid frameworks around and between which the loose skeletal sediment of the back-reef accumulates. Patch reefs are, therefore, major stabilizers (figs. 6.15, 6.16, and 6.22). Their orientation (on the Florida reef tract) as noted by Hoffmeister and Multer (1968) is not haphazard but in a general elongate fashion parallel to the seaward edge of the outer shelf (fig. 6.17). In this way, they receive the greatest amount of nutriment. Why a particular elongate band of patch reef started exactly where it did is an open question. Although not necessary, topographic highs having this general northeast-southwest trend due to earlier erosion (old shorelines) or deposition (old aeolian dunes) would be likely places for incipient patch reef growth. The belief that the present patch reefs represent the latest of many former patch reef belts on the Florida platform is discussed and illustrated in chapter 10 of this Guidebook.

Data on growth rates of typical back-reef corals in the Florida area has been described in various forms by several workers. Hoffmeister and Multer (1964) found a 10.7 mm/yr growth rate for living *Montastrea annularis* which if projected into the Pleistocene might indicate a time span of 250 to 500 years for the accumulation of a 2 meter vertical section of fossil back-reef rock as exposed along the Key Largo canal cut (see back cover photo of this

Figure 6.15. Single head of *Diploria* coral on bare rock-bottom environment, Florida reef tract. Note long spine *Diadema* echinoid (hidden beneath coral) and several varieties of Alcyonarians.

Figure 6.16. Top view of living, 1.5 m high *Montastrea annularis* with thriving *Halimeda tuna* filling small crevices. Top of knife handle is 4 cm in diameter. Patch reef south of Newfound Harbor Keys, Florida.

Figure 6.17. Aerial photograph showing from the northwest corner to the southeast, portions of Blackwater and Barnes Sound, central Key Largo, shoal area (subtidal to depths of approximately 2 m—shown by vertical lines), patch reef and outer-reef environments.

Guidebook). Shinn (1966) relates growth rates to environmental factors. Landon (1975) provides a detailed study of growth rates of various corals. She also notes that growth rates of *M. annularis* and *S. siderea* were higher in the back reef (off southeast shore of Boca Chica) than in the outer reef area. Hudson *et al.,* (1976) give growth rates for *M. annularis* and note two types of high-density growth bands at Hen and Chickens patch reef southwest of Tavernier. They describe 80-90 percent mortality of these corals during 1969-70 due to chilled waters.

Once established, the primary living coral framework is the site of a constant "battle" between organisms which can be classified as "constructive" and those that are "destructive" in their effects. These processes proceed simultaneously, and a great number of variables can affect the final outcome. Constructive agents can include encrusting coralline red algae, which form thick crusts actually replacing large portions of the framework and actions of the hydrozoan Millepora which encroaches on living corals encrusting external surfaces. Tube-dwelling polychaetes can fill crevices with cemented particulate material. Destructive organisms can include boring sponges (*Cliona*), pelecypods (*Lithophaga*), worms, aggressive neighboring corals, rasping reef fish (parrot fish) and echinoids. The combined effect of the above described destructive and constructive activities by various organisms can produce a "secondary framework" whose shape may roughly resemble that of the original primarily coralline framework but whose external and internal composition is altered to a significant degree. For a summary of these processes occurring in Bahaman reefs see Zankl and Schroeder (1972) and Schroeder and Zankl (1974).

Inorganic causes of coral destruction include siltation (believed by many to be the most common cause of coral mortality) producing "bald spots" on coral surfaces. In very shallow areas such bald spots may also be due to exposure during exceptionally low tide periods or exposure to freshwater surface lenses after heavy runoff from nearby shore sources. Whenever such bald areas occur they immediately become the site for competitive degrading or accreting organisms described above. For additional discussion on natural (and man-made) stresses on reefs of the Florida Keys see chapter 12 pages 289 to 313.

Once the dead base of a patch reef is weakened by burrowers, they can often be toppled by storm wave action. A schematic illustration showing a possible sequence involving early development, growth and final destruction of some Florida patch reefs is presented in figure 6.21. Illustrations of various

stages within this sequence as observed by the writer are shown in figures 6.18, 6.19, and 6.20.

Although not a submerged patch reef, the isolated emergent Soldier Key (7 km south of Key Biscayne—see index map 7) has aspects (such as corals) of both shoal and patch reef environments as defined in this Guidebook. Interested readers should refer to an excellent detailed ecological survey of this small area by Voss and Voss (1955) which includes zonation of organisms and a tabulation of fauna and flora.

Wiman and McKendree (1975) studied distribution of various *Halimeda* and adjacent sediments in a patch reef area 2 km east of Old Rhodes Key. They found that *H. opuntia* and *H. tuna* were controlled by availability of hard substrate while the distribution of other *Halimeda* species appears to be controlled by availability of light, water depth and temperature.

Dodd *et al.,* (1973) describe with good illustrations both biotic and sediment zonation along a patch reef south of Newfound Harbor Keys and Big Pine Key which they indicate compares favorably to the Key Largo limestone. A discussion/reply (publication pending) by Multer/Dodd will clarify statements relative to some aspects of this paper. The abstract of their paper is noted below.

POSSIBLE LIVING ANALOG OF THE PLEISTOCENE KEY LARGO REEFS OF FLORIDA
by
J. R. Dodd, D. E. Hattin, R. M. Liebe

A living, linear reef about 3.8 km in length has been discovered 0.8 to 1.2 km off Big Pine Key and the Newfound Harbor Keys in the lower Florida Keys. The reef is approximately 60 m wide, lies at depths from 4.5 to 7 m, and is growing on, and slightly landward of, a slope where Pleistocene bedrock rises with relative abruptness. A deeper, flat area on the seaward side of the reef is, in most places, a *Thalassia-Cymodocea* (sea grasses) meadow, which is usually separated from the reef by a narrow, barren belt of fine-grained sediment. The outer portion of the reef consists of small, scattered coral heads and octocorals separated by a patchy veneer of sediment. The main body of the reef includes large, commonly coalescent coral heads, some standing at least 2.5 m above the seafloor. Next landward is a zone of smaller coral heads and abundant octocorals, followed by an area of large heads of *Montastrea annularis*. The previously known patch reefs off the Newfound Harbor Keys lie farther landward in shallower water. Although the sediment veneer is thin across much of the linear reef, accumulations of skeletal sand and gravel up to 50 cm thick, are common in areas of greatest coral growth. Geometry and biota of this reef, as well as its location subparallel to the present shoreline, suggest that the reef may be a living analog of at least a part of the Pleistocene Key Largo Limestone.

Figure 6.18. Mature stage of "baldness" in 1.5 m high *Montastrea annularis* coral head south of New found Harbor Keys, Florida. Multilobular heads are undermined from edges of central bald area. See figure 6.19 for close-up view.

Figure 6.19. Close-up showing results of organic destruction of living *Montastrea annularis* coral head seen in figure 6.18. *Diadema* echinoids inhabit the niche beneath the multilobular living coral heads. Note "pedestals" left by selective organic destruction of the central dead core of this coral head.

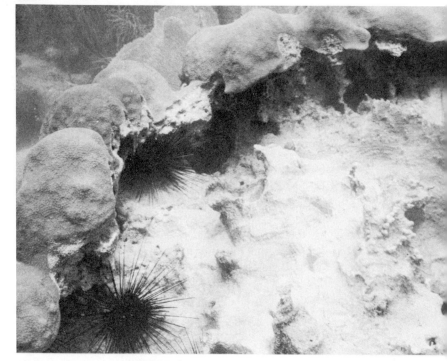

Figure 6.20. Rubble stage or final destruction stage of a *Montastrea annularis* coral head such as shown in figure 6.19. Cemented rubble has a relief of about .6 m. Photograph taken south of Newfound Harbor Keys, Florida.

1   EARLY STAGE
—incipient development of coral heads
  (commonly *Diploria*)

—no rubble

2   FLOURISHING STAGE
—large variety of corals flourish

—loose rubble and sand floors areas between individual
  coral heads, rubble flanks extend out into adjoining
  flats

3   RUBBLE STAGE
—living coral insignificant

—cemented rubble dominant

Figure 6.21. Schematic illustration showing an evolutionary sequence involving (1) early development, (2) flourishing growth and (3) final destruction of some Florida patch reefs. All stages in this series can be observed at different localities on the back-reef platform of the Flordia Keys today. Conditions permitting, these three stages can be successively repeated over the same topographic high. See also figures 6.18, 6.19 and 6.20.

For additional literature describing the living back-reef asemblages of the Florida and Bimini areas the reader is referred to Ginsburg (1964), Hoffmeister *et al.,* (1964), Kissling (1965), Smith (1948), and Squires (1958).

MOSQUITO BANKS (Approximately 5 km Northeast of Rodriguez Key, Key Largo): Mosquito Banks is one of a large number of typical patch reefs (fig. 6.22) readily accessible from either Largo Sound, Pennekamp Park or Tavernier (see figure 6.1 and "V" on index map 3). Mosquito Banks is within the boundary of the Pennekamp Coral Reef State Park (no sampling allowed) and is easily found by bearing toward the flashing 37-foot-high tower marked "35" along its western edge.

Other readily accessible patch reefs (outside the park) can be found such as at Triangles (southwest of the park boundary and northeast of Tavernier Key) and Hen and Chickens patch reef (off Plantation Key). A series of linear patch reefs can be found off Newfound Harbor Keys seaward of Coupon Bight (fig. 3.27). These patch reefs (figs. 6.18, 6.19, 6.20 and page 179 of this Guidebook) can be reached from boats chartered from marinas at the southeast end of Big Pine Key or along Newfound Harbor Channel.

The most comprehensive list of Florida patch reef organisms known to the writer is the unpublished list furnished by courtesy of D. L. Kissling for this Guidebook. This list, which was derived as a "partial list of organisms . . . from examination of patch reefs south of Boca Chica, Newfound Harbor Keys and at Mosquito Banks," is given below.

Algae

Goniolithon strictum
Halimeda opuntia
Halimeda tuna
Laurencia sp.
Penicillus dumetosus
Porolithon sp.
Rhipocephalus sp.
Sargassum sp.
Udotea sp.

Angiosperms

Thalassia testudinum

Sponges

Aplysilla sulferea
Halichondria melanadocia
Ircinia campana
Ircinia strobolina
Lissodendoryx isodictyalis
Anthosiqmella varians
Sphsciospongia vesparia

Corals, etc.,

Montastrea annularis
Montastrea cavernosa

Millepora alcicornis
Diploria strigosa
Diploria clivosa
Diploria labyrithiformis
Solenastrea hyades
Porites astreoides
Porite porites
Stephanocoenia michelini
Isophyllia sinuosa
Colpophyllia amaranthus
Favia fragum
Siderastrea radians
Siderastrea siderea
Dichocoenia stokesii
Agaricia agaricites
Mycetophyllia lamarckana
Oculina diffusa
Zoanthus sp.
Palythoa mammilatus
Gorgonia flabellum
Gorgonia ventalina
Briareum asbestinum
Plexaura flexuosa
P. homomalla
Muricea elongata
Plexaurella nutans
Erythropodium cariboreum
Pseudopterogorgia rigida
Eunicea tourneforti
E. knighti
E. asperula
E. fusca
Pseudopleaura flagellosa
P. porosa
Pterogorgia anceps
Pseudopterogorgia americana
Pseudopterogorgia acerosa
Plexaurella dictotoma

Molluscs

Arca umbonata
Arca zebra
Barbatia sp.
Brachiodontes recurvus
Isognomen alatus
Lithophaga antillarum
Codakia orbicularis
Astraea tectum
Strombus gigas
Vasum muricatum

Anthropods

Panulirus argus
stone crabs

Echinoderms

Clypeaster rocaceus
Diadema antillarum
Echinometra lucunter
Ophionereis reticulata
Ophiocoma echinata
Ophiocoma pumila
Ophiocoma wendti

Ophioderma appressum
Holothuria floridana
large serpent stars

Also

Numerous foraminifers, fishes,
ostracodes, bryozoans

BLANKET SEDIMENT ENVIRONMENTS

GENERAL HAWK CHANNEL AND WHITE BANK AREAS (Seaward of Key Largo, Florida): Large areas of the Florida inner shelf platform are blanketed with sediment (figs. 6.1, 6.2, and 6.17). Two common varieties of such sediment are (1) loose, often fairly well-sorted, clean, rippled sand with a paucity of obvious life, and (2) grass-held, less well-sorted sediment having a rich variety of living fauna and flora. In many ways, these two varieties may be considered as end members with gradations between them more commonly present on the back-reef platform than the end members themselves. In other words, patches of loose sand in grassy areas and some grass invading barren sand flats are very common occurrences. Generally speaking, the broad, loose, rippled sand areas occur more often on the seaward portions of the platform where stronger current action is present, and the grass-held sediment blankets are best developed on the leeward (bordering the emergent Keys) portion of the platform (see figure 7.28).

The area called "White Bank" is typical of the first type of sediment blanket noted above. This area averages about .8 km wide and lies as a northeast-southwest trending band along the southeastern edge of the back-reef platform just back (leeward) of the outer reef (see white band northwest of Key Largo Dry Docks outer reef on figure 6.17 and figure 7.28). This "underwater desert" is essentially devoid of life and its *Halimeda*-rich, medium-to-coarse grain, subrounded, clean sand has been brought in from the adjacent outer reef zone by strong current action. Ripples, bedding, mechanical abrasion of grains and lack of burrows are characteristic of these deposits (see figures 6.23 to 6.26). In describing the ecology and life habits of seventeen species of echinoids seaward from Key Largo, Kier and Grant (1965) note certain echinoids (*Leodia sexiesperforata, Encope michelini, Clypeaster subdepressus, Meoma ventricosa* and *Plagiobrissus grandis*) which burrow in the clean, grassless sand areas. Usually the only evidence of these echinoids are their burrows. A few conchs have also been observed by the writer on these

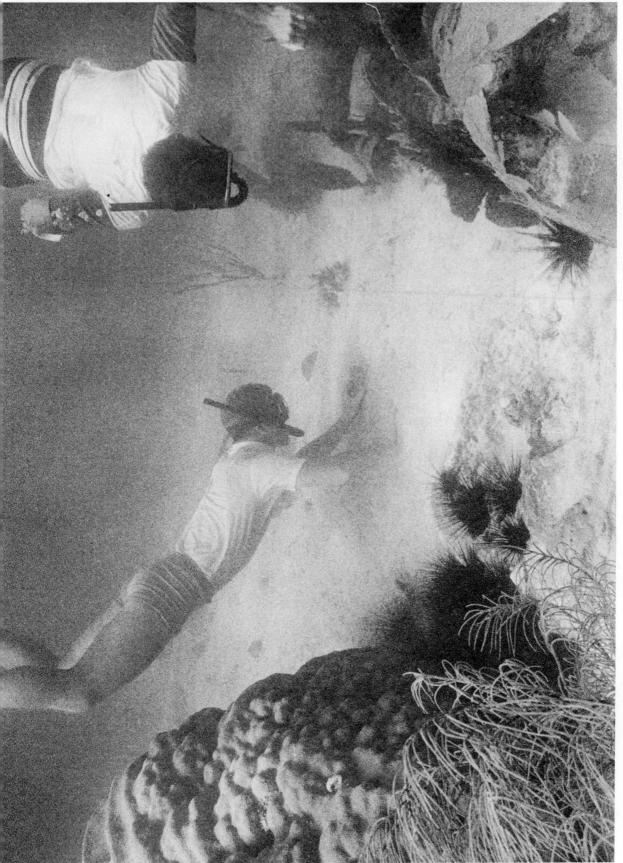

Figure 6.22. Living *Montastrea annularis* coral head and associated organisms surrounded by loose algal and molluscan sand at Mosquito Banks Patch Reef. **Note:** Compare with back cover showing outcrop of lithified Pleistocene equivalent of same coral species surrounded by calcarenite along the Key Largo Canal Cut.

Figure 6.23. Diver hovering over loose blanket sand, White Bank, Florida Reef Tract. Depth 3 m.

Figure 6.24. Close-up photograph of medium to very coarse *Halimeda*-rich sand taken from White Bank (see figure 6.23).

Figure 6.25. Rippled loose sand blanket near western edge of White Bank with three feet high topographic relief (left) developed by grass-held sediment patch.

Figure 6.26. Typical current-swept grass-held blanket sediment with local loose sand patch (foreground), Florida reef platform.

barren, rippled sand flats. Starck (1966, p. 730) in describing desertlike blanket sands of the back-reef platform (behind Alligator Reef light) notes (after use of paralyzing poison rotenone) the apparently sterile sand yields a crop of stunned shrimps, crabs, flounders, stargazers, lizardfishes, and various eels. Many of these animals, according to Starck, emerge only at night.

In sharp contrast to the clean, barren sand areas, the grass-held blanket sediment environment contains thriving communities of fauna and flora. A large variety of mollusks, echinoids (see Kier and Grant, 1965), foraminifera, sponges, arthropods, a few corals (*Porites* and *Manicina*) and algae inhabit these grass areas. The resulting debris from this organic milieu contributes a wide size range of skeletal debris which is retained by the effective stabilizing root and blade system of the grasses. In addition to this stabilizing effect the grasses (flat bladed turtle grass *Thalassia testidinum* and the round bladed grass *Syringodlum filiforme*) act as a baffle, trapping new current derived particulate material and reducing the winnowing out of contained particles (fig. 6.32). The effectiveness of this baffle mechanism is well illustrated by the high percentage of silt- and clay-size carbonate found in the grass-held sediment in the inner portions of the back reef, as noted in analyses graphically summarized below. Breakdown of skeletal constituents in the grass-held areas where the effect of mechanical abrasion and rolling by current action is minimal is due chiefly to biological activity of crustaceans, mollusks, sponges (*Cliona*) and boring algae. See Swinchatt (1965, p. 81-83) for a list of studies concerning biological breakdown of sediments. Fish, holothurians and worms may also contribute to the biological diminution of skeletal material.

The pioneer study by Ginsburg (1956) of the sediments found on the Florida platform is a classic paper which must be studied by serious students of carbonate rocks. The abstract of this paper is presented below (see also figures 6.27, 6.28, and 6.29).

ENVIRONMENTAL RELATIONSHIPS OF GRAIN SIZE AND CONSTITUENT PARTICLES IN SOME SOUTH FLORIDA CARBONATE SEDIMENTS
by
R. N. Ginsburg

In the southern extension of the Florida peninsula variations in the submarine topography, areal geography, and hydrography which control the distribution of sediment-producing organisms are reflected in the grain size and constituent particles of the calcareous sediments being deposited. Two major environments can be recognized: (1) a curving band-shaped reef tract with good water circulation, and (2) Florida Bay, a very shallow triangular area with semirestricted water circulation.

Florida Bay sediments have larger proportions of particles less than 1/8 mm than the sediments of the reef tract. The constituent particle composition of the fraction larger than 1/8 mm in Florida Bay is almost exclusively molluscan and foraminiferal, but in the same size fraction of the reef-tract sediments fragments of algae and corals are abundant. Similar distinctions in grain size and constituent particles for comparable environments can be derived from published data for the sediments around Andros Island, Bahamas.

In Florida Bay large local variations in physical environment obscure the expected effects of differences in environment from one part of the Bay to another, and no distinct subenvironments could be recognized from the gross grain size and constituent particle composition. However, in the reef tract local variations of environment are smaller, and the gradual but consistent changes in depth and water circulation effect differences in the fauna and flora, and thereby produce sediments which have recognizably different abundances of the major constituent particles as shown in figure 6.27. The three subenvironments, back reef, outer reef-arc, and fore reef are indicated by progressive changes in constituent composition, and in less degree by variations in gross grain size.

Because the estimates of constituent particle composition of the reef-tract sediments were made by point counts on standard petrographic thin sections this approach can be used to analyze ancient limestones.

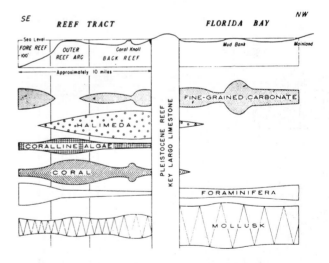

Figure 6.27. Generalized variations in sediment grain size and constituent composition of South Florida sediments. From Ginsburg (1956).

A paper by Swinchatt (1965) includes graphic summaries of sediment characteristics from samples col-

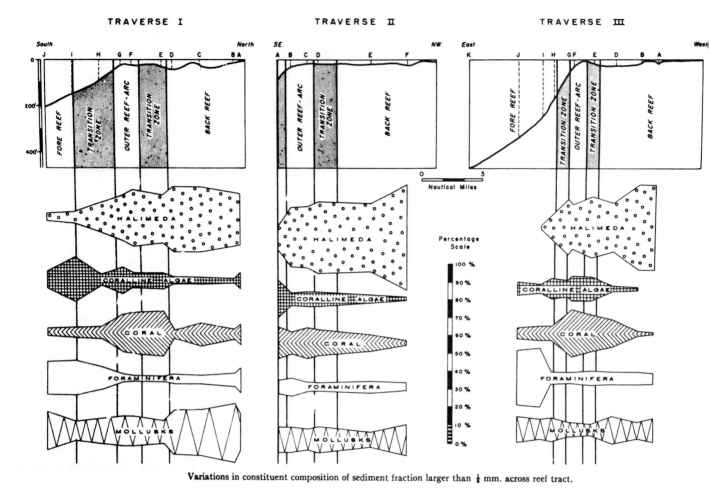

TRAVERSE I

TRAVERSE II

TRAVERSE III

Variations in constituent composition of sediment fraction larger than ⅛ mm. across reef tract.

Figure 6.28. Florida Reef Tract from Ginsburg (1956). For location of traverses see figure 3.1.

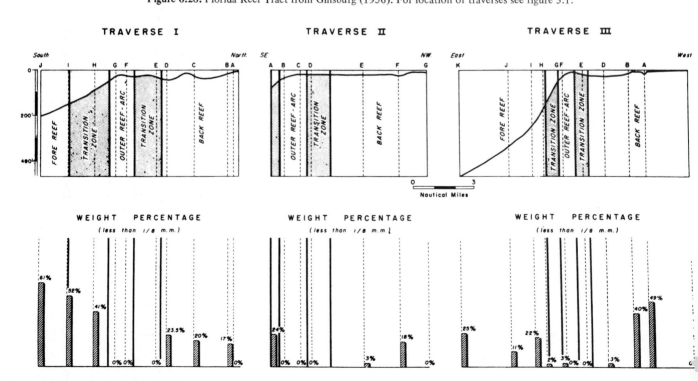

TRAVERSE I

TRAVERSE II

TRAVERSE III

WEIGHT PERCENTAGE
(less than 1/8 m.m.)

WEIGHT PERCENTAGE
(less than 1/8 m.m.)

WEIGHT PERCENTAGE
(less than 1/8 m.m.)

	Florida (This Paper)		Bahamas (Thorp, 1936)	
	Florida Bay (17 Samples)	Reef Tract (25 Samples)	West Side[1] Andros Is. (Fig. 10)	Reef Tract[2] Andros Is.
GRAIN SIZE				
Weight percentage less than ⅛ mm.	49	17	68	4
CONSTITUENT COMPOSITION[3]				
Algae	½	42	5	30
Mollusk	76	14	14	15
Coral	0	12	0	12
Foraminifera	11	9	23	26
Non-skeletal	3[4]	12	18[5]	9
Miscellaneous	½	9	10	8
Unknown	1	8	--	--
Ostracods	2	—	—[6]	—[6]
Quartz	6	—	—	—

[1] Calculated from the data given by Thorp (1936, pp. 110–12) for 7 samples—Nos. 176b, 177, 178, 179, 180, 181, 183—locations indicated in Figure 10.
[2] Calculated from the data given by Thorp (1936, pp. 110–13) for 11 samples—Nos. 145, 146, 147, 159, 162, 184a, 184b, 187, 188, 189, 195—locations shown in Figure 10.
[3] The estimates of constituent composition for Florida Bay samples were derived from numerical counts in each size grade less than ⅛ mm. cumulated according to weight percentage. The constituent composition of the reef-tract samples was estimated by point counts of thin sections of the fraction greater than ⅛ mm. as described in the Appendix. Thorp made visual estimates of the composition of each size grade. His data were averaged by the writer.
[4] The lack of pellets is due to disaggregation procedures as discussed in the Appendix.
[5] Includes 38% pellets.
[6] Included in "Miscellaneous."

Figure 6.29. Size grade and constituent particle composition data from Ginsburg (1956).

	Florida	Bahamas	
	Reef Tract This Paper (25 Samples)	Reef Tract[1] (Fig. 11) Thorp (1936) (10 Samples)	Marginal Shelf Eastern Bahamas Illing (1954, p. 17) (5 Samples)
GRAIN SIZE			
Weight percentage less than 1/16 mm.	9	4	0
CONSTITUENT COMPOSITION[2]			
Algae	42	30	39
Mollusk	14	15	18
Coral	12	12	12
Foraminifera	9	26	13
Non-skeletal	12	9	14
Miscellaneous	9	8	4
Unknown	8	—	0

[1] Calculated from the data given by Thorp (1936, pp. 110–13) for samples—Nos. 145, 146, 147, 159, 162, 184a, 184b, 187, 188, 195 whose locations are indicated in Figure 10 of this paper.
[2] The estimates of constituent composition were made as follows: this paper—volume estimates derived from point counts of thin sections of fraction greater than ⅛ mm.; Thorp and Illing visual estimates of numerical percentages in each size grade. Illing verified the estimates of coral and coralline algae by examination of thin sections.

Sample Number	Grain Size				Particle Constituents of Fraction Greater than ⅛ mm.							
	Wt. % Less than ⅛ mm.	Wt. % Less than 1/16 mm.	Md φ	Qd φ	Halimeda	Coralline Algae	Mollusk	Coral	Foraminifera	Non-Skeletal	Miscellaneous	Unknown
Traverse I												
I-A	0	0	1.15	0.60	29	7	20	10	13	3	8	10
I-B	16.9	9.5	1.30	1.10	30	1	33	5	7	16	3	5
I-C[1]	20.0	11.6	2.00	0.82	36	2	25	12	4	7	7	7
I-D	23.5	17.5	0.80	1.48	36	2	24	3	5	17	7	6
I-E	—	—	0.56	0.80	27	8	12	25	3	12	7	6
I-F	—	—	0.85	0.60	32	7	16	22	3	7	6	7
I-G[2]	—	—	0.36	0.44	27	13	16	21	3	8	8	4
I-H	67.7	45.5	3.85	—	16	.5	13	6	12	24	11	13
I-I	61.1	33.7	3.45	0.81	6	24	10	5	19	14	12	10
I-J	52.1	37.9	3.20	—	4	8	20	3	18	16	16	15
Traverse II												
II-A	24.0	13.9	0.65	1.62	13	21	14	17	8	11	12	4
II-B	—	—	0.98	0.56	34	6	14	12	10	12	5	7
II-C	—	—	0.85	0.52	34	7	11	17	6	13	4	8
II-D	—	—	1.50	0.42	41	7	14	14	4	11	2	7
II-E	2.4	—	2.15	0.50	38	4	8	9	6	18	10	7
II-F	17.7	7.3	1.35	1.00	60	1	17	2	5	5	4	6
Traverse III												
III-A	49.0	—[3]	—	—	46	0	13	1	6	14	5	15
III-B	39.8	13.0	2.85	0.56	46	2	9	3	7	12	14	7
III-D	3.0	—[3]	1.00	0.59	26	3	15	15	7	17	7	10
III-E	—	—	1.28	0.61	33	13	5	20	8	9	7	4
III-F	—	—	0.70	0.52	24	14	8	26	6	7	11	4
III-G	3.4	—	0.93	0.76	30	9	8	22	6	8	12	5
III-H[1]	1.7	—	0.25	1.16	15	11	13	8	7	22	18	6
III-I	22.0	8.1	2.35	0.81	2	5	11	8	32	9	19	14
III-J	11.2	4.3	2.18	0.68	0	7	14	6	30	5	23	15

[1] Particle constituent compositions are averages of the estimates from two different thin sections.
[2] All values are averages of two samples collected 50 feet apart.
[3] Not calculated because of incomplete analysis.

lected along 8 traverses across the back-reef platform (see figure 6.30). In these diagrams Swinchatt considers the outer third of the platform to be dominated by loose skeletal sand with little or no lime mud or grass present. The inner portions of the shelf are, in contrast, described as dominated by grass-held skeletal sand with 10-55 percent silt- and clay-sized carbonate. A study of Swinchatt's graphs (fig. 6.30) shows the specific variations between these two areas and indicates how the sediment of the back reef may be divided into two (outer and inner) general facies. Criteria noted for such subdivision may be used in evaluating ancient carbonate rock analogs. Swinchatt's abstract is noted below:

SIGNIFICANCE OF CONSTITUENT COMPOSITION, TEXTURE, AND SKELETAL BREAKDOWN IN SOME RECENT CARBONATE SEDIMENTS
by
J. P. Swinchatt

A study of sediments along a 35-mile length of the Florida reef tract indicates the presence of two subenvironments between the reefs and the Keys. Variational trends of mean phi, standard deviation, percent fines (less than 62 microns), roundness, and constituent composition indicate a distinct change in the sediment at a point about halfway between the reefs and the Keys. Trends show a general increase or decrease in the outer (reefward) one-half, but general uniformity within the inner one-half.

Bottom observations indicate a correlation between the area in which the change in sediment takes place, and the area in which density of growth of marine grass increases. Analysis of the sediments and the processes affecting them indicates that marine grasses not only modify the effect of current and cause the entrapment of fine sediment, but they also exert an indirect control on processes of skeletal breakdown and textural evolution of the sediment. For instance, in the outer (reefward) area, skeletal breakdown is dominantly a mechanical process and the sediment textures at least partially reflect current phenomena. In the inner (Keyward) area, however, skeletal breakdown is dominantly a biological process and the ultimate texture of the sediment reflects a complex of processes including those associated with growth and the life cycle of the organisms, the reaction of the shell carbonate to breakdown, and transport of fine sediment from the outer area. Of particular importance are the relative rates at which these processes take place; the variability of these rates strongly influences texture and constituent composition of the sediment.

Though little quantitative data concerning these processes is yet available, and their relative importance is poorly known, knowledge of their possible effects suggests caution in some aspects of interpretation of carbonate rocks.

Any discussion of blanket sediment environments of the Florida platform should include a reference to the extensive and classic work by Enos (1977), a por-

tion of which is included in this Guidebook on pages 229 to 234 (see also figures 7.32 to 7.35).

Frost (1974) describes the occurrence of Holocene algal stromatolites in subtidal (3-4 m depth) environments along the southwest part of White Banks. He differentiates these stromatolitic structures from the intertidal and supertidal varieties and emphasizes the type of collaborative evidence needed when identifying specific ancient rock environments containing algal stromatolites. The abstract of his paper is presented below:

SUBTIDAL ALGAL STROMATOLITES FROM THE FLORIDA BACK-REEF ENVIRONMENT
by
J. G. Frost

Blue-green algae form potential stromatolite structures in the Holocene subtidal Florida back-reef environment. The stromatolite structures are domal and circular to elliptical in plan view. The stromatolites occur randomly scattered among *Thalassia* sp. grass flats in about 3 m of water. The stromatolite structures are similar to many found in the Geologic record. Growth forms of the blue-green algae stromatolites (*Schizothrix*?) and their environmental occurrence are important when interpreting ancient depositional environments in terms of the water depth and relative position to the shoreline.

Wright and Hay (1971) describe the abundance and distribution of Foraminifera of the back-reef area along a traverse from Rodriguez Key to Molasses Reef. Their study includes a review of previous investigations in the Florida Keys area and an evaluation of sampling techniques. The abstract of their paper is noted below.

THE ABUNDANCE AND DISTRIBUTION OF FORAMINIFERS IN A BACK-REEF ENVIRONMENT, MOLASSES REEF, FLORIDA
by
R. C. Wright and W. W. Hay

Living and dead Foraminifers were examined from 53 samples collected along an 11 station traverse in a back-reef environment in the Florida Keys. Foraminiferal populations occurring in the sediment and on the vegetation were examined at each location.

The larger number of individuals counted and the closeness of the sampling locations yielded a high degree of precision. Samples taken only a few feet apart prove to be from statistically different populations in almost every case. However, the variation between adjacent samples is less than that between widely spaced samples. The variation in abundance between adjacent samples is due to the presence of microenvironments.

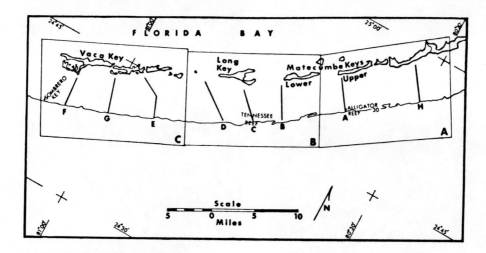

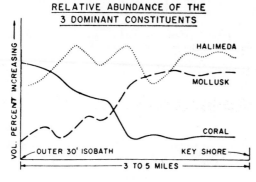

RELATIVE ABUNDANCE OF THE 3 DOMINANT CONSTITUENTS

Graph diagrammatically illustrating variation of relative abundance of the 3 major components of the sediment along a line from the reef to the Keys. This pattern is characteristic of all traverses to varying degree. Note the general decrease of coral and increase in mollusk debris in the outer one-half of the area and the more uniform values in the inner one-half.

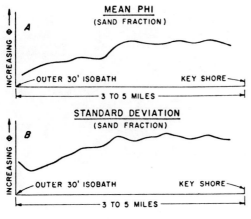

MEAN PHI (SAND FRACTION)

STANDARD DEVIATION (SAND FRACTION)

Graphs diagrammatically illustrating variation of $M\phi$ and standard deviation of the sand-size fraction along a line from the reef to the Keys. These patterns are characteristic of all traverses to varying degree. Note general increase of values over outer one-half of area, more uniform values in inner one-half.

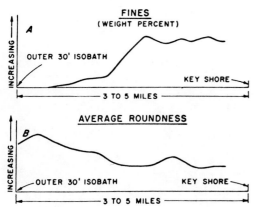

FINES (WEIGHT PERCENT)

AVERAGE ROUNDNESS

Graphs diagrammatically illustrating variation of weight percent fines (less than 62 microns diameter) and average edge roundness of the sand-size fraction along a line from the reef to the Keys. These general patterns are shown by all traverses to varying degree. Note increase of fines, decrease in roundness over outer one-half of area, and the general uniformity of values in the inner one-half.

Figure 6.30. Summary results of analyses of bottom samples from 8 traverses (top map) across the Florida reef platform. From Swinchatt (1965).

Figure 6.31. *(above)*: Two types of grass-held sediment. Left, thick mounds (over 2 m) of grass-held grass-stabilized sand near edge of White Bank. Right, Thin (less than 30 cm) accumulation of grass-held sediment in patches off Newfound Harbor Keys.

Figure 6.32. *(below)*: Close-up of dense baffle presented by flat-bladed *Thalassia testudinum* and round-bladed *Syringodium filiforme* grass. Note epiphytic growth on blades of grass and also burrow mound.

The Foraminifers are typical, shallow water, marine, tropical forms. The depth distribution of *Uvigerina peregrina* and *Pyrgo subsphaerica* is extended into shallow water. One new species is described.

The distribution of the total population and the distribution of certain species can be correlated with the grain size distribution of the sediments. Current movement is a major factor governing the distribution of the sediments and also the distribution of some of the foraminiferal tests.

In areas lacking a vegetative cover there are few living Foraminifers present in the sediment. In areas with a vegetative cover more living Foraminifers are found on the vegetation than on the bottom. Most of the Foraminifers live attached to the vegetation and settle to the bottom after death. There is a correlation between the distribution of living species and the empty tests.

The specimens are slightly smaller than is typical for these species due to the high temperature of the water. The species are well adapted to the environment as test structures and niche selection provide protection from water movements.

SUBTIDAL HARDGROUND ENVIRONMENTS

SHOALS BORDERING KEYS AND INNER SHELF AREAS (Florida Keys): Some areas of the

Florida inner shelf platform are not covered with loose sand, grass-held sediment or living patch reefs. These areas have hard uneven rock floors and/or cemented rubblelike pavement. Various organisms (alcyonarians, encrusting foraminifera, boring mollusks and worms, crustose coralline algae and *Halimeda tuna*) have been observed sometimes inhabiting these surfaces or their sediment-filled fractures.

The origin of the induration of these surfaces is not always clear. Subtidal hardgrounds in shoal areas bordering the emergent Florida Keys may represent former subaerially exposed and lithified sediment or Key Largo (Pleistocene) limestone bedrock highs (the latter sometimes found coated with subaerial crusts—see chapter 11 of this Guidebook). Other "hardground" areas may well involve submarine cementation and/or be the product of encrusting organisms such as coralline algae. Certain "lumpy" hardground or nodular pavement areas appear to be former rubble surfaces which have subsequently been welded together by one or both of the above mentioned processes.

See figures 6.33, 6.34 and 6.35 for illustrations of various hardground areas in the back-reef area of the Florida Keys.

Figure 6.33. Hard rock bottom near edge of shoal area (water approximately 2 m deep) off of Harry Harris Park, Tavernier, Florida. *Halimeda tuna* and Alcyonarians grow directly on the surface of the rock and turtle grass inhabits sand-filled crevices.

Figure 6.34. Area of hard rock-bottom floor developed as a low mound 50 m diameter with 1 m relief from cemented patch reef rubble. Remnants of rubble can be seen at "R." Small depressions "D" on top of mound support grass-held sediment patches. Example of advanced "Rubble Stage" of figure 6.21.

Figure 6.35. Hard rock bottom of rubble pavement made up of partially cemented reef rubble with small patch of living *Acropora cervicornis.* Southwest of Carysfort Light, behind outer reef.

RODRIGUEZ KEY, FLORIDA REEF TRACT:

1. How sharp a contact is actually present between the *Goniolithon* and *Porites* zones on the windward side of the Key? Why is there no *Porites* development along the western most edge of the Key (see figure 6.9)?
   Compare the development of windward and leeward organic zones and note how this could be used for interpreting similar ancient analogs.

2. Does the *Goniolithon-Porites* zone provide a protective function on the windward edge of this Key allowing the extra wide development of the "grass and green algae" zone (fig. 6.9) or did the latter zone build up independently?

3. Note how the flat-bladed turtle grass could cover, protect and stabilize the muddy sediments during low tide or storm periods.

4. How does this mud mound compare with ancient analogs?

5. What will happen to this mud mound if worldwide sea level rises 1 meter? Lowers 1 meter? Lowers 6 meters?

6. List on the cross section below the various fauna and flora that you observe along a section normal to the edge of the bank environment.

North   *South*

Fauna

Flora

Sample No.

Photo No.

RODRIGUEZ KEY DISCUSSION

The important role of calcareous red algae is very obvious to visitors at Rodriguez or Tavernier Keys. Paul Basan (Nov. 1963—personal communication) indicates that many people have

Sample No.

Photo No.

mistaken the low encrusting form of *Goniolithon* for *Porolithon*. Basan, together with Paul Brenckle, has found six growth forms for *Goniolithon* which appear to be controlled by water depth and agitation. Basan notes the following:

Basically, we found a gradation of growth forms from highly branched erect forms in the bank lagoon on Rodriguez Key through stubby-branched-erect to low encrusting forms at the same location. An interesting form can be found on the Gulf side of Spanish Harbor bridge near Don Quioxite Key. Here, many highly branched erect forms have coalesced to form very large "algal heads," a condition we found at no other location. These forms are found in long sand-filled washouts in water deeper than the surrounding area.

In the Barracuda Keys, *Goniolithon* can be found in a zone similar to that on Rodriguez, however, some interesting other forms can also be found. There is a limestone platform that outcrops on the Gulf side of these keys (probably Miami limestone). There are numerous erosional features on this limestone in the form of potholelike depressions in which small clusters of branching *Goniolithon* are growing. These algae, unlike the others seen at the other locations, are growing firmly cemented to the limestone. Furthermore, *Goniolithon* growing in areas of low *Thalassia* are consistently associated with *Porites porites* and display similar growth forms as the coral.

In the latter area, Basan also notes the occurrence of an unidentified, delicate, "puff-ball" articulate coralline alga.

WHITE BANK (North of Key Largo Dry Rocks Outer Reef) FLORIDA:

1. What is the composition and size grade of this sand? Record results on appropriate page in Appendix.

2. Is the sand really well sorted?

3. How many kinds of ripples are forming? Draw a sketch of each.

4. How easily do such ripples form? (Take off a flipper and using it as an underwater wave promoter, try to form ripples along a different trend than is now present on the bottom.)

5. Why are there no plants or coral heads here?

6. Are there many invertebrate animals? What kind?

7. How thick are these sand deposits (use a steel probe)? What will govern their thickness? What will govern their lateral distribution?

8. What is the origin of such large bodies of unconsolidated sand on this reef tract?

9. Note the constant shifting of sand grains. What influence does this have on shape of the grains and on paucity of life in the area?

Figure 7.1. *Acropora palmata* dominates portions of the outer shelf near Carysfort Light, northern Florida Keys. Depth approximately 4 meters.

OUTER SHELF AREAS

7

The term "Outer Shelf" as used in this Guidebook refers to that relatively narrow (about 1 1/2 km wide) area along the outer edge of the platform and behind the shelf break. This area contains what is commonly called the shelf "rim" or "lip." The bottom of such tidal-swept areas can be gently undulating and covered with relatively deep water or it can be shallow and the zone of high wave energy.

Carbonate environments associated with the outer shelf areas of the Florida-Bahama region can be described in many ways. For discussion purposes in this Guidebook and for ease in interpreting ancient analogs the following subdivisions will be used:

1. Intermittent flourishing outer reefs such as found on the rims of the Florida and Bahama platforms. See figure 7.1.
2. Essentially barren rubble flats and mounds as can readily be observed along the outer margins of the Florida platform.
3. Belts of loose permeable shelf-edge sand lying between established reef and rubble areas.

Warning: Although the outer edge of the Florida reef platform is only 7 or 8 kilometers from shore, persons not familiar with the area should realize that storms can develop quickly and motor failure can result in drifting northward with the Gulf Stream. Storms can also develop rather quickly off the east coast of Andros. For these reasons two methods of power (two motors, one motor and one sail or one motor and a pair of oars) are recommended. Chain at the end of the anchor line prevents chafing against coral and aids the holding power of anchors. Grapnels are used for anchoring in coral, and a powerful flashlight is useful if unexpectedly caught out after dark. It is always good to leave one's destination and estimated time of return with some reliable person on shore. Weather broadcasts (in Florida) are overly abundant for helping plan outer reef expeditions.

INTERMITTENT FLOURISHING OUTER REEFS

Illustrations of zoned "Marginal Reef Communities" as defined and described by Ginsburg and James (1974, p. 7.11) are found in many outer reef areas discussed in this Guidebook. A chart showing the abundance of principal organisms of these marginal reef communities is shown in figure 7.2. Newell and Rigby (1957) describe and illustrate similar reef communities. Gebelein (1974) describes outer barrier reefs of eastern Andros (Bahamas) and contrasts them with counterparts found in the Florida reef tract.

Growth rates for outer reef corals (*Montastrea annularis*) are given by Vaughan (1915) as 5-6.8 mm/yr and Hoffmeister and Multer (1964) as 10.7 mm/yr. Shinn (1966) relates growth rates to environmental factors. Shinn (1972) describes growth rates of *A. cervicornis, Diploria strigosa* and *M. annularis* in the Florida Keys and notes rapid regrowth after destruction by hurricane action. Landon (1975) gives data for growth rates of *M. annularis, P. porites* and *S. siderea* in various zones of the outer reefs of the Lower Keys. Her data, based on x-radiographs of coral skeletons, is summarized in figure 7.3. See also figures 7.4 and 7.5. For discussion of the history of Florida growth-rate studies, field and laboratory techniques and detailed tabulation of observed growth measurements including standard deviations and conclusions, see also Landon (1975). See Weber *et al.* (1975) and Hudson *et al.* (1976—also p. 179 of this Guidebook) for relationships of skeletal growth bands to mean annual water temperature. Graus and Macintyre (1975) describe new developments in coral growth studies including the reconstruction of ancient reef bathymetry.

The occurrence of flourishing outer reefs along the Florida Keys coast is not haphazard but rather

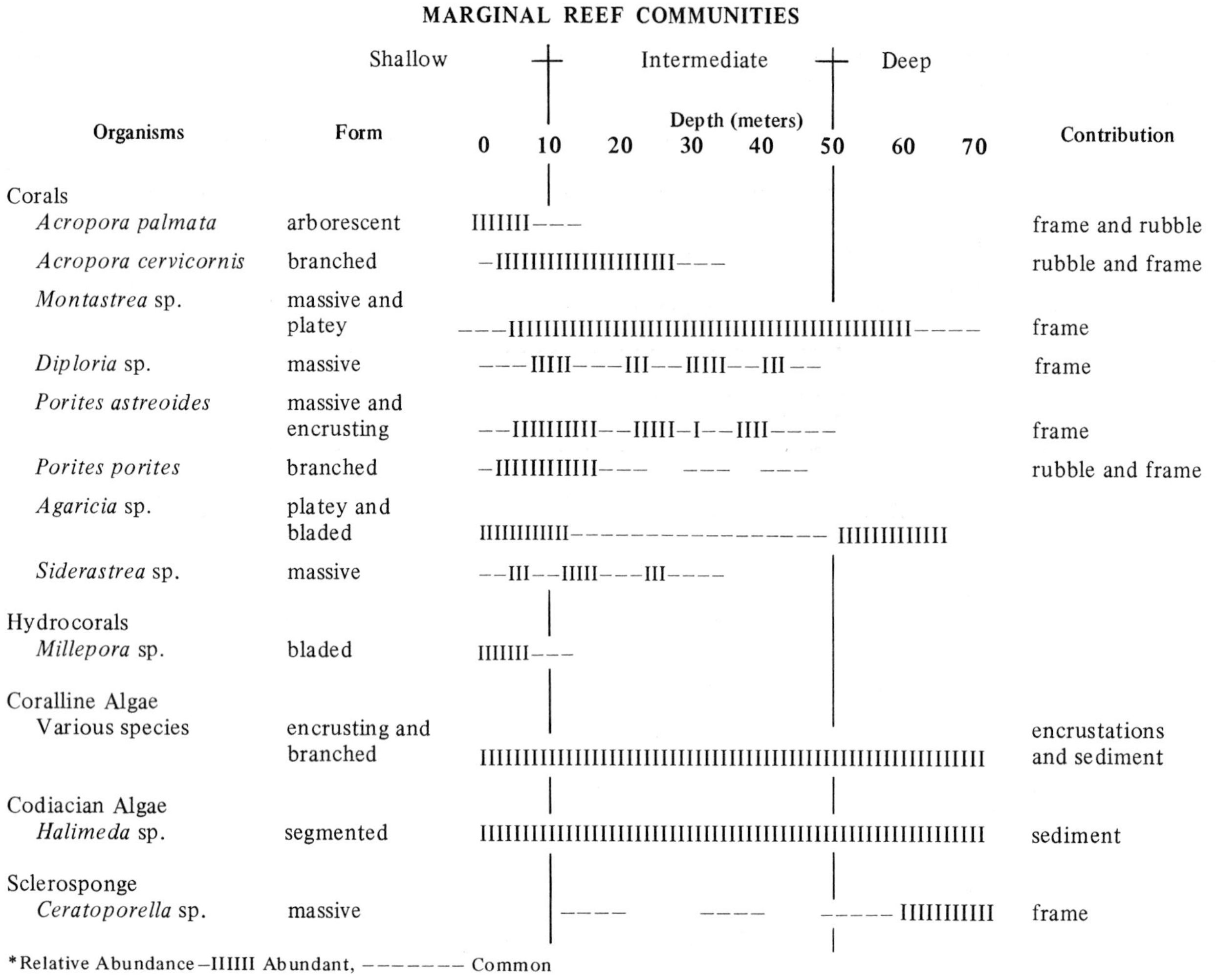

MARGINAL REEF COMMUNITIES

Organisms	Form	Shallow — Intermediate — Deep / Depth (meters) 0 10 20 30 40 50 60 70	Contribution
Corals			
Acropora palmata	arborescent	IIIIIII–––	frame and rubble
Acropora cervicornis	branched	–IIIIIIIIIIIIIIIIIIII–––	rubble and frame
Montastrea sp.	massive and platey	–––IIIIIIIIIIIIIIIIIIIIIIIIIIIIIIIIIIIIIIIIIIIIIIIIIIIII–––––	frame
Diploria sp.	massive	–––IIIII–––III––IIIII––III––	frame
Porites astreoides	massive and encrusting	––IIIIIIIIII––IIIII–I––IIII––––	frame
Porites porites	branched	–IIIIIIIIIIII––– ––– –––	rubble and frame
Agaricia sp.	platey and bladed	IIIIIIIIIII––––––––––––––––– IIIIIIIIIIII	
Siderastrea sp.	massive	––III––IIIII–––III––––	
Hydrocorals			
Millepora sp.	bladed	IIIIII–––	
Coralline Algae			
Various species	encrusting and branched	IIIIIIIIIIIIIIIIIIIIIIIIIIIIIIIIIIIIIIIIIIIIIIIIIIIIIIIIIIIIIII	encrustations and sediment
Codiacian Algae			
Halimeda sp.	segmented	IIIIIIIIIIIIIIIIIIIIIIIIIIIIIIIIIIIIIIIIIIIIIIIIIIIIIIIIIII	sediment
Sclerosponge			
Ceratoporella sp.	massive	–––– –––– ––––– IIIIIIIIII	frame

*Relative Abundance – IIIIII Abundant, – – – – – – Common

Figure 7.2. Chart showing variations in the relative abundance of the principal organisms of the marginal reef community with depth (generalized from Goreau, 1959, Logan *et al.,* 1969, and observations by Ginsburg and James). From Ginsburg and James (1974).

predictable. Flourishing reefs occur seaward of the emergent Keys and are generally absent seaward of the open channels between Keys, which distribute Florida Bay sediment and waters deleterious to active reef growth. The protecting presence of a solid island barrier (such as Key Largo) between Florida Bay and the reef tract is necessary for active reef growth. This relationship between channels and absence of flourishing reefs and vice versa may be applied to ancient carbonate analogs. This same principle holds true for the western edge of the Bahama platform where tides and easterly winds bring turbid and abnormally saline waters over the lip of the shelf hindering extensive coral development except in the lee of such islands as Bimini. Along the east coast of Andros turbid waters also locally hinder the otherwise flourishing reef development.

Three distinctive features of the flourishing Florida outer reefs are: (1) presence of *Acropora palmata,* (2) vertical coral zonation off the terraced reef front, and (3) spur and groove structures.

The moosehorn coral *Acropora palmata* is found in Florida waters *only* in shallow (usually 0-5 m depth) portions of the outer reef (as noted in chapter 10 this distinction is used in interpreting the paleoenvironment of the Pleistocene Key Largo limestone). Goreau and Wells (1967) note a depth range for *A. palmata* in Jamaican waters of 0-5 m with .3 to 1.5 m being the optimum depth. For data concerning this and other reef species see also Newell *et al.,* (1951), Storr (1955), Newell and Rigby (1957) and Newell (1958). The writer has actually found patches of *A. palmata* up to 2 km back from the average edge of the platform but only where the normal outer reef was absent and the full forward and backward surge of the waves was present.

A significant paper for those interested in outer shelf/platform margin areas is one by Wilson (1974) who after studying about 40 complexes of ancient carbonate facies has found three recurring patterns of shelf-margin facies, i.e., (1) downslope lime-mud accumulations, (2) knoll reef ramps, and

HABITAT		Montastrea annularis			Siderastrea siderea	Porites porites
		ALL FORMS	MASSIVE	COLUMNAR		
OUTER REEF	TOTAL	7.42			2.37	16.06
	FOREREEF DEEP				2.20 + .58	
	SHALLOW	7.29 + .39	8.37 + 1.52*	7.12 + 1.08	2.30 + .61	17.34 + 2.09
	BUTTRESS				2.58 + 1.88	15.97 + 4.29
	REEF FLAT		8.25 + 1.13**		2.43 + 1.68	14.41 + 4.77
PATCH REEF		9.00 + 2.18	7.98 + 2.35	9.66 + 1.40	2.74 + .38	13.80 **
KEY LARGO LIMESTONE		5.17 + 1.45	4.68 + 1.21	5.42 + 1.51	1.50 + .57	10.46 + 6.16

Figure 7.3. Coral growth rates in the Lower Florida Keys as determined by x-radiograph measurement of skeletal growth bands. (*above*) Average coral growth rates in mm, for three species in various reef habitats. Standard deviation of each sample population is also given. *Based on one specimen. **Based on two specimens. ‡Based on specimens collected approximately 500 m inshore of the patch reefs. (*below*) Mean growth rates with the standard deviation expressed by a bar for each species. The number of growth bands measured, followed by the number of colonies represented, are also given.

DFR–DEEP FOREREEF
SFR–SHALLOW FOREREEF

B–BUTTRESS
RF–REEF FLAT

PR–PATCH REEF
KL–KEY LARGO (Pleistocene) LIMESTONE

For habitat localities see figure 7.4. Figures and captions from Landon (1975).

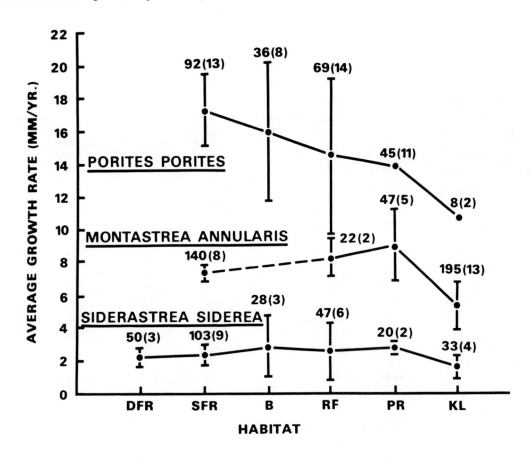

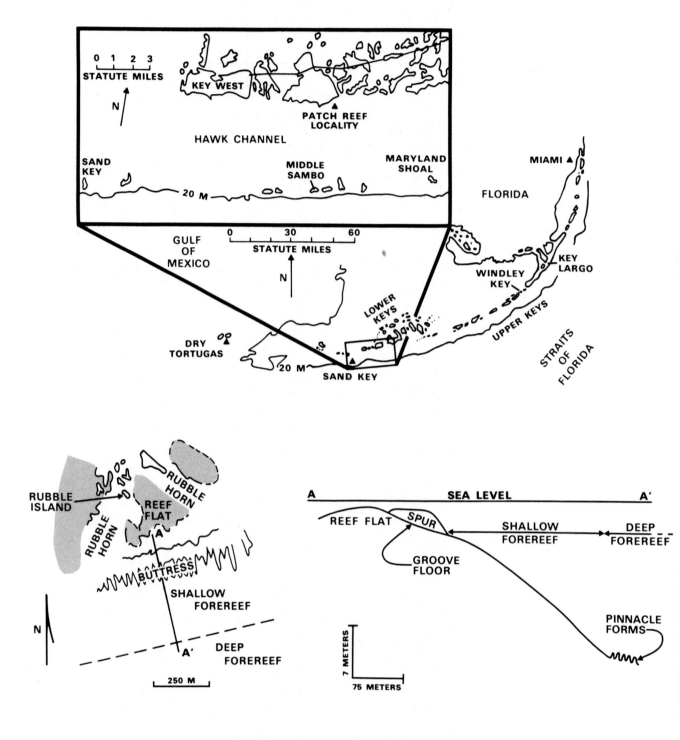

Figure 7.4. Index map of Lower Florida Keys. Plan view (lower left) of Middle Sambo reef shows habitats and grass cover (shaded areas). Profile (lower right) is also of Middle Sambo. For coral growth rates in these habitats see figure 7.3. From Landon (1975).

Figure 7.5. X-radiograph of a massive colony of *Montastrea annularis* (Ellis and Solander) from 6 m depth on shallow fore reef of Sand Key, X1. Bracket includes a pair of dense and less dense bands, one annual increment of growth. Arrow indicates a denser band within an annual increment due to poor conditions for skeletal deposition. Photo and caption from Landon (1975).

(3) frame-built reef rims. One of his interesting discussions (p. 821) relates various shelf-margin facies to wind and water movement and an excellent Appendix (p. 822) asks a wide variety of very appropriate questions. Although concerned with ancient carbonates this paper can also be used, of course, as a basis for letting past events help us interpret, classify and know what to look for in the Holocene too!

Although describing an area not covered by this Guidebook, the following abstract by Brooks (1962) illustrates how typical outer reef organisms may be found under special conditions at considerable depths.

OBSERVATIONS ON THE FLORIDA MIDDLE GROUND
by
H. K. Brooks

The Florida Middle Ground (28° 10´ to 28° 40´, 84° 5´ to 84° 25´), about 90 miles offshore from the Cedar Keys, was studied directly by diving with SCUBA gear. The study was made to determine the nature and origin of the relief features on the seafloor. Irregular rocky masses rise from the shelf of the Floridian Plateau at depths of 126 feet to summits 84 to 90 feet below sea level. Living scleractinians, hydrocorals, alconarians, coralline algae, and even the fish population are comparable to those on the reefs of the Florida Keys. The recurrence of these organisms at depths of 84 to 120 feet and this far north in the Gulf of Mexico has not been previously reported. Although a coral-reef fauna and flora exist in the area of the Florida Middle Ground, the recent growth has not contributed significantly to the development of the biohermal masses. The foundation is an older coral reef that must have developed during the last interglacial stage, the Sangamon.

UPPER FLORIDA KEYS (Carysfort Light, Key Largo Dry Rocks and Molasses Reef): There are many examples of readily available outer flourishing reefs in the northern Florida Keys. Carysfort Light can be reached from Garden Cove or Largo Sound. Key Largo Dry Rocks lies opposite Largo Sound. For those unfamiliar with the waters a local guide or one's own expertise using compass bearings and dead reckoning may be necessary to locate Key Largo Dry Rocks since no high marker is present. Molasses Reef at Molasses Light is available from Largo Sound or Tavernier. Actually, a transect involving several of these reefs as well as rubble and sediment blanket deposits (White Banks) can be readily visited in one good day's boat trip (see Field Itineraries in the Appendix of this Guidebook, index maps 3 and 5 and figure 7.8).

The present living outer reef at Carysfort Light (see figure 7.1) displays spurs and grooves on the seaward side of the reef and a roofed pillar structure (with roofs nearly emergent at low tide) in some central areas. The rigid reef framework, as with that at Looe Key and probably many other outer reefs, lies on an alternating sequence of unconsolidated shelly sand and indurated coralline rock all resting on Key Largo (Pleistocene) bedrock at depths of about 15 m. For a generalized block diagram showing the outer reef today see Hoffmeister and Multer (1968) and figure 10.4 of this Guidebook. See figure 7.6 for illustration of the Carysfort area.

Several growth forms of A. palmata can be observed at Dry Rocks. Figure 7.9 shows the unoriented flat palmate branch form which is common in the protected lee of the reef flat. According to Shinn (1963), eighty percent of all living A. palmata occurs in the zone of oriented growth form on the first terrace of the reef front (see Zone "A" in figures 7.10 and 7.11) which is approximately 60 m in width and which slopes down to a depth of 3 m.

Over most portions of this first terrace A. palmata branches are more circular and massive in cross section and are oriented pointing away from the open sea. A third growth form, nearly vertical branches, is found at the base of the first terrace where the wave thrust is weaker (see figure 7.12).

A second terrace (zone) extends for approximately another 30 m seaward and slopes to a depth of 8 m. Millepora complanata covers most of this zone. Below this level is a rubble zone made up predominately of coarse debris of Acropora and Millepora.

Spur and groove structures, as seen at Key Largo Dry Rocks as well as along many other Florida outer reef fronts, consist of seaward-pointing ridges 8 to over 30 m wide extending down to depths below 30 m. These ridges are separated by sand-floored channels (grooves) which are often 7 to 10 m below the tops of adjacent ridges. As noted by Shinn (1963), the various theories of the origin of spur and groove structures can be separated into three groups: (a) formation by erosional forces, (b) formation by coral growth, and (c) formation by combined erosion and growth. A study of the Florida spur and groove structures by Shinn (1963), which included examination of dynamite-blasted sections of these spurs, revealed that in situ colonies of A. palmata are the main component of most spurs and that biologic growth (and not erosion) is the responsible agent for these structures. Shinn's abstract is quoted below. See also figure 7.13.

Figure 7.6. (*above*) Shallow seaward edge of outer reef flat with rippled sand patches among branching *Acropora palmata* and round *Diploria* coral. Depth 1 meter. Near Carysfort Light, Florida.

Figure 7.7. (*below*) Multilobular head of *Montastrea annularis* surrounded by branching *Acropora palmata* (foreground) and *A. cervicornis* (upper right). Outer reef platform. Photo courtesy of E. A. Shinn.

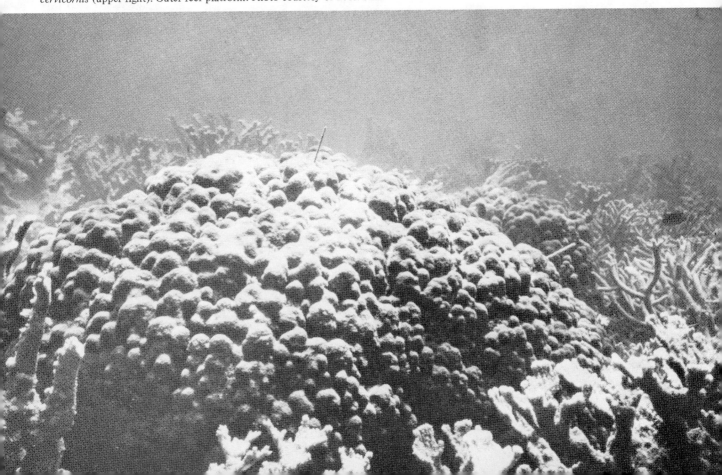

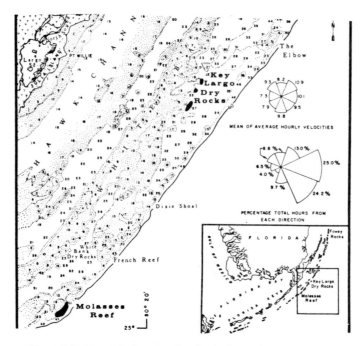

Figure 7.8. (*above*) Index map showing hydrography, wind velocity and wind direction data of area between Key Largo Dry Rocks and Molasses Reef area. From Shinn (1963).

Figure 7.9. (*below*) Palmate unoriented branches of *Acropora palmata* growing in protected lee side of Key Largo Dry Rocks. Finger points to round branching colony of *Acropora cervicornis; Montastrea annularis* head appears in upper left; sea fans and sea whips (Alcyonarians) are also present. Depth of water is 1.5 meters.

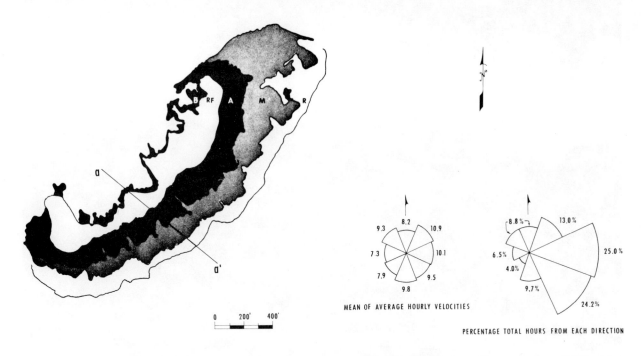

Figure 7.10. Map showing distribution of reef builders at Key Largo Dry Rocks. Shades of color and letters correspond to those in block diagram (figure 7-11) below. From Shinn (1963). *Key to Letters:* B-Back Reef, RF-Reef Flat, A-*Acropora* Zone, M-*Millepora* Zone, R-Rubble Zone. For section along a-a', see figure below.

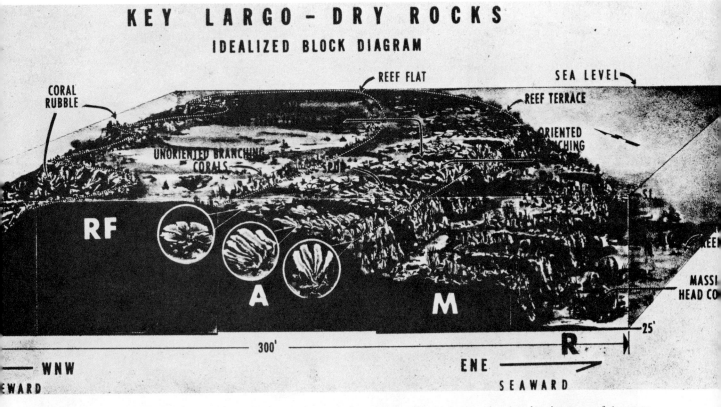

Figure 7.11. Block diagram of Key Largo Dry Rocks showing ecologic zones (see Key to Letters above) and major types of *Acropora palmata* growth forms. From Shinn (1963).

Figure 7.12. (*above*) Unoriented branched *A. palmata* (background) and bladed growths of stinging coral *Millepora complanata* (foreground). Sea side of Key Largo Dry Rocks at depth of about 3.0 m.

Figure 7.13. (*below*) Side view of a spur whose top is encrusted mostly with living *Millepora*. Floor of groove (lower left corner) contains loose sand and rubble and is 2.4 m below top of adjoining spur.

SPUR AND GROOVE FORMATION ON THE FLORIDA REEF TRACT
by
E. A. Shinn

On two Florida reefs, submarine reef spurs 10-12 feet high and up to 50 feet wide were dissected with explosives so that internal structures could be examined. *Millepora* and alga-coated spurs were found to be composed mainly of *in situ* coral (*Acropora palmata*). Comparison of living *A. palmata* with encrusted spurs suggests that a new interpretation of spur and groove formation is necessary.

A. palmata growing in less than 20 feet of water on the seaward slope of reefs which face prevailing seas modifies its growth form so that the branches can accommodate the forward thrust of impinging waves. The branches become oriented in the direction of wave movement, and degree of modification is proportional to the wave strength. Continued unidirectional growth causes individual colonies to coalesce into fingerlike spurs that project as much as 200 feet into oncoming seas. These "living spurs" die from crowding when they reach the surface and subsequently become completely masked with calcareous algae and *Millepora*. Moving sand, in the grooves between spurs, prevents coral attachment, and periodic hurricane seas remove accumulating debris derived from the overhanging walls of adjoining spurs.

For additional data dealing with Florida corals, the reader is referred to Agassiz (1880), Vaughan (1909, 1910, 1911, 1912, 1913, 1914, 1915a, 1915b, 1916a, 1916b), and Wells (1932). An illustrated summary of the Alligator Reef area is given by Starck (1966).

LOWER FLORIDA KEYS (Some 13 Outer Reefs Lying Offshore from Big Pine Key to Key West): Outer reefs of the Lower Florida Keys can be reached by boat from various land points (Big Pine Key to Key West) depending on which reef is to be visited. For those unfamiliar with the waters a local guide or one's own expertise using compass bearings and dead reckoning may be necessary.

Summarized specifically for the 1975 edition of this Guidebook (and augmented for the 1977 edition) by D. L. Kissling is the following excellent summary of some results of his 1971-1974 study of 9 of 13 outer reefs of the Lower Keys. Very little work has been done on these reefs although, as Kissling puts it, they "perhaps represent the best developed coral reefs in the continental United States." Of particular interest is Kissling's reference to internal reef cement, data concerning rampart islands, textural analyses of reef-derived sediment, hydrologic monitoring results and benthic organism census. Much of this data is the first of its kind from the Florida area.

CORAL REEFS IN THE LOWER FLORIDA KEYS: A PRELIMINARY REPORT
by
D. L. Kissling

Introduction

The thirteen "outer-arc" reefs lying west of Big Pine Key form a discontinuous southern barrier to the lagoonal Inner Reef Tract of the Lower Florida Keys (fig. 7.14). These perhaps represent the best developed coral reefs in the continental United States. Yet surprisingly, except for some seasonal water temperatures included in Vaughan (1918), no published investigations of these reefs exist. Several specific studies conducted in part on the generally smaller and less contiguous "outer-arc" reefs in the Upper Florida Keys deal with reef structure (Shinn, 1963), recent sediments (Ginsburg, 1956; Swinchatt, 1965), foraminiferal families (Moore, 1957), echinoid ecology (Kier and Grant, 1965) and reef fishes (Starck, 1968; Starck and Davis, 1966). Despite their accessibility, the bank-edge or "outer-arc" reefs of the Florida Keys have not been subject to that rigorous kind of comprehensive investigation already accorded other Caribbean coral reef systems.

In an attempt to achieve an integrated view of the many sedimentological and ecological elements of these coral reefs, I—together with eleven graduate and undergraduate field assistants—have completed 190 days of underwater field studies on nine of the thirteen reefs during the time spanning June 1970 to January 1974. Field studies were supported by grants awarded by the National Science Foundation (GA-30548), the Link Foundation and the SUNY Research Foundation. Analysis and synthesis of data gathered are incomplete at present. This report simply endeavors to convey the scope of our investigation along with a few observations which may help to characterize various aspects of these reefs.

Reef Structure

Major features of reef structure were delineated from habitat maps prepared from enlarged aerial photographs and from nine bathymetric profiles traversing the structural trends of eight reefs. Water and sediment depths were measured at five-meter intervals along profiles extending up to 650 m in length. Triplicate profiles were prepared across each buttress zone (see figure 7.15).

From their lee (or north) side seaward the reefs consistently display three principal physiographic units (fig. 7.16). Water depths cited are corrected to mean low tide.

But there is little conformity beyond these major units, as each reef exhibits considerable individuality. Distal sand-grass flats are not developed in some, and rubble flats are replaced by extremely shallow reef-rock pavements in others. Buttress zones vary in pattern and spur lengths (generally characteristic for any reef) may range from 50 to 130 m. Shallow fore reefs may be relatively featureless, or they may possess massive *Montastrea annularis* mounds up to 5 m high and 6 m diameter. Others display a second, less distinct buttress

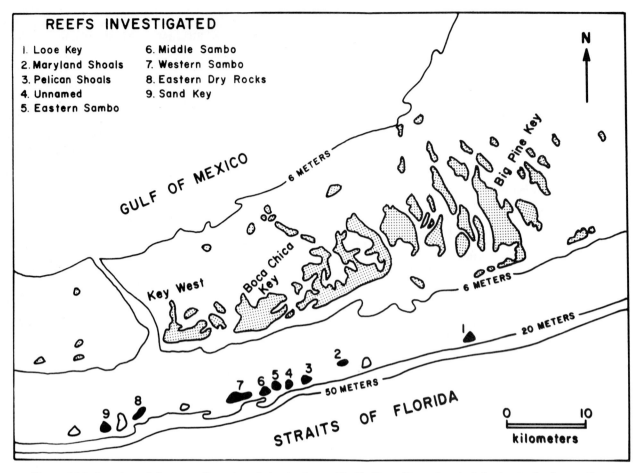

Figure 7.14. Location of "outer-arc" coral reefs in the Lower Florida Keys. Unnumbered circles locate the four reefs not investigated. Figure courtesy of D. L. Kissling.

zone of low spurs and narrow grooves at 9 to 12 m depth. Some reefs are fronted by an outer ridge such that the deep fore reef ends in a barren muddy plain formed in the intervening trough (fig. 7.16), but the seaward limit for others is necessarily arbirary. Much of the individuality reflects various stages of reef development. Those possessing narrow buttress zones and pavementlike reef flats supporting rampart islands are apparently undergoing active erosion and may be relatively young.

Internal reef structure—specific skeletal identity and orientation mapped where exposed by erosion—consists essentially of displaced coral rubble, weakly to firmly bound by melobesioid algae and calcite cement. Rubble intertices are filled with fine sediment only on reef flats. Transported coral rubble, particularly both species of *Acropora,* also dominates internal spur structure, although in places spur framework is constructed by *in situ* colonies.

Rampart Islands

One ephemeral and three permanent rampart islands, measuring up to 170 m long, were each mapped at six-month intervals spanning two years. The 16 maps record low tide and high tide island outlines, distribution patterns of major sediment-texture classes, relative surf force and drift directions, storm ridge positions, and are supplemented by 113 transverse profiles with measured elevations. Quantitative study of intertidal and supratidal faunas were made on two islands.

Most rampart islands lie along one side of the reef flat and are simply supratidal parts of mounded, crescent-shaped rubble horns which curve leeward and inward from buttress extremities, creating sharp lateral boundaries to reef flats. Rubble horns and rampart islands are formed by wave refraction around the reefs, which effectively restricts the lateral migration of coral rubble and sand supplied by storm destruction to the reefs. Islands tend to be teardrop-shaped and concave toward the reef flat. Their shapes and sizes are modified greatly by major storms, but they soon reestablish forms consistent with prevailing winds and wave refraction characteristic of summer and winter seasons, thereby quickly cancelling out storm effects. Cobbles and boulders comprising the islands are well rounded and are dominated by those corals most readily broken during storms (species of *Acropora, Millepora complanata, Porites porites*). See figure 7.17.

Reef Sediments

Fine sediments, including carbonate sand and mud, were sampled by three procedures. Eighty-four samples

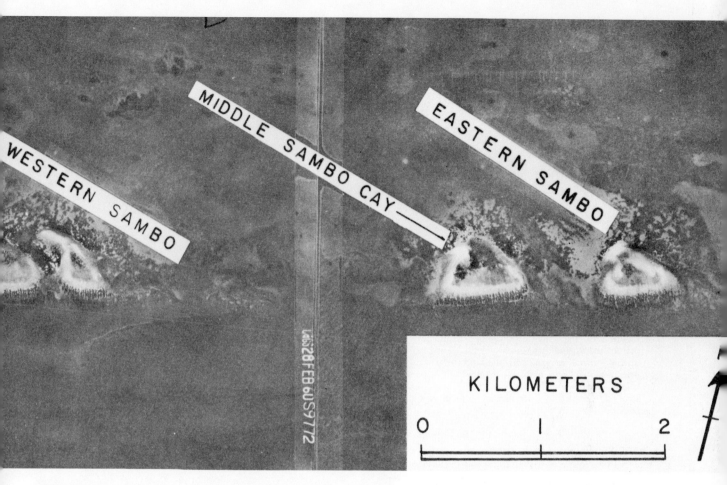

Figure 7.15. Aerial view of Western Sambo, Middle Sambo and Eastern Sambo reefs. Refer to figure 7.14 for location. Photo courtesy of D. L. Kissling.

were taken by hand at 30-meter intervals along traverses across eight reefs to 23 m depth. Fine sediments on reefs are typically left-skewed, coarse sands progressing toward better sorted, medium sands in all directions away from the reef flat. Seventy samples were taken by grab sampler along four traverses extending across the Inner Reef Tract to 55 m depth seaward of the reefs. Nearshore and deep substrates at distances more than one to two kilometers from reefs are dominated by muddy fine sands or sandy muds. Thirty-one samples from intertidal and supratidal parts of rampart islands invariably display better sorting than subtidal reef samples. Textural analyses for 185 samples have been completed and thin sections from epoxy-impregnated samples have been prepared for point count analyses of skeletal constituents.

Fine sediments on reefs result not only from the death of foraminifers, echinoderms, and calcareous algae (especially *Halimeda*), but also from mechanical and biogenic breakdown of coral and mulluscan rubble. Actually, coral rubble in cobble and boulder sizes represents the vast bulk of reef sediment—not sand. If distribution vectors for textural and generic compositions of reef rubble were accurately known in relation to existing sources (living and inert) for rubble, then the relative importance of present-day erosion and accretion of reefs might be ascertained. Accordingly, 54 rubble

samples (all surface cobbles and boulders within one-square-meter quadrats) were taken at 30-meter intervals across five reefs to 12 m depth. Sample populations range from 60 to 380 cobbles and boulders per square meter. Textural analyses were accomplished by measuring three axes on each clast and generic (or specific) identity was based on fine-skeletal structure. Fourteen rubble samples from one traverse were retained to obtain dry weights to evaluate mass contribution by various coral genera and species for all samples. Five surface rubble populations from three rampart islands were also analyzed.

Coral cobbles and boulders are at maximum size and number on rubble flats just leeward of reef buttresses. These parameters decrease markedly leeward into sand-grass flats (which overlie earlier rubble accumulations) and seaward from midway through grooves. Rubble is absent on spurs. The notably sparse rubble assemblages on fore reef surfaces (dominantly *Acropora cervicornis,* many *Porites porites,* fewer *Agaricia* and massive corals) closely match the living coral populations there if one assumes that most massive corals don't contribute to the rubble, but continue as solid substrate even after death (see figure 7.18). Rubble assemblages found in buttress grooves (many *A. palmata* and *Millepora complanata,* fewer *P. porites, Agaricia* and massive corals) would correspond well to the endemic fauna, excepting the con-

North   South

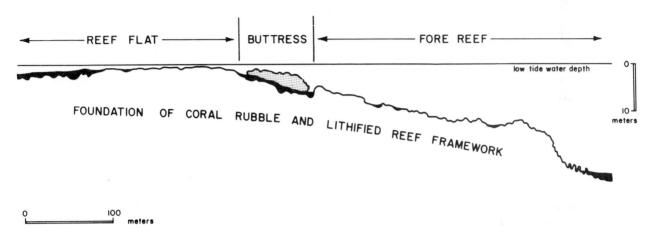

Figure 7.16. Bathymetric profile across Middle Sambo Reef. Stippled pattern represents one spur of the buttress zone. Solid pattern shows the distribution and thickness of fine sediments on the reef foundation. (1) *Reef Flat*—0.4 to 2.5 m deep; usually divisible into distal sand grass flats and (to seaward) shallow, sparsely vegetated rubble flats. (2) *Reef Buttress*—0.6 to 7.5 m deep; divisible into several habitats including spur tops densely populated by corals and zoanthiniarians, sheer spur walls up to 4 m in height, caverns and fissures beneath spurs, and deep intercalated grooves floored by sand, rubble or lithified reef rock. (3) *Fore Reef*—4.0 to 24.0 m deep; divisible into the shallow fore reef to 12.0 m depth, and the more steeply sloping deep fore reef at the seaward extremity of each reef. Figure courtesy of D. L. Kissling.

tinued dominance of *A. cervicornis.* An even poorer match exists between the living coral populations of reef flats and the abundant rubble dominated by species of *Acropora.* These observations point to two active processes: (1) leeward storm-wave transport of coral rubble from fore reef and buttress zones to reef flats, and (2) erosion of inert framework of spurs and groove heads to supply part of the existing surface rubble. Without radiometric dating it is quite hopeless to distinguish coral boulders only a few years old from those centuries old.

Following a major storm (June 1972) large quantities of freshly broken *Acropora palmata* fragments were collected. After rejecting those organically bored or encrusted, the remaining 72 large coral fragments and ten *Strombus gigas* shells were drilled to receive an anchor wire, bleached in sodium hyposulphate solution, washed, oven dried, individually weighed and labelled. These boulders and shells, together with eight tile pipes, were anchored onto substrates representing eight different reef environments in July 1972. One-third of the samples were collected and preserved in formalin after each of three six-month intervals (January 1973, July 1973, January 1974). By microscopic examination, identification, x-radiography and remeasurement of these boulders and tiles considerable information should be forthcoming concerning rates of boring by endolithic organisms, rates of encrustation by other organisms, successional patterns in colonization, and rates of initial breakdown of coral and molluscan rubble. Encrustations are numerous after only six months and many boulders anchored for 18 months have a surprisingly "old" appearance because of numerous borings.

Taxonomic study of endolithic organisms has been carried out on natural rubble samples.

Reef Waters

Hourly measurements of hydrologic properties, spanning 24-hour periods, were taken alternately at reef flat and fore reef stations during summer, autumn and winter seasons on two reefs. These included tide level, wind conditions, wave heights, current directions and velocities, incident illumination, air and water temperatures, salinity, pH, dissolved oxygen, and plankton samples. Many additional measurements were recorded for specific habitats on several reefs. Tidal amplitudes were recorded continuously over a six-week period.

Maximum and minimum amplitudes for the mixed, semidiurnal tides are 80 cm and 20 cm, respectively. Current velocities on most reefs commonly range between 10 and 30 cm/sec, but twice that where entire reef buttresses are deeply submerged. Surface water temperatures very from 23 to 30° C and dissolved oxygen content from 5.2 to 8.4 milligrams/liter; both changing with hour of day and season. Salinity is relatively uniform at 36 to 38°/oo and pH varies from 8.1 to 8.5. Gale force winds spawned by Hurricane Agnes (June 1972) provided an opportunity to observe storm effects on specific underwater sites and on rampart islands.

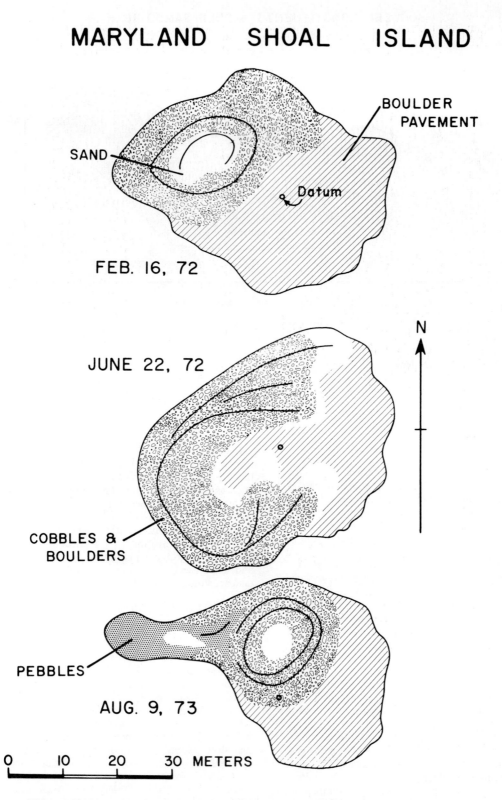

Figure 7.17. Successive change in rampart island morphology, Maryland Shoal reef. Top map is typical for mid-January to mid-July, bottom map for mid-July to mid-January. Central map was prepared four days after prolonged gales associated with Hurricane Agnes occurred. Heavy lines denote ridge crests. Figure courtesy of D. L. Kissling.

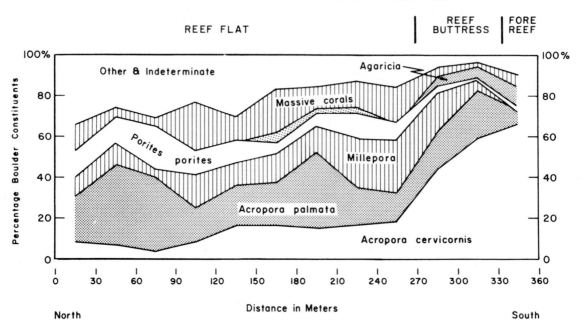

Figure 7.18. Taxonomic composition of rubble assemblages across Middle Sambo reef. Composition of rubble assemblages corresponds to endemic living assemblages most closely on the fore reef, least closely on the reef flat. Leeward proportional increase of indeterminate category reflects progressive biogenic destruction in that direction. Figure courtesy of D. L. Kissling.

Benthic Organism Distribution

Zonation of benthic organisms with respect to specific habitat conditions (e.g., depth or substrate) has been a major task in field observation and identification. Non-numerical censuses of macrobenthos were made at alternate 10-meter intervals along measured traverses crossing six reefs to 23 m depth. An average of six man hours of underwater observation was required to record an estimated relative abundance for each species encountered within the 400 m² area covered at each of the 104 census sites. Quantitative population censuses of cryptic biota (organism assemblages living beneath boulders) were conducted for 20 random boulders at each of 65 sites spaced at 20-meter intervals across five reefs. Twenty-one grided quadrats, encompassing a total area of 1820 m², were laid out in various habitats on five reefs to 18 m depth. In some quadrats all surface-dwelling and cryptic macrobenthos were identified (or described) and counted; in others only stony corals and octocorals were counted. Infaunal organisms in sand substrates were sampled from groove floors and sand-grass flats at 16 sites. Specimens sieved from 800 liters of sediment were counted and identified. The largest, most diverse infaunal assemblages inhabit *Thalassia*-covered sandy substrate.

Benthic microorganism populations were sampled at 30-meter intervals across five reefs to 12 m depth. Additional samples were taken from seagrass and algal substrates. The 54 samples were treated with Rose Bengal to stain live microorganisms (largely foraminifers and ostracods) and preserved. The foraminiferal assemblage includes more than 40 species.

Probably more than 550 macrobenthic species were recorded; approximately 350 have been identified. These include benthic macroalgae, marine grasses, sponges, anemones, scleractinian and hydrozoan corals, octocorals, flatworms, nemertine worms, sipunculids, bryozoans (Cuffey and Kissling, 1973), oligochaete and polychaete annelids, chitons, gastropods, scaphopods, bivalves, cephalopods, crustaceans, pycnogonids, tunicates, asteroids, ophiunoids, echinoids, holothurians and crinoids. Distribution patterns for a few of these taxa are summarized below.

Stony Corals

Stony corals (largely scleractinians), represented by 42 species on these reefs, encrust all hard substrates in great profusion. Species diversity and skeletal mass/area increase successively from sand-grass flats seaward to shallow fore reefs; both attaining maxima at 8 to 10 m depth. Because the bathymetric limits for several species became exceeded with increasing depth, diversity and skeletal mass/area decrease slightly into the deep fore reef despite the appearance of a few species not found in shallower waters. Coral distribution seems to be influenced most by intensity of illumination, substrate conditions, wave intensity, and displacement by storms.

Reef zonation, traditionally based on compositions of variously situated coral assemblages, is applicable to these reefs only if the less common species and rare species (especially) are given quantitatively disproportionate weighting to discriminate zones. But if the numerically dominant species only are considered, zonal differences become much more subtle. Thus, in descending order of abundance for each zone, the reef flat is characterized by *Porites astreoides, Favia fragum, Siderastrea radians, Porites porites, Agaricia agaricites* and *Millepora complanata;* the reef buttress by *A. agaricites, F. fragum, P. astreoides, M. complanata, P. porites, Siderastrea siderea* and *Acropora palmata;* and the fore reef, having greater equitability in species representation, is characterized by *Acropora cervicornis, A. agaricites, P. astreoides, S. siderea, Montastrea cavernosa, Millepora alcicornis, Montastrea annularis, P. porites* and *Dichocoenia stokesii. Agaricia agaricites* is by far the numerically dominant coral on these reefs.

Octocorals

Patterns for diversity and abundance for the octocoral assemblages of 41 species are much the same as stated above for stony corals. Greatest diversity and abundance are found on fore reefs; the least on sand-grass flats. On the whole, octocorals are somewhat more narrowly adapted than stony corals, as they are virtually absent on shallow spur tops, in unlighted crevices and on thick sand substrates. Their unjustified neglect in coral reef studies relative to stony corals probably stems from the necessity of microscopic spicule identification rather than any diminished utility. The neglect is unfortunate, for octocorals are no less diagnostic for distinguishing various reef habitats than stony corals.

In descending order of abundance, the dominant octocoral species on reef flats are *Gorgonia ventalina, Plexaura flexuosa, Pterogorgia citrina, Pterogorgia quadalupensis* and *Eunicea mammosa.* Dominant species of the reef buttresses are *G. ventalina, Plexaura homomalla, P. flexuosa, Erythropodium caribaeorum* and *Pseudopterogorgia americana.* Dominant species on fore reefs are *E. caribaeorum, Pseudopterogorgia bipinnata, Briareum asbestinum, P. flexuosa, P. americana, G. ventalina* and *Eunicea calyculata.*

Ophiuroids

After stony corals and octocorals, ophiuroids (brittle stars) are perhaps the most abundant large invertebrate on the reefs. Because of their retiring habits and because the systematic literature is barely adequate, ophiuroids remain poorly known and unappreciated by most coral reef investigators. A paper describing the 21 species encountered on these reefs and their habitat distribution is under preparation jointly with Gordon T. Taylor. Except for the basket star, *Astrophyton muricatum,* all species belong to the cryptic fauna; most living beneath coral rubble or live corals. Three species dwell commensally within certain sponges and one species is infaunal. As recesses beneath coral boulders constitute their principal habitat, ophiuroids are most abundant and most diverse on rubble flats and leeward ends of grooves where boulders are most abundant and where the assemblage is characterized by three species of *Ophiocoma, Ophioderma appressum* and *Ophiomyxa flaccida.* They are somewhat less common on sand-grass flats with *Ophiothrix orstedii* and *Ophiocnida scabriuscula* (infaunal) most characteristic and on sand-floored grooves and shallow fore reefs typified by *Ophionereis reticulata* and *Ophiozona impressa.* They are relatively sparse on deep fore reefs where characterized by the coral and sponge dwelling species of *Ophiothrix* and

Ophiactis quinqueradia. Species distribution among ophiuroids is apparently regulated by bathymetry, substrate texture and availability of inert cover or commensal hosts. Many mollusks and decapod crustaceans on the reefs share similar habitats and distribution patterns with ophiuroids.

Reef Fishes

Most coral reef visitors are astonished by the wonderful variety of color, shape, and motion among the fishes. Probably more than 300 species inhabit these reefs. Like sedentary corals, many fish species are characteristic of particular reef zones or habitats, while others are ubiquitous. Estimated relative abundances were recorded for approximately 200 identified species encountered at 98 locations, encompassing all habitat types to 22 m depth on eight "outer-arc" reefs and two patch reefs. Specimen counts were made of each species occupying seven of the quadrats mentioned above. Rotenone fish poisons which would insure more thorough surveys were not used. Several hundred live specimens were measured *in situ* in order to estimate biomass distribution of various trophic categories among fishes over the several reef zones.

Species diversity for reef fishes increases progressively from the sand-grass flats seaward to the shallow fore reef, then decreases again as one crosses the deep fore reef. In general, greater environmental heterogeneity enhances the local diversity and abundance of fish, many of which are just as responsive to substrate conditions as are benthic invertebrates.

Sheetlike encrusting bryozoan assemblages represent an important but neglected component of the outer reef fauna. Such assemblages are described in the Lower Florida Keys by Cuffey and Kissling (1973) in the abstract given below.

ECOLOGIC ROLES AND PALEOECOLOGIC IMPLICATIONS OF BRYOZOANS ON MODERN CORAL REEFS IN THE FLORIDA KEYS
by
R. J. Cuffey and D. L. Kissling

Investigation of modern barrier reefs along the Lower Florida Keys indicates that bryozoans represent an important but neglected component of the reef fauna. Because of geographic proximity to fossil reefs of similar character, these modern reef-dwelling bryozoans should furnish comparative data useful for interpreting fossil reef assemblages.

As on other western Atlantic reefs, sheetlike encrusting cheilostomes are ubiquitous and dominate the diverse bryozoan assemblages. Aeteid and tuftlike cheilostomes are moderately common, while lichenoporid and idmoneid cyclostomes, reteporid cheilostomes, and boring ctenostomes are relatively rare and less widely distributed. Crisiid cyclostomes have not yet been seen here.

The bryozoans function as "hidden encrusters" and are opportunistic in habitat selection. They are most abundant in rubbly, shallower parts of the reef-flat and buttress zones, where they join many other organisms in encrusting undersides of semistable coral boulders and concavities of molluscan shells. Where sandy substrates predominate, and on fore reef slopes to 10 m depth where boulders are rare, bryozoans inhabit inert undersides of living corals and millepores. They are virtually absent on unstable boulder ramparts adjacent to reef islands, and on walls of open reef cavities where predation and space competition exclude them. The bryozoans contribute little to the reef detritus, and seemingly do not trap or bind sediment.

DRY TORTUGAS (96 km West of Key West):
Field trips to Dry Tortugas (Ft. Jefferson National Monument) require advance permission from the Office of the Superintendent, Everglades National Park, P.O. 279, Homestead, Florida 33030. Transportation usually can be chartered from Key West. Water, food and lodgings have in the past been the responsibility of the visitor. Check with the Park officials for an update on these and other important requirements.

General Statement

Dry Tortugas (fig. 7.19) represents an isolated cluster of submarine carbonate sand shoals, reefs and low emergent islands resting on submerged Pleistocene (?) bedrock platform. This area, lying some 96 km west of Key West, represents the most westerly of the known Keys along the southern border of the Florida Platform.

The area provides examples of all three carbonate environments associated with Outer Shelf Areas as defined in this Guidebook, namely, (1) intermittent flourishing reefs, (2) rubble flats (see page 223 of this Guidebook), and (3) blanket deposits of sediment (see page 229 of this Guidebook).

In addition to its fascinating human historical record, Dry Tortugas is unique in at least three ways for scientific study: (a) it was the site of the former Tortugas Laboratory of the Carnegie Institution of Washington (Washington, D.C.) and thus an area for many early investigations and subsequent publications, (b) it is one of the best geographically located areas for the study of tropical storm and hurricane effects on reef growth, shoal morphology and sediment distribution, and (c) it is the site of the only known beachrock in the Florida area (see page 33).

A general geologic description of Dry Tortugas is given by Vaughan (1914) who refers to Tortugas as an atoll. Osburn (1914) describes the bryozoa of the Tortugas. Growth-rate measurements of local corals are described by Vaughan (1915); the role of alcyonarians in reef formation, by Cary (1918). A paper describing ecology and topography of the "Sand Keys" of Florida (including the Dry Tortugas area) is given by Davis (1942). A paper by Ginsburg

(1953) describes beachrock on Loggerhead Key (fig. 7.20). Brooks (1964) refutes the "atoll" concept for Dry Tortugas and presents a summary of the origin of the banks and shoals in the following abstract.

REEFS AND BIOCLASTIC SEDIMENTS OF THE DRY TORTUGAS
by
H. K. Brooks

Many miscomprehensions exist relative to origin of the Florida Reef tract and, in particular, the southwestern extremity—the banks, shoals, and reefs known as the Dry Tortugas. They are not an atoll as stated by Vaughan (1914). The component physiographic features rise from a shallow limestone platform 80 to 100 feet below sea level. Relief features are predominately banks and shoals of bioclastic sands. Their genesis and circular distribution is related to the prevailing seasonal storm patterns.

Large patches of *Acropora cervicornus* (Lamarck) are widely distributed throughout the area in water less than 60 feet deep. Live coral on these patches is sparse. Proliferation of the staghorn corals is slow, but cumulative growth has produced a magnitude of skeletal remains. The coralla are preserved and are ultimately indurated into a porous rocky mass by the luxuriant growth of *Lithothamnion* and its cognate encrusting associates. The shallow reefs of Garden and Loggerhead keys, populated by calcareous algae, alcyonarians, and scleractinians, etc., originate upon a foundation of the remains of these organisms. This can be seen where erosion in surge channels has exposed the underlying materials.

Reef organisms, reef rock, bioclastic sands (i.e., coralline sands, algal sands, pseudo-oolite, and beachrock), and bioclastic muds will be illustrated and discussed with the aid of a color underwater film.

O'Neil (1976) notes the radical change in size of East Key over a 200 year span. He notes that its simple strand/dune plant community is a current source for short-term stability of the central core of this Key.

Ft. Jefferson, on Garden Key (fig. 7.21), is constructed on loose calcareous bioclastic sand and rubble underlain by the "Key Largo (Pleistocene) limestone" found at a depth of 9 m (Multer, 1967—unpublished drill record). **Note:** exact age and correlation of this bedrock unit will have to await results of radioactive dating. In 1967 the writer also visited all 7 of the Keys in the Tortugas group and found no bedrock exposed (although beachrock and encrusted littoral rubble was present on some Keys). It is interesting to note that the bedrock beneath Garden Key was found at an elevation considerably higher than the bottom of channels between all Keys of the Tortugas group, thus indicating the strong possibility that bedrock relief can itself be one of the various factors responsible for the location of today's emergent Keys.

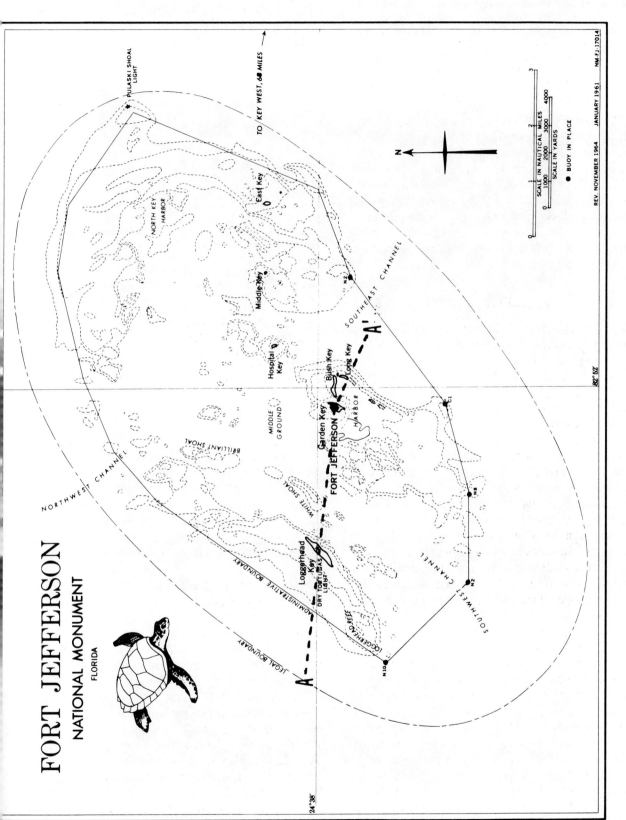

Figure 7.19. Sketch map of Dry Tortugas showing location of seven presently emergent Keys. From U.S. National Park Service, rev. Nov. 1964. Added dashed line indicates location of cross section by Jindrich (1972) shown in figure 7.22.

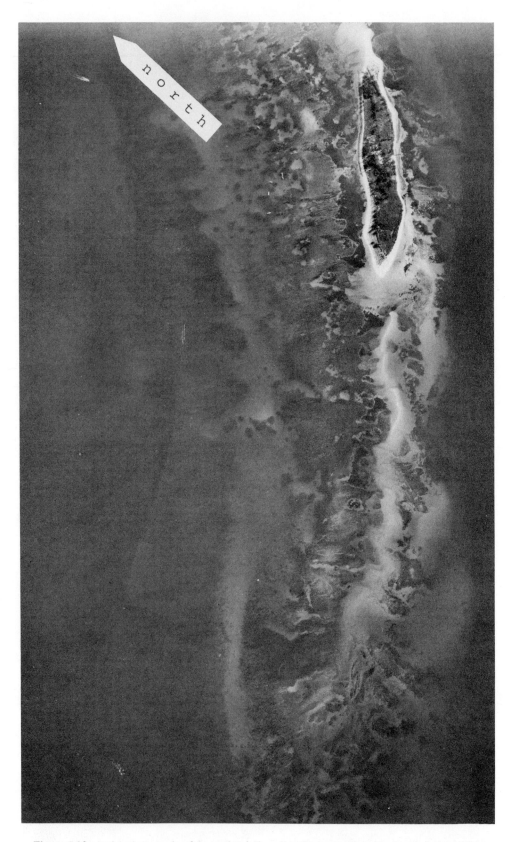

Figure 7.20. Aerial photograph of Loggerhead Key, Dry Tortugas. Note beachrock in intertidal zone and shifting sand at either end of the emergent Key. Evidence for active major recurving of spit at northern end was observed in March of 1967 by the author. Submerged shoals and reef extend southwestward to Loggerhead Reef (bottom of photo). Photo from U.S. Coast and Geodetic Survey (28Feb60 S9669).

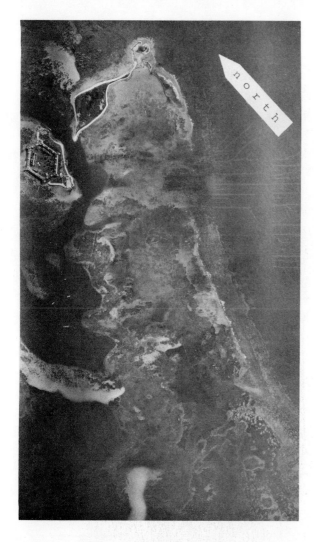

Figure 7.21. Aerial photograph of Garden Key (with Ft. Jefferson) and Bush and Long Key to the east. Submerged sand shoals and reefs extend southward. Photo from U.S. Coast and Geodetic Survey (28Feb60 S9608).

A comprehensive study by Jindrich (1972) describes three basic biogenic facies found on the Tortugas platform lying both adjacent to one another and in vertical succession. The abstract of Jindrich's dissertation is noted below. See also figures 7.22 and 7.23.

BIOGENIC BUILDUPS AND CARBONATE SEDIMENTATION, DRY TORTUGAS REEF COMPLEX, FLORIDA
by
V. Jindrich

The Dry Tortugas, a horseshoe-shaped complex of carbonate banks and coral reefs, is located at the southern terminus of the Florida limestone shelf. The complex rises to the surface waters from a drowned Pleistocene surface that forms a circular platform having a general depth of 17-21 m.

Three basic biogenic buildups (facies) comprise the reef complex: (1) detrital lagoonal bank, (2) Montastrea reef bank, and (3) *Acropora palmata* reef. These facies lie adjacent to one another and are also present in vertical succession as individual growth stages of varying thickness and lateral extent. A zone of *Acropora cervicornis* is developed as a transition between the *Montastrea* and *A. palmata* growth stages.

The present organic assemblages and topography bear evidence of dominantly lateral progradation and cumulative storm effects that are linked to the slow eustatic sea level rise for the past several millenia. Long-continued storm degradation is manifested by (1) continuous removal of *A. palmata* and its replacement by storm-resistant coralline algae and *Millepora* sp. to produce truncated rocky surfaces, (2) abundant reef rubble, (3) erosion of spur-grooves, and (4) development of intertidal rubbly reef flats.

Sediments ranging from cobble-sized rubble to medium silt are composed of *Halimeda,* coral and mollusc grains; coralline algae and foraminifers are present in minor amounts. Variations in texture and constituent particle composition are interpreted to be mainly a result of mode of sediment transport and effect of grain shape. Broadly-defined grain size populations produced by three modes of transport have characteristic assemblages of constituent particles. The populations include a gravel-sized surface creep population, sand-sized saltation population, and very fine sand- to silt-sized suspension population. Strong mixing occurs between the gravel and sand population on the storm-degraded shoals, and between the sand and silt population on the lagoon bottom. Sands flanking the reefs and reef banks show minimum mixing hence good degree of sorting. Incongruous mixtures of the in-place fraction and varying proportions of the transported populations constitute detrital lagoonal banks as a substrate stabilized by seagrass and coral growth. The gravel-sand and sand-silt mixtures are related to deposition under highly variable energy conditions. Variability in energy conditions does not cause strong population intermixing on beaches. From the same reason, beach sediments show a high degree of sorting in all size grades from cobbles to fine sand.

WINDWARD REEFS EAST AND SOUTH OF LONG KEY (Dry Tortugas): Living reefs found flourishing in the Dry Tortugas areas today must be viewed as "survivors" in that they, more than in the case of reefs in many other areas, represent remnants of abnormally intense and frequent storm activity over a shallow carbonate platform.

As shown in the cross section by Jindrich (1972) in figure 7.22, reef development in the Long Key areas can be described in terms of reef wall and reef flat.

1. Windward facing reef walls (see figure 7.23) are capped by rubble pavement and along the upper seaward portions contain only a limited amount of small size living coral (*Acropora palmata, Diploria* sp., *Porites asteroides* and *Montastrea annularis*). According to Jindrich (1972, p. 8) the upper reef wall is dominated by such forms as abundant soft algae

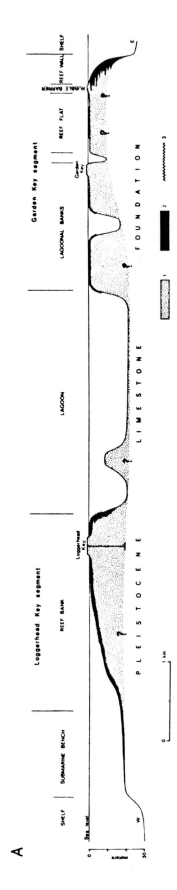

Figure 7.22. Section across the Dry Tortugas reef complex showing its physiography and inferred internal facies. (1) Unconsolidated sediments of the lagoonal-bank facies, including rubble beds and detrital aprons. (2) Rigid frame of the *Montastrea* reef-bank facies and *Acropora palmata* reef facies, possibly with intercalated beds of cemented reef rubble. (3) Distribution of abundant coral growth (dominantly *A. cervicornis*). Figure and caption from Jindrich (1972). See figure 7.19 for location of this cross section A-A'.

Figure 7.23. Upper reef wall slope on Garden Key segment of cross section (figure 7.22). The picture shows two of a few large *Acropora palmata* colonies found preserved on the storm-degraded reef crest. Reef rubble containing *A. palmata* fragments testifies to the former extensive growth of this coral species now being replaced by brown algae, alcyonarians, and *Millepora* sp. Photo and caption from Jindrich (1972, p. 72).

(*Pocockiella variegata, Turvinaria tubinata* and *Padina* sp.) which are replaced by a variety of alcyonarians and *Millepora* sp. with depth. Limited spur and groove structures can be observed extending seaward from the reef wall displaying truncated and coralline algae-encrusted surfaces.

2. The *reef flat* lies directly behind (leeward) of the reef wall and is a shallow skeletal debris-covered surface which extends toward and intergrades with the lagoonal banks (see figure 7.22). In some areas aprons of coarse skeletal sand display thickets of *A. cervicornis* (fig. 7.24) and *Porites porites*. Thalassia-stabilized sand and large numbers of *Diadema* are found on other portions of the reef flat surface. A variety of *Halimeda* are present. Coarse rubble can be encrusted with coralline algae.

REEF BANKS OF EAST KEY, SOUTHERN LOGGERHEAD KEY AND SOUTHERN

GARDEN KEY (Dry Tortugas): Large areas (2 × 17 km along East Key) of carbonate shoal banks can be gullied and indurated with encrusting coralline algae and covered with various coral (*Diploria* sp., *Porites porites, A. cervicornis* and *Montastria* sp.) which commonly increase in density on the flanks of these banks. Tremendous quanities of *Halimeda opuntia* grow on hard substrates such as dead *A. cervicornis.* Millepora and alcyonarians are also very common. Where these banks are commonly current swept, only coarser rubble remains as a veneer with skeletal sand being confined to cavities and pockets (such as provided by *A. cervicornis* thickets—see figure 7.24). For further details see Jindrich (1972).

EASTERN ANDROS ISLAND BARRIER REEF
(Bahamas): This magnificent barrier reef is readily viewed by chartered plane from Nassau or Miami.

Figure 7.24. Dense thickets of living *Acropora cervicornis* growing on and between coralline algae-encrusted thickets of dead *A. cervicornis*. Such thickets provide excellent baffles for sediment accumulation during storms. Photo taken on reef flat approximately 540 m southwest of the southern end of Long Key in 60 cm of water. Similar but more widespread thickets occur on various Dry Tortugas reef banks.

Andros Town (at Fresh Creek) is also the site of a commercial Bahaman airline terminal. Chartered boats from Nassau and local Andros sources are usually available for more detailed investigations.

The one-half to one km wide "barrier reef" found along the eastern coast of Andros commonly has an *Acropora palmata* crest with several successive seaward fore reef terraces or platforms displaying coral development and commonly spur and groove structure. At the edge of these terraces the platform break or rim (1 to 2 km seaward of the reef crest) is usually abrupt and can have a notched clifflike face. Along its north-south trend the barrier reef is fragmentary in its middle and southern portions where patch reefs and scattered coral commonly occur.

See Newell *et al.* (1951) and Newell and Rigby (1957) for comprehensive reports on this reef. The latter report includes illustrated discussions and also gives a foldout map showing over 170 km of barrier and associated patch reefs along the east coast of Andros. Purdy and Imbrie (1964) summarize some of the pertinent features of this barrier reef. For a succinct description of the distribution, morphology, zonation of biotas, internal structures and sediment and cementation of this reef see Gebelein (1974) who also provides illustrations and discussion of windward lagoon patch reefs. Cuffey and Gebelein (1975) have investigated the role of bryozoans as "hidden-encrusters and cavity-dwellers" along the barrier reefs of eastern Andros. The abstract of their paper is noted below.

REEFAL BRYOZOANS WITHIN THE MODERN PLATFORM-MARGIN SEDIMENTARY COMPLEX OFF NORTHERN ANDROS ISLAND

by

R J. Cuffey and C. D. Gebelein

The eastern margin of the Great Bahama Bank provides unusual opportunity to expand knowledge of reef-dwelling bryozoans, not only for the Bahamas specifically, but also for general use in paleoecology.

East of northern Andros Island, a narrow barrier reef dominated by *Acropora palmata* lies just islandward inside the bank margin. The barrier's back-reef slope yields many bryozoans, functioning as hidden-encrusters and cavity-dwellers most common are encrusting cheilostomes, but some tuftlike and aeteid cheilostomes and idmoneid and lichenoporid (*Disporella*) cyclostomes also occur. In the shallow windward lagoon, islandward from the barrier, are scattered patch reefs (with small headlike and branching corals), occupied by abundant but noticeably less diverse, hidden-encrusting and cavity-dwelling (but not sediment-forming) cheilostomes, mostly sheetlike, a few tuftlike and aeteid. Immediately adjacent to the island shore, the only bryozoans encountered are a few encrusting cheilostomes on branching *Porites* in grassy shallows.

North of Andros, barrier reefs are absent, and elongate oolite-sand shoals lie bankward from the platform margin. Bryozoans are absent from the active oolite shoal and the stabilized oolitic sand flat bankward behind the shoal. However, rocky mounds in tidal channels are small patch reefs whose frame consists of encrusting cheilostomes (comprising up to 75% of some reef rock) accompanied by red algae and milleporines. Bryozoans functioning as reefal frame-builders are known from very few modern situations; these reefs are thus exceptionally interesting, because of their similarity to Paleozoic bryozoan reefs.

Author's Note: Although not in the areas covered by the present edition of this Guidebook, three important Bahaman studies of flourishing outer reefs should be noted for the interested reader.

1. A thought-provoking and important study of reefs off North Eleuthera by Zankl and Schroeder (1972) and Schroeder and Zankl (1974) describes reefs as various types of pillars. These pillars, in turn, can be composed of primary and secondary frame builders. The authors emphasize and illustrate various types of organism growth and destruction, sedimentation and cementation as well as mechanical breakdown which determines the final shape and composition of a reef.

2. Atolls in the Caribbean are rare. An excellent, detailed and well-illustrated study of an atoll in southeastern Bahamas (Hogsty Reef) is given by Milliman (1966, 1967a and 1967b).

3. The Hopetown Reef off Elbow Cay (northeastern Great Abaco) has been described in more than usual detail by Storr (1964). His 98 page well-illustrated account can in many ways serve as a guide to others interested in similar open ocean-facing reefs which Storr defines as lying somewhere "between the concept of a 'near fringe' and a 'bank-barrier' reef."

RUBBLE FLATS AND MOUNDS

A second type of environment found associated with Outer Shelf Areas in the Florida-Bahama region is that of essentially barren rubble flats and mounds. In one sense "flats" and "mounds" can be looked at as two end-members with a variety of gradational forms in between. Broad sheets of rubble can be worked up into "horn shaped" submerged mounds by tropical storms which in turn can be remodified by long-term tidal action. Storms of hurricane intensity can build rubble into emergent islands.

The source of most rubble is the result of a variety of mechanical and biological activity. Intense storms can break up even the thicker branching coral (see figure 7.25) but the ever present internal and external destruction of reef frame builders by various organisms weakens erect reef forms and of course reduces the size of fragments once they are mechanically broken and fall to the seafloor. For a discussion concerning the speed of this destructive process by organisms see a description of research in progress by Kissling on page 212 of this Guidebook. Rubble may also become encrusted and cemented as discussed under Hardground Environments on page 191 of this Guidebook.

The distribution and topographic form of rubble flats and mounds relative to the shelf edge varies and is worthy of study for interpreting ancient analogs.

FRENCH REEF (Outer Florida Reef Tract): French Reef can be found most easily by going approximately 3 km northeast from Molasses Light along the seaward edge of the shallow outer reef platform (see figure 7.8 and index map 5 in the Appendix).

French Reef represents one example of a particular bottom environment found on the outer reef platform. This is an area which formerly had flourishing coral but is now merely a rock-floored, barren, rubble-veneered topographic high (fig. 7.26). The barren state in which this area is now found is due to the destructive forces of hurricane and/or Pleistocene sea fluctuations. Another example of a barren outer reef is seen at Molasses Reef, 3 km to the southwest, where a more extensive development of new living coral can be observed coating the old barren reef surfaces. In many respects, the sequence seen in French Reef, Molasses Reef and Key Largo

Figure 7.25. Overturned *Acropora palmata* (result of tropical storm) on back side of outer reef, 1 mile southwest of Carysfort Light, Florida. Depth of water 2 m.

Dry Rocks, respectively, represents a series of progressive stages in living reef development on barren outer reef platforms.

Only a very little incipient coral growth is found in the French Reef area. Evidence for a former flourishing coral reef in this area is readily found by examination of the cemented coralline rock floor (fig. 7.27) and the loose coral rubble (fig. 7.26) which veneers large portions of particularly the leeward side of the area.

The destructive nature of tropical hurricanes on outer reef platforms is well described by Ball *et al.*, (1967) who note the development of rubble "horns" or leeward pointing, often curving projections composed of great quantities of hurricane-derived coarse coral rubble. Such "horns" are shown by Ball *et al.*, (1967) on the back (leeward) edge of French Reef (see "X" on figure 7.28) as well as many other outer reefs.

The effectiveness of hurricanes in the production and spread of such rubble zones is well documented by Ball *et al.*, (1967) who note the leeward movement of rubble some 60 to 150 m distance behind Molasses Reef as a result of a single hurricane. They also describe and picture the leeward western edge of one such coarse hurricane-derived rubble horn as having a relief of 6 m and a slope of 30-35 degrees.

One of the most important aspects of these leeward rubble horns and sheets is the fact that they represent a greater accumulation of coarse debris than that found on the windward slopes of the outer reef plat-

form. This relationship stands in sharp contrast to the more classic idea of rubble being confined to the areas in *front* of the outer reef. This inconsistency is explained by Ball *et al.*, (1967) as due, in the present Florida situation, to lack of a combination of (1) materials transported seaward by storm-ebb tidal currents over topographic lows in the platform edge, (2) materials slumped from the platform edge, and (3) subaerially eroded material off the exposed platform edge during past emergences. Whatever the reason, however, the fact that the majority of the coarse rubble found today along the Florida reef tract is located *behind* and not in front of the outer reef is significant for interpretation of ancient reef analogs.

LITTLE MOLASSES ISLAND (Molasses Outer Reef, Florida Reef Tract): About .8 km northeast of Molasses Light is a small emergent island composed of reef rubble called Little Molasses Island (fig. 7.29). This island rises to a height of approximately 1.3 m above sea level and is subject to change in shape with each successive tropical storm. Little Molasses represents the cumulative product of past storm activity which has piled up rubble on this leeward side of Molasses outer reef. Aerial photographs illustrating the reshaping of this island by Hurricane "Donna" in 1960 are shown by Ball *et al.*, (1967). These photos also show the extension of a submerged rubble "horn" to the northwest.

Rubble fragments up to 1 m in longest dimension can be found on the emergent portion of this island

Figure 7.26. Loose coral rubble on top of French Reef, outer Florida reef tract. *Halimeda opuntia* grows in great profusion on some of this rubble.

Figure 7.27. Current swept hard rock bottom made up of organically cemented reef rubble ornamented with various Alcyonarians. French Reef, Florida outer reef tract; depth of water approximately 1 meter.

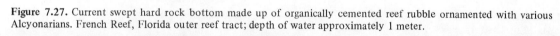

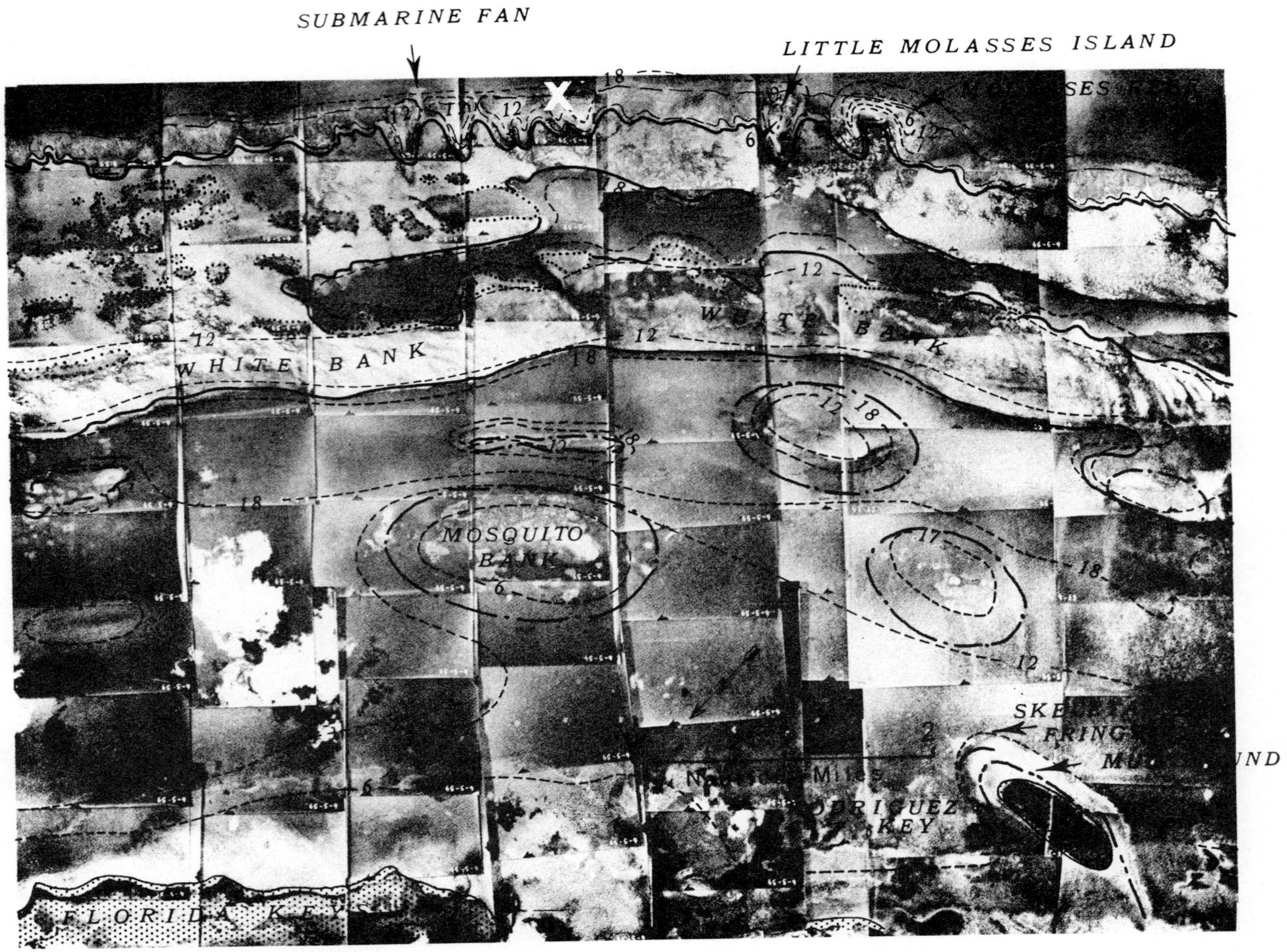

Figure 7.28. Aerial mosaic of part of the Florida reef tract. X = French Reef area. Approximate bottom contours (- - - - -). The numbers on these lines are depths in feet. Zone of platform-edged reefs (———). Very little living coral is found in this zone relative to the amount of dead-coral rock and rubble. The contours in this zone delineate several horns of coral debris that are readily visible on the aerial photograph. Linear patches of dead coral (. . . . .). These patches are incrusted with organisms such as calcareous algae. Sand shoals behind the reefs and linear rock patches (———). The largest and farthest west of these shoals is White Bank. The sand shoals are white where they have not been overgrown with marine grass. Areas of patch-reef growth (. - . - .). Skeletal zone on the flanks of the sediment mound, Rodriguez Bank (- - - - -). The arrow points to a submarine fan of bedded sand building from the platform edge into deeper water. From M. M. Ball, *et al.*, "The Geologic Effects of Hurricane Donna in South Florida," JG 75:5 (1967). Copyright © 1967 by the University of Chicago Press.

Figure 7.29. (*above*) The emergent island, Little Molasses, completely made up of storm rubble like the rounded 60 cm long *Acropora palmata* branch at "A." View is looking northeast.

Figure 7.30. (*below*) Submerged flank of Little Molasses showing (left) "D" rounded *Diploria* head (60 cm in diameter) in water over 1 meter deep and (right) "A" rounded branch of *Acropora palmata* in 2 meters of water.

Figure 7.31. View on part of the rubbly spit projecting seaward from Bush Key, Dry Tortugas. Note the frontal rampart of well-sorted *A. cervicornis* rubble overlying the coarse rubble of massive coral heads containing slabs of *A. palmata*. Photo and caption from Jindrich (1972).

and down its submerged flanks (fig. 7.29 and 7.30). Shallow digging on the island reveals coarse sand alternating with rubble zones. A cursory inspection reveals no cementation taking place today although conditions for it (salt spray and evaporation, rain water) certainly exist.

Little Molasses Island today is important since it demonstrates (1) how wave activity can transport and pile rubble into a wave resistant island some 1.3 m *above* sea level, (2) how subsequent cementation would result in the preservation of a ''reeflike'' mound on the outer platform composed of welded coarse coral fragments and intermixed sand (an ancient analog of the latter might easily be confused with true bioherms formed by organic growth of corals infilled with much rubble and sand), and

(3) how a very high porosity/permeability mound may originate in the outer reef platform.

Banks (1959) describes the emergent rubble pile at Little Molasses Island as a ''summit conglomerate,'' notes its potential as an excellent future reservoir rock, and compares it with the Cretaceous Sunniland coquina pay zone of the Forty Mile Bend oil field in south Florida.

MIDDLE SAMBO EAST TO MARYLAND SHOAL REEFS (Lower Florida Keys): The outer reef area of the Lower Florida Keys, reached by boat from local marinas, is the site of extensive rubble deposition. The most detailed investigation known to the writer of outer reef rubble and their associated horns and rampart islands (some emergent) is by Kissling (see pages 210 to 212 and figures 7.15

through 7.18 of this Guidebook). This preliminary report by Kissling describes rubble on both the back, crest and fore reef zones as well as details of techniques for investigating these unique deposits.

BUSH KEY AND GARDEN KEY (Dry Tortugas): Access to this area is described on page 216 of this Guidebook.

As an area "overworked" by hurricanes (see page 219 and figure A.4 in the Appendix), Bush Key and Garden Key display much evidence of earlier destructive work by hurricanes to such branched forms as *A. cervicornis* and *A. palmata*. Remnants of these corals are found as sticks and irregular plates forming rubble barrier cappings and leeward rubble flats (see cross section figure 7.22).

According to Jindrich (1972, p. 22), "The highest elevations of the rubble barrier emerging as islands display crude bedding with a coarse rubble of massive corals and larger slabs of *A. palmata* at the base topped by ramparts of A. cervicornis sticks" (see figure 7.31). He also notes (p. 23) that "All significant rubble deposits point to leeward transport. No rubble accumulations were encountered seaward from the windward fronts."

SHELF EDGE SANDS

The third type of environment found associated with Outer Shelf Areas in the Florida-Bahama region is actually nothing more than the extension of skeletal or oolitic sand found as blanketlike deposits on the inner shelf platform (chapter 6). In other words, in addition to rubble and intermittent thriving reefs, the platform edge areas in the Florida-Bahamas are very commonly found covered with either loose rippled or grass-held sand.

CAT CAY OOLITIC SHOALS (Northwest Great Bahama Bank): Oolitic sand originates today along various areas of the Bahama shelf edge as described on pages 151 to 155 of this Guidebook and the reader is referred to these discussions for further information. Conditions favorable for the formation of similar oolitic sands on the Florida shelf are not found today although oolitic sand did form along the Florida platform during the Pleistocene (see chapter 9).

CARYSFORT LIGHT TO TENNESSEE REEF LIGHT SKELETAL SANDS (Northern Florida Keys): The outer shelf of the northern Florida Keys contains intermittant flourishing outer reefs which are the "factories" responsible for the production of large quantities of *Halimeda*-rich skeletal sand. This sand is carried back onto the platform where it forms large blanket like deposits (such as White Banks) on both the inner and outer shelf areas. See page 182 of this Guidebook for a discussion of this sand found on the inner shelf. Elongate belts of this same Halimeda-rich porous sand, stabilized by marine grasses or in loose, rippled deposits, can be found between established reefs and rubble areas along the outer shelf zone.

One of the most comprehensive studies ever made in the Florida Keys deals with these skeletal sands of the Inner and Outer Shelf. This study was made by Paul Enos who first provided the data given below for the 1975 edition of this Guidebook. The complete paper will be published *in* Enos and Perkins (1977-in press). See also Enos (1974a and 1974b). Enos examined patterns of Holocene carbonate accumulation on the south Florida shelf in three dimensions. Extensive use of a high-resolution reflection profiler, custom designed for shallow carbonate shelves as well as improved coring equipment allowed him to evaluate the wedge of Holocene sediments covering the rock-floored shelf to a greater extent than has ever been done before (see figure 7.32). Examination of this previously inaccessible third dimension allowed recognition of (1) accumulation patterns and their variations (figs. 7.33 and 7.34), (2) sediment sources, (3) controls on accumulation, and (4) pore space characteristics.

In discussing reefs along the shelf edge, Enos notes that there are many steep-sided internal reflections in the general vicinity of the shelf edge subsurface which are interpreted as actual reefs which have been covered over with sediment (see figure 7.35). Enos also notes that since the Pleistocene, reef growth at the shelf edge has been very intermittent and interspersed with periods of skeletal grainstone accumulation. Cores taken by Enos indicated that patch reefs colonized in generally loose sediment (usually wackestone), indicating that in some cases the "hard surface" usually required for attachment must be as small as a large shell or piece of debris.

According to Enos, "The history of reef development at the south Florida shelf break has involved continuous changes in the location of active reefs so that no large segment of the shelf break lacks some evidence of former reef development. Evidence seems to favor a long-term reduction in reef biomass or perhaps a steady state" (implying neither increase nor decrease in biomass but more shifting of loci).

With regard to the origin of White Bank, Enos notes its presence is due to the presence of the outer reefs which act as barriers breaking the competence of deep water swells. "The sediment in White Bank is located in the area of maximum outer reef produc-

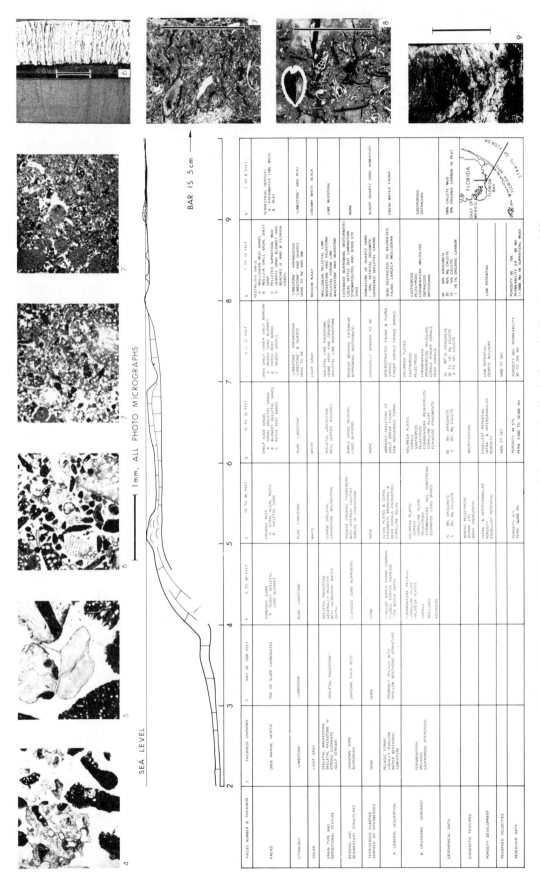

Figure 7.32. Summary of Holocene sediments of the south Florida shelf margin. From Enos (see p. 229).

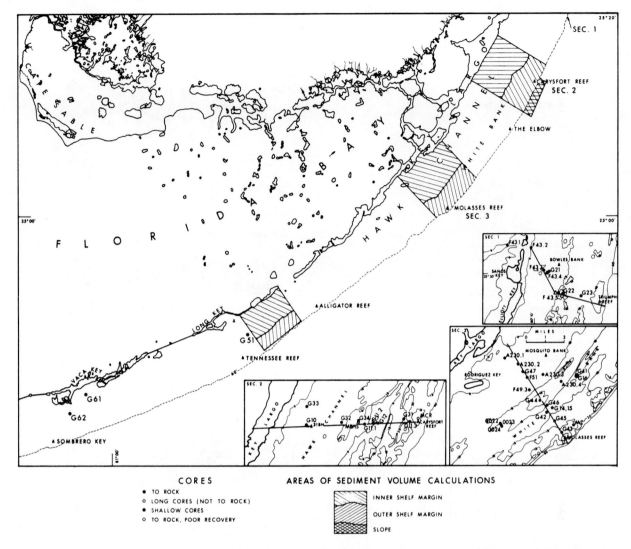

Figure 7.33. Location of cores, cross sections (see figure 7.34) and areas of sediment volume calculations. Core locations marked by open circles were unimpregnated cores examined through the courtesy of H. Gray Multer. From Enos (see p. 229).

tivity, in-place sediment production is very low, skeletal sand transport is dominantly landward, and some sediment parameters show gradational trends from the reefs across the outer shelf margin. These arguments apply to the origin of shelf margin skeletal sand in general." Enos indicates that a 15 km long elongate rock high beneath the middle part of White Bank and the detection of several patch reefs within White Bank are of secondary importance in its origin.

In discussing patch reefs, Enos notes that "most patch reefs developed on an antecedent bank whose location was presumably influenced by the rock floor features."

Numerous flood tide deltas and less common ebb tide deltas are described by Enos in the Upper Florida Key area where fine sediment from the inner shelf (such as Florida Bay) and the shelf margin can deposit significant packstone and wackestone deltas. Such deltas represent the only mappable sediment feature that trends normal to the shelf edge and therefore are useful when interpreting ancient reef analogs. Enos describes the largest (primarily ebb) delta at Caesar Creek (5.6 km long). Another complex of deltas can be found at Matecumbe (7.2 km wide).

The abstract and summary of the paper by Enos are given below.

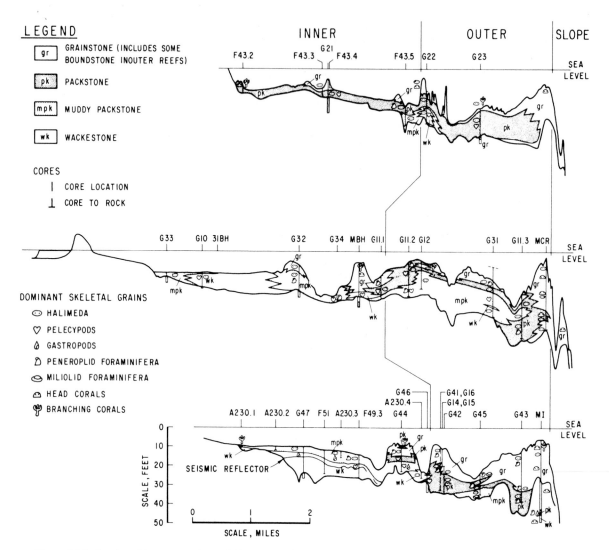

LEGEND

gr	GRAINSTONE (INCLUDES SOME BOUNDSTONE INOUTER REEFS)
pk	PACKSTONE
mpk	MUDDY PACKSTONE
wk	WACKESTONE

CORES

I	CORE LOCATION
⊥	CORE TO ROCK

DOMINANT SKELETAL GRAINS

- HALIMEDA
- PELECYPODS
- GASTROPODS
- PENEROPLID FORAMINIFERA
- MILIOLID FORAMINIFERA
- HEAD CORALS
- BRANCHING CORALS

Figure 7.34. Stratigraphic cross sections of Holocene sediments, south Florida shelf margin. Section and core locations are shown in figure 7.33. Sections are constructed from slightly smoothed, velocity-corrected relfection profiles with 170x vertical exaggeration. From Enos (see p. 229).

CARBONATE SEDIMENT ACCUMULATIONS OF THE SOUTH FLORIDA SHELF MARGIN
by
Paul Enos

Abstract

The narrow south Florida Holocene carbonate shelf contains four of the facies recognized as particularly favorable for reservoir development in ancient carbonate shelves: reefs and shelf edge sands (outer shelf margin), open shelf (inner shelf margin), and restricted (inner) shelf. Distinct organic communities are recognizable in the shelf margin and shallow slope environments reflecting primarily type of substrate and secondarily degree of circulation. Inner shelf, inner shelf margin, outer shelf margin, and shallow slope environments of deposition may be distinguished by composition of skeletal grains, depositional textures, and sedimentary structures.

Reflection profiling of the shelf margin shows that the thickest accumulations of Holocene sediments, as much as 45 feet, lie in parallel belts along the outer reefs, along the shelf edge shoals and along the shallow fore reef slope. These accumulations also have the most favorable pore space arrangements with porosities of 40-45 percent and permeabilities up to 35 darcies. Second order accumulations with much lower permeabilities are tidal deltas, nearshore wedges of sediment and a few banks, all of the inner shelf margin.

The belts of thick sediment vary laterally along the margin, particularly the sand shoals which disappear southwestward and are replaced northward by discontinuous patch reef banks. The shoals are derived primarily from mechanical concentration of skeletal debris produced in the outer reefs. In general, skeletal productivity is the primary control on accumulation patterns. Sediment transport, the configuration of the lithified Pleistocene rock floor, and the evolving Holocene depositional topography also influence accumulation patterns.

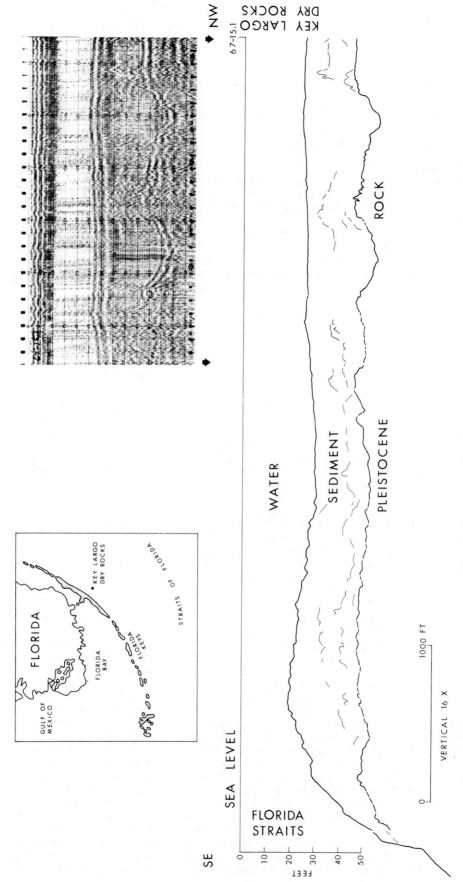

Figure 7.35. Internal reflectors in Holocene sediment near the shelf break, interpreted as buried reefs. Part of the actual record is reproduced at the same scale as the tracing. From Enos (see p. 229).

The sedimentary record of the present cycle is transgressive. Depositional environments have shifted progressively shoreward. The Holocene shelf margin is essentially stacked on a late Pleistocene one but the Holocene belts of grainy sediment are much narrower and the Holocene topographic crest is at the shelf break.

Summary and Conclusions

The Holocene depositional environments in the south Florida shelf are summarized (fig. 7.32) following the conceptual scheme for carbonate facies patterns of Dooge and Raasch (1966). Four facies having reservoir potential are well developed on the south Florida shelf: organic reef, shelf edge sands, open marine shelf, and restricted shelf.

The shelf margin, in particular and outer 2 miles, is the site of the thickest (typically 25 feet or more) and most permeable (to 35 darcies) sediments in the Holocene package. The porosities of the outer shelf margin sediments (about 45 percent) are some of the lowest encountered among Holocene sediments, but because the sediment is grain-supported sand, most of the porosity is potentially preservable. The shelf margin sediment package is generally muddier toward the base.

The thickest sediment accumulations lie in belts parallel to the shelf edge. The trends of the belts are quite predictable but the sediment thickness and physical properties vary along the trend. The outer reef belt is the most nearly continuous and the thickest. It is about a mile wide and located immediately behind the shelf break. The shallow slope sediment blanket is thickest within a mile of the shelf break at a water depth of about 100 feet. Slope sediments are packestone grading into wackestone with greater water depth. Shelf edge sands form prisms of thick sediment as much as a mile wide, 1 to 1 1/2 miles behind the slope break. The largest sand shoal is 25 miles long and as much as 25 feet thick. Where sand shoals are lacking a belt of discontinuous patch reef banks may occupy the same position. The patch reef belt is also about a mile wide, trends parallel to the shelf, and may be more than 25 feet thick, but continuity is poor and the sediments are generally muddy. A bedrock depression with a thin layer of muddy sediment isolates the patch reef belt of the lower Keys fron the outer reef belt.

The inner shelf margin is not nearly as "full" of sediment as the outer margin. Sediment accumulations are muddier (packstone and wackestone), generally thinner, and less predictable than those of the outer shelf margin. Wedges of sediment piled against the Pleistocene rock of the Keys reach more than 15 feet in thickness. These wedges are elongate parallel to the shelf edge. Patch reef banks of the inner shelf margin tend to be elongate parallel to the shelf edge, but most are small, less than a mile long, and not readily predictable. Tidal deltas oriented perpendicular to the shelf edge are developed where passes between rock floor highs enter restricted basins of the inner shelf. Most of the sediment in tidal deltas is wackestone and muddy packstone, generally less than 15 feet thick.

Changes in sediment thickness and total volume along trend are most closely tied to the amount of tidal exchange with the restricted shelf. Where water from the inner shelf and the broad Gulf of Mexico shelf flows across the shelf margin, sediment accumulation is reduced, probably owing to decreased productivity. Tidal deltas and inner shelf

margin sediment wedges are lacking here, too, because they are dependent on current constriction by the Florida Keys.

The primary controls on sediment distribution patterns, in apparent order of importance, are productivity, mechanical redistribution, preexisting rock topography, and contemporary sediment topography. The relatve importance of these factors may be qualitatively evaluated for the various types of accumulations.

The Holocene history of the south Florida shelf sedimentation is a continuous transgression, with only local regressive pulses. Sedimentary facies reflect less restriction and higher energy levels toward the top of the sequence. The Holocene sequence differs from the Pleistocene shelf sediments in that it contains little quartz or nonskeletal carbonate and the distribution of grainstone is much more restricted. The differences can be accounted for by the relatively small degree of submergence of the Florida Plateau and low supply of terrigenous material from the Florida peninsula during the Holocene.

Boyer (1972) describes various types of grain accretion as being more abundant and varied along the platform edge in comparison to areas of sand back toward the center of the platform. The abstract of his paper is given below.

GRAIN ACCRETION AND RELATED PHENOMENA IN UNCONSOLIDATED SURFACE SEDIMENTS OF THE FLORIDA REEF TRACT
by
Bruce W. Boyer

Abstract

Accretionary features (grain coatings, intragranular void fillings, internal and external cements) of Florida reef tract calcarenites are morphologically diverse, but are all products of nonskeletal submarine carbonate precipitation and lithification. Such features are most abundant in sands along the platform edge near the outer margin of the reef tract where flushing of sediment by ocean water is most intense and generally become less abundant on the back reef toward the Keys.

Accretionary features and lithified micritic aggregates (mostly fecal pellets) common in clean sands are rare in muddy calcarenites of the inner back reef. Presumably, even piecemeal cementation is inhibited in muddy environments, for two possible reasons: (1) The flux of dissolved carbonate to and within impermeable muddy sediments is relatively small, and (2) the muddy sediments are generally poorly aerated and relatively rich in organic matter; failure of oxidative decomposition of organic films on carbonate particles may prevent welding of groups of particles or the formation of overgrowths on particles.

Well-developed superficial ooids and clear microcrystalline cements essentially free of entrapped detrital micrite are found only near the platform edge, where water turbidity is low. A systematic change in abundance and type of accretionary phenomena, like that observed on the present Florida reef tract, may be a useful paleogeographic indicator.

FIELD QUESTIONS
Outer Shelf

KEY LARGO DRY ROCKS (Florida Outer Reef Tract)

1. How many geologically significant "zones" (equivalents of which may be recognized in the fossil record) can you find normal to the elongate trend of this outer reef? What are the chief constituents of each zone?

BACK   TOP   FRONT   FORE REEF AREA

2. What percentage of the exposed hard rock surface is covered with:

   a.   living coral   _____

   b.   living coralline algae   _____

   c.   living green algae   _____

   d.   rubble   _____

   e.   loose sand   _____

   f.   other   _____?

3. What are the chief types of corals present that are not seen in the back-reef areas?
   Are there any peculiar shapes of the corals in this outer reef as compared with shapes of corals in the back-reef area?

4. Does this outer reef act as a "baffle" for sediment accumulation?
   How does the loose sediment in back (leeward) of the outer reef compare in size and composition with the sand in channels (grooves) between the reef spurs—and the sediment of the fore reef (windward) area?

	Size Grade	Composition
Leeward Sand		
Channel Sand		
Windward Sand		

   Run sieve analysis and plot up on sieve analysis sheet in appendix.

Sample No.

Photo No.

5. Do the grooves always have loose sand bottoms? Do they always have open ends?

6. What effect would a hurricane have on this outer reef? Would such effects be preserved in the fossil record?

7. What would happen if sea level dropped an average of 1 m? 3 m? 100 m? raised 3 m?

8. Ancient reefs often have micrite associated with them to some extent—where is the mud fraction on this outer reef? If present—what is its origin?

9. How does the rubble on the windward side of the outer reef compare in amount, size grade and distribution with the rubble on the leeward side of the reef?

	Lee Side	**Windward Side**
Amount—		
Size—		
Distribution pattern—		

FRENCH REEF (Florida Outer Reef Tract—3 km Northeast of Molasses Light):

10. Examine the rubble "horn" on the leeward side of this reef. What is the composition of the rubble?

11. Why is there such a paucity of living attached forms on top of or in front of French Reef?

LITTLE MOLASSES ISLAND (Florida Outer Shelf—a Little Less Than 1 km Northeast of Molasses Light):

12. Break apart some rubble (e.g., piece of *Acropora palmata*) and note effect of boring sponge (orange) and algae. On the emergent Little Molasses rubble pile look for aragonite recrystallizing to calcite (spar).

13. Coral rubble on the outer shelf accumulates in several shapes—elongate mounds (sometimes emergent), blanketlike sheets and horns or pointed rubble sheets. Question: If lithified, how would each of these three be classified in the terms of "biostromes" and "bioherms"?

14. Examine the submerged rubble "horns" extending out leeward from Little Molasses Island and from adjoining Molasses Reef (Molasses Light) area (see figure 7.28). Contrast and compare these horns with those observed in back of French Reef.

FIELD NOTES

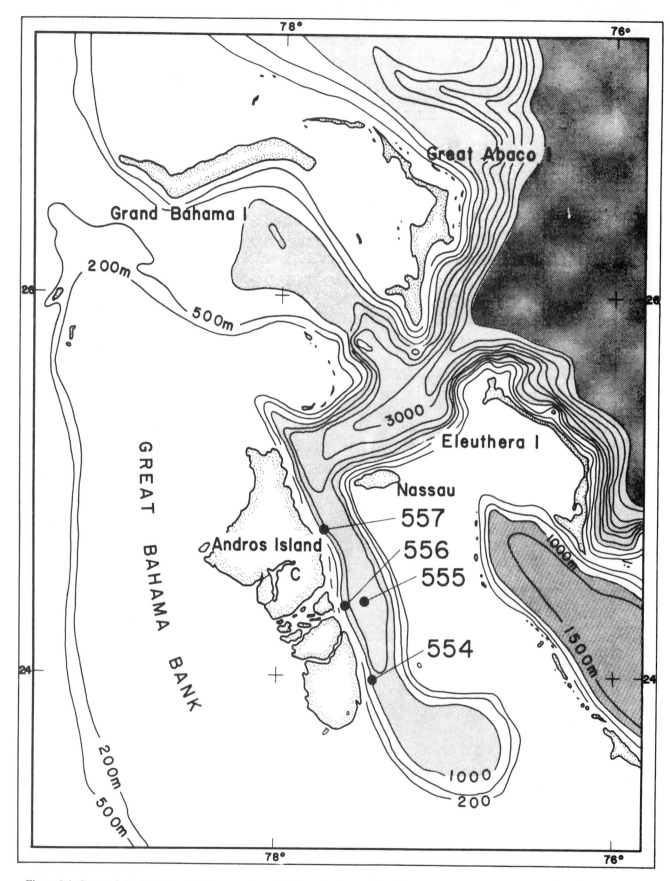

Figure 8.1. Some of the deepest and best examples of shelf margin slopes off carbonate platforms exist in the Bahamas. The development of submersibles and new geophysical techniques has increased our accessibility and knowledge of this heretofore almost unknown area. The above bathymetric map from Schlager *et al.*, (1976) shows sites of four submersible (Alvin) dives. Contour interval is 500 m. The 200 m contour is added to show the edge of the banks.

SHELF MARGIN SLOPE AREAS

8

The term "Shelf Margin Slopes" as used in this Guidebook relates to those submarine slopes that extend from the seaward edge of the outer shelf down to the floor of the bordering basin.

This seaward extension of the edge of carbonate platforms has been the subject of increasing interest during the last few years. A significant reason for this is the development and availability of research tools such as submersibles and new geophysical techniques.

Some of the reasons for the study of carbonate shelf margin slopes are noted below.

1. They represent the "highway" over which a tremendous load of known shelf-produced skeletal carbonate must pass on its way to the basin. Thus a study of basin slopes should reveal various "processes" responsible for large-scale transport of carbonate sediment.
2. They represent a slope that can evidently (a) grow by various types of accretion extending the platform edge seaward or (b) be cut back by various erosional processes thereby reducing the size of the platform at its outer edge. More information concerning such accretion/degradation processes is highly desirable.
3. They contain various environments with many characteristic (some hopefully preservable) sedimentary structures. These can be useful in identifying the edges of ancient carbonate basins.

The purpose of this chapter is *not* to present a comprehensive review but rather a "peek" at some of the environments and associated processes that are currently being evaluated along a few submarine slopes which extend from the platform rim down to the basin floor in the Florida-Bahama area. Obviously only a few readers of this Guidebook will be fortunate enough to be able to either directly or indirectly observe or measure features of this environment; therefore it is hoped that this chapter will provide a glimpse of some of the environments and their active processes and aid those interested in interpreting ancient carbonate shelf margin slopes. For additional

information, the reader is encouraged to check bibliographies given in most of the literature cited.

Five localities which have been the site of investigations along today's shelf margin slopes in the Florida-Bahama area are noted below:

1. shelf margin slopes that border the Straits of Florida off Miami and Bimini,
2. the Pourtales Terrace area off the platform edge bordering the lower Florida Keys,
3. shelf margin slopes off the platform edge bordering Dry Tortugas in the Tortugas and Agassiz valley areas,
4. shelf margin slopes that drop off the eastern Andros outer shelf into the Tongue of the Ocean, and
5. shelf margin slopes of the Little Bahama Bank.

A foldout map showing the bathymetry of four out of five of the above-mentioned areas is given in a paper by Malloy and Hurley (1970) who describe the geomorphology and geologic structure of the Straits of Florida. Included in their paper is a discussion of the slopes leading down to the bottom of the Florida Straits from both the Bahaman and Florida shelf edges. Reproductions of actual seismic reflection profiles are given indicating some subbottom characteristics beneath these shelf margin slopes. A review of previous works and pertinent bibliography is included.

SUBMARINE SLOPE ENVIRONMENTS BORDERING THE STRAITS OF FLORIDA OFF MIAMI AND BIMINI

One of the most detailed accounts of today's shelf margin slope morphology, sedimentary structures and organisms is given by Neumann and Ball (1970) from observations aboard the submersible *Aluminaut* on two dives—one along each of the shelf margin slopes bordering the Straits of Florida off both the Miami and Bimini areas (see figure 8.2). The abstract of their paper and two profiles describing some of the characteristics of these slopes are noted below.

SUBMERSIBLE OBSERVATIONS IN THE STRAITS OF FLORIDA: GEOLOGY AND BOTTOM CURRENTS
by
A. C. Neumann and M. M. Ball

The submarine slopes that border the Straits of Florida off Miami and Bimini were traversed by the submersible *Aluminaut*, in August and September of 1967. The Bimini escarpment is characterized by a three-part zonation consisting of relatively strong northerly bottom currents in both deep and shallow zones, and an intermediate zone of low northerly bottom current velocities. From 538 m (the bottom of the traverse) to 222 m, there is a sloping, smooth, rock surface veneered with sand, ripple-marked by northward bottom currents of 50 cm/sec or more. The middle, low velocity zone, from 222 m to 76 m, exhibits a muddy slope of largely bank-derived material. The sediment surface exhibits tracks, burrows, and mounds, which indicates that currents here are never as strong as in the rippled zones above and below. Observed current velocities in the middle zone were only 5 to 10 cm/sec. Above 76 m, a steep, vertical to overhanging cliff with large talus blocks at its base rises to a crest at 30 m. Currents in the upper zone are northward at 50 to 150 cm/sec. The inverted situation of higher energy bottom current conditions and associated sedimentary features and textures existing in the same area, but at a greater depth than low energy surface features and fine sediments, is of significance to the stratigraphic interpretation of ancient rocks.

On the western side of the Straits, at the base of the Miami Terrace, is an elongate trough 825 m deep. The bottom here is characterized by ridges and mounds of muddy sand capped by thickets of living deep-water branching coral. The eastward-facing escarpment of the Miami Terrace exhibits ledgelike outcroppings of dark phosphatic limestone from depths of 719 m to the crest at 457 m where the traverse ended.

An interesting finding of the dives in the Straits of Florida is the observation, based on both current measurements and sedimentary structures, that the bottom current on the western side of the Straits flows *southward* at observed velocities of 2 to 50 cm/sec. This southerly bottom flow is opposite to either the northerly Florida Current above or to the bottom current on the Bahama side of the Straits. The nature and orientation of the sedimentary structures, plus the combined observations of several *Aluminaut* dives in the same area, indicate that the southward bottom counterflow is persistent and not a temporary tidal reversal. An extensive sedimentary anticline in the west-central sector of the Straits may have been built by this bottom countercurrent bringing material from the north.

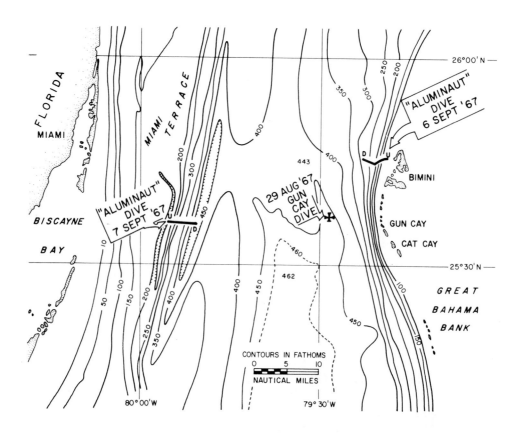

Figure 8.2. Map of Straits of Florida showing location of 1967 *Aluminaut* dives. The May 1969 dive was at 25°38.6'N./80°01.7' W., or near the "V" in "DIVE" on the left side of the figure. "D" and "U" refer to the down and up locations, respectively. From Neumann and Ball (1970).

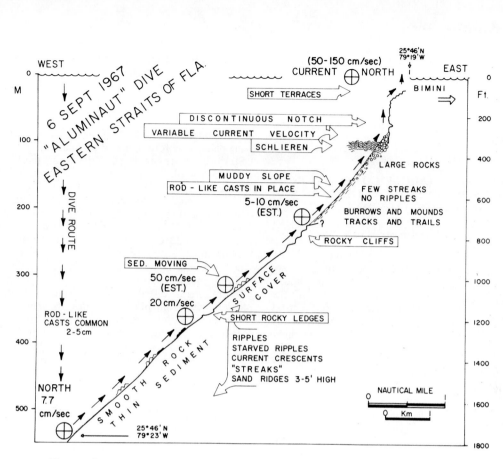

Figure 8.3 Schematic profile of submersible dive traverse off Bimini, Bahamas, indicating major observations. Bottom current direction is north. From Neumann and Ball (1970).

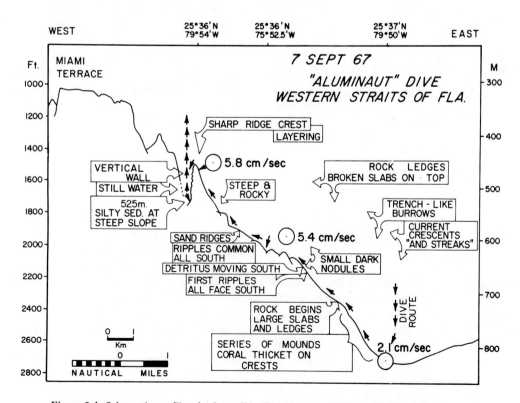

Figure 8.4. Schematic profile of submersible dive traverse off Miami, Florida, indicating major observations. Bottom current direction is south. From Neumann and Ball (1970).

A very limited amount of information is available concerning the nature of the sediment which moves down the shelf margin slopes from both the Florida and Bahaman shelf edges into the Straits of Florida. An abstract by Gassaway (1970) describes some aspects of these sediments.

MINERAL AND CHEMICAL COMPOSITION OF SEDIMENTS FROM THE STRAITS OF FLORIDA
by
J. D. Gassaway

The three main sources of sediment in the Straits of Florida are, (i) the Florida Current contributing a low-magnesian calcitic biogenic ooze; (ii) the Bahama Banks, where the highly aragonitic sediments arise; and (iii) the "relict" sediments which are the source for the detrital silicates and phosphates. A possible fourth source is the Miami Terrace Reef which has a relatively large high-magnesian calcite content. Offshore currents carry the detritals southward along the coast of Florida until they are blocked by the Florida Current. Aragonitic sediments move westerly from the Bahama Banks, but to the west the Florida Current strongly curtails their cross channel movement. The mineral suite of the Straits of Florida sediments is composed principally of biogenic carbonates except to the northwest where quartz comprises more than half of the sediment, and reflects the strong faunal control of sediment mineralogy.

Trace elements found in the carbonate fraction of the sediments of the Straits of Florida are divided into three groups. The first group, strontium and chromium, has a positive correlation with aragonite and similar quadratic trend surfaces east of the axis of the Florida Current. West of the axis chromium behaves differently than does strontium and correlates with the insolubles to the northwest. Statistical evidence indicates that strontium is not significantly enriched in the carbonate sediments after deposition. Manganese and cobalt, the second group, vary directly with each other and statistical analysis shows that they are present in the same phase. From the trend surfaces and linear statistical treatment of manganese and cobalt with respect to the carbonate minerals and insolubles, these two elements occur in low magnesium calcite. Therefore, the Florida Current is a principal source for the distribution of manganese and cobalt in the Straits of Florida. The third group of elements, iron, zinc and chromium, are correlated with the "relict" sediments to the northwest where their most probable source is the carbonate-fluorapatite.

Planktonic Foraminifera contribute to the sediment of the shelf margin slopes along both sides of the Florida Straits. Jones (1971) documents the temporal and vertical population distribution of living planktonic Foraminifera in the Straits of Florida. The paper includes a description of faunal-environmental relationships (temperature, salinities, nutrients, oxygen, light characteristics of the Florida Straits) and 27 plates of planktonic Foraminifera. The abstract of his paper is given below.

THE ECOLOGY AND DISTRIBUTION OF LIVING PLANKTONIC FORAMINIFERA IN THE STRAITS OF FLORIDA
by
J. I. Jones

Seasonal, diurnal, and depth distribution patterns of living planktonic foraminiferal populations from the Florida Straits have been analyzed to determine their relationships to measured environmental parameters. Maximum seasonal foraminiferal concentrations were found in January and July. Maximum populations were measured in the 50 and 150 meter depth range, commonly associated with the 26°C isotherm, and restricted to a temperature range from 22°C to 30°C. A vertical diurnal population migration was observed, with maximum concentrations occurring during daylight hours in the nearsurface waters. No apparent relationship was found to exist between foraminiferal distribution and nutrient concentration, determined from phosphate measurements. Phytoplankton concentrations and foraminiferal distribution patterns coincided. Seasonal, depth, temperature and salinity preferred ranges have been established for the foraminiferal species occurring in this study.

POURTALES TERRACE AREA ALONG THE SHELF MARGIN SLOPE BORDERING THE LOWER FLORIDA KEYS

Shelf margin slopes are sometimes interrupted by submarine terraces. One of the most investigated and largest in the Florida-Bahama area is the Pourtales Terrace (approximately 200×30 km). This terrace interrupts the normal shelf margin slope forming a 200-400 m floor. The inner (200-500 m deep) portion of the terrace is partially covered with sediment; however, most of the Pourtales surface appears to be karstlike and composed of an exposed hardground pavement which ends seaward in a steep pre-Miocene shallow water limestone escarpment which drops off into the Florida Straits. A considerable area of this terrace can be seen on figure 8.5.

Malloy and Hurley (1970) present a detailed bathymetric map of the whole terrace and discuss various theories for the origin of its karstlike surface including one involving development of the solution and collapse topography taking place while the terrace was still submerged. Other illustrated papers (with extensive bibliographies) which discuss the origin of the karst topography and other aspects of the terrace, such as step fault or slump-slip origins, are Jordan (1954), Jordan and Stewart (1961), Jordan (1962), Jordan *et al.,* (1964), Uchupi (1968), and Gomberg (1974). Of perhaps special significance to geologists wishing to interpret ancient analogs of basin slope terraces is a study by Gomberg (publication pending) which describes crusts, borings, cavity fillings and lag deposits of the current-swept surfaces. He defines characteristics of a "condensed sequence" which he indicates as diagnostic of drowned shelves. An abstract giving preliminary results of these investigations by Gomberg (1973) is given below.

DROWNING OF THE FLORIDAN PLATFORM MARGIN AND FORMATION OF A CONDENSED SEDIMENTARY SEQUENCE
by
D. N. Gomberg

The southeastern part of the Floridan carbonate platform seaward of the Florida Reef Tract has been drowned to a depth of 200-400 meters—too deep for accumulation of shallow marine carbonate sediments. The rocky surface of the Pourtales Terrace is exposed at these depths, and is kept free of fine-grained sediments by the swift Florida Current. Slow accumulation of coarse residual sediments and severe alteration of preexisting rocks characterize this depositional setting.

Dredging on the terrace and its seaward-facing escarpment shows that it is composed of several hundred meters of pre-Miocene, shallow water limestone. A deep water hardground on the uneven, partly-exposed surface of this Tertiary limestone is distinguished by ferromanganese crusts and replacements, conglomerate and breccia formation, superimposed borings, and pelagic sediment cavity-infill. Scattered about this hardground is an unconsolidated lag of Miocene marine-vertebrate bones, sharks teeth, coprolites and phosphatized limestone cobbles.

On the shallower parts of the terrace, a thin (less than a meter) discontinuous cap of Quaternary-Recent bioclastics is accumulating and being cemented. Unlike the Tertiary units, these younger rocks are poorly consolidated, and composed of the unaltered skeletons of organisms now living on or above the terrace. Little or no material derives from the adjacent areas of shallow water carbonate accumulation.

The upper meter of the Pourtales Terrace is a "condensed sequence," with the general succession "shallow water deposits—hardground—deep water deposits" most diagnostic of drowning. Intensified circulation and minimal terrigenous input appear necessary for the formation of this sequence, and for similar ones in the geologic record.

SUBMARINE SLOPE ENVIRONMENTS OFF THE PLATFORM EDGE BORDERING DRY TORTUGAS AND AGASSIZ VALLEY AREAS

Jordan and Stewart (1961) in a definitive paper describe a swarm of valleys incising the shelf margin slope for more than 30 nautical miles along the 400 fathom contour south of Dry Tortugas (see figure 8.5). These smaller valleys converge into two princ-

iple (Tortugas and Agassiz) valleys which are tracable to depths of over 900 fathoms. Above the heads of these valleys are smooth even slopes from 25 to 300 fathoms which the authors believe represent aprons of sediments gradually burying older valley heads. The source of these sediment aprons is described as probably from Florida Bay and the western Florida coast during northeastern storms. In a latter paper, Malloy and Hurley (1970, p. 1961) confirm, by sub-bottom profiling, the existence of these buried channels.

Jordan and Stewart (1961) believe that the Tortugas and Agassiz valleys have an origin unrelated to processes presently active in the area, suggesting former turbidity currents, and emphasizing the fact that multiple theories are often necessary to explain shelf margin slope features shown in figure 8.5.

Minter *et al.,* (1974) report additional information based on new bathymetric data and actual observations from the submersible *Alvin*. They indicate that these valleys are being filled shoreward of 600 m and the two main valley systems may have different origins. They also note sluggish current and sediment transport within the valleys, which themselves show terracing and possibly local slumping. Numerous escarpments are also mentioned.

SUBMARINE SLOPE ENVIRONMENTS OFF THE EASTERN ANDROS OUTER SHELF—BAHAMAS

Some submarine shelf margin slopes extending from the platform edge down onto adjoining basin floors in the Bahamas (fig. 8.1) have created the world's highest canyon walls (either submarine or subaerial) according to Andrews *et al.,* (1970). The shelf margin slopes of a somewhat shallower extension of this 'greatest of all deep canyons' forms the Tongue of the Ocean (TOTO) off eastern Andros and has been the site of investigations by the U.S. Navy and various other groups. Interested readers should refer to Gibson and Schlee (1967), Busby and Merifield (1967), Markel (1968) as well as Andrews *et al.,* (1970). Most of the papers summarize previous investigations and describe in various detail (often with photo showing slope conditions) the character of the TOTO shelf margin slopes. Busby (1962) and Andrews *et al.,* (1970) also give foldout bathymetric maps showing configurations of the shelf margin slopes. In addition to papers which include discussion of TOTO basin slopes per se, there are many publications dealing with the origin of the TOTO and sediments found on and beneath its floor. A paper by

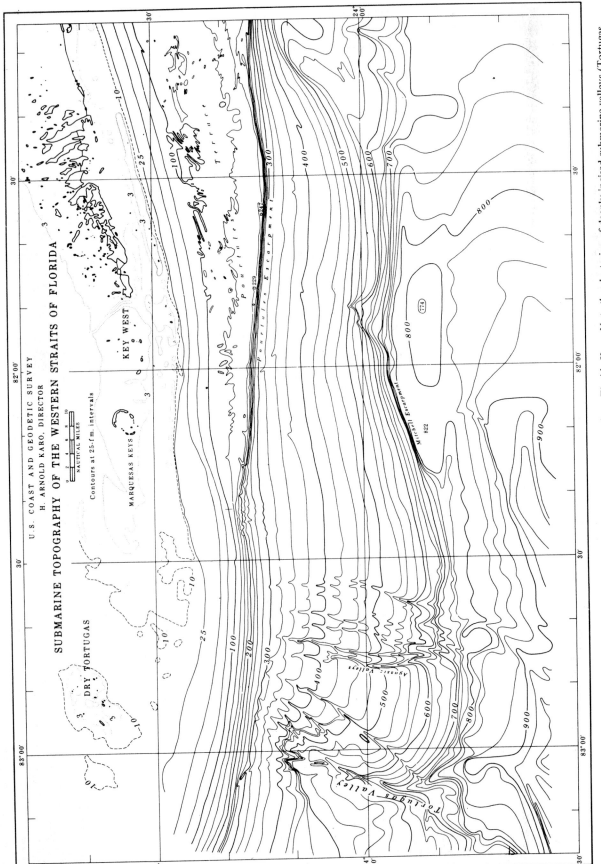

Figure 8.5. Bathymetry of a portion of the basin slope off the platform edge of the southern Florida Keys. Note the clustering of deeply incised submarine valleys (Tortugas and Agassiz Valleys) in the southwest portion of the map compared to relative smooth slopes to the east. A portion of the basin slope is interrupted by the Pourtales Terrace with its steep southern escarpment in the upper right area of the map. From Jordan (1962).

Lynts (1973) summarizes many of these investigations and gives an extensive bibliography.

Schlager *et al.,* (1976) present an excellent summary of the amazing variety of past published theories for explaining the origin of the steep slopes and adjoining basins commonly observed in the Bahamas. More important, however, the paper reports on actual observations made down the steep basin slope off the east coast of Andros (see figure 8.1) during 4 dives in the submersible *Alvin.* They subdivide this basin slope into three morphologically distinct units:

1. a **reef wall** from 0-400 m with slopes of 35-90° and a talus slope at its base,
2. **gullied slopes,** 400-1,400 m with inclinations of 5-25°, and
3. **flat bottom** or floor of the basin.

Although not observed, the **reef wall** was believed to probably be under aggradation. In contrast the **gullied slopes** and **flat bottom floor** were observed to show abundant evidence of erosion (scours, sediment removal, erosional terraces, undercutting, V-shaped valley profiles). Processes responsible for this erosion, according to Schlager *et al.,* (1976), are thought to be not continuous but episodic in nature (i.e., at one place perhaps several important erosional events in the past 200,000 to 300,000 years). In summary, observations by Schlager *et al.,* (1976) stress the importance of erosion (and not faulting) for the steep canyons and scours of the shelf margin slopes below the reef walls. They further state . . . "we believe the high and steep slopes of the Tongue of the Ocean to be the result of an interplay of upbuilding of the platforms and erosion of the troughs, as Andrews and others (1970) suggested." They indicate that, in the troughs, deposition partly fills these depressions forming a flat bottom. A schematic illustration of this interplay is given in figure 8.6. As pointed out by the authors of the above paper, the model presented by a sequence of erosional surfaces within slope deposits may be a useful tool for interpreting ancient analogs. The abstract of the paper by Schlager *et al.,* (1976) is given below.

EPISODIC EROSION AND DEPOSITION IN THE TONGUE OF THE OCEAN (BAHAMAS)
by
W. Schlager, R. L. Hooke and N. P. James

During four dives in the Tongue of the Ocean in the deep submersible *Alvin,* we observed extensive evidence for erosion, including cliffs several tens of meters high of horizontally bedded pelagic chalk, crescentic scour marks around boulders in channel bottoms, and bands of freshly exposed rock at the bases of cliffs that were otherwise stained dark brown by manganese oxide. In the central part of the Tongue of the Ocean, the change from a depositional to an erosional regime must have occurred only recently, as the axial valley is cutting headward into bedded chalks of mid-Pleistocene age. Such changes may have occurred several times in the history of the Tongue of the Ocean, and the steep walls may be the result of an interplay between upbuilding of the platforms and erosion of the troughs.

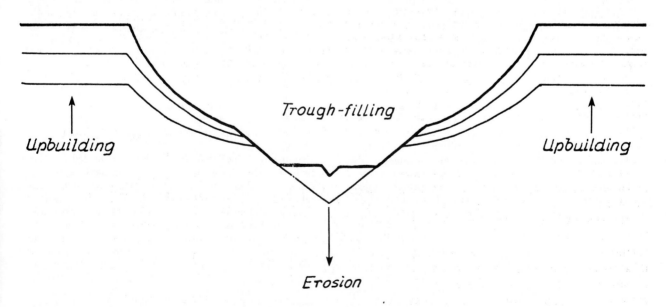

Figure 8.6. Schematic cross section of the TOTO showing the projected result of the interplay of upbuilding of the platforms and episodic erosion and refilling of the trough. From Schlager *et al.,* (1976).

SHELF MARGIN SLOPE ENVIRONMENTS OFF THE LITTLE BAHAMA BANK

Submarine slopes bordering the Straits of Florida were investigated in 1971 by submersible (*Alvin*) and some of the results were discussed in abstract form by Neuman, Keller and Kofoed (1972). A more detailed discussion of these dives and the concept that the deep shelf margin slopes off carbonate platforms of the Little Bahama Bank are accretionary forms building outward is given by Neuman (1974a) and reproduced below.

CEMENTATION, SEDIMENTATION AND STRUCTURE ON THE FLANKS OF A CARBONATE PLATFORM, NW BAHAMAS
by
A. C. Neumann

As part of a NOAA program in November 1971, 5 dives were made on the research submersible, *Alvin,* to depths of 600 to 700 m in the northeastern Straits of Florida at the base of Little Bahama Bank. In an area previously contoured as east-west trending ridge and valley topography (Malloy and Hurley, 1971), we encountered instead an irregular pattern of north-south oriented, elongate, rocky mounds between which hardgrounds occasionally cropped out beneath a thin veneer of sandy mud. The largest of these rocky mounds were 30 to 40 m high and 100's of meters long (Neumann, *et al.,* 1972).

Rocks and sediments from the mounds were sampled and photographs were taken from *Alvin.* During the operation the WHOI research vessel *Gosnold* ran a grid of 10 in³ air gun profiles over the area. The R/V *Gilliss* returned to the area in May 1973 and obtained dredge rock and sediment samples as well as more air gun profiles.

Preliminary results from these observations suggest that the mounds are forming by a combination of processes. Sediments are being transported along the bottom by bottom currents. In some areas the sediments are sandy and extensively rippled. More commonly, however, a thin veneer of loose dark material is gathered into long-narrow bands which parallel the south to north bottom current direction and give the bottom a streaked appearance. Close observation through the viewing port revealed that this phenomenon is a result of sediment and organic matter in loosely bound aggregates of a few millimeters in diameter which roll and drift over the bottom under the influence of currents 5 to 10 cm per second or less. This process was nick-named "bed fluff transport." The sediment collects behind objects and irregularities as leeside accumulations.

Submarine lithification of the carbonate sediment must occur in this region at or just below the sediment surface. Its products include small irregular fragments, hummocks, large mounds and flat-topped hardgrounds. Organisms such as sponges, crinoids, alcyonarians and aheratypic corals attach to and grow upon these rocky substrates. Growth of the mounds is achieved by the organisms trapping sediment by baffling, contributing to the buildup by their faecal products, and lending their skeletal remains upon death. Boring and burrowing organisms help by pro-

viding a honeycomb of holes within the rock which also serves as a sediment trap. Several generations of boring, infilling, and cementation can be seen in rock sections.

The sediment which accumulates is cemented by subsea lithification, a process still poorly understood. The buildup cycle is one of cementation, organism attachment, sediment entrapment, and more cementation. The validity of this sequence of events is enhanced by the observation on some mounds of an onionskinlike layering of several concentric shells of lithified crust each about 10 cm thick.

Subsequent laboratory analyses of the rock and sediment further document the significance of subsea cementation in this area and its role in mound building. A radiocarbon date of one whole rock sample is 26×10^3 yrs and contained coral fragments (*Dendrophyllia* sp. and *Lophohelia* sp.) yield a date of 27×10^3 yrs. Rocks exposed at comparable depths on the Miami Terrace on the other side of the Straits are Miocene in age (Bush, 1951). This suggests that the surficial rocks on the Bahamian flank are products of modern processes relative to this outcropping across the Straits. Data from X-ray diffraction, stable isotopes of carbon and oxygen, electron microprobe analysis, scanning electron microscopy and rock thin section studies also indicate that the rock is largely aragonitic and contains pelagic forams and pteropod tests. In the process of subsea lithification, this sediment is transformed to a rock that can be quite dense and is composed largely of micrite of magnesian calcite which is about 14 mole percent $MgCO_3$. Calcite tests of pelagic forams are abundant, and microprobe scans show little Mg in the tests relative to the micritic matrix. The originally aragonitic pteropod valves are scarce in the rock relative to the sediment. Aragonite, common in the sediment, is reduced to trace amounts in the rock except for the occasional large coral fragments. The ratio of oxygen 16 and 18 indicates cementation at temperatures of 6 to 8°C, close to the 7 to 10°C bottom temperature presently recorded for the Bahamian side of the Straits (Moore, 1965). Thin section studies and scanning electron microscopy of the rocks show features commonly associated with other examples of documented deep-sea lithification such as "dog tooth" crystals of Mg-calcite, micritic rinds, a matrix of Mg-calcite micrite, pelletal micrite, and fibrous cement. The latter two features have been previously associated with shallow water environments of cementation (Bricker, 1971, p. 49), and this might now be reconsidered.

Submarine lithification of carbonate sediments has now been documented for both deep and shallow marine environments (i.e., Bricker, 1971), and Teichert (1958) has summarized deep-sea coral banks and discussed the geological significance of bioherand buildups in environments deeper than those usually regarded for these features. The significance of the abundant, apparently oriented, elongate rocky mounds observed on the dives reported here is that they are lithified bioheral buildups capable of considerable upward growth and associated, not with bank margin environments of carbonate platforms, but instead with the deep (600-700m) bank flank environment at the base of the platform. Some of the subbottom reflection profiles indicate that slumping and megabreccia transport may have implaced material at some places along the base of the bank. It is conceivable that some of the irregular topography observed was originally a result of slumping rather than cementation. Perhaps the lithified rinds develop on megabreccia blocks.

The evidence for slumping, however, is restricted to local areas. The evidence against slumping as a major mode of emplacement of the mounds is the widespread occurrence of hardgrounds between the mounds, the fact that the mounds are elongate in the current direction, and that they exhibit concentric crusts of deep water material lithified in place. To distinquish these deep water, lithified, fossiliferous mounds from shallow water bioherms, and to signify the important role that subsea lithification plays in their origin and growth, the term "lithoherm" is proposed for the type of feature described here.

It is proposed that the deeper flank environment at the seaward base of carbonate platforms be considered a site of biohermal growth as well as the shallow bank margin or lagoonal environment. Heckel (1974) reviews the concepts behind the definition of bioherms, but does not discuss lithified bioherms of the deep, bank flank environment.

There is much yet to be known concerning these curious and significant features. Some subbottom profiles suggest megabreccia emplacement of an original irregular topography; others suggest *in situ* development and revealed buried lithoherms growing from a planar basement and mounds exhibiting a concentric shell-like inner structure. More detailed profiles with better resolution are needed, but are difficult to obtain from surface profiling of these features which are of small relief relative to their depth.

Several profiles reveal that the mounds seem to grow upward from any one of a series of nearsurface planar reflectors. It is tentatively concluded that these buried reflectors are the same as the lithified surfaces, or hardgrounds, observed to crop out between the mounds. A similar surface was earlier reported from similar depths off Bimini by Neumann and Ball (1970) diving in *Aluminaut*. It appears that the development of hardgrounds and other products of subsea cementation is important in the maintenance and outgrowth of steep, carbonate platform margins and also in the accretion of the flanking beds. The source of the flank deposits is in part from the overlying water column as evidenced by the abundant pelagic foram and pteropod tests. The aragonitic mud fraction is probably derived from the prolific calcareous algae of the platform surface (Neumann and Land, 1968). The fact that lithification is active in some areas and not in others is a function of the chemical oceanographic regime. The Straits of Florida is a ramp sloping upward to the north. The bottom flow on the west side is to the south (Neumann and Ball, 1970) and downhill; in contrast, the flow on the eastern side is uphill and to the north. The effect of this on the carbonate saturation is that the partial pressure of CO_2 increases in the southward flowing water and decreases in the northward flowing water. Thus, the western side would be one of solution or at least nonlithification; whereas the eastern side would be one of accretion by lithification. The different rocks on each side of the Straits document such a regime, but more studies of the carbonate chemistry of the waters are needed. Not only might ascending waters in straits and seaways be associated with concurrent subsea lithification, but upward movement of water along the sides of carbonate platforms may induce subsea lithification and thus help to maintain the steepness of these platform margins.

We must modify the oversimplified view of carbonate platforms as largely lagoonal deposits held behind a steep reef and/or oolite wall and fronted seaward by coarse reefal talus. Our observations of deep carbonate platform flanks suggest that carbonate platforms, like the Bahama Banks, are accretionary forms building outward as well as upward, with the shallow marginal deposits playing a volumetrically minor role compared with the bulk of lagoonal material contained behind the barrier rim and the wedge of flanking deposits piled without.

Neumann (1974b) describes the results of high-resolution profiling and submersible observations around the northern margin of Little Bahama Bank. He defines three areas of the northern shelf edge of the Little Bahama Bank and their offshore slopes in the abstract given below. See also figure 8.7 for graphic illustration.

SHALLOW AND DEEP BANK MARGIN STRUCTURE AND SEDIMENTATION LITTLE BAHAMA BANK
by
Conrad A. Neumann

High-resolution profiling around the northern margin of Little Bahama Bank, plus air gun profiling and submersible observation of the deeper flanks, has revealed a complex of features that extends our knowledge of carbonate platform construction beyond generalizations derived from ancient analogs. Shallow margin growth is determined by sea level history, degree and direction of physical energy, reef growth, and sediment supply. Antecedent topography exerts a feedback effect. Deep flank growth is largely influenced by circulation, lagoonal and pelagic sediment production, and tectonics. Subsea lithification plays a major role in slope maintenance, hardground development and growth of downslope bioherms. The three sectors of the northern margin are each different. On the east, a reef complex of coalescing coral pinnacles rises from a sloping bedrock shoulder; seaward talus is thin; bankward, a belt of grass-bound muddy sand fronts active, flood-oriented oolite shoals. Northward, drowned reef masses shelter leeside talus wedges. Deep flanking deposits here are a leeside accumulation of the northerly oceanic circulation. To the west, a drowned Holocene reef lies buried beneath a shoulder of sand swept westward by the bank-top drift. The deep flank here is dissected by a canyon complex. Hardgrounds and lithified bioherms, "lithoherms," are observed at 2000'. Subbottom profiles of these features suggest *in situ* growth, stacked sequences and in places, antecedent control of slump topography. Upwelling and regional circulation determine the extent of subsea cementation. The accreting platform builds upward by subsidence and lagoon filling, and outward by shallow marginal growth plus deep flank deposition from pelagic, benthic and lagoonal sources.

See Wilber and Neumann (1976) for an abstract describing results of a petrological examination of samples from subsea carbonate mounds (lithoherms) from the lower shelf margin slopes off Little Bahama Bank. See also Hine (1975) for shallow bank margin structure and depositional processes, northwestern Little Bahama Bank.

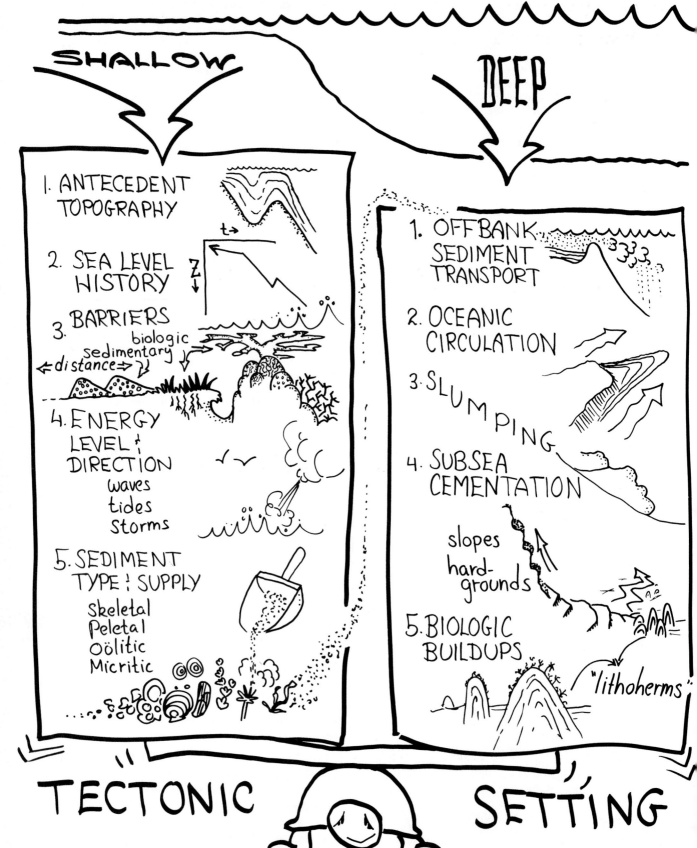

Figure 8.7. Schematic illustration of some variables affecting growth and related activities on shelf margin slopes of carbonate platforms. Drawing courtesy of A. C. Neumann.

FIELD QUESTIONS
Shelf Margin Slope

1. What are some of the methods by which shelf margin slopes may grow by accreting seaward? How would you go about documenting such growth in the field?

    Methods   **Documentation**

    a.

    b.

    c.

2. What are some of the methods by which shelf margin slopes may be cut back and reduced? How would you go about documenting such erosional processes in the field?

    Methods   **Documentation**

    a.

    b.

    c.

3. List the various processes involved in transporting large quantities of shelf-produced carbonate down or over the shelf margin slopes. How would you go about quantifying each of these processes in the field?

    Processes   **Techniques**

    a.

    b.

    c.

    d.

    e.

    f.

4. What would you look for in an ancient carbonate rock section to identify a fossil shelf margin slope?

Figure 9.1. Cross-bedded oolite of oolitic facies at Silver Bluff, South Bayshore Drive, Cocunut Grove, Miami.

9

PLEISTOCENE SHELF EDGE SAND BELT AND ADJACENT LAGOONAL ENVIRONMENTS

In addition to a wide variety of unconsolidated Holocene carbonate sediments, the south Florida-western Bahamas area provides excellent opportunities for the study of ancient indurated limestones which have been mapped and interpreted as to their origin and paleoenvironment. Such rocks are of particular interest to the carbonate student since their contained constituent particles are often identical to the Holocene sands and reefs described in earlier chapters of this Guidebook. In other words, by an examination of local bedrock, it is possible to see the indurated equivalents (including same species of organisms) of many present-day Florida-Bahaman environments.

Indurated carbonate rocks which represent Pleistocene age (a) shelf edge sand belts, and (b) lagoonal environments (adjacent to and landward of the later sand belts), are widespread in south Florida. Shelf edge sands are represented by the oolitic facies of the Miami Limestone formation (fig. 9.2). Indurated lagoonal carbonate rocks can be found directly underlying this oolitic facies and extending westward into the Everglades. This latter blanket sheet of lagoonal rock is called the bryozoan facies of the Miami Limestone formation.

Both of these facies are of particular interest as their modern counterparts can be seen and studied in the process of formation some 50 miles to the east along the western edge of the Bahamian platform.

MIAMI LIMESTONE

SHELF EDGE OOLITIC FACIES (Bayshore Drive, Coconut Grove, Florida): During the late Pleistocene, conditions were present for the development of unstable oolite sand belts just back from the outer edge of the Florida Platform. These deposits are now lithified and for the most part elevated above sea level. This fossil marine sand unit has been recently described and interpreted by Hoffmeister, *et al.,* (1967) as the oolitic facies of the Miami Limestone formation. This oolitic facies occupies large areas of what is called the Atlantic Coastal Ridge bordering the east coast from south of Boca Raton to a point southwest of Homestead (see figure 9.2). It also covers the Key Largo Limestone from Big Pine Key south to Key West in the southern Florida Keys.

Its maximum known thickness of 11 meters occurs on the mainland where it is found as a ridge (see cross section, figure 9.2) which thins abruptly to the east and gradually westward. One of the best exposures of this cross-bedded (dips up to 30°) oolitic rock is found along Bayshore Drive (see index map 1) in Coconut Grove .8 km south of Viscaya where it occurs as a wave-cut bench (Silver Bluff) about 2.5 m above mean sea level (fig. 9.1).

In addition to ooids, various amounts of shell debris and pellets are found in this oolitic facies. Many aragonite ooids have been replaced by calcite. Large areas of the rock found at or below sea level show an "oomoldic" texture of spherical and ellipsoidal cavities. Sinkholes are common features on top of this rock unit. Northwest-southeast trending channels, often filled with marl and quartzose sand, dissect the oolitic ridge found on the mainland, and similar trending tidal channels dissect this oolitic facies in the lower Florida Keys from Big Pine Key to Key West (fig. 9.3).

A modern equivalent of this Pleistocene oolitic rock can be found along the western edge of the Bahaman Platform in the Cat Cay area (see figures

9.4 and 5.15). Similarities between this Holocene oolitic belt and the Florida Pleistocene oolitic facies include composition, cross-bedding, transverse tidal channels, geometry and position on the edge of a platform. The northwest-southeast trending channels and overall change in shape of the Florida Keys south of Big Pine Key are due to the fact that this area was at one time the site of a submerged belt of oolitic sand similar to that found today at Cat Cay (see figures 9.3 and 5.15). For further details the reader is referred to Hoffmeister *et al.,* (1967), Robinson (1967) and White (1970). For a short illustrated discussion of characteristic features of drusy mosaic

growth in the Miami Limestone see Multer (1971b). Comments and illustrations concerning cementation and recrystalization of Miami Limestone ooids are noted by Multer (1971c). Halley (1976a) notes how the cemented ooid sand barrier bar of the Pleistocene Miami formation southwest of Miami could serve as a model for ancient ooid sand belt exploration. Halley *et al.,* (1976b) describe natural casts of microborings in ooids of the Miami Limestone and their process of formation, comparing them to endolithic algae borings found penetrating outside lamellae of Holocene ooids at Joulters Cays, Bahamas, today.

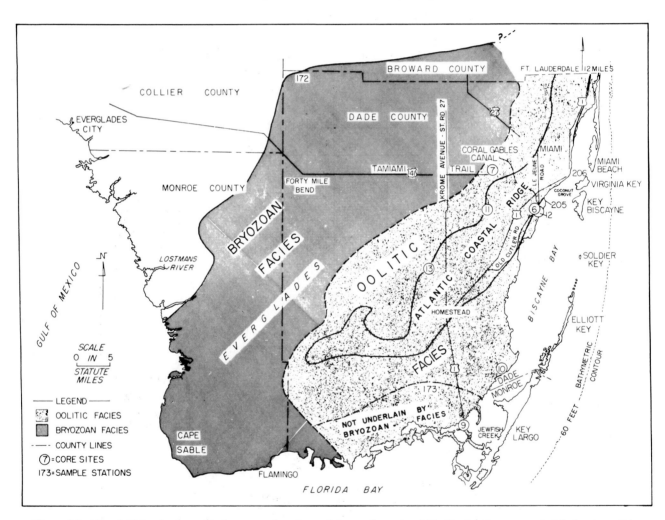

Figure 9.2. (*above*) Main physiographic features and geology of the southern tip of Florida showing area underlain by the oolitic facies and bryozoan facies of the Miami Limestone. The oolitic facies covers the bryozoan facies except for the southern border outlined. (*below*) Generalized east-west cross section of Miami Limestone, Florida, showing stratigraphic relationship between the two facies of the Miami Limestone and interfingering of the upperpart of the Key Largo coral reef limestone of the Florida Keys with the bryozoan facies of the Miami Limestone. From Hoffmeister *et al.,* 1967.

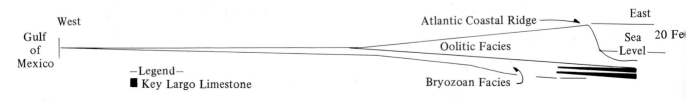

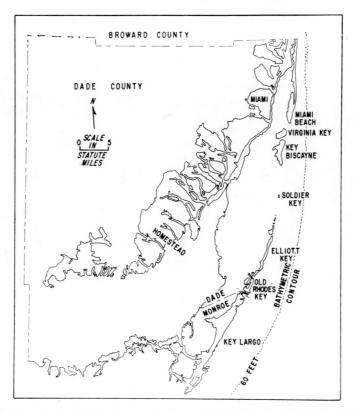

Figure 9.3. (*above*) Map showing transverse valleys or "glades" in the Atlantic Coastal Ridge. The valleys are comparable to the tidal channels of the unstable oolite ridge of the Bahama area (see figure 5.15). (*below*) Geologic map of the Lower Keys of Florida composed of the oolitic facies of the Miami Limestone. The shape and orientation of these Keys owe their origin to tidal currents which cut transverse channels in the unstable oolitic ridge which formerly occupied the area. (Both figures from Hoffmeister *et al.*, 1967.)

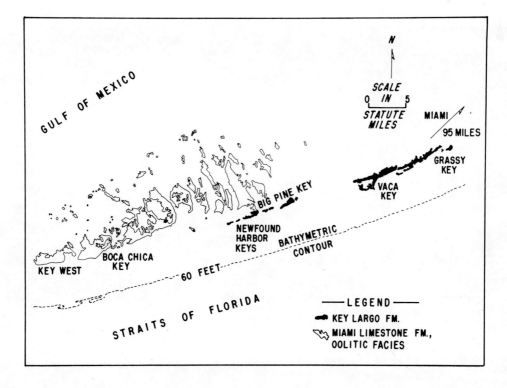

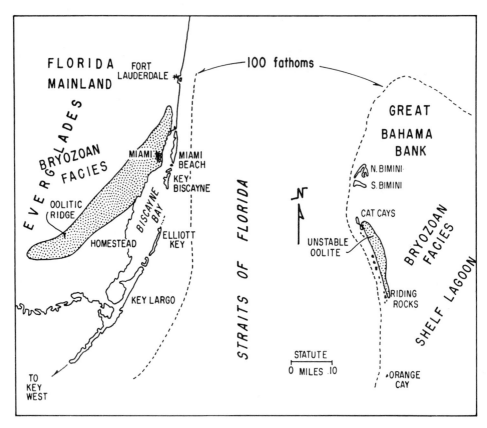

Figure 9.4. (*above*) Map of southeastern Florida and the northwestern section of the Great Bahama Bank showing the chief topographic and stratigraphic features of southern Florida and the mirror image relationship to their Recent Bahaman counterparts. From Hoffmeister *et al.*, (1967).

Figure 9.5. (*below*) Typical knobby specimen from bryozoan facies of Miami Limestone. Knobs and tubes commonly hollow. From Hoffmeister *et al.*, (1967).

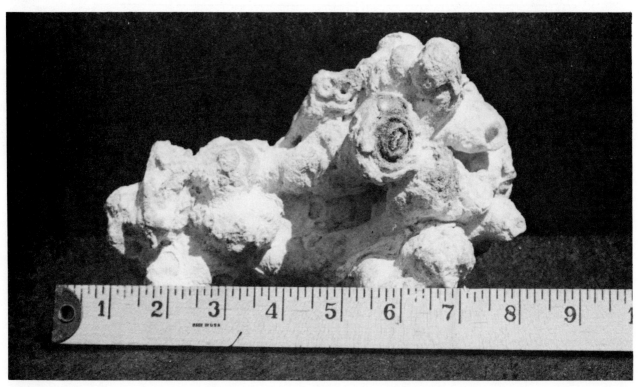

LAGOONAL BRYOZOAN FACIES (Old Cutler Road, Miami, Florida): Lying with a gradational contact directly beneath the oolitic facies described above is the bryozoan facies of the Miami Limestone formation. Good exposures can be found either south of Old Cutler Road (see index map 1) from dredged piles adjacent to canals or housing developments (the bryozoan facies is below sea level here) or north of the Atlantic Coastal ridge along dredged piles bordering drainage canals of the Everglades swamp. This blanketlike unit covers an area of over 5,200 sq km from the coast northward into the Everglades (see figure 9.2). It averages 3 m thick under the coastal ridge and thins westward. The most characteristic feature of this facies is the abundance (up to 70% by volume) of massive compound colonies of cheilostrom bryozoa belonging to a single species (*Schizoporella floridana* Osburn). See figure 9.5. Two growth forms have been identified by Hoffmeister *et al.,* (1967, p. 180) for this encrusting bryozoa, and as many as 46 separate laminae can be found around a central void which probably represents the former position of some marine grass. Mixed with the bryozoa are ooids, pellets, calcareous worm tubes and various amounts of skeletal sand. *Schizoporella floridana* can today be found in the Florida Bay area environment encrusting *Halimeda*, gorgonian fronds and *Thalassia* (fig. 9.6).

A Holocene environment having many of the characteristics of this Pleistocene lagoonal deposit can today be found behind the Cat Cay oolitic sand belt along the western edge of the Bahaman shelf (see figure 5.15). The pelletoid sand of this shelf lagoon contains *Schizoporella floridana*. For more details concerning both the Florida bryozoan limestone and its Bahaman counterpart see Hoffmeister *et al.,* (1967). See Steinker and Floyd (1975) for additional data on dredged samples containing *Schizoporella floridana* found in the lower Florida Keys.

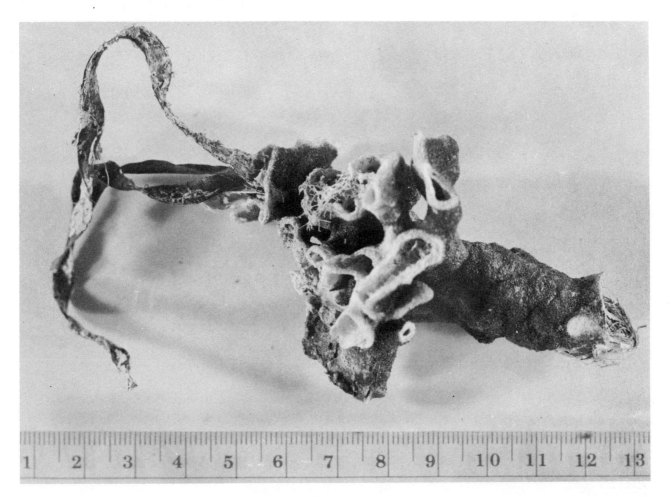

Figure 9.6. Living bryozoan (*Schizoporella floridana*) encrusting *Thalassia* in 1 meter of water, southwest of Crane Key, Florida Bay.

FIELD QUESTIONS
Pleistocene Shelf Edge Sand and Lagoonal Rocks

SOUTH FLORIDA AREA

1. What are the paleoenvironmental implications of such a widespread (5,200 square km area) single species bryozoan-rich limestone?

2. What is the source of the quartz found as nuclei in some of the ooids of the oolitic facies in the Miami area?

3. What are the criteria for differentiating between submarine and aeolian origins for oolitic sand such as found in the Miami Formation? List criteria below:

    a.   a.

    b.   b.

    c.   c.

    d.   d.

4. Why is oolitic sand not found forming today along the shelf edge of the Florida Platform as it did in the Pleistocene and as it is doing today along the Bahaman shelf some 80 km away?

5. What is the role of encrusting bryozoans today? List examples of what they encrust.

Figure 10.1. Back-reef facies of the Key Largo (Pleistocene) Limestone as seen exposed along the face of the Windley Key quarry wall. Corals of this 100,000 year-old-rock unit can be observed here in growth position. Note *Diploria* coral below hammer. Dense, subaerial, laminated crusts as well as karst breccia fillings can also be found along the top of the quarry walls.

10

PLEISTOCENE BACK-REEF ENVIRONMENTS

PLEISTOCENE ENVIRONMENTAL SETTING—FLORIDA KEYS

The Pleistocene depositional setting during which various Key Largo facies evolved (including the back-reef facies shown in figure 10.1) is described in a paper by Hoffmeister and Multer (1968). See also figure 10.2. The abstract, conclusions and some pertinent illustrations from this paper are noted below.

GEOLOGY AND ORIGIN OF THE FLORIDA KEYS
by
J. E. Hoffmeister and H. G. Multer

Abstract

The Florida Keys form a crescentic chain of small limestone islands which extend from near Miami to Key West, a distance of about 150 miles. They are made of two main formations of Pleistocene age—the Key Largo Limestone and the Miami Limestone. The former, named and described by Sanford, is an elevated coral reef rock, and the latter, also described by Sanford, is an oolitic limestone in this area. The Key Largo Limestone is the surface rock of the Upper Keys, and the Miami Limestone covers the Lower Keys. A contact found at Big Pine Key shows that the oolitic limestone overlaps the Key Largo, and core borings show that the Key Largo underlies the oolitic cover for the entire area of the Lower Keys. The Key Largo, therefore, extends for the total length of the Keys. A core placed at Dry Tortugas, 70 miles west of Key West, encountered it at 30 feet below sea level. The Key Largo has a maximum thickness of over 200 feet.

The fact that the Keys are located about 5 miles from the seaward edge of the Florida coral reef platform creates an interesting problem. They may be (1) remnants of an ancient outer reef similar to that found today bordering the living coral reef tract, or (2) they may have become established as a line of patch reefs in the back-reef zone of a preexisting platform which extended seaward some distance, possibly as far as the present platform. It is believed that the answer to this problem can be found through an understanding of what was happening seaward and leeward of the old reef at the time its corals were flourishing. Consequently, the composition and structure of the reef platform seaward of the Keys and the geologic history of the rocks of the Florida mainland immediately behind the Keys were studied. Information thus obtained plus observations on the composition, especially the coral content, of the Keys themselves have led the authors to believe that the Keys were formed as a line of patch reefs in a back-reef area which was bordered on its seaward edge by an outer reef which has since been lowered, chiefly by erosion, and covered by more recent material.

Conclusions

During the last interglacial period, about 95,000 years ago (Broecker and Thurber, 1965, p. 58-60), the coral reefs which today make up the Key Largo Limestone of the Florida Keys were flourishing as a line of patch reefs in the back-reef area of a broad-reef platform similar to the Florida reef tract of today. Seaward of them the platform was occupied by parallel lines of other patch reefs and edged by an outer reef (fig. 10.3).

Subsequent marine and subaerial erosion following the withdrawal of the sea during the Wisconsin, possibly accompanied by a structural downward tilting or faulting of the area, or both, resulted in the lowering of the platform to a depth of about 75 feet at its seaward edge and progressively less farther inland. With the return of the sea, new reef growth began on the eroded platform and continued to the present (fig. 10.4).

Some of the main observations upon which it is concluded that the Key Largo Limestone of the Keys originated as a line of patch reef in a back-reef environment are the following. (1) The species of corals and other organisms found fossil in the old reef are identical with those of the living patch reefs. (2) There is an absence of *Acropora palmata*, a coral species found commonly in the turbulent waters of living outer reefs. (3) There is a favorable proportion of other species common in all zones. (4) Since the community of coral species found in the Key Largo requires an environment of low-level energy and since it was determined that the water in which the corals grew was shallow, it becomes clear that they must have developed in the protected area of a back-reef zone.

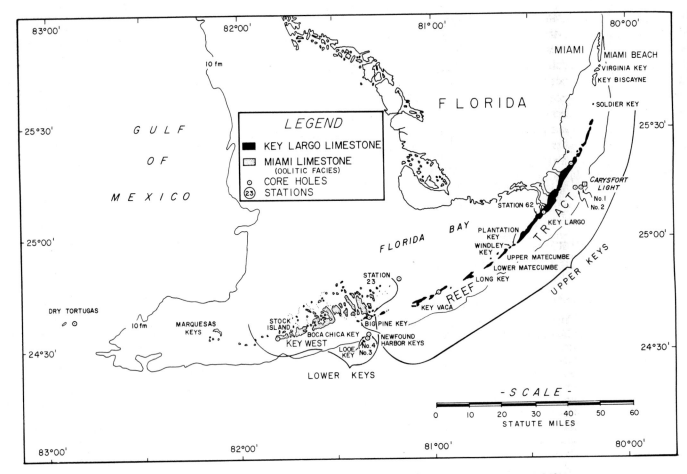

Figure 10.2. Main features and geology of the Florida Keys. From Hoffmeister and Multer (1968b).

(5) The discovery of *Acropora palmata* in the Key Largo Limestone, underlying the most recent material of the outer reef, demonstrates its presence at the time the Keys were being formed and indicates that its absence in the rocks of the Keys is due to the unfavorable environment of a back reef. Its presence also indicates the existence of an outer reef at the edge of the platform during Key Largo time. (6) The elongated crescentic shape of the Keys, running approximately parallel to the outer edge of the platform, is similar to the alignment of present-day patch reefs.

See Landon (1975) for discussion of growth rates of *Montastrea annularis* and *Siderastrea siderea* from the Key Largo Limestone based on x-radiographs of coral skeletons. She found growth rates of *S. siderea* and *M. annularis* to be less than half those recorded for present-day living reef forms and annual band thicknesses for both corals showing low standard deviation. Landon (p. 50) interpreted this slower growth rate as indication of less favorable Pleistocene environment for calcification and the low standard deviation as indicative of fewer environmental fluctuations. See figure 10.5 for x-radiograph of *M.*

annularis from the Key Largo Limestone and compare it with figure 7.5 showing x-radiograph of same (living) species from the Florida reef tract today. To the writers knowledge, these are the first published x-radiographs documenting such variations.

GUIDE TO KEY LARGO LIMESTONE OUTCROPS

A summary of additional information (published earlier than the above described paper) which may be of particular interest to visiting geologists was published by Hoffmeister and Multer (1964b) in the Guidebook 1 for the 1964 G.S.A. annual meeting held in Miami. The text of most of this discussion is quoted below.

The Key Largo is an elevated coral reef of Pleistocene age. Its horizontal extent is now fairly well known by aerial examination and core borings. It underlies Miami Beach in the north, comes to the surface at Soldier Key and is again submerged beneath the Miami Oolite from Big Pine through Key West. In addition it has recently been found a few feet beneath sea level along the eastern shore of the Florida mainland from Miami southward for at least 40 miles.

It varies considerably in thickness throughout. At Key West and Big Pine Key it is at least 180 feet thick, at Grassy Key 170 feet, at middle Key Largo 70 feet, and at the northern tip of Key Largo 145 feet. Wherever its base has been located it rests on an unconsolidated quartz and calcareous sand.

Its composition is that of a typical coral reef with large, massive coral heads, many in place, surrounded by smaller coral colonies, shells and shell fragments of all sizes of common marine organisms. Reef building corals are found from top to near the bottom of the formation, but, in general, are more prolific in the upper two-thirds than in the lower third. An indurated calcarenite of varied organic components is probably the most important rock by volume.

Probably the dominant coral species of the formation are *Montastrea annularis, Diploria clivosa, D. Strigosa, D. labyrinthiformis* and several species of *Porites.* In addition *Acropora cervicornis* is prolific at several localities. Practically all the coral species found living today on the Florida reef tract can be recognized in the Key Largo. One notable exception is *Acropora palmata.* This species, commonly known as the Elkhorn coral, is one of the most prolific in the living reefs and as yet has never been located in the Key Largo.

Author's Note: The front cover of this Guidebook illustrates a living back-reef environment which is readily compared to a Pleistocene Key Largo counterpart displayed on the back cover (same coral species and sand composition in both photos). The fossil coral photo on the back cover is taken along the Key Largo canal cut which is the second most exposed section known of the Key Largo Formation.

The upper part of the Key Largo can be seen in many cuts, canals, and boat slips found throughout the Keys. One of the best exposures is on Key Largo—where a canal has been cut entirely through the Key from the Atlantic side to the Bay of Florida exposing a maximum thickness of about 10 feet of the formation. Certainly one of the most striking features here is the presence of several huge colonies of *Montastrea annularis* which have been bisected exposing the internal structure of the specimens. These are undoubtedly in their original position of growth.

One of the best preserved of the colonies stands 7 feet high and has a branch which measures 8 feet in length. It is interesting to estimate the time required to grow a specimen of this size. The writers for several years have been conducting a series of growth-rate experiments on Florida corals and have determined that the average upper growth of *Montastrea annularis* amounts to 10.7 mm a year (1 foot in about 28.5 years). On this basis a specimen which has a branch 8 feet in length and which stands 7 feet high would take about 230 years to be produced. The next question concerns the time required to produce the thickness of 7 feet of coral reef material, including the coral. On the basis of the fact that corals, particularly large branching types, are relatively easily overturned by wave action, and destroyed by boring organisms after death, a period of 250 to 500 years has been estimated.

A quarry of the coral reef rock on Windley Key offers probably the best opportunity to examine the material at close range. The great bulk of the Key Largo Limestone is greatly altered and recrystallized. Some excellent specimens of well-preserved corals can be found here. From these it has been determined, on the basis of the Thorium-Uranium ratio that the apparent age of the upper part of the Key Largo Formation is about 100,000 years.

One of the most interesting types of lithology of the Key Largo is what has been called for want of a better name "holey limestone." This rock displays an unusual framework structure in which numerous large and irregularly shaped holes, which comprise 40 to 60 percent of the total volume, are present. The rock is found chiefly a few feet below sea level and is brought to the surface in large quantities by dredges engaged in making cuts for boat slips and canals.

The origin of this rock has posed a difficult problem. Although it is believed to have been formed in more than one way, it is now known that the accumulation of tremendous amounts of fragments of a thin branched *Porites* accounts for much of it.

The stratigraphic relation between the Key Largo Limestone and Miami Oolite can be seen at a contact at the southeastern point of Big Pine Key. Here the Miami Oolite gently overlaps the old coral reef in a southern direction. The contact appears to be of a transitional character. No other surface contact has been seen in the lower Keys. However it is known that the oolite cover of these Keys is relatively thin, particularly along their southern borders. For example at Boca Chica and Stock Island it has a thickness of only 3 to 6 feet. The oolite cover appears to thicken gradually to the north; at the center of Key West it reaches 20 feet.

Most writers have considered the Key Largo coral reef as having formed entirely during the Sangamon interglacial period of the Pleistocene. Parker, *et al.,* (1955), however, are inclined to the belief that it began its formation during the Aftonian and was added to during Yarmouth and Sangamon times. An unconformity found near sea level on Key Largo strengthens the possibility of a prolonged history.

Papers not discussed in this chapter that do deal with aspects of the Key Largo Limestone include Cooke (1929), Ginsburg (1953c), Hoffmeister and Multer (1964a) and Dodd *et al.,* (1973—see page 179 of this Guidebook). The stratigraphy and geologic history of the Key Largo Limestone and Miami Limestone are summarized by DuBar (1974, p. 227-231).

PETROLOGY AND SIGNIFICANCE OF THE KEY LARGO LIMESTONE

The Key Largo Limestone most frequently seen (and discussed) is that found cropping out along the emergent Florida Keys. However, the reader should be aware that the Key Largo Formation has been found by the writer (drilling from a floating platform over the reef tract and Florida Bay areas) to occupy an area up to 20 km or more wide and at least 376 km long. A study of cores from these wells is being conducted by the author. Some data, based on early

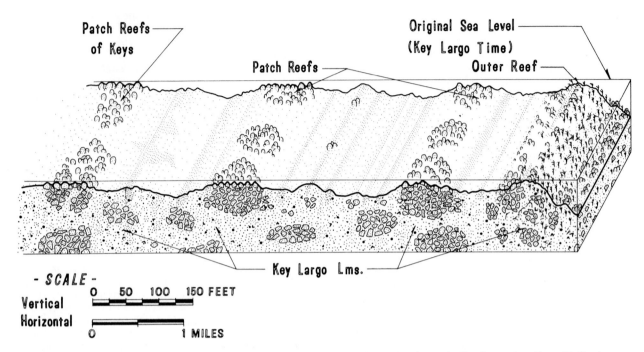

Patch Reefs of Keys

Patch Reefs

Original Sea Level (Key Largo Time)

Outer Reef

Key Largo Lms.

- SCALE -

Vertical

Horizontal

| O | 50 | 100 | 150 FEET |

O   1 MILES

Figure 10.3. Generalized block diagram showing conditions across Florida coral-reef tract in Key Largo time. From Hoffmeister and Multer (1968b).

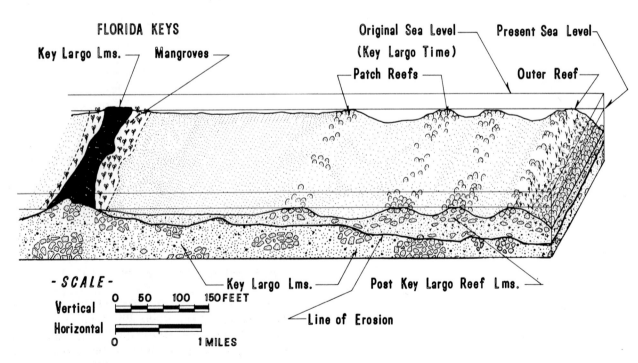

FLORIDA KEYS

Key Largo Lms. — **Mangroves**

Original Sea Level (Key Largo Time)

Present Sea Level

Patch Reefs

Outer Reef

Key Largo Lms.

Post Key Largo Reef Lms.

Line of Erosion

- SCALE -

Vertical

Horizontal

| O | 50 | 100 | 150 FEET |

O   1 MILES

Figure 10.4. Generalized block diagram showing conditions across Florida coral-reef tract at present time. From Hoffmeister and Multer (1968b).

Figure 10.5. X-radiograph of *Montastrea annularis* from the Key Largo (Pleistocene) Limestone collected from dredged material at Tavernier Creek south of U.S. Route 1 at east end of Plantation Key. X 1. From Landon (1975). Compare with figure 7.5.

analysis of core data, is noted in the abstract presented below by Multer and Hoffmeister (1968b).

PETROLOGY AND SIGNIFICANCE OF THE KEY LARGO (PLEISTOCENE) LIMESTONE, FLORIDA KEYS
by
H. G. Multer and J. E. Hoffmeister

The study of exposed bedrock and 13 strategically placed diamond core borings, drilled both offshore and along the emergent Florida Keys, demonstrates that the Key Largo (Pleistocene) Formation extends in a 235-mile, apparently continuous band from Dry Tortugas to North Miami Beach and that it can be divided petrologically into three major facies.

Thin sections and polished slabs permit distinction of the following facies: (1) an outer reef facies (2 to 4 miles seaward of the present Keys) containing 4 common rock types, including encrustate *Acropora palmata* boundstone, (2) a back-reef facies (2 miles seaward and approximately 1 mile lagoonward of the Keys) containing 6 common rock types, and (3) a lagoonal facies in the approximate site of modern Florida Bay containing five common rock types.

Vertical persistence of major facies and similarity of each with overlying Holocene sediments indicate a general continuity of environments for at least the last 100,000 years. The Key Largo Limestone provided a major control for late Pleistocene sedimentation in southern Florida. The area can be used today as a model for both modern and ancient shelf-edge carbonate deposition.

BACK-REEF FACIES OF THE KEY LARGO LIMESTONE (Area Along the Emergent Florida Keys):

The thickness of the Key Largo beneath the emergent Keys as determined by widely spaced core borings varies from 20 to over 50 meters. An examination of such cores as well as overlying surface exposures has led the writer to classify the Key Largo found along the emergent Florida Keys into six common rock types which together characterize the back-reef facies of the Key Largo Formation. These six

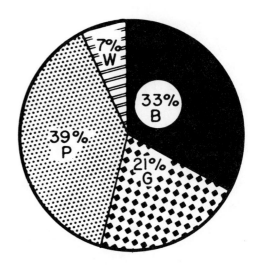

B = BOUNDSTONE
G = GRAINSTONE
P = PACKSTONE
W = WACKESTONE

Figure 10.6. A summary of the relative abundance of rock types along the emergent Keys (back-reef facies) obtained by averaging the data from six cored wells taken along the emergent Keys (see figure 10.2). Coralline boundstones were found to be most abundant in the upper portions of the formation with the deepest coral zone found at a 45 m depth below Big Pine Key.

rock types (rock terms used are after Dunham, 1962) are listed below. See figure 10.6 for a graphic summary.

Back-Reef Facies

1. Incrustate Coral Boundstone
2. Algal-pelletoid Grainstone
3. Mollusk Grainstone
4. Coral-mollusk Packstone
5. Mollusk Packstone
6. Foraminiferal-mollusk Wackestone

The best exposure of this back-reef facies is at the Windley Key quarry (just north of Route 1 on the high part of the Key) where many quarry walls (fig. 10.1) provide excellent exposures of various rock types. See Pasley (1972) and figure 10.7 for the results of a species distribution survey of the Key Largo Limestone along the south wall of this Windley Key quarry.

A detailed account of this quarry and related diagenetic processes is given by Stanley (1966) who describes (p. 1939) the Key Largo of the exposed Keys as a former deep open water reef rather than a series of coalescing patch reefs growing in back of a seaward barrier. Since Stanley's paper, however, drilling of 12 core holes by the writer as well as other data (see Hoffmeister and Multer, 1968) has provided evidence indicating that the emergent Florida Keys do represent a rather shallow Pleistocene back-reef environment similar to the series of patch reefs which can be seen today growing within a mile seaward of the Keys. In other words, evidence to date seems to indicate that if today's sea level were

lowered, exposing offshore living patches until they stood some 17 feet above sea level (as at Windley Key quarry) and the tremendous land-producing activities of mangrove and man were allowed to take place, one would have another series of emergent Keys similar to those connected by U.S. Route 1 today.

"SHELL AND MANGROVE FACIES" (Key Largo Formation): Bain and Teeter (1975) describe two carbonate facies associated with the upper part of the Key Largo Limestone on Key Largo, Florida. The data as presented would appear to indicate the possibility of lagoonal conditions during the interval described. This is certainly a possibility, consistent with the known transgressive/regressive nature of the Pleistocene shoreline. However, the writer (who has reviewed the complete paper for the authors) has indicated that rock units similar to what they describe as "mangrove facies" have also been found to be composed of *Porites* coral and/or formed by the removal of less resistant carbonate surrounding indurated burrow tubes. (See discussion of "holey limestone" on p. 261 of this Guidebook.) In any case, this would not necessarily change the significance of the paleoenvironmental interpretation by Bain and Teeter. The abstract of the paper by Bain and Teeter is given below; illustrations from their paper are noted in figure 10.8.

PREVIOUSLY UNDESCRIBED CARBONATE DEPOSITS ON KEY LARGO, FLORIDA
by
R. J. Bain and J. W. Teeter

Recent dredgings on Key Largo have exposed previously undescribed deposits within and possibly above the Key

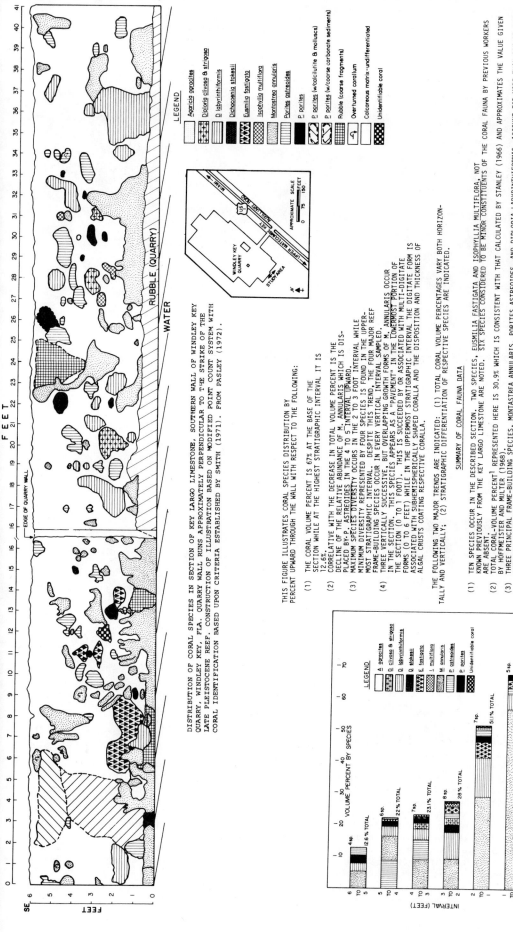

Figure 10.7. Distribution of coral species in a section of the Key Largo Limestone, southern wall of Windley Key quarry. From Pasley (1972).

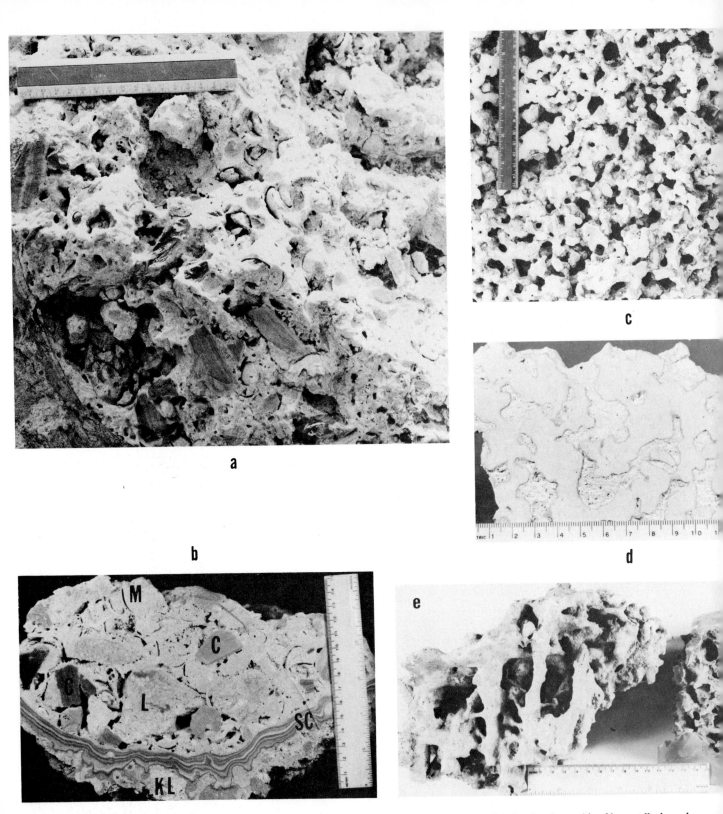

Figure 10.8. Carbonate facies described by Bain and Teeter (1975). (a) *Chione cancellata* shell facies showing molds of bay mollusks and clasts of subaerial crust with chaotic texture. Key Largo Canal cut bay side float from new marina dredgings. (b) Slabbed block of *Chione cancellata* shell facies showing weathered Key Largo Limestone (KL) beneath subaerial laminated crust (SC). Conglomeratic deposit overlying the crust contains molds of *Chione cancellata* (M) and clasts of subaerial crust (C) and Key Largo Limestone (L) all in chaotic arrangement. Note crust and conglomerate occupy a depression in underlying limestone. (c) Dredged block of black mangrove root facies. View parallel to bedding surface and perpendicular to vertically oriented limestone elements. Note circular voids which mangrove roots once occupied and dark rind on margins of limestone. (d) Slabbed sample of black mangrove root facies with soft sediment infilling between well-lithified vertical limestone elements. Note dark rind on margins of well-cemented limestone elements. (e) Sample of mangrove reef rock (Hoffmeister and Multer, 1965) from Key Biscayne on left and bioplastic model of voids within the black mangrove root facies on right. Note similarities in scale, pattern of dominant vertical elements and branching smaller horizontal elements.

Largo Limestone. Associated with the frequently discussed *Montastrea-Diploria* coral patch reef facies are a *Chione cancellata* shell facies and a black mangrove root facies.

The *Chione cancellata* shell facies is characterized by a molluscan fauna, typical of present-day Florida Bay, consisting predominantly of molds of *Chione cancellata*. It occurs as a well-lithified quartzose biocalcirudite containing fragments of subaerial crust which it overlies. The *Chione cancellata* deposit represents a storm accumulation either: (1) deposited on a lower subaerial crust within the Key Largo Limestone during a lower stand of sea level or (2) the most recent deposit on Key Largo postdating the latest subaerial crust. The nature of the preservation indicates subaerial leaching after deposition.

The black mangrove root facies consists of a grey biocalcarenite with resistant vertical elements separated by roughly circular voids which are commonly filled with loosely consolidated sediment. The calcarenite contains a fossil fauna characteristic of modern Florida Bay and is believed to have been deposited between the roots of a black mangrove thicket. Later decay of plant roots caused the cementation of the surrounding calcareous sediment and created voids within the deposit which were later partially filled by loose carbonate sediment.

Steinker and Floyd (1975) present a published "Comment" relative to the above paper by Bain and Teeter indicating additional pertinent field observations. They also "disagree with the designation of the *Chione cancellata* shell-mold deposits and the black mangrove root-mold deposits as facies." Referring to a paper by Steinker and Steinker (1972, p. 52) they note that "*Stratigraphic facies* refer to different aspects of rocks of essentially the same age."

REAPPRAISAL OF PLEISTOCENE STRATIGRAPHY OF SOUTH FLORIDA

The Pleistocene stratigraphy of southern Florida appears (fortunately!) to be entering a new era. Although the recognition of unconformities in the Pleistocene of south Florida has been known for many years (i.e., Parker and Cooke, 1948, figure 4, p. 89), the significance and correlation of these surfaces and the intervening "packages" of sediments have not always been clear. For those interested in the Pleistocene of the Florida Keys a clear *regional* picture including these disconformable surfaces is necessary.

A reappraisal of the Pleistocene of southern Florida was held at a conference in 1968 bringing together prominent workers for discussion and re-evaluation of field evidence. Persons interested should read the Guidebook for this conference edited by Perkins (1968). One worker, for example, indicated six marine units (7 including the Silver Bluff) with disconformities between them (see Brooks, 1968 and 1974).

Results of the above reappraisal conference together with imaginative regional field studies such as those of Brooks (noted above) and Perkins (noted below) and new techniques such as described by Mitterer (noted below) are all providing a new and welcomed framework for south Florida stratigraphic studies.

Perkins (1974) describes the regional recognition of five discontinuity surfaces based on data from a large number of drilled core holes with excellent recovery. These wells cover an area from Lake Okeechobee to Big Pine Key in the Florida Keys. The abstract of his study is given below. The complete paper will be published *in* Enos and Perkins (1977-in press).

DISCONTINUITY SURFACES AS A STRATIGRAPHIC TOOL: THE PLEISTOCENE OF SOUTH FLORIDA
by
R. D. Perkins

Detailed stratigraphic analysis indicates that the Pleistocene rocks of south Florida contain evidence for subaerial exposure along well-defined discontinuity surfaces. These discontinuity surfaces are often found to be intraformational when related to formal stratigraphic units. Stratigraphic units delineated by regional discontinuity surfaces are proposed as an alternative approach to more clearly understand the depositional history of these rocks. Discontinuity surfaces separate the Pleistocene into five major marine units. Marine units are correlated with eustatic high sea level stands and discontinuity surfaces with subaerial exposure during low sea level stands. Evidence for subaerial exposure includes: laminated crusts, paleosols, root structures, vadose sediment, bored surfaces, corrosion zones, and freshwater limestones.

Each marine unit so delineated records a lower transgressive sequence grading upward into a regressive sequence capped by a discontinuity surface. Paleotopography, revealed by contouring discontinuity surfaces, strongly influenced isopach thickness and lithofacies patterns within individual units. An integration of lithofacies with ecologic facies provides the basis for reconstruction of depositional environments for each marine unit. Subdivision of the Pleistocene of south Florida into workable stratigraphic units on the basis of discontinuity surfaces, suggests that such an approach may be useful in the study of other eustatically controlled sedimentary sequences.

Mitterer (1974) describes a promising new application involving diagenetic alteration of amino acids in fossil shells for use in Pleistocene stratigraphic analysis. He indicates that this method can be used to identify and correlate Pleistocene deposits of south Florida where he has found evidence for seven interglacial cycles. Each marine unit, according to Mitterer, was deposited during a different interglacial event, and each is separated by unconformities from adjacent units. Mitterer also indicates (personal communication) that Lake Okeechobee is the most southerly point he has been able to sample to date although the technique is applicable to the Florida Keys. The abstract of Mitterer's (1974) paper is noted below.

PLEISTOCENE STRATIGRAPHY IN SOUTHERN FLORIDA BASED ON AMINO ACID DIAGENESIS IN FOSSIL *MERCENARIA*
by
R. M. Mitterer

Diagenetic conversion of isoleucine to alloisoleucine in shells of *Mercenaria* provides a quantitative geochemical aid for interpreting the Pleistocene sequence in southern Florida. At least seven different cycles of marine sedimentation, reflecting high sea levels, and seven unconformities, indicating low sea levels, are recorded. The technique can be applied to other Pleistocene deposits in the Atlantic Coastal Plain and could be a useful adjunct in Quaternary biostratigraphic interpretations. In addition, the alloisoleucine content provides an independent check on the validity of radiocarbon dates.

Author's Note: The possibilities of a new "stratigraphic framework" as described above by various workers is particularly appreciated by those of us who are involved with the Pleistocene on the Florida Keys edge of the Florida Platform. For example, the numerous successive regional transgressions/regressions described above could be employed in explaining the evidence for a lagoonal environment within the Key Largo Limestone described by Bain and Teeter (1974, see page 264 of this Guidebook). Likewise, many subsurface features previously observed by the writer in the Key Largo Limestone of the Florida Keys area become more tenable. Included in such would be the subsurface subaerial laminated crusts and root cast zones (Multer and Hoffmeister, 1968a, illustrated in figure 11.6 of this Guidebook), solution zones, and changes in mineralogy (calcite changing abruptly upward into aragonite zones similar to those core changes described by Gross and Tracey, 1966, and Schlanger, 1963, in Pacific island cores). These have all been found by this writer in the subsurface of the Key Largo Limestone.

In other words, what has previously been only locally described and suspected can now be regionally documented and confirmed. Many thanks are due such workers as Brooks, Mitterer and Perkins for their efforts.

FIELD QUESTIONS
Pleistocene Back-Reef Bedrock

WINDLEY KEY QUARRY (Windley Key):

1. What percentage of the bedrock is coral?

2. What are the chief types of coral present?

3. What types of coral are missing and of what significance is this?

4. What percentage of the coral is essentially whole and in growth position? What does the lack of coral in growth position signify as to original depositional conditions and rate of sedimentation?

5. What percentage of the coral is original aragonite and what percentage has changed to drusy calcite?

6. What are the other common (noncoral) rock constituents?

7. How many different kinds of carbonate rocks can you find in this quarry? Classify according to:

               Folk   **Dunham**

   a.

   b.

   c.

   d.

   e.

   f.

Sample No.

Photo No.

8. Are there any unconformities or is there evidence of emergence in this quarry?

9. Do you see evidence of "storm layers" or ancient hurricanes?

10. Locate an "average" type of back-reef limestone and make a detailed sketch below of a 2 m × 2 m wall indicating all constituents.

2 m

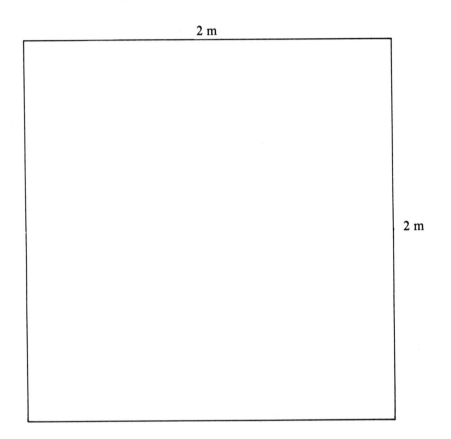

2 m

11. Is there much micrite present in this quarry? If so, what is its source?

12. How does this Pleistocene (100,000 years old) limestone compare with more ancient (Paleozoic) equivalents that you have examined as to:

Porosity/Permeability—

Color—

Preservation of fossils and structures—

Diagenesis—

cementation

interstitial solution

interstitial filling

replacement

13. Can you observe any geopetal structures?

14. Estimate, using method employed by Hoffmeister and Multer (1964a), the time for the exposed rock in this quarry to have formed.

15. How efficient is the solution-reprecipitation process of aragonite going to low-magnesium calcite . . . is the efficiency greater than 90 percent as reported by Harris and Matthews (1968) for the emergent Pleistocene coral reef rock in Barbados? Check to see if there is more drusy calcite in the older quarry faces than in the freshly cut quarry walls.

16. Shinn (1968b) has described birdseye structure found in both ancient and recent sediments as indicators of supratidal and intertidal environments. The muddy carbonate tailings that spill out on this Windley Key quarry floor were used by Shinn who said that bubblelike vugs were made by gas bubbles and planar vugs were made by skrinkage resulting from desiccation of exposed sediments. These vugs tend to resist compaction and hence are readily preserved. Cut a clean face in this leathery mud and look for vugs. Do you see any similar vugs in the Key Largo? Have you seen similar vugs in ancient carbonates? Sketch below the types of vugs and bubbles you see in the Key Largo and in the mud on the quarry floor.

17. All of the Pleistocene bedrock in this quarry has gone through at least one period of being exposed to a littoral zone. The upper surface of the quarry has been subjected to more than 10,000 years of subaerial action. Do you see evidence for, or effects of, the following processes:

    a. littoral physical/chemical processes:

    b. littoral biological processes:

    c. subaerial processes:

KEY LARGO CANAL CUT (Key Largo—See Index Maps 3 and 5):

1. What is the percentage of coral in this canal cut? Does the percentage change as one walks along the canal (across the strike or elongate trend of the reef zone)?

2. What is the origin and relative age of the red cavity fillings in the bedrock along the top of the canal cut?

3. Is there an overall seaward slope or any topographical configuration on top of the Key Largo Limestone along this canal cut? If so, what is the origin of such configuration?

Sample No.

Photo No.

Figure 11.1. Subaerial crusts in the Florida Keys. From Multer and Hoffmeister (1968a).

11

LATE PLEISTOCENE SUBAERIAL FEATURES

Exposed Pleistocene bedrock surfaces in the Florida Keys-Western Bahamas area today display many results of subaerial action (solution, reprecipitation, soil formation, storm action, etc.). These phenomena are relatively recent, having evolved after the elevation of the Pleistocene bedrock above sea level. Three of these features are (1) subaerially formed crusts, (2) solution hole breccias, and (3) the influence of karst topography on sedimentation and biota. These phenomena are of particular importance to the student of ancient carbonate rocks as their proper identification in ancient rock sequences could indicate the presence of subaerial conditions, unconformities and clues to ancient climatic conditions as well as sea level fluctuations.

SUBAERIAL CRUSTS

The abstract and text of a paper by Multer and Hoffmeister (1968) concerning subaerially formed crusts (fig. 11.1) and their implication is given below.

SUBAERIAL LAMINATED CRUSTS OF THE FLORIDA KEYS
by
H. G. Multer and J. E. Hoffmeister

Abstract

Exposed Pleistocene marine limestones of the Florida Keys are often coated by laminated 1-to-6-cm-thick calcitic crusts. Heretofore these crusts have locally been identified as indurated marine algal stromatolites similar to the soft, marine, living algal stromatolitic mats of the Florida Keys, which border and occasionally even coat the encrusted bedrock; such juxtaposition is now considered merely coincidental.

C^{14} dating of five different crust samples reveals a time of formation (within the last 4395 ± 90 years) during which the land surface was above sea level. Field relationships and laboratory evidence also indicate subaerial origin. Three general types of crusts are: (1) microcrystalline rind, (2) dense laminated, and (3) porous laminated.

Similar laminated crusts found in subsurface cores suggest emergence followed by submergence of the Key Largo reef in late Pleistocene time.

Proper identification of such subaerially formed laminated crusts, to distinguish them from similar-appearing crusts formed in marine environments, is necessary for correct interpretation of paleoenvironments and former sea level fluctuations. Thin crusts may be the only evidence for recognizing some ancient unconformities.

Introduction

Exposed Pleistocene limestones of the Florida Keys, the Key Largo Limestone, and the Miami Oolite are in many places coated by laminated 1-to-6-cm-thick calcitic crusts and rinds which conform to the irregularly weathered limestone surface (fig. 11.1). Such crusts occur from Soldier Key in the north to Key West in the south (fig. 11.2), and heretofore have locally been identified as indurated marine algal stromatolites similar to the soft, marine, living algal stromatolitic mats (fig. 11.3) which border and occasionally even coat the encrusted bedrock. The primary purpose of this paper is to describe the crusts in some detail and to demonstrate that although they occasionally underlie and often resemble the marine algal stromatolites of the intertidal zone, they are in fact, the result of subaerial weathering, often closely associated with organically rich soils. Therefore, the juxtaposition of living algal stromatolitic mats and indurated crusts examined by the authors is now considered merely coincidental.

The ability to recognize evidence of relatively short periods of subaerial exposure within a normal marine sequence of sedimentary rocks is of obvious importance to the geologists. The second purpose of this paper, therefore, is to call attention to the application of the above-mentioned subaerial crusts in the identification of Pleistocene sea level fluctuations in late Key Largo time resulting

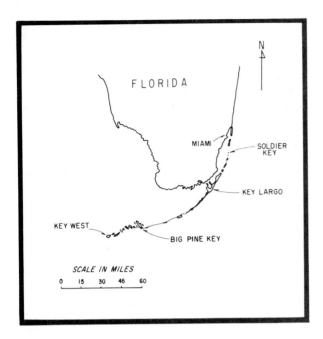

Figure 11.2. Index map of the Florida Keys. From Multer and Hoffmeister (1968a).

Figure 11.3. Soft, algal, stromatolitic mats over supratidal mud, Cluet Key, Florida Bay. From Multer and Hoffmeister (1968a).

in alternating emergence and submergence of the coral reef rock. Such emergence is identified by subsurface core samples which reveal a laminated crust layer similar to the subaerially-formed crusts coating the surface of the Key Largo Formation.

Although both surface and subsurface indurated Pleistocene crust samples described in this paper are relatively young, their identification and principles of formation are believed to have broad universal application to rocks of all ages.

Samples of the crust were collected throughout the Florida Keys, and more than 50 thin sections and 100 polished sections were studied by the senior author. In addition, laboratory experiments were conducted which resulted in the artificial production of laminated crusts similar to those in nature.

Of the previous workers in the Florida Keys, Sanford (1909) was the first to describe the general geology of the bedrock. R. N. Ginsburg has published a series of papers dealing with different aspects of the geology of the Keys including excellent descriptions of the soft algal mats (1954-unpublished; 1955; 1960). Hoffmeister and Multer (1961; 1964) determined the extent and thickness of the Key Largo Limestone and its stratigraphic relationship to the Miami Oolite, and estimated its original rate of sedimentation at one locality. The present paper represents the first published work attempting to describe the indurated crusts of the Keys and their origin.

Acknowledgments

Appreciation must be expressed to the following persons who aided and encouraged this investigation. R. N. Ginsburg checked thin sections and, along with R. J. Dunham, provided stimulating discussions. C. O. Oppenheimer and F. R. Fosberg provided helpful chemical and biological information. Dating of samples was done by the C^{14} Dating Laboratory of the Institute of Marine Sciences, Miami, Florida, under direction of Göte Ostlund. All research was supported by a grant from the National Science Foundation. Particular thanks are due E. A. Shinn for accompanying and encouraging the senior author in the field and for critical reading of the manuscript.

Geologic and Climatic Setting

The Florida Keys represent elevated remnants of a Pleistocene coral reef tract extending in a southeast-southwest arcuate band for more than 150 miles from Miami to Key West, Florida (fig. 11.2). The Key Largo Limestone is exposed as a chain of keys extending from Soldier Key south to Big Pine Key. From Big Pine Key to Key West this coralline reef arc is covered by a veneer of Miami Oolite which reaches at least as much as 30 feet in thickness in some places.

The climate in the Florida Keys area is subtropical, with an average temperature of 75°F. Rainfall averages about 60 inches per year, with 75 percent of it falling during the hurricane and tropical storm season extending from June through October.

The numbers and types of subaerial processes affecting today's exposed or soil-covered limestone bedrock surfaces in low-lying tropical environments near the sea are many and complex. Long and short periods of tropical rainfall alternating with similar periods of drought and intense heat produce solution and reprecipitation, fluctuations in the shallow groundwater table, and changes in microbial activity. Decaying lush vegetation provides humic acid and plant detritus. Animals yield pelletoid and skeletal debris. Hurricanes and lesser tropical storms frequently mix detritus from adjacent environments and occasionally saturate the soil and underlying vadose zone with salt water. Vegetation of topographically low islands furnishes abundant baffles for entrapping wind- and wave-carried detritus during such storms.

Distribution and Thickness of Crusts

Bedrock is found randomly coated by laminated crusts from Soldier Key to Key West. Such crusts are either exposed, or they are covered by thin soils. The color of the laminae varies from light brown (5YR6/4) to grayish-brown (5YR3/2) and medium-dark gray (N4). These crusts, harder than the underlying rock, usually occur as smooth dome- and basinlike coatings on the bedrock (fig. 11.1), and break into hard, tough, platy fragments under the hammer.

The thickness of the crust is highly variable; it averages 2 cm, with a maximum thickness of over 5 cm. Crusts coating the Key Largo Formation north of Big Pine Key are usually thinner than those over the Miami Oolite, which crops out on Big Pine Key and southward to Key West.

Composition

Particles composing the crust vary from one area to another, but generally reflect the nature of the underlying bedrock, and are augmented with additional particles from adjacent and distant sources. Calcified roots, pellets, shells, occasional quartz grains, and even egg cases of flies are also found to be constituent particles of the crusts. Laminae vary from 0.2 to 6.0 mm in thickness, depending on the type of crust involved (see below).

In thin section the crust is composed of stained, microcrystalline calcite, pelletlike structures and all the various inclusions noted above. Voids are often coated or filled with blocky, prismatic calcite. Under high magnification the boundaries of the laminae are not sharp and are usually not due to changes in grain size of constituent particles, but rather to thin discontinuous bands and streaks of organic staining. When slowly heated to 450° C in air, these brown laminae turn black, with only rare, oxidized, reddish-brown pellets indicating the presence of iron. When treated with a Chlorox solution, the laminae turn light gray. Traces of algae are rare and are believed to be remnants of ubiquitous, filamentous blue-green forms which commonly bore into calcareous deposits.

Types of Crusts

Three general forms of crust may be recognized, each varying in texture and thickness. All three forms may be found at a single outcrop, although such an occurrence is rare. The three forms recognized, in decreasing order of volumetric abundance, are: (1) porous laminated crusts, (2) dense laminated crusts, and (3) nonlaminated microcrystalline rind. Porous and dense laminated crusts differ in density and constituent particles. They may occur separately or together. When occurring together, porous layers may overlie dense layers or vice versa. The microcrystalline rind, on the other hand, is generally at the bottom of the crust layers and is in direct contact with the underlying bedrock; it also coats vugs and cracks within the limestone. In rare instances the microcrystalline rind is found coating a laminated crust. Characteristics of each general type are given in table 11.1.

TABLE 11.1
Characteristics of Indurated Crusts Coating the Florida Keys

	Porous Laminated Crust	Dense Laminated Crust	Microcrystalline Rind
Laminations usual thickness persistency	sharp to vague 1.0–5.0 mm poor	sharp 0.5–1.0 mm fair to good	sharp 1.0–4.0 mm poor
Constituent particles	*5–20 percent* horizontal root tubes, organic debris, mollusk fragments, oöids, pellets, rock and coral fragments *0–5 percent* black limestone grains and subangular quartz grains (0.25–0.5 mm diam.)	*0–5 percent* organic debris, mollusk fragments, oöids, pellets, rock fragments, coral fragments, *rare* black limestone grains, sub-angular quartz grains (0.25–0.5 mm diam.)	occasional ghost structures of pellets and oöids in a dense, tan, microcrystalline calcite mosaic.
	microcrystalline calcite matrix		
	crinkled, undulating, horizontal solution cracks		
Porosity types	common solution holes 0.10–0.80 mm diam. coated with prismatic calcite	rare to common solution holes 0.10–0.40 mm diam. coated with prismatic calcite	rare pinpoint solution holes 0.10 mm diam. coated with prismatic calcite
Surface appearance	rough-surfaced domes and basins	smooth-surfaced domes and basins	smooth; rare mammillary surface
Usual nature of underlying bedrock	permeable; friable to firm	relatively impermeable; solid	all types
Comments			fills low relief areas on top of bedrock; fills deep cracks and coral calices.

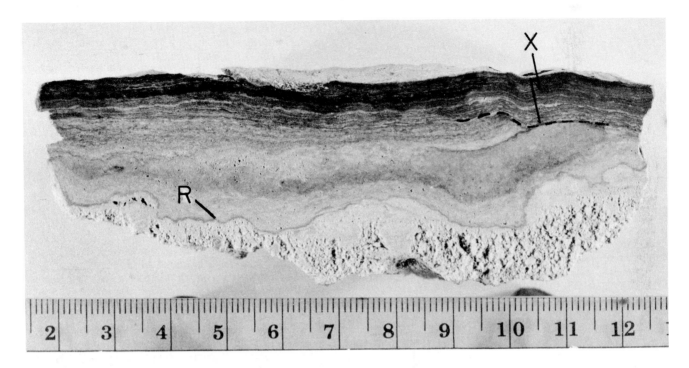

Figure 11.4. (*above*) Dense laminated crust overlying porous laminated crust. Unconformity along surface "X". Thin microcrystalline rind crust (R) coats coralline limestone bedrock. Key Largo, Florida. (*below*) Porous laminated crust with root cast holes (H) and thin basal microcrystalline rind (R) crust on top of Miami Formation (oolitic facies) bedrock, Stock Island, Fla. Scale in centimeters. From Multer and Hoffmeister (1968a).

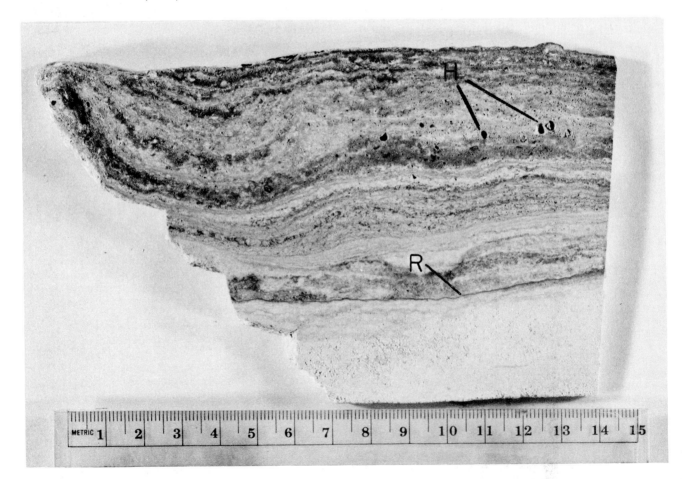

Origin of Crusts

Two major interpretations for the origin of the crusts will be considered here. The first concerns the possibility of their formation by means of algae—either marine or fresh water. The second deals with a subaerial origin.

Algal origin. The origin and structure of laminated soft marine algal mats in the littoral zones of the Florida Keys by accretion of alternate layers of mucilaginous algae and detrital particles is described by Ginsburg (1953; 1954; 1955; 1960). Comparison of such algal mats with the indurated crusts described by present writers indicates both similarities and dissimilarities. Both mats and crusts show banding, undulatory surfaces, and porous and dense textures. The mats may be found directly overlying the indurated crusts in the littoral zone. Dissimilarities, however, outweigh similarities and are listed below:

(1) Crusts lack the common desiccation features of mats. (2) Bands in crusts are due dominantly to differential staining; bands in mats are due dominantly to textural changes. (3) Crusts have been found to contain fossil roots of plants presently living only above the saline intertidal zone. (4) Crusts are found deep within vertical bedrock cracks and cavities—in both cases outside the realm of sunlight, which precludes the presence of light-requiring algae. (5) In today's intertidal zone there is no observable transition from soft algal mats to hard indurated crusts lying directly beneath the mats. (6) Crusts of the Keys do not occur as extensive blanket deposits as do algal mats of Florida Bay. (7) Hardened algal sediments often lose or do not contain many laminations, but can contain Recent dolomite; indurated crusts are laminated and have not been found (as yet) to contain dolomite. (8) Mats contain abundant algae; algae, although often not prone to fossilization, are absent or rare in crusts. (9) Crusts are found intimately associated with subaerial solution features which must have been formed above the littoral zone. The presence of both rind and laminated crusts as coatings on vertical sinkhole walls, as coatings on pebbles within filled sinkholes, and as coatings over the above-mentioned sinkholes some eight feet deep, requires an origin not only above sea level (precluding marine algae), but within the zone of aeration above the groundwater table.

The weight of present evidence, therefore, seems to eliminate marine algae as responsible for the indurated crusts under discussion.

The influence of freshwater algal mats, as described by Black (1933, p. 169), however, should not be underestimated. The exact role, if any, of freshwater algae in the Florida Key crust is hard to evaluate. No filaments of the common Bahamian algae *Scytonema* or radial, algal head structures as described by Black (1933, Pl. 22), were observed. The typical Florida Key crust with banding due primarily to organic staining is unlike the common algal mat texture involving alternate coarse-grained detrital laminae with finer-grained organic laminae. Present evidence therefore indicates a limited role for freshwater algae in the formation of crusts coating the Keys.

Subaerial origin. Of the three forms of crust found in the Keys (table 11.1), the microcrystalline rind type must be dealt with somewhat differently than the laminated types. The latter are dominant and will be considered first.

In many upland areas of the Florida Keys, laminated crusts can be found coating both low and high topographic surfaces, implying the possibility of a uniform mechanism of origin such as might occur beneath and adjacent to soil cover. The so-called "soil" of the Florida Keys should more accurately be termed "duff," as it does not possess a typical zonation and is composed dominantly of decayed plant litter. In the interest of simplicity, however, the term "soil" is maintained. In many places laminated crusts are today found underlying or adjacent to soil layers which are supporting dense vegetal cover. Examination of samples of some of these soils and the underlying encrusted bedrock reveals a close similarity of constituent particles in both soil and indurated crusts. Such field observations led to further examination of the relationship between soils and laminated crusts.

Plastic impregnation and thin sectioning of samples of undisturbed soil overlying encrusted bedrock reveals that today's soils can contain material both from the local bedrock and from distant sources (i.e., black limestone fragments of unknown origin and coral are found over oolitic bedrock). All such particles found in the soil can also be identified in thin sections of the indurated laminated crust directly underlying the soil. Such data led the authors to develop a hypothesis involving crust formation under and adjacent to aggrading soil deposits.

Figure 11.5 illustrates some of the mechanisms believed responsible for the formation of such laminated crusts.

Dissolution of mollusk fragments and carbonate grains in today's soil layers, as well as measured pH ranges (maximum recorded range being pH of 5.6 in peaty soil overlying calcarenite with a pH of 8.2), demonstrate the presence of conditions for solution and precipitation of carbonate on top of bedrock by percolating charged solutions draining through overlying soils. In addition to layers being built up on top of bedrock beneath soils, it is also believed that layers may be built up on exposed bedrock surfaces bordering soil-covered areas during rainy seasons when organic- and $CaCO_3$-rich waters from under soil patches flood the exposed adjacent bedrock surfaces. If such exposed deposits survive subsequent solution and erosion, they will build up an indurated laminated crust coating.

Recrystallization is also believed to have taken place as shown in thin section by "ghost structures" of former constituent particles and inversion of originally aragonite ooids to calcite. Both subaerial cementation and recrystallization processes are also described by Kornicker (1958) as responsible for laminated crusts coating portions of the Bimini Islands in the Bahamas.

Another indication of a pedologic origin for the crust is its intimate relationship as both a lining and a surficial covering of former soil and rock-filled sinkholes (some more than 8 feet deep), now found as indurated breccia pits on the surface of both the Key Largo Limestone and the Miami Oolite. Such field relationships demand the presence of a soil, as well as wind and wave mechanics strong enough to bring detritus up to boulder size across some of the highest portions of the Keys.

Evidence for the dominance of aggradation over degradation is: (1) presence of plant fibers and organic grains from soil within indurated crusts; (2) presence of superimposed, distinctly different laminated crusts separated by unconformable surfaces which demonstrate past solution of underlying crusts and availability of different source material for building up of next overlying crust sequence; (3) admixtures of detrital rock particles within crust distinct from those available in underlying bedrock.

As previously mentioned, the banding in laminated crusts is due chiefly to organic staining and concentration

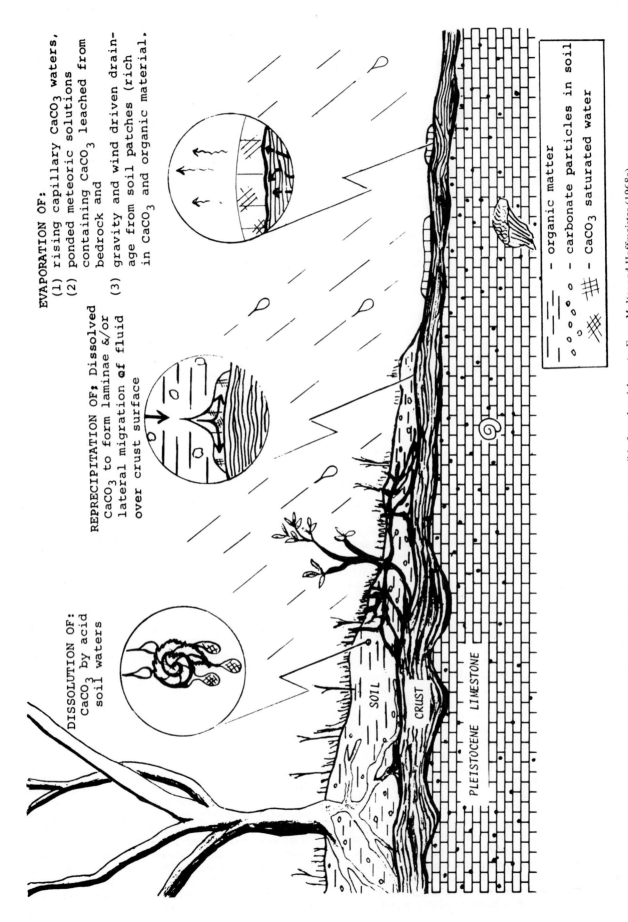

EVAPORATION OF:
   (1) rising capillary CaCO₃ waters,
   (2) ponded meteoric solutions
       containing CaCO₃ leached from
       bedrock and
   (3) gravity and wind driven drain-
       age from soil patches (rich
       in CaCO₃ and organic material.

REPRECIPITATION OF: Dissolved
   CaCO₃ to form laminae &/or
   lateral migration of fluid
   over crust surface

DISSOLUTION OF:
CaCO₃ by acid
soil waters

— organic matter
— carbonate particles in soil
— CaCO₃ saturated water

SOIL

CRUST

PLEISTOCENE LIMESTONE

Figure 11.5. Schematic drawing of some mechanisms responsible for subaerial crusts. From Multer and Hoffmeister (1968a).

of minute organic particles. One explanation of the origin of such bands could be the washing down of such minute soil particles (product of microbial decomposition) during excessively long wet periods. Light bands would represent merely the normal accumulation of fine-grained, chiefly nonorganic detritus during intervening drier periods when bacterial decay activity is minimized. Drier periods lasting as long as seven consecutive years have been recorded in the neighboring Homestead area.

Abundant "micro-unconformities" (fig. 11.4) reflect past changes in the local pH where solution dominated over precipitation. In some instances changes in constituent particle composition of encrusted layers over the unconformable surface indicate probably erosion and covering by new veneer.

To test the above theory, tubes of various soils from Key Largo were subjected to percolating waters in a laboratory experiment in an attempt to produce an artificial crust. After 3.5 months a crustlike veneer of microcrystalline calcite containing fine particulate rock and organic matter (washing down naturally through a fine plastic screen at the base of each tube) was formed on underlying bedrock surfaces from freshwater percolation only. No recognizable veneer was formed from saltwater percolation.

Early field investigations indicated that when both types were found together the dense crust was always below the porous crust, which led to a belief that the dense variety was merely an older and more recrystallized (denser) portion of the whole crust coating. Subsequent field and laboratory work indicated that, although the dense crust is more common as a basal member, it can be found overlying the porous crust (fig. 11.4) and the contact between the two is usually sharp. The compositions of the two crusts were also found to be distinctive.

The exact conditions responsible for both the dense and porous laminated crust types are not known. It is quite possible that a variety of mechanisms may permit the development of each type. Present evidence leads the authors to suggest that the porous laminated crust with its root casts, vegetal debris and coarser-grained detritus, was produced beneath a relatively thick, aggrading tropical forest cover. In addition to porosity caused by solution of organic debris, elongated plant fibers and roots tend to bridge over underlying grains, providing protection from filling and complete compaction. In such protected zones, permeability and solution are often more highly developed. The porous laminated crusts also contain aragonitic ooids-especially in the southern Keys where it overlies the Miami Oolite. Under solution these aragonitic grains tend to dissolve more readily leaving a more porous structure than the nonaragonitic dense crust more often found over the Key Largo Limestone.

The dense laminated crust, on the other hand, is believed to reflect environmental conditions different from those producing a porous crust. The dense laminated crusts lack the abundance of detrital constituents which would indicate possible origin under a thin soil or on an exposed bedrock surface receiving periodic drainage from adjacent soil patches. Neither a thin soil nor an exposed bedrock surface would be as capable of producing detritus or entrapping transported debris by its baffle effect as would a dense forest cover over the porous crust. Such thin soil or exposed rock surface would also allow more frequent wetting-drying (solution-precipitation) processes, which would produce a denser (more recrystallized) crust.

Environments for producing both types of crusts may exist at the same time in the same area. In time, one environment may migrate across another (that is, thick forest cover might replace thin soil cover or exposed bedrock surfaces; this might be why porous crusts commonly overlie dense crusts when the two are found together, as noted above).

The microcrystalline rind type of crust is believed to originate independently of tropical soil patches. The location of the rind below the laminated crust also indicates that it was formed first—perhaps before the availability of soil and debris cover. Conditions permitting the formation of a rind type crust are discussed below.

Wide ranges in pH, demonstrating the presence of conditions capable of dissolving and precipitating $CaCO_3$, are found on exposed and covered bedrock surfaces of the Florida Keys. Oppenheimer and Master (1965, and personal commun.) show diurnal fluctuations in alkalinity (pH from 7.4 to 9.2 in littoral marine environments and pH from 6.5 to 10.0 in freshwater environments) from experimental and natural observations of material in the Florida Keys. During the day, high temperatures and low CO_2 (taken up by plants in photosynthesis) yield a high pH, and $CaCO_3$ is precipitated; at night temperatures drop and CO_2 is given off by plants, lowering the pH. Kellerman and Smith (1914) describe three types of biological processes found responsible for bacterial precipitation of $CaCO_3$, using samples from the Florida Keys area. Littoral solution and precipitation are also discussed by Emery (1946).

Above the littoral zone microbial activity, together with evaporation of ascending capillary water and ponded meteoric water, is thought to be responsible for deposition of the microcrystalline rind type crust. Whenever local precipitation is equal to or in excess of solution, a microcrystalline rind is built up. Field observations indicate that such rind tends to fill in depressions in the bedrock surface, indicating a localization whenever ponding is allowed. Percolating, carbonate-charged solutions also leave microcrystalline rind deep within cracks and fissures of exposed bedrock.

In summary, it is believed that microcrystalline rind is produced in the subaerial (including upper littoral) zone by the leaching and repricipitation processes described above. In the Florida Keys today erosion and solution are dominant on exposed bedrock surfaces, allowing only random protected areas for build-up of rind crust. The discontinuous distribution of the rind preserved under laminated crusts may well indicate similar sporadic deposition in the past.

Age and Additional Evidence for Subaerial Origin of Crusts

Bulk radiocarbon[14] analyses were run on five samples from three localities. The ages indicated by these analyses show that the crusts have a maximum age ranging from 880 to 4395 ± 90 years. These age determinations probably exceed actual ages because of included carbonate derived from the underlying Pleistocene limestone. Contamination from Recent material, although less likely, would produce a younger age.

Published data by Shepard (1963), Ginsburg (1964, p. 5), and Scholl (1964) indicate that sea level has been lower than at present throughout the past 4,000 years. Thus the surface crusts described in this paper could never have been immersed in sea water. Since there is no evidence of radical climatic or tectonic change during these same past 4,000 years, it is assumed by the writers that subaerial processes like those operating today (see, Geologic and Climatic Setting, above) were responsible for the origin of these crusts.

Subsurface Crust

Cores from a diamond core test hole drilled on Grassy Key revealed, at a depth of 14 feet below sea level, a sequence of laminated crusts within marine limestone of the Key Largo Formation. These subsurface crusts are interpreted as subaerial because of their close similarity to those found coating the exposed surface of the Key Largo described in this paper. A description of this cored interval is given in figure 11.6.

Effects of solution and recrystallization are more obvious with this subsurface crust than with more recent counterparts now coating the exposed Key Largo. An encrusted cobble approximately 6 inches above the top of the crust indicates the possibility of tropical storms having accompanied transgressing seas which deposited the overlying biosparite.

Interpretations

The above example of an ancient (the age of the upper part of the Key Largo Formation [at Windley Key] has been determined, on the basis of thorium-uranium dating, at 95,000 years by Broecker and Thurber [1965, p. 59]) subsurface laminated crust serves to show how such features can be used to indicate former subaerial exposure in a coralline reef marine sequence. Although the crust is thin, it is the only evidence of an unconformity.

Many different valid hypotheses exist for the origin of laminated crusts on limestone surfaces. Careful examination of all field and laboratory evidence is therefore necessary before applying any particular hypothesis for paleoenvironmental interpretation of ancient limestones.

The writers of this paper have described three types of crusts which formed in a dominantly subaerial (some rind may have formed in the upper littoral) environment. Correct identification of similar crusts in a sequence of ancient marine carbonate rocks also containing marine (algal) crusts may make it possible to predict periods of regression and transgression of seas, as illustrated in figure 11.7. It should also be realized that a single stromatolitic zone in ancient rocks may contain both subaerially formed laminae and similar-appearing laminae of marine algal origin superimposed upon one another. Such a sequence will eventually develop over the subaerially encrusted bedrock

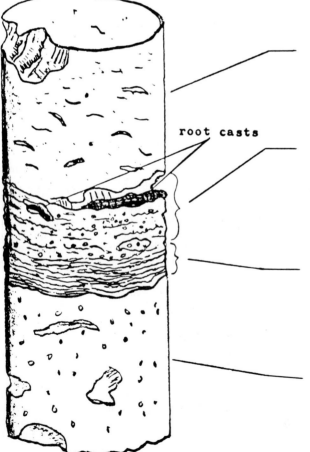

root casts

UNSORTED BIOSPARITE - very pale orange (10YR8/2) common mollusks, superficial oölites, forams, algal debris; up to 3% subrounded quartz.

POROUS CRUST - dark brown (5YR3/2) common ghost grains, horizontal discontinuous voids partially filled with spar. Upper contact shows open solution crack up to 5 mm wide with frequent root-like casts; trace of subrounded quartz.

DENSE CRUST - dark yellow-brown (10YR4/2) dense with ghosts of pelletoid grains, superficial oöliths, thin horizontal plant fibers and grains, trace of subrounded quartz, random voids partially filled with spar.

OÖSPARITE - grayish-orange-pink (5YR7/2) superficial oölith rims often with void or spar filled centers; trace of mollusk and algal fragments; subrounded quartz rare.

Figure 11.6. Schematic illustration of subsurface core from Grassy Key showing laminated crust. From Multer and Hoffmeister (1968a).

of the Florida Keys if the present soft marine algal mat covering becomes fossilized and preserved as the sea level continues to rise.

Observed field and thin-section evidence indicating differences between the subaerially formed *crusts* coating the Florida Keys and marine algal *mats* observed by the writers or described in the literature may be summarized, with the reservation that there are exceptions to all criteria listed. The following statements, therefore, must be considered generalizations.

1. Sediment-binding algal mats tend to thicken over domal relief features; subaerial crusts described in the Florida Keys do not.

2. Laminae in mats tend to alternate in size-grade of constituent particles; laminae in cursts of the Florida Keys tend to remain the same size-grade with banding due to staining.

3. Mats have abundant algae; crusts of the Florida Keys rarely show evidence of algae.

4. Mats show presence of intense boring and mixing by littoral organisms; crusts covering the Florida Keys do not.

5. Supratidal mats show desiccation features; crusts of the Florida Keys do not.

The existence of subaerial crusts which may superficially closely resemble marine algal stromatolites is established. Caution, therefore, should be observed in the interpretation of similar laminated deposits in ancient limestones.

Postscript (by H. G. M.—9/69)

Since the above article was published several other generalizations have been noted which may aid in the identification of ancient subaerial crusts as distinct from ancient algal mats.

6. Banding in sediment-binding algal mats tends to be sharp; bands in crusts tend to be diffused.

7. Bands which appear to coat deep cavities and undersides of former surfaces are more likely to be crusts for algae need light (photosynthesis) and sediment does not tend to accumulate as readily on vertical or inverted surfaces.

8. Total outcrop thickness of successive laminations can be of considerable magnitude from algal mats; mechanics of crust formation tends to usually limit great thickness of laminated rock.

9. Birdseye structures as described by Shinn (1968) are intertidal and supratidal in origin and should therefore be found predominantly in algal mats and not in subaerial crusts.

10. Crusts have a common intimate relationship as both a lining and surficial covering of karst features such as sinkholes. Algal (fresh water) mats would be expected to play a more limited role in this capacity.

11. Algal mat laminations tend to accumulate and thicken on one (windward) side of local topographic highs (C. D. Gebelein—personal communication); crusts do not.

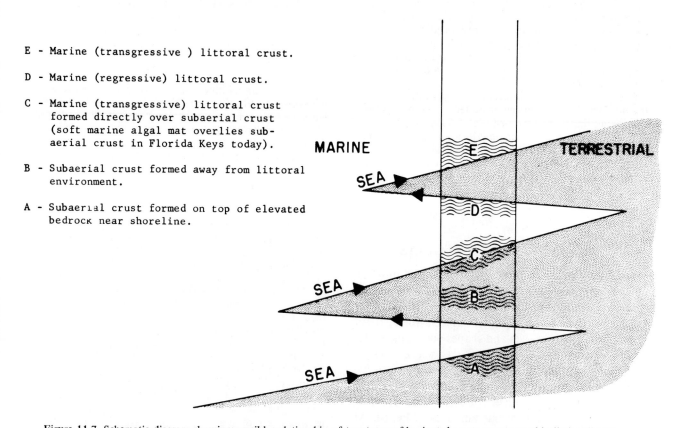

E - Marine (transgressive) littoral crust.

D - Marine (regressive) littoral crust.

C - Marine (transgressive) littoral crust formed directly over subaerial crust (soft marine algal mat overlies subaerial crust in Florida Keys today).

B - Subaerial crust formed away from littoral environment.

A - Subaerial crust formed on top of elevated bedrock near shoreline.

Figure 11.7. Schematic diagram showing possible relationship of two types of laminated crusts on topographically low limestone coastlines subjected to fluctuating sea levels. From Multer and Hoffmeister (1968a).

Kahle (1974) describes the relationship between the micrite of the crusts overlying the Miami Limestone and presence of fungal and algal borings. His abstract is given below.

ALGAE, FUNGI, AND TRANSFORMATION OF SPARRY CALCITE TO MICRITE
by
Charles F. Kahle

The Pleistocene Miami Limestone in the lower Florida Keys contains numerous calcareous crust (caliche) profiles. Crust, dominantly micrite, may or may not be gravitationally oriented, and evolves during diagenesis of different rock types, but mainly oolite, when the rocks are exposed subaerially. Diagenesis involves various processes including micritization of allochems, plus a previously unreported process, the degrading recrystallization of sparry calcite cement to micrite.

A direct correlation typically occurs between the location and abundance of fungal and algal borings infilled by micrite, unicellular algae, and organic matter, and the location and abundance of micrite within sparry calcite cement. Continuous gradation in textures can be demonstrated between unaltered oolite, partly altered oolite, and crust which contains relics of allochems and sparry calcite grain boundaries.

Repeated boring of sparry calcite by fungi or algae, and infilling of the borings by micrite, leads ultimately to transformation of the original sparry calcite to micrite. Both fungal borings into or through grain boundaries and unicellular algae concentrated along grain boundaries are associated with all stages of recrystallization of sparry calcite to micrite. Each stage of recrystallization appears to be controlled by reactions between pore fluids and unicellular algae, or organic matter derived from decay of algae and fungi, or both.

The present study reinforces and extends work by Bathurst and Purdy who demonstrated a relation between micrite infilling of borings, or the abundance and localization of organic matter and micritization of allochems. Degrading recrystallization of sparry calcite cement to micrite may explain the origin of at least some micrite in a variety of rock types that are not associated with calcareous crust.

SOLUTION HOLE BRECCIAS

Quarrying and canal cutting operations into the Pleistocene age Key Largo Limestone and oolitic facies of the Miami Limestone of the Florida Keys has revealed reddish and grayish, breccia-filled, fossil sinkholes of various sizes and shapes occurring at elevations up to 18' above present-day sea level (figs. 11.8 to 11.13). Such "solution breccias" can be found in all stages of induration and composed of various angular to rounded fragments of local bedrock. Also found are various amounts of Fe, Si, Al and Mn enriched silt- and clay-sized matrix as well as angular to rounded fragments of distant bedrock and

coral which are believed to be of hurricane origin. Subaerially laminated crusts and rinds are often found coating some of these solution features.

Such solution hole breccias are excellent indicators of subaerially exposed surfaces and hence are of particular interest to students of ancient carbonate rocks. Although only preliminary studies have been made of these features, the area is particularly well suited for such studies due to the fact that, in addition to well-exposed fossil solution hole breccias showing varying degrees of induration, there are many modern counterparts (fig. 11.14). The latter are in the form of enclosed depressions found today on the bedrock surface of the Florida Keys filled with loose rock debris, vegetal material with or without a covering layer of water.

Persons interested in such breccias and their various origins should refer to Woodring *et al.,* (1924), Bretz (1940), Shrock (1945), Fry and Swineford (1947), Hohlt (1948), Van Houten (1948), Ginsburg (1953), Kohout (1960) and Krawiec (1963).

In a study of the solution breccias found in the Florida Keys Krawiec (1963) reached the following conclusions.

SOLUTION PITS AND SOLUTION BRECCIAS OF THE FLORIDA KEYS
by
W. Krawiec

Conclusions

The Key Largo Limestone and the Miami Oolite were formed in the Pleistocene during a high stand of the sea. During a following low water stand, created by a period of Pleistocene glaciation, lithification of these two formations occurred in the zone of meteoric waters. At the same time solutions aided by organic acids produced by the decay of plant material, created pits or holes in the surface of the limestone.

As enlargement progressed pits of various shapes corresponding to the texture of the limestone being dissolved were produced. The coarse textured lithologies, composed of irregularly placed coral heads and fragments of other reef organisms developed pits which were roughly V-shaped and possessed irregularities in their outline. Those in the fine grained, calcarenitic material were, in general, bowl shaped with smooth sides.

As the pits developed they became the natural collecting place for loose material of all available types. Soils produced in the vicinity and enriched in Fe found their way into the holes, as well as did fragments of associated woody materials. Irregular fragments of the host rock tumbled into the depressions and became embedded in the finer grained matrix. It is also believed that storm waves played an important part in the filling of the cavities. The gray and red, friable matrices which are found in layers as well as fragments of black limestone were carried by storm waves from nearby shores and mangrove swamps and eventually settled in the depressions.

Figure 11.8. (left) Typical v-shaped solution hole breccia found in Key Largo Limestone. Canal cut is about 2 1/2 m high. Seven Acres, Key Largo, Fla.

Figure 11.9. (right) Close-up of the corner of the wall seen in figure 11.8. Note variety and shape of included fragments.

Figure 11.10. (left) Red cemented breccia showing rounded to angular Key Largo Limestone fragments. Note laminated crust on one (formerly exposed) surface of many of these fragments. Grassy Key, Fla.

Figure 11.11. Bowl-shaped depression typically found on top of Miami Formation (oolitic facies). Fragments are poorly cemented within cavity. Northern Big Pine Key, Fla.

Figure 11.12. V-shaped breccia filling at Seven Acres, Key Largo, Fla.

Figure 11.13. Looking down on weathered surface of Key Largo Limestone showing solution pits and common rubble. Bedrock beneath hammer is a well-cemented red breccia locally coated with a dense laminated crust. North Key Largo, Fla.

Figure 11.14. Modern sinkhole filled with loose rock debris and carbonate sand and mud. North Key Largo.

Solutions passing through the pits, during and after the accumulation of the breccias, dissolved calcium carbonate and carried it downward to mingle with the groundwater. Here it was carried away by tidal currents working in the pore space beneath the Keys. Some of the solutions, however, precipitated the calcium carbonate at the interface of the breccias and the host rock, and here a tan, cryptocrystalline limestone crust was formed by chemical neutralization. Many of the breccia fragments were also coated by the tan limestone. The interfingering nature of the crust and evidence of replacement indicates that precipitation by solution has taken place. The greater thickness of the crust in the higher part of the pits is due to the greater amount of calcium in solution in this area. The latter was brought about by the greater quantity of organic matter here.

The solutions that passed through the unlithified matrix concentrated Fe, Mn, and Al as they dissolved the calcium carbonate. As the Fe concentration increased, it was spread throughout the pit creating the red matrix and obliterating the layering of the unlithified matrix. As this action continued the red color became deeper. The loss of organic matter from the matrices resulted in the raising of the pH and lowering of the calcium concentration. The solutions could then cement the matrix into a solid breccia. When this occurred the solutions could no longer percolate easily through the breccias.

The sequence of events in the history of the breccias may be summarized as follows:

1. formation of the pits by solution
2. filling of the pits with unconsolidated material of various sizes, constituent compositions and from various sources
3. precipitation of the limestone crust lining the pits and the fragments
4. creation of the red breccia by Fe concentration
5. lithification of the friable matrices to produce the indurated breccia as organic constituents were eliminated

INFLUENCE OF KARST TOPOGRAPHY ON SEDIMENTATION AND BIOTA

Exposed karst topography (figs. 11.1 and 11.14) has a distinct effect on drainage and sediment collection (soil formation) in the present Florida Keys. Terrestrial biota distribution are in turn influenced by the presence or absence (and thickness) of soil and debris cover.

In like manner, *buried* karst topography beneath a shallow marine environment may well be at least partially responsible for initiating certain topographic forms (sediment accumulations) which, in turn, are attractive for certain types of biota. Thus there can be a relationship between buried karst topography and presence of plants (i.e., mangroves and/or grasses with their baffling capabilities) and presence of skeletal producing calcareous animals associated with these habitats.

In other words, there may in some instances be a reason for a skeletal mound initiating in one area (over a buried karst bedrock low) rather than another. One reason for this may be due to the following sequence: (a) deposition of sediment in protective depression, (b) stabilizing and upbuilding (by plant baffling) of sediment into a mound rising above the edge of the depression, and (c) enrichment of the skeletal component of the sediment by invertebrate epifauna and infauna which prefer the grass-held mound habitat. A skeletal hash mound which might have had some of the same history is described on page 76 of this Guidebook in Florida Bay.

Sample Description	Location	Fe$_2$O$_3$	MnO	Al$_2$O$_3$	SiO$_2$
White recrystallized calcarenite	Vaca Key	0.15	0.001	0.5	1.0
White lithographic limestone	Vaca Key	0.15	0.12	0.25	1.0
White lithographic limestone	Windley Key	0.035	0.001	0.22	1.0
Light tan calcarenite	Vaca Key	0.22	0.0035	1.2	1.2
Tan calcarenite	Key Largo	2.5	0.013	8.5	7.5
Brownish matrix (unlithified)	Big Pine Key	2.2	0.017	7.5	11.0
Brownish calcarenite	Vaca Key	0.8	0.005	1.8	1.3
Reddish tan lithographic limestone	Vaca Key	0.45	0.012	2.2	1.0
Reddish brown calcarenite	Key Largo	1.7	0.012	6.8	13.0
Red calcilutite	Windley Key	1.8	0.033	6.0	6.2
Red calcilutite	Key Largo	4.0	0.020	7.5	12.0
Red calcilutite	Key Largo	4.8	0.025	12.0	18.0

(Analyses by Dr. O. Joensuu and Mrs. S.K. Johnston, Institute of Marine Sciences, Miami, Fla.)

Figure 11.15. Weight % of Fe$_2$O$_3$, MnO, Al$_2$O$_3$ and SiO$_2$ in samples of the Key Largo Limestone and the Miami Oolite as determined by spectrochemical analyses. From Krawiec (1963).

This possible relationship between buried karst topography and types of resulting sediment and/or biota has been observed by many workers and provides an interesting challenge for those interpreting ancient carbonate analogs. Some of the investigators working in Holocene sediments who have noted these relationships in the Florida Keys-Bahamas area are listed below with page reference as to where more details can be found in this Guidebook: (1) Boardman (1976), Bight of Abaco, page 93, (2) Gray (1974), Florida Bay, page 76, (3) Kissling (1968), Coupon Bight lagoon, p. 97, (4) Multer (unpublished), Coupon Bight lagoon, p. 94, (5) Siemers and Dodd (1969) and Dodd and Siemers (1971), Bahia Honda and Big Pine Keys, p. 75, and (6) Turmel and Swanson (1964 and publication pending), Rodriquez Bank, p. 167.

Author's Note: For those interested in karst features in the Bahama area, Bourrouilh (1974) describes the results of a geomorphic study of the Fresh Creek, Andros, region including a description of fossil megaripples, protective calcitic crusts, blue holes, karst "doline" high-magnesium rich depressions and a comparison with Yucatan karst topography.

FIELD QUESTIONS
Post Key Largo Subaerial Features

KEY LARGO (Florida Keys):

1. Examine the rind and various types of laminated crusts over the bedrock at Seven Acres, Marion Park or other exposed sites. Is the rind always found directly on top of bedrock and beneath the laminated crust layers? If so why?

2. Compare crusts found overlying the Key Largo Limestone in the northern Keys with those overlying the oolitic facies of the Miami Formation in the lower Keys. Is the crust over the later really usually more porous and thicker than the former? If so why? What is the role of roots and burrows in the porous crust?

3. At low tide find some encrusted bedrock overlain by a soft mat of blue green algae. Try and break away a fragment which includes both mat and crust and examine same. Is the contact sharp? Are the laminations and/or constituents similar? Are there any significant characteristics of the mat distinct from those of the crust and vice versa?

4. Examine a cross section of the laminated crust. What is the origin of the microunconformities and the discontinuity of individual dark and light bands?

5. Can you locate a place where subaerial crusts are in the process of formation today? If so—what type of crusts are forming and by what mechanics?

6. Examine as many exposures as possible of solution breccia filling cavities in the top of the Key Largo Limestone and the Miami Limestone. Such exposures can be found at places such as Seven Acres or Marion Park on Key Largo and at various places off the road on north Big Pine Key.

    What is the source of the red color?

    What is the source of iron on an emergent coralline limestone island?

    Why is the matrix sometimes grayish and other times red? What is the significance of the brown colored matrix?

    Does the presence of microcrystalline rind and laminated crust lining some cavities vary with the type (porosity) of the adjacent bedrock? Does it thin opposite more impervious bedrock?

    How do you tell if the encrusted pebbles within the matrix were formed subaerially or are they coated with algal laminations formed in a marine environment and washed in during storms?

    Why is there such a difference in induration of the various breccia fillings?

    Is there active solution and breccia forming processes going on today? If so, where and to what extent?

Figure 12.1. Man-made changes, such as the extension of individual Keys out into Florida Bay by dredging and filling as seen above, can have severe detrimental impact on the ecosystem if proper protective measures are not implemented.

12

SOME NOTES ON ENVIRONMENTAL QUALITY

INTRODUCTION

There is growing concern about how the natural environment of the Florida Keys can realistically accomodate the intense environmental pressures brought about by man's increasing use and development of the area (for a review of adjacent Dade County growth management challenges see Carter, 1976). This is obviously a "people problem" and good land use planning and continued management are urgently needed. *To accomodate growth while maintaining environmental quality is the challenge.* It is hoped that the present chapter, which documents some of the stresses on the animals and plants of the Florida Keys and compiles reference sources on organisms under man-made stress in various marine areas of the world, will be useful to those involved with these problems.

The impact of man on the ecosystem can be described and/or measured in many ways. This can include the effect of drainage, dredging and filling, dumping of heated waters, toxic metals or petroleum products, introduction of exotic plants and animals, reduction of vegetation (including timber) and overfishing. For examples of various impact studies see Avon (1973), Chesher (1971, 1973, 1974), Clark (1972), Greenfield and Rich (1972), Larsen and Michael (1972), McNulty (1970), Michel (1972a, 1972b, 1973), Pitt (1972) and Thomas (1970).

Many ecosystems in the Florida Keys area contain animals and plants that are being subjected to a variety of man-made pressures. Those that will be illustrated in this chapter are:

1. Thermal Stress
2. Sewage pollution
3. Siltation/turbidity
4. Heavy metal pollution.

In addition to man-made stresses noted above, animals and plants of the Florida Keys also undergo a wide variety of *natural* stresses. These include thermal stress, siltation from storms, and predation from other organisms. Determining whether the cause of the stress is natural or man-made is an important although sometimes difficult task. For this reason, a brief discussion of some local natural stresses will follow the main topic of this chapter—that which concerns man-made stresses on the environmental quality of the Florida Keys.

SOME MAN-MADE STRESSES ON ORGANISMS OF THE FLORIDA KEYS

MAN-MADE THERMAL STRESS (Turkey Point, Lower Biscayne Bay/Card Sound): According to the U.S. Department of Interior (1974) the two Turkey Point fossil fuel power plants require 1,270 cfs of cooling water and raise the water temperature 12° F. The one nuclear power plant at the same site requires 4,250 cfs and raises the water temperature 16° F. The report by the Department of Interior (1974) also indicates that this released heated water has significantly affected the plankton, benthic algae, sea grasses, benthic invertebrates and fish near the discharge area. To reduce thermal stress use is being made of a large-scale system of cooling canals. The results of studies of *thermal stress* on a wide variety of marine plants and animals is given by Bader and Roessler (1972) in a description of a large-scale ecological study of the south Biscayne Bay and Card Sound area. This area, the site of the Florida Power and Light Co. Turkey Point power plant, was monitored from July 1970 to February 1972 to determine baseline data including assessment of hydrography, water chemistry, vegetation, planktonic and benthic organisms. Of particular interest are recommendations (excerpts from hearings and court statements) made by various scientists as they interpret their research concerning thermal stresses on various marine organisms. Recommended temperature limits (maximum) are also given.

Excerpts from the summary of this comprehensive volume are given below. Portions of each author's conclusions bearing on pertinent thermal data is given. *Persons wishing to read the whole summary should procure a copy of the original report.*

R. G. Bader, M. A. Roessler, G. L. Voss, and D. P. deSylva*

Our observations indicated that areas which average 4.5° F above the ambient bay temperature (as measured on Pelican Bank) have reduced standing stock of benthic plants and animals during the summer months in an area of about 200 acres; however, in other seasons the production of the normal biota is enhanced by the higher winter temperature and the result is that on an annual basis no decrease in production occurred. At stations which averaged 6.3° F above the ambient temperature the plants and animals were severely depleted during the summer months in an area of about 75 acres. Some recovery is noted in winter and early spring but the net effect is lowered annual production. The recovery in winter does; however, indicate that if the discharge were terminated this 75 acre area would recover.

For example with an average ambient bay temperature in summer of about 89 to 90° F the addition of the temperature anomaly would develop temperatures of 95 to 96° F. Although 95° F is biologically detrimental, it appears that if the macrobenthic flora and fauna in Card Sound which is similar to that in Biscayne Bay is exposed to no more than 95° C it will recover after the discharge is curtailed.

We have observed that in an area of about 30 acres elevated 8.1° F or to an average summer temperature of 98° F, there is virtually no recovery in winter. The high temperature effects added to the increased current velocity at the outfall has caused a barren area and this area will probably only recover slowly if it recovers at all.

The proposed values of temperature and flow rate for the interim period are very close to those which appear as critical for permanent damage and every precaution should be taken to insure that a temperature of 95° F and a discharge of 2,750 cfs in Card Sound and 1,500 cfs in Biscayne Bay are not exceeded.

The discharge temperature of 90° F and flow of 1,200 cfs recommended after construction of the cooling system will still produce some change in the Card Sound environment but the impact of these changes on the use of the Sound by sport and commercial fishermen as well as recreational users should be minimal.

Martin A. Roessler**

As a marine ecologist I would like to urge the rapid development of gas cooled reactors, MHD reactors or other developments which eliminate water as a coolant for power stations. Since these will not be developed for some time another approach is indicated.

During 41 months of fieldwork at the Turkey Point Power Plant station on Biscayne Bay, Florida, we have observed significant damage to the invertebrate fauna at water temperature of 33° C (91.4° F) and even lower temperatures have reduced the productivity of *Thalassia* (turtle grass) and macroalgae. (*Penicillus, Halimeda, Batophora,* etc.) in the summer months.

Thus a maximum discharge temperature of 90° F or +1.5° F above the monthly average of the maximum daily temperature in summer appears to be a reasonable guideline for Biscayne Bay, but a number of scientists at the Rosenstiel School of Marine and Atmospheric Sciences suspect that this single control temperature probably should not be imposed statewide for the following reasons.

In the first place we live in a state with approximately 1,400 miles of coastline. In the southern portion, the coastal and marine conditions are subtropical in nature, while in the more northern areas, temperate conditions exist. Secondly, some of our bays and estuaries, particularly those on the west coast, appear to be dependent on phytoplankton as a source of primary productivity while the Everglades region appears to be principally a detritus food chain ecosystem depending on red mangroves and sea grasses and algae for its source of energy. Some areas, for example Biscayne Bay, appear to depend on the benthic flora (sea grasses, macroalgae and microalgae) as the major energy source. Of course all three (phytoplankton, detritus and benthic plants) contribute somewhat in each of these areas but the majority producer in each region is different.

To a large extent the source of food regulates the types of animals which will be found in each area. And the type of animals present and their tolerances should dictate the temperature limits and the cooling method to be used.

In some areas where phytoplankton is the dominant food producer, the volume of water passing through the power plant is critical because of the potential for killing plankton which in turn feeds oysters and other filter feeding specimens. In these regions a small volume of water discharged at a high temperature and then rapidly cooled by mixing with a diffuser system has an advantage of passing less water through the plant where thermal and mechanical shock can kill plankton.

In areas dependent on detritus from the shoreline and bottom plants, mixing with a larger volume of cool water may be preferable, or perhaps if the water is sufficiently deep, a surface discharge of hot water which does not affect the bottom organisms would be best.

We feel that state regulatory agencies and the power industry in the state should openly get together and discuss the realistic increases in power demands for the next 20 years and then allow the environmental divisions to recommend sites for the location of power plants sufficient to meet the projected power demands. Only through the use of statewide or better yet regional planning can the best allotment of water resources of the state be affected.

If numerical standards either absolute temperatures or allowable temperature increases, are set, they should not

---

*State of Florida Department of Pollution Control, Tallahassee, Florida, September 8, 1971.

**State of Florida Department of Polution Control, Orlando, Florida, November 22, 1971.

be inviolate. If the state agencies or environmentalists envision damage at any site at 90° or + 4° F (if these criteria are accepted) they should conduct the necessary research to determine the allowable temperatures for that site. If the power industry believes higher temperatures will not damage the existing populations they should support the research necessary to prove that no damage will result. Preferably this research should be done by local unbiased agencies such as state or private universities.

Anitra Thorhaug*

To begin with it must be said that the temperature limits of tropical estuarine organisms have had scant attention in the past. Because of this error in both judgment and emphasis, a vast amount of basic information is now essential if we are to have any capabilities in predicting the effects of elevated temperatures, either by natural means or by man-made heated effluents. In field conditions many factors such as salinity, organic composition, alkalinity, reduction potential and metallic concentration may be changing simultaneously, making it difficult at least, or impossible in the extreme, to separate with confidence, the sole effect of elevated temperature.

Despite the complexity, the problem that has plagued the ecologist is one of lack of information on the precise point at which death by heat occurs. It is still not perfectly clear whether the lethal temperature observed in the field is the *mean* temperature over a selected period of time or the *highest* temperature encountered over a decided period of exposure. For practical purposes these obstacles can be overcome and thermal limits *can* and *must* be defined. The question does not revolve about whether there will be a limit, but rather what that limit will be and what variances will be permitted—if any—and for what reasons.

In an effort to overcome the many problems associated with man-made thermal additions to the coastal marine environment, laboratory investigations were conducted to examine the effect of *temperature alone* on the viability of selected tropical and subtropical organisms. The experiments concerned various organisms which were important at selected trophic levels; this allowed a very close integration with readily observable field information.

The organisms investigated in the laboratory were selected from the many found living on Biscayne Bay and Card Sound. Over 18,000 individuals were examined in order to determine the temperature effect on their viability. Lethal temperature limits were determined for 27 different species and life stages. This composite included nine species of important benthic plants, selected life stages of the stone crab (*Menippe mercenaria*) and the pink shrimp (*Penaeus duorarum*). Five species of caridean shrimp, very important members of the food chain and thus the entire ecological system of the South Florida Bays were also studied.

The results of the laboratory studies are given in the included illustration (see Section IX, Figure IX.1).** The most important observation to be gained from the experimental data is that tropical organisms are living very close to their thermal limit. Temperature increases, by natural means or otherwise, can cause an elimination of species and thus an interruption of the ecological balance.

Field investigations substantiated the laboratory data. Our observations indicated that areas which average 4.5° F (2.3° C) above the ambient bay temperature (as measured on Pelican Bank) have reduced standing stock of benthic plants and animals during the summer months in an area of about 200 acres; however, in other seasons the production of the normal biota is enhanced by the higher winter temperature and the result is that on an annual basis no decrease in production occurred. At stations which averaged 6.3° F (or 3.3° C) above the ambient temperature the plants and animals were severely depleted during the summer months in an area of about 75 acres. Some recovery is noted in winter and early spring but the net effect is lowered annual production. The recovery in winter does, however, indicate that if the discharge were terminated this 75 acre area would recover. For example with an average ambient bay temperature in summer of about 89 to 90° F the addition of the temperature anomaly would develop temperatures of 95 to 96° F. Although 95° F is biologically detrimental, it appears that if exposed to no more than 95° F macrobenthic flora and fauna in Card Sound and South Biscayne Bay will recover after the curtailment of discharge.

We have observed that in an area of acres elevated 8.1° F or to an average summer temperature of 98° F, there is virtually no recovery in winter. The high temperature effects added to the increased current velocity at the outfall has caused a barren area and this area will probably only recover slowly.

In summary, laboratory and field observations indicate that the discharge of 90° F water into the Bay waters will be acceptable. The impact of any changes on the use of the waters by all parties should be minimal.

Gilbert L. Voss*

In deriving the average temperature tolerances of marine organisms we confined ourselves to those organisms which are found below mean low water and which belong to groups of organisms important because of numbers, ecological indications, or commercial importance and important in the general biological economy of the tropical shallow waters. We eliminated from our study those species which normally live intertidally or on exposed flats at the water margin. These organisms are adapted to live in regions of wide temperature range and can exist under high temperatures. However, they are not important in the general ecology of tropical waters, are not of importance commercially or recreationally, and do not contribute a significant role in the ecology of the sublittoral zone. Similarly, certain estuarine fishes are found in high temperature water for part of the tidal cycle but cannot survive without cyclic relief from these conditions.

Our recommendations relating directly to the temperature problem were the following: (1) The maximum permissible temperature at the outfall must not exceed 90° F. (2) While the maximum permissible temperature must not exceed 90° F, the cooling of the water in the colder months is critical for spawning of most animals. The recommended temperature limits by month are shown in table 12.1.

(3) We recommend that in consideration of all future developments, industrial, urban, and rural, competent

*State of Florida Department of Polution Control, Miami, Florida, October 13, 1971.
**Author's Note: This figure is reproduced as figure 12.2 of this Guidebook with the permission of Dr. Thorhaug.

Organism	Time of Exposure	Upper Lethal Limit in °C	No. of Organism
I. Plants			
1. *Halimeda incrassata*	8 days	32.9-34.8	152
acclimated 15°C	15 days	34.7-36.6	41
acclimated 30°C	15 days	32.7-34.6	40
2. *Penicillus capitatus*	8 days	31.5-34.7	159
3. *Acetabularia crenulata*	4 days	38.1-39.1	600
4. *Valonia ventricosa*	72 hrs.	30.0-31.5	
5. *V. macrophysa*	42 hrs.	32.0-33.6	
acclimated 30°C	72 hrs.	32.6-34.2	9,725*
acclimated 15°C	72 hrs.	33.2-34.7	
6. *V. utricularis*	120 hrs.	31.0-31.4	
7. *V. aegrophilia*	72 hrs.	31.4-33.0	
8. *V. ocellata*	72 hrs.	32.8-34.0	
9. *Laurencia poitei*	10 days	31.7-34.9	144
II. Invertebrate Larvae			
1. *Penaeus duorarum* nauplii	22 hrs.	30.5-31.5	2,159
2. *P. duorarum* 1st protozoea	18 hrs.	36.0-37.6	
3. *P. duorarum* 3rd protozoea	17 hrs.	36.8-37.8	
4. *P. duorarum* 3rd mysis	72 hrs.	36.8-37.8	
5. *P. duorarum* 1st postlarvae	1 hr.	37.9-40.7	
6. *P. duorarum* late juvenile	40 hrs.	36.3-38.5	
7. *Menippe mercenaria* eggs	40 hrs.	36.3-38.5	3,886
8. *M. mercenaria* 1st zoea	24 hrs.	34.4-36.0	
9. *M. mercenaria* 2nd zoea	91 hrs.	33.1-34.2	
10. *M. mercenaria* 5th zoea	44 hrs.	34.7-35.5	
11. *M. mercenaria* megalopa	16 hrs.	36.0-37.0	
12. *M. mercenaria* zoea/mega.	24 hrs.	30.5-31.4	
13. *M. mercenaria* mega./juve.	24 hrs.	28.9-30.5	
III. Caridean Shrimp			
1. *Tozeuma carolenensis*	72 hrs.	32.3-33.9	768
2. *Palaemonetes intermedius*	72 hrs.	36.2-37.8	570
3. *Paraclimenes* sp.	168 hrs.	36.1-37.6	190
4. *Hippolyte* sp.	48 hrs.	31.0-32.8	66
5. *Leander tenuicornia*	24 hrs.	34.4-35.5	24
IV. Mollusca (intertidal)			
1. *Nassarius vibex*	72 hrs.	37.5-40.2	48
		Total	18,572

*Valonia of all species

Figure 12.2. Upper temperature limits of selected tropical estuarine organisms in the laboratory.

professional ecologists-biologists be consulted by the county prior to approval of such plans. We recommend the formation of such an advisory board on an unpaid, honorary basis with financial support for expenses of the committee paid for by industry. On the recommendation of such an advisory board, a paid consultive committee may be constituted for the study in depth of particular proposals, the expenses paid for by the applicant (industrial or local government). (4) Care must be taken that after permission has been given for plant construction which meets the above requirements, unforeseen bulkheading, causeway construction, canals, dikes, beach replenishment and other developments are not permitted within the area without prior consultation with the county's advisory board. Such later developments might well so change the physical and chemical features of the region as to completely negate all recommendations approved for industrial uses.

The findings of the committee were based upon a variety of types of data: (1) about 40 published papers (therefore reviewed and open to contradiction or criticism by the authors' scientific peers) pertinent to the area and subject; (2) fieldwork on ecological problems in which natural temperatures were involved; (3) laboratory studies of temperature tolerances of a wide range of animals and plants both in mass culture and individually; (4) tissue culture temperature tolerances of a variety of local fishes and (5) the considered opinion of established and experienced scientists other than those on the committee.

In interpreting the data three problems were immediately apparent. I will address myself to these separately. The first is that in the study of the long range effect of widespread and long duration high temperature its effect only upon adult organisms cannot be the sole criterion. Over the long haul, the effect of elevated temperatures at the weakest and most susceptible link in the life history of the organism must be the point to consider. In all marine organisms with which I am familiar, this point is either in the maturation of the gonads, spawning, or the first few days

TABLE 12.1.
Recommended Temperature Limits

January	70°F	July	90°F
February	72°F	August	90°F
March	73°F	September	90°F
April	77°F	October	81°F
May	81°F	November	77°F
June	83°F	December	73°F

of larval life. The temperatures affecting these stages may and usually do lie as much as 10-15° C lower than the thermal death point of the adult. Deleterious effects upon adults occur several degrees below the thermal death point with a corresponding general shift downward all along the life history. Therefore, in the long range assessment of the problem, the weakest link should be the point for consideration as death at any period has the same effect upon the standing crop.

Another problem is the one of duration of high temperatures. In estuarine conditions it is normal for the temperature to rise during the heat of the day, especially during periods of low tide and to fall with the flood tide or during the night when cool waters from further offshore are brought in or the sun's rays are no longer heating the shallow waters. Estuarine animals have evolved adaptations for this cycling temperature condition and can survive the higher temperatures if given relief by the low temperature cycle. If, however, the water is maintained at high temperatures through heat pollution, this cyclic relief is not provided and damage results. This cyclic relief is also necessary on a broader scale at the annual level. It is absolutely necessary in the tropics and warmer temperate regions for the water to cool down sufficiently for spawning to occur. Therefore, care must also be taken that the waters can cool down to the region of about 17-18° C during the winter months. If they are elevated much beyond this, gonad development and spawning do not occur in many South Florida animals with resultant failure of that year's production.

The third main point was that the presence of even large amounts of certain plants and animals in waters with elevated temperatures cannot be taken as evidence that the temperature increase is beneficial to the environment. Certain temperature tolerant or even thermophyllic species of plants and animals are well known. Among these are certain green algae, many blue-green algae and certain so-called fouling organisms such as barnacles, etc. Some of these, especially the blue-greens, are harmful to the biota and their presence is harmful to the environment. They are not part of the beneficial food chain.

In summation, we can now say with confidence that (1) temperatures above 32° C are harmful to marine life, with few exceptions, if sustained over the natural tidal cycles; (2) that this is a general rule throughout the world oceans and may be considerably lower in higher latitudes; (3) that certain ranges of temperature on an annual basis are necessary for spawning, growth and survival; (4) that tropical organisms being subject to less annual and daily temperature ranges than temperate organisms are more susceptible to even small temperature changes than are temperate species; (5) that this reaction to high temperatures is due to basic physiological and biochemical factors within the cell.

See Cole (1974) for the effect of thermal stress conditions on benthic foraminifera in Biscayne Bay.

MAN-MADE POLLUTION (Northern Biscayne Bay): The effect of pollution (untreated domestic sewage) in the northern Biscayne Bay area is described by McNulty (1970). Studies of various organisms were made before and after pollution abatement. The paper lists species and quantities of animals and plants involved on both hard and soft bottoms of the bay. A concise general review of the history of pollution and the marine environment is presented. The abstract and summary of McNulty's (1970) paper is given below together with a few of his illustrations.

EFFECTS OF ABATEMENT OF DOMESTIC SEWAGE POLLUTION ON THE BENTHOS, VOLUMES OF ZOOPLANKTON, AND THE FOULING ORGANISMS OF BISCAYNE BAY, FLORIDA
by
J. K. McNulty

Abstract

Various elements of the biota of northern Biscayne Bay, Florida, were studied before and after abatement of pollution. The population consisted of 136 to 227 million liters per day of untreated domestic sewage. Four years after removal of the population certain changes had taken place. At distances of 100 to 740 meters seaward from outfalls, in water depths of one to three meters in hard bottom, populations of benthic macroinvertebrates had declined from abnormally large numbers of species and individuals to normal numbers of each, while soft-bottom populations had changed qualitatively but not quantitatively. Adjacent to outfalls, populations had increased in numbers of species and numbers of individuals in hard sandy bottoms only. Volumes of zooplankton had decreased to about one-half the preabatement values in poorly flushed waters; elsewhere, they remained about the same. Dissolved inorganic phosphate-phosphorus decreased similarly. Abundance of amphipod tubes had declined markedly, a change not shared by the quantities of other fouling organisms (including barnacles), which remained about the same. There was no evidence of improved commercial and sport fishing following abatement; this is interpreted to mean that long-lasting detrimental effects have resulted from pollution and dredging.

Summary

1. A domestic sewage disposal plant built by the city of Miami went into operation between September 18 and December 15, 1956, with the result that some 30 to 50 million gallons of waste per day were discharged into the disposal plant instead of into northern Biscayne Bay.

2. Ecological field studies by the author and others were conducted in the preabatement period between November, 1953, and September, 1956. Many of the field observations were repeated between November, 1960, and July, 1961. Results of the restudy are reported here, and comparisons of ecological conditions before and after abatement are made.

3. The study includes comparison of preabatement with postabatement distribution and abundance of benthic macroinvertebrates, concentration of dissolved inorganic phosphate-phosphorus, displacement volumes of zooplankton, numbers of barnacles, numbers of amphipod tubes, and volumes of all fouling organisms.

4. The effect of domestic sewage on the benthic macroinvertebrates is analyzed by attempting to show the changes in the clean water bottom communities brought on by pollution. Thorson's method of naming bottom communities is used (Thorson, 1957).

5. Two bottom communities are described, one in soft bottom and the other in hard bottom. Their major characterizing species are, respectively, the ophiuran, *Ophionephthys limicola,* and the venerid lamellibranch, *Chione cancellata.*

6. Pollution affected the soft-bottom community less than it affected the hard-bottom community, possibly because animals of the soft sediment are adapted to higher organic content of the environment than are animals in hard bottom.

7. Near outfalls, in relatively deep water and soft bottom under polluted conditions, there are few species and few individuals per species. Under clean water conditions, the species of polychaetes were different, but the fauna remained impoverished both qualitatively and quantitatively.

8. A few hundred meters seaward from the outfalls, in soft bottom under polluted conditions, about half of the normally common species were missing, while several immigrant species normally not in the community were present. The total quantity of animals before and after abatement remained about the same.

9. Near an outfall, in shallow water and hard bottom under polluted conditions, the fauna was depressed. After pollution abatement, it increased markedly both qualitatively and quantitatively.

10. A few hundred meters seaward from outfalls, in shallow water and hard bottom under polluted conditions, the fauna was unusually diverse and abundant. Many more species were found and were at much higher population levels than under any other conditions.

11. The diversity and abundance of attached vegetation declined between 1956 and 1960, contrary to expectations. It is possible that turbdity and sedimentation from hydraulic dredging for construction of the Julia Tuttle Causeway may have been a major factor in the decline. Another event of possible significance was Hurricane Donna, which, on October 10, 1960, brought sustained easterly winds of 89-101 kilometers per hour with gusts to 129-132 kilometers per hour.

12. The concentration of dissolved inorganic phosphate-phosphorus dropped markedly following abatement of pollution. Mean concentrations in Miami River water dropped to about half the preabatement levels, while decreases to about one-quarter the preabatement level occurred in the northwestern part of the study area.

13. Volumes of zooplankton decreased to about half the mean volumes recorded in preabatement work in the northwestern part of the study area. In other parts of the study area, however, volumes remained about the same, indicating that pollution-related factors stimulated production of zooplankton only in the poorly flushed, nutrient-rich, northwestern study area.

14. The abundance of amphipod tubes declined markedly following abatement of pollution, as shown by the monthly settlement of fouling organisms on glass panels exposed throughout the study area. Neither the total displacement volume of all fouling organisms nor the numbers of barnacles settling per month show a clear relationship to pollution.

15. The literature on pollution in the marine environment is reviewed.

16. A summary of the geology of Biscayne Bay is presented.

17. The history of dredging and filling in the area of interest is reviewed briefly.

MAN-MADE SILTATION/TURBIDITY (Upper Florida Keys): Excessive suspended sediment from dredging-filling operations can kill many organisms by reducing necessary sunlight requirements and by sedimentation effects.

It is commonly thought that excessive siltation (from natural as well as man-made sources) is the most common killer of coral reefs. Griffin and Antonius (1974) have reported significant additions to the turbidity of the Pennekamp Park area waters of the Florida Keys. This 18-month-long study indicates that although turbidity has not (to date) harmed corals within the park, damage to reefs outside the park has taken place. An abstract of their investigation is given below.

TURBIDITY AND CORAL REEF HEALTH IN WATERS OF PENNEKAMP PARK, UPPER KEYS

by

G. M. Griffin and A. Antonius

Turbidity from dredge-fill and other sources was measured for 18 months and effects on water clarity and Pennekamp State Park coral reefs assessed. A typical dredge project, monitored for 12 months, yielded 2×10^9 mg suspended sediment per working day, usually into a 0.3×0.3 inshore area. This is 5% of the total natural load of the *entire* 21 mile inshore park area, a significant local addition. Unless coastal mangroves and sea grasses, the natural sediment traps/water clarifiers, are preserved, clarity will continue to decline. Our survey indicates that all coral reefs in the park remain healthy and, so far, have not been adversely affected by increased sedimentation. However, continued monitoring is needed as development progresses because of damage noted on some reefs outside the park.

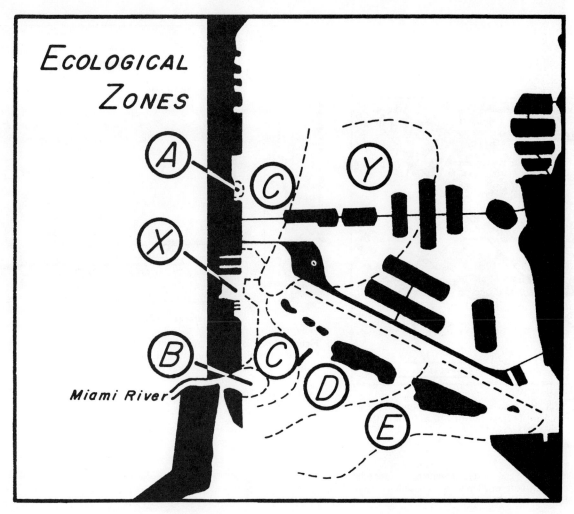

Figure 12.3. Ecological zones. X and Y—soft bottom; A through E—hard bottom. From McNulty (1970).

| | Zone A 20 m from outfalls | | Zone B 100-370m from outfalls | | Zone C 185-740m from outfalls | | Zone D 925-2,220m from outfalls | | Zone E 1,300-4,100m from outfalls | |
| | Sta. 401W | | Stas. 703, 704, 704E 501, 16 | | Stas. 401, 402, 402E, 501, 16, 706 | | Stas. 604, 605, 606, 707, 20 | | Stas. 607, 608, 609, 708, 20E, 73W, 73 | |
Plant Group	1956 ml/m^2	1960 ml/m^2	1956 ml/m^2	1960 ml/m^2	1956 ml/m^2	1960 ml/m^2	1956 ml/m^2	1960 ml/m^2	1956 ml/m^2	1960 ml/m^2
Spermatophytes	–	–	–	–	81.8 *	–	100.5	–	292.0	106.1
Red algae	trace	–	12.0	–	60.8	–	0.9	–	18.0	3.2
Other	–	–	4.5	–	15.0	–	–	–	9.1	–
TOTALS	–	–	16.5	–	157.6	–	101.4	–	319.1	109.3

*Displacement volume in milliliters.

Figure 12.4. Comparison of volumes of major groups of benthic plants in hard-bottom areas (median particle-size 0.05-0.22 mm) in northern Biscayne Bay, Florida, for July-August, 1956 (polluted) and November-December, 1960 (unpolluted), in zones A through E. From McNulty (1970).

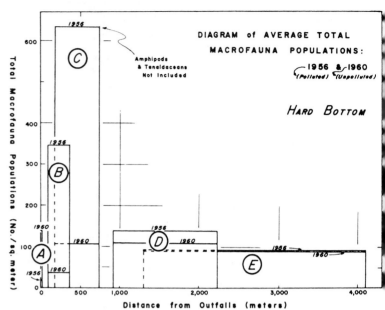

Figure 12.5. Pre- and postabatement populations in 1/4 m² of hard bottom 185 to 740 m from outfalls, zone C. From McNulty (1970).

Figure 12.6. Diagram of mean total numbers of individuals per m² in hard-bottom areas plotted against the distance from outfalls, zones A, B, C, D, and E, for the years 1956 and 1960. From McNulty (1970).

Figure 12.7. Pre- and postabatement populations in 1/4 m² of hard bottom 20 m from an outfall, zone A. From McNulty (1970).

Figure 12.8. Diagram of mean total volumes of sea grasses in hard-bottom areas plotted against distance seaward from outfalls, zones A, B, C, D, and E, for the years 1956 and 1960. From McNulty (1970).

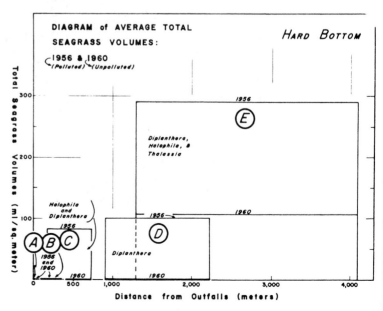

Barnard *et al.,* (1974) document the preservation of noncarbonate detrital fractions in skeletal cavities of live tropical corals. They indicate the usefulness (when detrital fractions are tied to dated growth increments) of such data for baseline studies of environmental stress due to excessive suspended sediment (see figure 12.9). They note that in one Panamanian specimen (shown in figure 12.9) the presence of gibbsite correlates with the start of a local housing development! Their data also includes one *Solenastrea hyades* specimen from Florida Bay with a detrital content including illite, talc, kaolinite-chlorite, gibbsite, quartz and feldspar.

See Baron (1976) for a discussion of the source and dynamics of suspended sediments in nearshore regions of southwestern Florida.

Additional data concerning suspended matter in waters of the Florida Reef tract is noted below by Griffin (1974).

NATURAL VARIATIONS IN SUSPENDED SEDIMENT CONCENTRATIONS, DIVER'S HORIZONTAL VISIBILITY, AND COLOR RECOGNITION LIMITS—UPPER FLORIDA KEYS—1972-1974
by
G. M. Griffin

The concentration of suspended matter in the water column controls the limits of diver's horizontal vision and ability to correctly identify colors. Under some conditions it can also be a significant ecological stress factor. Both of these consequences of turbidity are potentially important in the waters east of the Keys. Accordingly, suspended sediment concentrations were monitored from October 1972 to February 1974 along six traverses and at other spot stations throughout the area. The method utilized a recording 1-meter transmissometer and experimentally determined relationships between total light attenuation and particulate concentrations of < 44 micron lime mud.

Along the traverses, concentrations increased from the outer reefs toward a maximum in Hawk Channel, usually followed by a small decline toward shore (figs. 12.10 and 12.11). In the zone within 100 m of the shore, concentration varied as a function of percentage of bottom covered by sediment-trapping vegetation (mainly *Thalassia*), and proximity to the mangrove fringe.

Overall, the lowest concentrations occurred in the summer, onshore southeast trade wind regime. On some summer traverses, there was little variation from the outer reefs to shore (e.g., figure 12.11, June 15 and August 15, 1973). However, in winter, under a varying wind regime, often including a significant northerly component, overall concentration increased from 3 to 10 fold, and turbid water covered a broad area including Hawk Channel and White Bank. The outer edge of this winter maximum is abrupt and quite visible to the eye, marking a decline from 5-10 mg/1 over White Bank to <1 mg/1 over the outer reef tract. Often the turbid edge is within 0.5 n mile of the outer reefs, and on occasion it overruns parts of the coral area (e.g., figure 12.10, Jan. 23 and Dec. 17, 1973). At these times, reef divers sometimes experience visibilities to white objects of 2 m or less, rather than the more typical 10-20 m.

Figure 12.12 lists mean concentrations at a number of

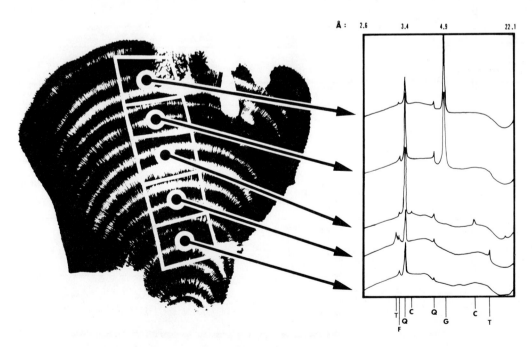

Figure 12.9. X-radiograph of *pavona gigantea* showing variation in composition of noncarbonate detrital fraction relative to annual paired light and dark density bands. Note the appearance of gibbsite (4.85 A) approximately six years ago. C, chlorite; T, talc; G, gibbsite; Q, quartz; F, feldspar. From Barnard *et al.* (1974).

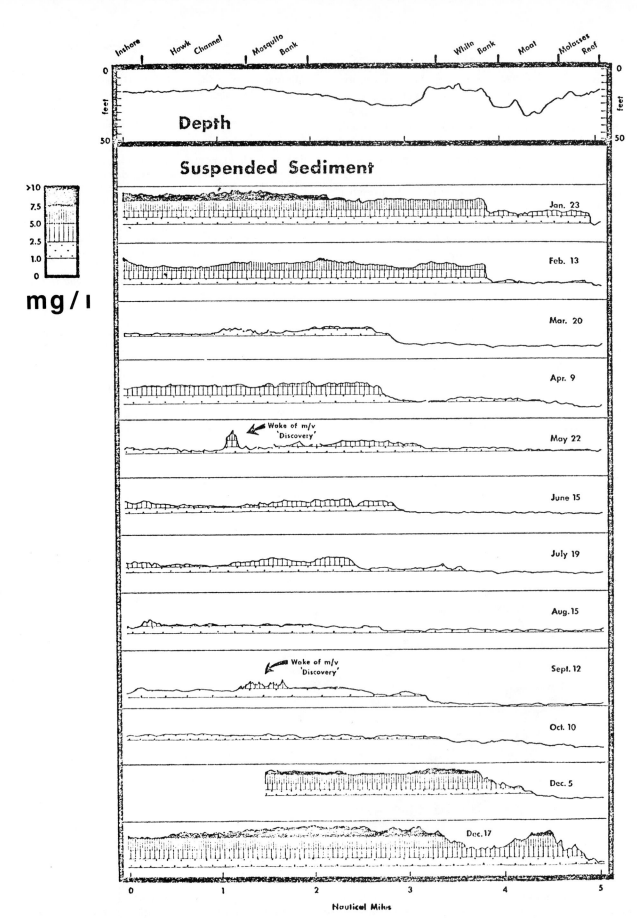

Figure 12.10. Suspended sediment concentrations and bathymetry along a repeated traverse from South Sound Creek to Molasses Reef—1973. From Griffin (1974).

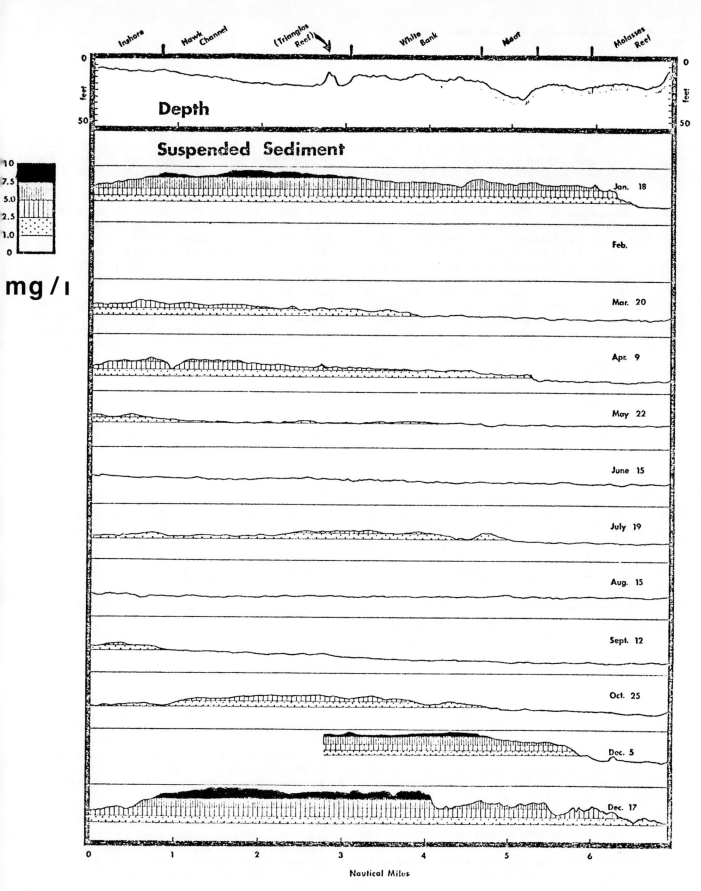

Figure 12.11. Suspended sediment concentrations and bathymetry along a repeated traverse from Tavernier Creek to Molasses Reef—1973. From Griffin (1974).

reef stations in the study area, and the approximate diver's horizontal limits of visibility of black or white 30 cm discs held 2.5 m below the sea surface. The horizontal limits at which divers could correctly identify the color of randomly displayed color discs are also indicated. These values are for 30 cm discs held against the natural hazy white water background, 2.5 m below the surface, with a clear sun nearly overhead (10 am-3 pm). They are averages of several divers, varying from novice to highly experienced. The discs were held so that the direct reflection of the sun was minimized.

Concentrations were lowest in the Florida Straits, averaging 0.50 mg/1 at the locations indicated in the figure 12.12, this increased to 0.80 over the outer reefs, and to 4.4 over the inner reefs. Visibility varied in a reciprocal manner, and the recognition distance for longer wave lengths (red-orange) gradually improved relative to the shorter wave lengths (blue) as suspended sediment concentration increased. On the relatively turbid inner reefs, orange and yellow were recognized at the greatest distance. Because the natural background over the outer reefs is a hazy white ("Ganzfeld"), the black disc produced greater contrast and was visible at slightly greater distances than the white. But, in the more turbid inner reef waters, there was little difference in black or white visibility distances.

Mineralogically, the suspended material over the outer reefs averaged 39% aragonite, 42% high-mg calcite, and 42% normal calcite. Other types of analyses are in progress, and the diver's observations are being refined.

The effects of dredging on water clarity around a Key Largo dredge-fill site (Basin Hills Dredge Project, northern Key Largo) was monitored for a 390 day period using three independent methods: (1) traverses by a towed optical transmissometer, (2) monitoring by sediment traps, and (3) x-ray diffraction of dredged-generated sediment which differed from that of the natural (Hawk Channel) environment. The techniques and results of this comprehensive study, which might well serve as a model for other areas, were presented orally in a paper by Griffin and Manker (1974—see references cited for availability of the complete text of this unpublished report). The conclusions of this report included a series of recommendations concerning future dredge projects in the Florida Keys which are reproduced, with the author's permission, below.

RECOMMENDATIONS CONCERNING FUTURE DREDGE PROJECTS IN THE KEYS*
by
G. M. Griffin and J. P. Manker

Because dredging of the entrance canal at Basin Hills appears to have had no detectable impact on the coral patch reef 0.48 n miles to the NNE or on the remaining grass flat areas, it seems reasonable that future dredging regulations in the Keys could use this project as a minimum model, at least until it is proved that the system can tolerate greater stresses. Based on this general philosophy, it is suggested that future regulations include consideration of these criteria prior to approval:

(1) Significant reefs composed of hermatypic corals, and more than 20 percent alive, within one n mile of the proposed canal must be located and mapped. Canals and related temporary or permanent spoil areas should be positioned so as to approach no closer than 0.5 n mile to such reefs in order that they be protected from excess sedimentation. The more or less continuous linear zones of low (less than 1 foot high) nonreef forming *Porites divaricata* and other similar corals that occur within several hundred yards of shore should not be included in this restriction.

(2) Locations where the surface of the nearshore bottom is composed predominantly of bare limestone bedrock should be favored for entrance canals, and areas of significant *Thalassia* beds should be avoided. In this way, the sediment trapping ability of the *Thalassia* will continue to aid in water clarification.

(3) Also to aid in sediment trapping and water clarification, a fringe of red mangrove should be preserved along the shoreline and care must be taken to preserve its vitality during and after dredging. The width of this zone should be determined by future research; for the present it is suggested that it be at least 100 feet, or no less than the preexisting width if that should be less than 100 feet. (The natural width of the mangrove fringe along Key Largo varies from approximately 60 feet to several hundred yards, and is easily discerned on color aerial photos.)

All spoil shall be deposited no closer to the coastline than the width of this fringe. There should probably be no objection to stilt or catwalk structures, or piers over parts of this fringe zone, so long as they do not involve clearing of vegetation or otherwise interfere with healthy growth of the mangrove.

(4) The number of dredged entrance canals should be limited so as to avoid excessive turbidity during dredging, and also to avoid the low-level turbidity that persists after dredging. A periodicity averaging one entrance canal per linear mile of coast seems reasonable, with the actual canal site being selected so as to avoid live coral reefs and grass flats, as described above.

(5) Between entrance canals, perimeter canals, separated from the coast by the mangrove fringe described above, seem on the whole to be a desirable alternative to an excessive number of entrance canals. However, legislation seems necessary to force property owners to connect into them. Perhaps entrance and perimeter canals should be dedicated for public use in the same way as streets in inland subdivisions.

The maximum depth of perimeter canals should be limited to whatever depth will allow for adequate water exchange with adjacent natural open water bodies. Otherwise the perimeter canals quickly become oxygen depleted, with resulting fish mortality and diminished recreational usage. Also, adequate vents to open water must be provided for oxygen ventilation. It is suggested, in lieu of further research, that vents be provided every 200 linear feet of perimeter canal, and that these be open channels 3 feet eep and 10 feet wide to allow limited passage of small boats. These vents should not extend more than approximately 50 feet seaward of the mangrove fringe.

(6) No additional artificial "cross-key" waterways

---

* This represents only portions of an oral report presented by the authors.

should be allowed between the Atlantic side of the Keys and the Florida Bay, Barnes Sound, Card Sound, side. This restriction would prevent greater influx of the more turbid bay waters into the reef tract area. In addition to higher turbidity, the bay waters also undergo much greater seasonal temperature and salinity fluctuations than the Hawk Channel waters, and all of these factors are detrimental or even lethal to growth of coral and other sensitive organisms of the reef tract area.

(7) The hard-rock dredge techniques described earlier, as employed at Basin Hills, produce much less turbid water than hydraulic dredging. Therefore, it is recommended that no other type of dredging be permitted in the Keys.

Also, bscause the rate of effluent generation and dispersal is important in assessing its effect on water clarity and possible biologic damage, it is recommended that, in lieu of further research, the rate of dredging in the Keys be restricted to that at Basin Hills, i.e., approximately 570 cubic yards per 8-hour working day. In addition, the total rate of fallout should be monitored by sediment traps 100 feet away on both sides of the canal extension, and limited to a maximum 200 mg/cm²/day averaged over a one-week period. If the total fallout exceeds this amount, dredging should pause for one week, to allow the natural forces to clear the organisms of sediment.

(8) Turbidity diapers seem beneficial only if the dredge operator repositions them frequently, so as to close gaps. Attention to this seems especially necessary, in the final phase, when one of the parallel spoil fingers has been completely removed, leaving a large potential opening. Also, gross leaks were frequently observed at anchor points on the corners of the diaper. This suggests that a redesign of diapers is needed to eliminate the depression of the corners.

The diaper allows suspended matter to settle to the bottom instead of being dispersed immediately as a turbid plume. However, no permanent benefit is obtained from this unless the canal is dredged deeply enough to form an effective sediment trap; otherwise, natural waves and currents and boat wakes will resuspend the fines whenever the diaper is removed. Therefore, it is suggested that regulations requiring a diaper, to be effective in reducing turbidity permanently, must be coupled with a requirement that that canal be dredged to several feet below the effective base of the expected disturbances. The minimum required depth would have to be determined by further research, but is probably on the order of 8 to 10 feet. This depth would exceed the maximum of 6 feet previously recommended by the Department of Pollution Control (1973) for all canals. Perhaps the DPC recommendation should be reexamined and possibly applied only to perimeter and other interior canals.

Locations	Susp. Sed. Conc. (optical mg/1)			Approx. Horiz. Visibility Limit (m)*		Approx. Horiz. Limit of Color Recognition (m) *			
	No. of Obsv.	Mean Conc.	Std. Dev. ±	White	Black	Blue	Yellow	Orange	Red
A. *Atlantic Ocean* (Fla. Straits)									
6.3 mi SE of Molasses Rf.	1	< 0.1		> 27	> 27	> 27	> 27	> 21	> 14
1 to 2 mi SE of reefs	3	0.29	0.20	12	13	9.7	8.3	7.7	5.4
1/2 to 1 mi SE of reefs	1	0.38		11.5	12	9.3	8.0	7.4	5.3
1/4 to 1/2 mi SE of reefs	7	0.86	0.46	8.8	9.8	7.5	6.9	6.1	4.7
< 1/4 mi SE of reefs	15	0.50	0.53	10.5	11.3	8.6	7.6	6.8	5.0
All above	26	0.50	0.71	10.5	11.3	8.6	7.6	6.8	5.0
B. *Outer Reef Coast*									
Pacific	11	0.93	0.58	8.4	9.6	7.4	6.7	6.0	4.6
Carysfort	11	0.83	0.48	8.8	9.9	7.6	6.9	6.1	4.7
Elbow	14	0.56	0.45	10.5	11	8.6	7.6	6.8	5.0
Molasses	14	0.80	0.51	9.1	10.1	7.8	7.1	6.3	4.7
Crocker	8	1.03	0.87	8.1	9.3	7.2	6.6	5.9	4.6
All Outer Reefs	58	0.80	0.55	9.0	10.1	7.8	7.1	6.3	4.7
C. *Inner Reefs*									
Mosquito Bank	14	5.53	3.3	2.0	1.5	< 1.0	< 1.0	< 1.0	< 1.0
Triangles	9	3.27	3.0	5.2	5.3	4.2	4.4	4.4	3.3
Hen and Chickens	15	3.90	3.0	4.7	4.2	3.6	3.9	4.2	2.6
All Inner Reefs	38	4.4	3.1	3.9	3.2	2.7	3.3	3.2	2.4

* = 30 cm disc, 2.5 m deep, bright sun, 10 AM to 3 PM.

Figure 12.12. Suspended sediment concentrations, driver's horizontal visibility and color recognition limits—east of the Upper Keys, October 1972-December 1973—all season averages. From Griffin (1974).

(9) Lastly it is recommended that research into the technology of dredging and its potential effects continue. At present there is insufficient quantitative knowledge of at least five points: (a) the tolerance limits of organisms to increased sedimentation and turbidity; (b) the width of mangrove fringe and/or *Thalassia* beds necessary to provide adequate natural suspended sediment traps (i.e., natural water clarification); (c) the ultimate depositional site of the excess particles generated by the dredge; (d) the optimum methods of providing oxygen bearing water to the perimeter and other interior canal systems; and (e) the size distribution of the dredge effluent and the effects of abnormalities in size distribution on the respiration of some of the important organisms of the inshore area.

MAN-MADE STRESS FROM HEAVY METAL INPUT (Upper Florida Keys):

Manker (1975) describes dispersal of heavy metals from areas of great human activity to the Florida reef environment (see figure 12.13). In a significant detailed study Manker measured heavy metal levels in (a) bulk bottom sediment samples, (b) four-micron fractions of bottom sediment, (c) suspended sediment, and (d) living coral specimens—all from the same area of the northern Florida Keys. These various levels of control provided an unusual amount of information. The summary of his paper is given below.

DISTRIBUTION AND CONCENTRATION OF MERCURY, LEAD, COBALT, ZINC, AND CHROMIUM IN SUSPENDED PARTICULATES AND BOTTOM SEDIMENTS— UPPER FLORIDA KEYS, FLORIDA BAY, AND BISCAYNE BAY
by
J. P. Manker

Summary

Concentrations of Pb, Hg, Cr, Co, and Zn have been determined in bottom sediments, the four-micron fraction of bottom sediments, and in suspended particulates from the Upper Florida Keys/Biscayne Bay area. Highest concentrations are found in the four-micron and suspended fraction and may be related to more surface areas being provided by fine particulates for adsorption/absorption of toxic metals. Organics associated with fine particulates also allow additional toxic metals (especially Hg) to be chelated or taken up in organic combinations.

Toxic metal concentrations in suspended versus four-micron fractions are similar as are their particle assemblages indicating the possible existence of equilibrium conditions (chemical and physical) between these fractions. Such equilibrium conditions could be disrupted by increased wave and/or current conditions which would place large volumes of bottom-derived material (4u to 20u) into suspension with associated toxic metals. A great potential exists during high winds for dispersal of this mobile fraction of bottom sediments concentrated in toxic metals to areas of low concentration. Because lagoonal environments contain large amounts of fine material, this
process could take place to a greater extent. Even under normal wind conditions (≡5 kts) two to three times as much fine material is in suspension in bay areas as compared to the reef tract areas.

The mineralogy/chemistry of finer particles may play a role in toxic metal concentration and distribution. In Biscayne Bay, which is mainly a sand-size quartz environment, high metal concentrations were predicted but low values were obtained. Immediately outside the quartz environment (i.e., on the reef tract) where $CaCO_3$ dominates, sediments high in concentrations of toxic metals were recorded. Such a discrepancy may be caused by calcium carbonate particles allowing toxic metals to be adsorbed into the carbonate structure, whereas the SiO_2 structure does not permit such substitution. Other reasons for low concentration of these metals in Biscayne Bay may be caused by the lack of fine particulates in this dominantly sand-size quartz environment for adsorption or absorption of metals, or perhaps most of these metals originate from outside the bay.

In general, toxic metal concentrations in the study area can be correlated with areas of dense population with associated high automobile and boat traffic and improperly monitored and maintained sewage disposal systems. In addition the Turkey Point nuclear/fossil fuel power plant may be linked with some metal concentrations. Pollutants (sewage and toxic metals) introduced in northern portions of the study area are able to move southward toward less populated areas by means of longshore drift and countercurrents present at the shelf margin.

Highest toxic metal concentrations come from sediments receiving effluent from a storm sewer system (Tavernier Key), a marina on Key Largo, and a lagoonal area in Florida Bay (station 8; figure 12.13). The latter two locations should be expected to yield high toxic metal concentrations because they are restricted basins in populated areas where these metals could accumulate. Metal concentrations are also noted near the Turkey Point generating facility for Hg, Zn, Cr, and Co. Build up of toxic metals in the outer reef tract area can be correlated with adjacent regions of maximum population and development. Carbon-12 enrichment in outer reef corals may indicate that sewage, which also carries toxic metals, is reaching the reef environment in the study area.

It is considered that concentration of Pb and Hg in bottom sediments is reaching sufficient levels above background in certain areas of the Upper Keys to be of environmental concern. Such areas are those proximal to sewage outfalls (i.e., Tavernier Key) and in restricted basins (i.e., Pennekamp marina and station 8; figure 12.13). Values are also high in bottom sediments for Cr, Co, and Zn, but little research has been done on the environmental impact of these metals. Therefore, predictions cannot be made for the study area ecosystem concerning the adverse effects of these metals.

As previously indicated, toxic metals are more concentrated in suspended particulates and in the four-micron fraction of bottom sediments. The influences of such concentrations on the environment are unknown. However, adverse effects may be expected in bottom-dwelling organisms that feed on silt and clay fractions of sediments where toxic metal concentrations are high. Organisms that ingest suspended particulates (including plankton) in areas of elevated metal concentrations may be subject to the adverse effects of these toxins.

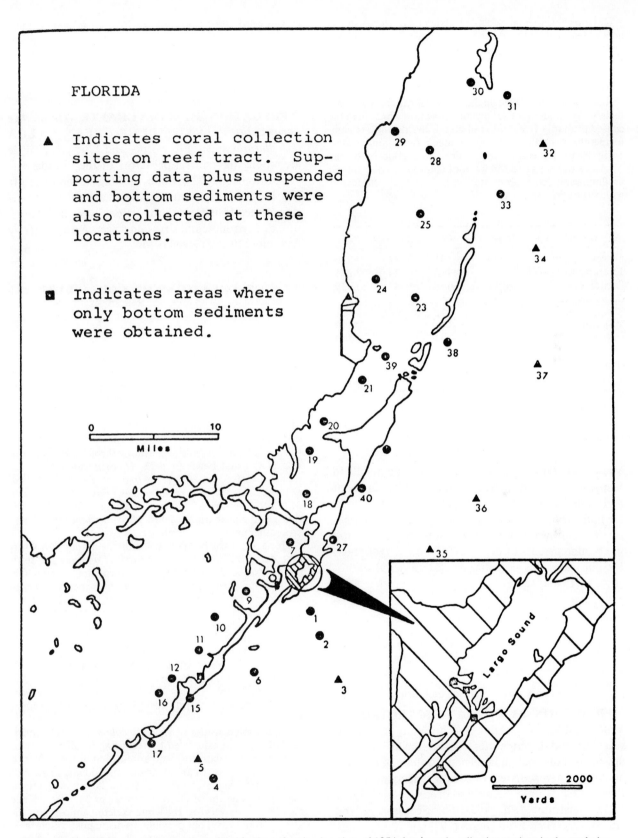

Figure 12.13. Index map of the northern Florida Keys showing location of 1974 data/sample collection stations in the study by Manker (1975). From Manker (1975).

Obvious detrimental effects of toxic metals in the study area are elusive and difficult to define. A reduction in sea grass and green algae populations near industrialized/populated areas has been reported in areas of high toxic metal concentration in this paper and by other workers.

Reefs which displayed highest concentrations of toxic metals have undergone serious deterioration which may be caused primarily from influx of cold water during extreme temperature conditions during winter months and secondarily by toxic metal contamination. Such concentrations of metals may serve to forewarn of continuing toxic metal accumulation commensurate with rapid development within the study area.

Location	Hg	Cr	Co	Zn
Fowey Rocks	549	361	882	1343
Triumph Reef	77	482	124	1285
Pacific Reef	52	678	84	929
Carysfort Reef	36	1381	58	1153
Elbow Reef	104	1272	168	891
Molasses Reef	46	496	73	2329
Hen and Chickens Reef (alive)	37	767	60	2949
Hen and Chickens Reef (dead)	114	694	183	4767

Figure 12.14. Concentration (ppb) of Hg, Cr, Co, and Zn in corals from northern Florida Keys (see location map, Figure 12.13). Data from Manker (1975).

Manker (1975) expresses concern for toxic metals in suspension being dispersed southward from areas of high population to the outer reef tract environment. He indicates one possible mechanism as being plumes carried southward in southwest longshore drift and in countercurrents at depth off the outer reef tract. He illustrates this potential using two aerial photos (figs. 12.15 and 12.16).

SOME NATURAL STRESSES ON ORGANISMS IN THE FLORIDA KEYS

Some Natural Stresses on Corals

PREDATION BY FISH AND SELECTED INVERTEBRATES: Grazing and browsing by certain reef fish and invertebrates are responsible for the destruction of coral and other substrates and the production of new sediment (an important by-product). Figure 12.17 summarizes these processes and figure 12.18 illustrates fish predation. See also Glynn (1973), Gygi (1975) and Odgen (1977).

AGGRESSIVE INTERACTIONS BETWEEN REEF CORALS: Stresses can come from within an individual group of corals. One coral can extrude its mesenterial filaments and feed on an adjacent coral as described by Lang (1971) in Jamaica. This process can also be observed in the Florida Keys area. The result of this aggressive behavior leaves patches of white bare coral skeleton.

PREDATION BY POLYCHAETE (*Hermodice carunculata*) AND BLUE-GREEN ALGA (*Oscillatoria* sp.): Fireworms and blue-green algae are among the natural predators of the Florida corals. Antonius (1973), while working in the reefs off Key Largo, found that the importance of these species seemed to increase proportionally with the pressure of environmental stress on a reef. An abstract by Antonius (1973) is given below.

NEW OBSERVATIONS ON CORAL DESTRUCTION IN REEFS
by
A. Antonius

Hermodice carunculata

A considerable number of invertebrates which feed directly on living stony corals have been discovered over the last decade. One of those is the amphinomid polychaete *Hermodice carunculata* (Pallas). First observed by J. Marsden and P. Glynn in 1962 to feed on *Porites porites* and *Acropora palmata*, the list of known prey species has substantially increased since. Mainly through the contribution of Ott and Lewis in 1972, *H. carunculata* has been reported to feed on: *Porites astreoides, Deploria clivosa, D. strigosa, Montastrea annularis, Favia fragum,* and *Millepora,* besides off nonreef-building actinians, zooanthids, and alcyonarians. Thus, *H. carunculata* appears to be a rather unselective carnivore.

However, it has been observed by the author that wherever *Acropora cervicornis* is present, *H. carunculata* feeds almost exclusively on this species. Surprisingly, the white, tissue-stripped ends of *A. cervicornis* branches, familiar to any experienced diver, have so far not been brought into connection with *H. carunculata* as predator. A theory of natural pruning in *A. cervicornis* has even been published (Shinn, 1972).

Feeding behavior of *H. carunculata* in *A. cervicornis* fields has been observed in 1969 and 1970 at Los Roques, Venezuela; in 1971 at Acklins Island, Bahamas; in 1972 at Carrie Bow Cay, British Honduras; and in 1973 in the John Pennekamp Coral Reef State Park, Key Largo, Florida.

Population densities of *H. carunculata* in *A. cervicornis* fields have been calculated by recording feeding behavior patterns in the laboratory and application of the results to field observations. In an aquarium, individuals of *H. carunculata* of any size between 5 and 20 cm will eat an average of one tip of *A. cervicornis* per 24 hours. Since algae begin to overgrow the white skeleton almost immediately, it is not easy, but possible, to distinguish yesterday's prey from today's. Thus, every brilliant white tip of *A. cervicornis* equals one individual of *H. carunculata*. These white ends of branches were counted inside a 1 m² frame, contiguously moved along transect lines on the Grecian Rock Reef and Key Largo Dry Rocks. This method yielded

Figure 12.15. ERTS imagery of Upper Keys/Biscayne Bay area. *Upper photo:* indicates turbid mass flowing out of Biscayne Bay and from the Virginia Key area onto the reef tract and then south. Low tide conditions prevailed at time photo was taken and winds were from NNE at 13 kts. From Manker (1975).

Figure 12.16. *Lower photo:* indicates turbid masses restricted to lagoonal areas. High tide conditions prevailed at time photo was taken and winds were from NNE at 15 kts. From Manker (1975).

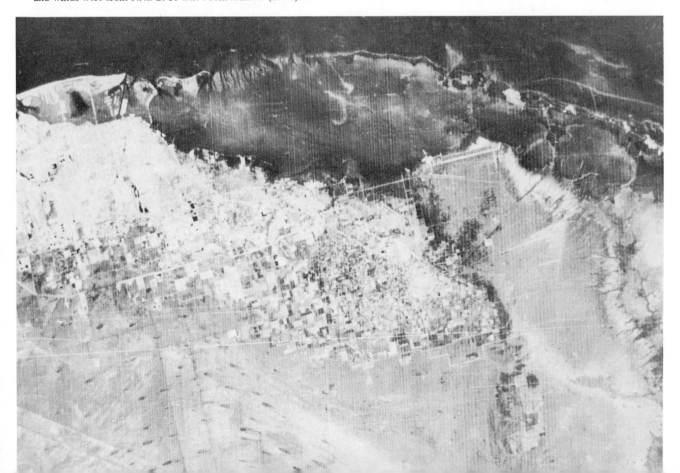

RATES OF EROSION AND TURNOVER BY SOME INVERTEBRATE AND VERTEBRATE POPULATIONS

Species	Locality	Population Density or Biomass	Mass Substrate Removed or Turnover Rate/Yr.	Substrate	Source
Grazing and browsing fishes	Bermuda	55 kg/hectare	0.23 kg/m^2/year turnover, 0.1 kg/m^2/year sediment production.	Coral reef	Bardach, 1961
Stoplight parrotfish *Sparisoma viride*	Bermuda	151 indiv./hectare	2,088 kg/hectar/year turnover 1.2 m uncompacted sediment produced/10,000 yrs/per fish	Calcarenite	Gygi, 1975
Striped parrotfish, *Scarus croicensis*	Panama Caribbean coast	47.2 kg/hectare	1.3 kg/m^2/year turnover, 0.5 kg/m^2/year sediment production.	Coral reef	Ogden (personal communication)
Grazing and browsing fish	Pacific atolls		0.4-0.6 kg/m^2/year production of sand and fine gravel.	Coral reef	Cloud, 1959
Long-spined sea urchin, *Diadema antillarum*	Caribbean, U.S. Virgin Islands	8.7 individuals/ m^2 average	9.3 kg/m^2/year turnover, 4.5 kg/m^2/year sediment production.	Coral reef	Ogden (personal communication)
Urchin *Echinometra*	Caribbean, U.S. Virgin Islands	100 indiv./m^2	3.9 kg/m^2/year	Algal ridge reef	Ogden, 1977
Rock-boring sea urchin, *Echinometra lucunter*	Bermuda	25 individuals/m^2	7.0 kg/m^2/year	Eolianite	Hunt, 1969
Chiton tuberculatus	Puerto Rico	22/m^2	394 gm rock removed/m^2/year	Coral rubble	Glynn, 1970 and unpubl.
Acanthopleura granulata	Puerto Rico	8/m^2	478 gm rock removed/m^2/year	Coral rubble	Glynn, 1970 and unpubl.
Nerita tesselata	Barbados	220/m^2	154 gm rock removed/m^2/year	Beachrock	McLean, 1967
Clinona lampa	Bermuda	Abundant	2.6 × 10^4 gm rock removed/m^2/year	Calcarenitic	Neuman, 1966

Figure 12.17. Some rates of erosion and turnover by some grazing fish, and invertebrates illustrating both the destruction of coral and other substrates and the production of new sediment. Sources noted.

numbers from 0 to 13 white tips per m², or, all presently available data put together, a preliminary average of almost exactly three individuals of *H. carunculata* per m² of *A. cervicornis* field.

Eaten tips of *A. cervicornis* never recover. They are overgrown by a succession of algae, usually blue-greens, sometimes filamentous greens. Parrot fish will then, by browsing, completely remove the overgrown tips. Living tissue later overgrows the stump and heals the injury. However, by removing a noticeable number of tips of living branches, *H. carunculata* may well amount to a controlling factor in a healthy *A. cervicornis* field.

Under adverse conditions, though, the role of *H. carunculata* might become more important. It has been ovserved in the laboratory that *A. cervicornis* colonies which are under stress (e.g., collected, transported, put into an aquarium with different water conditions) will die upon the feeding impact of *H. carunculata*. Subsequent to predation, the living tissue of *A. cervicornis* begins to dissolve just below the eaten tip; the process continues toward the base of the colony and finally kills most or all of the branches. Especially in areas of man-made environmental stresses, this process could become an important factor limiting the growth and development of *A. cervicornis* populations.

Oscillatoria sp.*

Compared with our present knowledge of predators on stony corals (e.g., Robertson, 1970), relatively little is known about biodestruction other than predation. A striking case of coral killing by the blue-green alga *Oscillatoria*. sp. was first discovered by the author in 1972 in British Honduras and reconfirmed in 1973 in the reefs off the Florida Keys.

*Author's note: More recent identification of this blue-green alga by Stan Eiseman, Harbor Branch Lab, has resulted in changing the identification originally used by Antonius (*Schizothrix mexicana*) to *Oscillatoria* sp. The latter has been substituted, with permission of Dr. Antonius, throughout the above abstract.

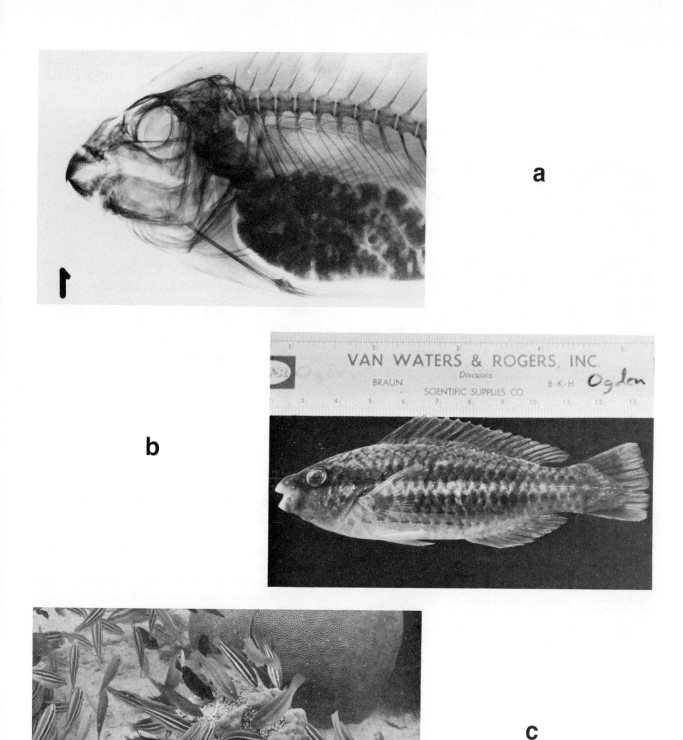

Figure 12.18. Fish can be aggressive predators of reefs and produce large quantities (see figure 12.17) of sediment as a by-product. (a) X-ray photo of Queen Parrotfish *Scarus vetula* showing beak, pharyngeal mill and resultant carbonate sand (in gut), (b) *Scarus croicensis* striped parrotfish showing beak, (c) feeding group of *Scarus croicensis*. All photos courtesy of J. C. Ogden.

Oscillatoria sp. Gomont 1892, family Oscillatoriacea, is very common in shallow marine and freshwater environments of tropical or temperate climate. It grows on a vast variety of substrates, one of which appears to be the living surface of at least four coral species: *Siderastrea siderea, Montastrea annularis, Diploria strigosa,* and *Dichocoenia stellaris.*

The initial stage of infection by *Oscillatoria* sp. has not yet been observed in the field; the smallest alga pitch found so far had a diameter of about 10 cm. However, from this stage on to total destruction of the coral, the succession of events is well documented. The alga on coral heads always appears in the form of a band or belt, 1-3 cm wide, encircling an area of tissue-stripped coral.

MORTALITY BY EXCESSIVE TEMPERATURE CHANGE: Hudson *et al.,* (1976) describe coral mortality due to excessive chilling at Hen and Chickens patch reef in the Florida Keys.

Some Natural Stresses on Mangroves

PREDATION BY WOOD-BORING ISOPOD CRUSTACEAN (*Sphaeroma terebrans*): Mangroves play a key role in lagoonal and littoral environments of the Florida Keys- Bahama area. These plants are an extremely efficient sediment stabilizer and land builder besides providing protection and food for many organisms (see page 61). A threat to the mangrove environment has recently been discussed by Rehm and Humm (1973) who describe the *wood-boring isopod crustacean Sphaeroma terebrans* (fig. 12.19a) as responsible for the destruction of extensive areas of the Ten Thousand Island and in Northern Florida Bay and White Water Bay in the Cape Sabel area. Fortunately the Florida Keys from Key Largo to Key West appear to be free, so far, from this threat.

Adult *Sphaeroma* enter the red prop roots by boring 5 mm size holes (fig. 12.19d). Reproduction within the root increases the amount of boring until the intertidal segment of the root is destroyed (fig. 12.19c). Without these roots, waves undercut the unprotected substrate (fig. 12.19b) and the trees fallover. Shrinkage of mangrove dominated environments along the southwest coast of Florida has been going on due to this mechanism since 1950 according to Rehm and Humm. This reversal in the usual (agressive land building) function of the red mangrove is not only reducing land, environmental nitches and food supply, but can destroy the wave barrier effect that protects many man-made roads and shore developments.

It is interesting to note that Humm (personal communication) believes that some type of intertidal predation may keep the *Sphaeroma* from being so destructive in the Florida Keys. He also notes that red mangroves containing numerous nesting birds seem to be exempt from *Sphaeroma* attack!

Rehm (1976) describes the destructive power of *Sphaeroma terebrans* and notes their distribution (Sapelo Island, Ga., south around southern Florida and then north and westward to Louisiana). He indicates that where boat traffic is heavy and *S. terebrans* active (i.e., Whitewater Bay), erosion of the no longer stabilized sediments destroys mangrove islands.

RECOVERY FROM ENVIRONMENTAL STRESS

Reports on recovery from specific environmental stress are not common. Shinn (1972) discusses coral recovery from natural environmental stress (hurricanes and death by chilling). He also discusses briefly "coral farming" as a recovery method applicable to the Florida reef tract in case of man-induced or natural destruction. Wayne (1974) describes reducing wave energy up to 20 percent by emplacement of sea grass beds such as *Thalassia testudinum.* Collier (1975) discusses natural beach changes and gives details on construction (including costs) of shoreline protection structures for Florida homeowners. Charlton (1976) describes mangroves in Florida in applied terms for those interested in transplanting or using mangroves as stabilizers of coastal sediment. He notes experimental projects and case histories in various Florida areas including the Florida Keys. He describes pruning as a successful method of utilizing mangroves and notes that black mangroves may be more useful than red mangroves as shore stabilizers.

SUMMARY AND SELECTED LIST OF REFERENCES

Marine plants and animals of the Florida Keys are undergoing a wide range of environmental pressures. It should be emphasized that easy accessibility to a great variety of organisms living in various stress situations make the Florida Keys an ideal area for study.

The evaluation of the effect of environmental stress on tropical marine animals and plants has hardly begun. Most environmental stress studies, to date, have been conducted in fresh water and for temperate marine organisms. A selected listing of papers dealing with investigations of marine

Figure 12.19. (a) a 10 mm adult specimen of *Sphaeroma terebrans*, (b) undercutting of the peat substrate after mangrove prop roots have been destroyed by *S. terebrans*, (c) intertidal prop root destruction zone, (d) bored holes from *S. terebrans* action. From Rehm and Humm (1973). Copyright 1973 by the American Association for the Advancement of Science.

organisms under stress and related topics has been compiled for the interested reader and is provided below.

Albertson, H. D. (1972) Comparison of the Upper Lethal Temperatures of Fifty Common Species of Organisms of Biscayne Bay, Florida. Masters Thesis, Univ. of Miami, Coral Gables, Florida.

Alexander, W. B., Southgate, B. A., and Bassindale, R. (1936) Summary of Tees Estuary Investigations. Survey of the River Teers. Part II. The Estuary, Chemical and Biological. *J. Mar. Biol. Ass. U.K.*, v. 20, no. 3, pp.. 717-24.

Anderson, L. (1972) On the Economics of Thermal Pollution. *Quart. Jour. Fla. Acad. Sci.*

Antonius, A. (1969) The Distribution of Stony Corals in the Gulf of Cariaco, an Area of Extreme Environmental Conditions (abst). 8th Meeting Assoc. Island Marine Lab. Carib., p. 24.

———. (1971) The *Acanthaster* Problem in the Pacific (Echinodermata). *Int. Revue Ges. Hydrobiol.*, v. 56, no.2, pp. 283-19.

Bader, R. G., Roessler, M. A., and Thorhaug, A. (1970) Thermal Pollution of a Tropical Marine Estuary. FAO World Conf. on Pollution in the Ocean, FIR:MP/70/E-4: 1-6.

Bader, R. G., Roessler, M. A., and Voss, G. L. (1972) Power Production, Heat Dispersion and Possible Solutions. *Quart. Jour. Fla. Acad. Sci.*

Bardach, J. E. (1961) Transport of Calcareous fragments by Reef Fishes. *Science*, v. 133, pp. 98-99.

Barlow, J. P., Lorenzen, C. J., and Myren, R. T. (1963) Eutrophication of a Tidal Estuary. *Limnol, Oceanogr.*, v. 8, no. 2, pp. 251-62.

Barnard, J. L. (1958) Amphipod Crustaceans as Fouling Organisms in Los Angeles-Long Beach Harbors, with Reference to the Influence of Seawater Turbidity. *Calif. Fish Game*, v. 44, pp. 161-70.

Barnard, J. L., and Reish, D. J. (1959) Ecology of Amphipoda and Polychaeta of Newport Bay, California. *Occ. Pap. Allan Hancock Fdn.*, no. 21, 106 pp.

Basu, A. K. (1965) Observations on the Probable Effects of Pollution on the Primary Productivity of the Hooghly and Mutlah Estuaries. *Hydrobiologia*, v. 25, no. 1-2, pp. 302-16.

Baughman, J. L. (1948) An Annotated Bibliography of Oysters with Pertinent Materials on Mussels and Other Shellfish and an Appendix on Pollution. The Texas A & M Research Foundation, College Station, Texas, 794 pp.

Bellan, G. (1964) Contribution a l'Etude Systematique, Bionomique et Ecologique des Annelides Polychaetes de la Mediterranee. *Rec. Trav. Sta. Mar. d'Endoume, Bul.*, v. 33, no. 49, 371 pp.

Berkely, S. (1972) Some Factors Affecting the Distribution of *Cerithium muscarum* and *Neopanope packardii* (Kingsley) in South Biscayne Bay, Florida. Masters Thesis, Univ. of Miami, Coral Gables, Florida, 54 pp.

Blegvad, H. (1932) Investigations of the Bottom Fauna at Outfalls of Drains in the Sound. *Rep. Danish Biol. Sta.*, no. 37, pp. 1-20.

Butler, P. A., and Springer, P. F. (1963) Pesticides—A New Factor in Coastal Environments. *Trans. N. Amer. Wildl. Conf.*, v. 28, pp. 378-90.

California State Water Pollution Control Board (1956) An Investigation of the Efficacy of Submarine Outfall Disposal of Sewage and Sludge. Publs. Calif. St. Wat. Pollut. Contro Bd., no. 14, 154 pp.

———. (1960) Summary of Marine Waste Disposal Research Program in California. Publs. Calif. St. Wat. Pollut. Control. Bd., no. 22, 77 pp.

———. (1964a) An Oceanographic Study Between Points of Trinidad Head and Eel River. Publ. Calif. St. Wat. Pollut. Control Bd., no. 25, xii + 135 pp.

———. (1964b) An Investigation of the Effects of Discharged Wastes on Kelps. Publs. Calif. St. Wat. Pollut. Control Bd., no. 26, xiv + 124 pp.

———. (1965a) An Oceanographic and Biological Survey of the Southern California Mainland Shelf. Publs. Calif. St. Wat. Pollut. Control Bd., no. 27, xiv + 232 pp.

———. (1965b) An Investigation on the Fate of Organic and Inorganic Water Discharged into the Marine Environment and Their Effects on Biological Productivity. Publs. Calif. St. Wat. Pollut. Control Bd., no. 29, xi + 117 pp.

Davies, J. (1969) Epidemiologic Studies of the Effects of Pesticides on the General Population. Report of the Secretaries Commission on Pesticides and their Relationship to Environmental Health, U.S. Dept. of Health, Education and Welfare, Washington, D.C.

Dean, D., and Haskin, H. H. (1964) Benthic Repopulation of the Raritan River Estuary Following Pollution Abatement. *Limnol. Oceanogr.*, v. 9, no. 4, pp. 551-63.

deSylva, D. P. (1969a) Theoretical Considerations of the Effects of Heated Effluents on Marine Fishes. Pp. 229-93 in *Biological Aspects of Thermal Pollution*, P. A. Krenkel and F. L. Parker (eds), Nashville, Tenn., Vanderbilt Univ. Press.

———. (1969b) The Unseen Problems of Thermal Pollution. *Oceans Mag.*, v. 1, no. 1, pp. 37-41.

deSylva, D. P., and Hine, A. E. (1970) Ciguatera—Marine Fish Poisoning—A Possible Consequence of Thermal Pollution in Tropical Areas. FAO World Conf. on Pollution in the Ocean, mimeo. rep., 8 pp.

Drost-Hansen, W. (1969) Allowable Thermal Pollution Limits—A Physico-Chemical Approach. *Ches. Sci.*, v. 10, no. 3-4, pp. 281-88.

Drost-Hansen, W., and Thorhaug, A. (1967) Temperature Effects in Membrane Phenomena. *Nature*, v. 215, p. 506-508.

Feltz, H. R., and Culbertson, J. K. (1972) Sampling Procedures and Problems in Determining Pesticide Residues in the Hydrologic Environment. *Pesticide Monitoring Jour.*, v. 6, no. 3.

Flice, F. P. (1958) Invertebrates from the Estuarine Portion of San Francisco Bay and Some Factors Influencing their Distributions. *Wasmann Jour. Biol.*, v. 16, no. 2, p. 159-211.

———. (1959) The Effect of Wastes on the Distribution of Bottom Invertebrates in the San Francisco Bay Estuary. *Wasmann Jour. Biol.*, v. 17, no. 1, pp. 1-17.

Foyn, E. (1960) Chemical and Biological Aspects of Sewage Disposal in Inner, Oslofjord. P. 279-94 in *Waste Disposal in the Marine Environment*, E. A. Pearson (ed), New York, Pergamon Press, 569 pp.

Fraser, J. H. (1932) Observations of the Fauna and Constituents of an Estuarine Mud in a Polluted Area. *J. Mar. Biol. Assoc. U.K.*, v. 18, no. 1, 69-85.

Fraser, T. H. (1967) Contributions to the Biology of *Tagelus divisus* (Tellinacea: Pelecypoda) in Biscayne Bay, Florida. *Bul. Mar. Sci.*, v. 17, no. 1, pp. 111-32.

Gaufin, A. R. (1957) The Use and Value of Aquatic Insects as Indicators of Organic Enrichment. Pp. 136-43 in *Biological Problems in Water Pollution*, C. M. Tarzwell (ed), U.S. Public Health Service, Robert A. Taft Sanitary Engineering Center, Cincinnati, Ohio, 272 pp.

Gerchakov, S. M., Rooth, C., Segar, D. A., and Stearns, R. D. (1972) Sediment Temperatures in the Vicinity of a Thermal Discharge. *Bul. Mar. Sci.*

Gerchakov, S. M., Segar, D. And Stearns, R. D. (1971) Chemical and Hydrological Investigation in the Vicinity of a Thermal Discharge into a Tropical Marine Estuary. Univ. of Miami, School of Marine and Atmospheric Science, Miami, Florida, mimeo. rep., 26 pp.

Gilet, R. (1960) Water Pollution in Marseilles and its Relation with Flora and Fauna. pp. 39-56 in E. A. Pearson (ed) *Proceedings of the First International Conference on Waste Disposal in the Marine Environment*. New York, Pergamon Press, 569 pp.

Glynn, P. W. (1963) *Hermodice carunculata* and *Mithraculus sculptus*, Two Hermatypic Coral Predators (abst). 4th Meeting Assoc. Island Marine Lab. Carib., pp. 16-17.

Greenfield, L. J. (1952) The Distribution of Marine Borers in the Miami Area in Relation to Ecological Conditions. *Bul. Mar. Sci. Gulf & Carib.*, v. 2, no. 2, pp. 448-64.

Greenfield, I. J., and Kalber, F. A. (1954) Inorganic Phosphate Measurement in Seawater. *Bul. Mar. Sci. Gulf & Carib.,* v. 4, no. 4, pp. 323-55.

Greenleaf/Telesca Planners, Engineers, and Architects (1972a) Engineering and Economic Report on Solid Waste Collection and Disposal for Metropolitan Dade County, Miami, Florida.

———. (1972b) Summary Report, Solid Waste Collection and Disposal, Miami, Florida.

Hartman, Olga (1955) Quantitative Survey of the Benthos of San Pedro Basin, Southern California; Part 1. Preliminary results. Allan Hancock Pacific Expeditions, 19, pp. 1-185.

———. (1960) The Benthonic Fauna of Southern California in Shallow Depths and Possible Effects of Wastes on the Marine Biota. pp. 57-81 in Proceedings of the First International Conference on Waste Disposal in the Marine Environment, E. A. Pearson (ed), New York, Pergamon Press, 569 pp.

Hohn, M. H. (1959) The Use of Diatom Populations as a Measure of Water Quality in Selected Areas of Galveston and Chocolate Bay, Texas. *Publ. Inst. Mar. Sci. Univ. Texas,* v. 6, pp. 206-12.

Hood, D. W., Duke, W., and Stevenson, B. (1960) Measurement of Toxicity of Organic Wastes to Marine Organisms. *Jour. Wat. Pollut. Control Fed.,* v. 32, no. 9, pp. 982-93.

Hynes, H. B. N. (1960) *The Biology of Polluted Waters.* Liverpool, Liverpool University Press, 202 pp.

Hyperion Engineers (1957) *Ocean Outfall Design.* Los Angeles, Calif., Holmes & Narver, Inc., 851 pp.

Ingram, W. M., and Wastler, T. A. III (1961) Estuaries and Marine Pollution. U.S. Public Health Service, Robert A. Taft Sanitary Engineering Center, Cincinnati, Ohio, Tech. Report W-61-4, 30 pp.

International Atomic Energy Agency (1960b) Disposal of Radioactive Wastes. v. I, Vienna, K. Ring, 603 pp.

———. (1960b) Disposal of Radioactive Wastes., v. II, Vienna, K. Ring, 575 pp.

Jeffries, H. P. (1962) Environmental Characteristics of Raritan Bay, a Polluted Estuary. *Limnol. Oceanogr.,* v. 7, no. 1, pp. 21-31.

Klein, H., Armbuster, J. T., McPherson, B. F., and Freiberger, H. J. (1973) Water and the South Florida Environment. U.S. Department of the Interior Spec. Publ.

Kolipinski, M. C., and Higer, A. L. (1969) Some Aspects of the Effects of the Quantity and Quality of Water on Biological Communities in Everglades National Park, U.S. Geol. Survey Open-File Report, Tallahassee, Fla.

Kolipinski, M. C., Higer, A. L., and Yates, M. L. (1971) Organochlorine Insecticide Residues in Everglades National Park and Loxahatchee Wildlife Refuge, Florida. *Pesticides Monitoring Jour.,* v. 5, no. 3.

Lackey, J. B. (1960) The Status of Plankton Determination in Marine Pollution Analysis. pp. 404-12 in *Proceedings of the First International Conference on Waste Disposal in the Marine Environment,* E. A. Pearson (ed), New York, Pergamon Press, 569 pp.

Leach, S. D., Klein, H., and Hampton, E. R. (1972) Hydrologic Effects of Water Control and Management of Southeastern Florida. Florida Dept. of Natural Resources, Bureau of Geology, Report on Investigation 60.

Lee, T. N. (1972) Effects of Power Plant Discharge on Exchange Processes in Shallow Estuaries. *Quart. Jour. Fla. Aca. Sci.*

Liebman, E. (1940) River Discharges and Their Effect on the Cycles and Productivity of the Sea. *Proc. Sixth Pacif. Sci. Congr.,* v. 3, pp. 517-23.

Lynn, W. A., and Yang, W. T. (1960) The Ecological Effects of Sewage in Biscayne Bay, Oxygen Demand and Organic Carbon Determinations. *Bul. Mar. Sci. Gulf & Carib.,* v. 10, no. 4, pp. 491-509.

MacGinitie, G. E. (1939) Some Effects of Fresh Water on the Fauna of a Marine Harbor. *Amer. Midl. Nat.,* v. 21, pp. 681-86.

Macintyre, I. G., and Pilkey, O. H. (1969) Tropical Reef Coral: Tolerance of Low Temperatures on the North Carolina Continental Shelf. *Science,* v. 166, pp. 374-75.

Mackenthum, K. M. (1965) Nitrogen and Phosphorus in Water. U.S. Public Health Service, Government Printing Office, Washington, D.C., 111 pp.

Marsden, J. R. (1962) A Coral Eating Polychaete. *Nature,* v. 193, no. 4,815, p. 598.

Mauchline, J., and Templeton, W. L. (1964) Artificial and Natural Radioisotopes in the Marine Environment. *Oceanogr. Mar. Biol.,* v. 2, pp. 229-79.

Mayor, A. G. (1916) The Lower Temperatures at which Reef-Corals Lose Their Ability to Capture Food. Cargnegie Inst. Wash., Year-Book no. 14, p. 212.

McCulloch, D. S., Conomos, T. J., Peterson, D. H., and Leong, K. (1971) Distribution of Mercury in Surficial Sediments in San Francisco Bay Estuary. U.S. Geol. Survey Open-File Report.

McNulty, J. K. (1957) Pollution Studies in Biscayne Bay During 1965. Progress Report to Nat. Inst. Health, The Marine Laboratory, Univ. of Miami, Coral Gables, Fla., Mimeo. Report Series, no. 57-8, 15 pp., unpublished manuscript.

———. (1961) Ecological Effects of Sewage Pollution in Biscayne Bay, Florida: Sediments and the Distribution of Benthic and Fouling Macroorganisms. *Bul. Mar. Sci. Gulf & Carib.,* v. 11, no. 3, p. 394-447.

McNulty, J. K., Reynolds, E. S., and Miller, S. M. (1960) Ecological Effects of Sewage Pollution in Biscayne Bay, Florida: Distribution of Coliform Bacteria, Chemical Nutrients, and Volume of Zooplankton. pp. 189-202 in Biological Problems in Water Pollution, C. M. Tarzwell (compiler), U.S. Public Health Service, Robert A. Taft Sanitary Engineering Center, Cincinnati, Ohio, Tech. Report W 60-3, 285 pp.

McNulty, J. K., Work, R. C., and Moore, H. B. (1962a) Level Sea Bottom Communities in Biscayne Bay and Neighboring Areas. *Bul. Mar. Sci. Gulf & Carib.,* v. 12, no. 2, pp. 204-33.

———. (1962b) Some Relationships Between the Infauna of the Level Bottom and the Sediment in South Florida. *Bul. Mar. Sci. Gulf & Carib.,* v. 12, no. 3, pp. 322-32.

Meyer, R. W. (1971) Preliminary Evaluation of the Hydrologic Effects of Implementing Water and Sewerage Plans, Dade County, Florida. U.S. Geol. Survey Open-File Report 71003.

Meyers, S. P. (1953) Marine Fungi in Biscayne Bay, Florida. *Bul. Mar. Sci. Gulf & Carib.,* v. 2, no. 4, pp. 590-601.

———. (1954) Marine Fungi in Biscayne Bay, Florida II. Further Studies of Occurrence and Distribution. *Bul. Mar. Sci. Gulf & Carib.,* v. 3, no. 4, pp. 307-27.

Michel, F. (1970a) Addendum to Technical Report Dated May 1970 Analysis of the Physical Effects of the Discharge of Cooling Water into Card Sound by the Turkey Point Plant of FPL. Univ. of Miami, School of Marine and Atmospheric Science, Miami, Florida, mimeo. Report ML 71001, 4 pp.

———. (1970b) An Analysis of the Physical Effects of the Discharge of Cooling Water into Card Sound by the Turkey Point Plant of FPL. Univ. of Miami, School of Marine and Atmospheric Science, Miami, Florida, mimeo. Report, 23 pp.

Minkin, J. L. (1949) Biscayne Bay Pollution Survey May-October, 1949. Florida State Board of Health, Bureau of Sanitary Engineering, Jacksonville, Fla., 78 pp. mimeographed.

Mohr, J. L. (1953) The Relationship of the Areas of Marine Borer Attack to Pollution Patterns in Los Angeles-Long Beach Harbors. p. I-1 through I-5 in Marine Borer Conference, Miami Beach, Fla., 1952, The Marine Laboratory, Univ. of Miami, 144 pp., mimeographed.

———. (1960) Biological Indicators of Organic Enrichment in Marine Habitats. pp. 237-39 in Biological Problems in Water Pollution, C. M. Tarzwell (compiler), U.S. Public Health Service, Robert A. Taft Sanitary Engineering Center, Cincinnati, Ohio, Tech. Report W60-3, 285 pp.

Moore, D. R. (1963) Distribution of the Sea Grass, *Thalassia,* in the United States. *Bul. Mar. Sci. Gulf & Carib.,* v. 13, pp. 329-42.

Moore, H. B., and Frue, A. C. (1959) The Settlement and Growth of *Balanus improvisus, B. eburneus* and *B. amphitrite* in the Miami Area. *Bul. Mar. Sci. Gulf & Carib.,* v. 9, no. 4, pp. 421-40.

Moore, H. B., and Gray, M. K. (1968) A Pilot Model of a Programmer for Cycling Temperature or Other Conditions in an Aquarium. Univ. of Miami, School of Marine and Atmospheric Science, Miami, Fla., mimeo. Report ML 68161, 4 pp.

———. (1969) A Temperature Control System for Running Sea Water. Univ. of Miami, School of Marine and Atmospheric Science, Miami, Florida, Mimeo. Report ML 69090, 8 pp.

———. (1970) A Pilot Model of a Salinity Control System for Running Sea Water. Univ. of Miami, School of Marine and Atmospheric Science, Miami, Florida, Mimeo. Report ML 70073, 9 pp.

Moore, H. B., and Lopez, N. N. (1969) The Ecology of *Chione cancellata. Bul. Mar. Sci.,* v. 19, no. 1, pp. 131-48.

National Academy of Engineering, The Environmental Studies Group for the (1969) Environmental Problems in South Florida. Washington, D.C.

National Research Council (1957) The Effects of Atomic Radiation on Oceanography and Fisheries. National Acad. of Sciences, Washington, D.C., publ. no. 551, ix + 137 pp.

———. (1959) Radioactive Waste Disposal into Atlantic and Gulf Coastal Waters. National Acad. of Sciences, Washington, D.C., viii + 37 pp.

Nelson, T. C. (1960) Some Aspects of Pollution, Parasitism and Inlet Restriction in Three New Jersey Estuaries. pp. 203-11 in *Biological Problems in Water Pollution,* C. M. Tarzwell (compiler), U.S. Public Health Service, Robert A. Taft Sanitary Engineering Center, Cincinnati, Ohio, Tech. Report W60-3, 285 pp.

Newell, G. E. (1959) Pollution and the Abundance of Animals in Estuaries. pp. 61-69 in *The Effects of Pollution on Living Material,* W. G. Yapp (ed), London, The Institute of Biology, 154 pp.

Nugent, R. S., Jr. (1970a) The Effects of Thermal Effluent on Some Macrofauna of a Subtropical Estuary. Ph.D. Dissertation, Univ. of Miami, Coral Gables, Florida, 198 pp.

———. (1970b) Elevated Temperatures and Electric Power Plants. Trans. Amer. Fish. Soc., v. 99, no. 4, pp. 848-49.

Odum, H. T. (1960) Analysis of Diurnal Oxygen Curves for the Assay of Reaeration Rates and Metabolism in Polluted Marine Bays. pp. 547-55 in Waste Disposal in the Marine Environment, E. A. Pearson (ed), New York, Pergamon Press, 569 pp.

Odum, H. T., DuRest, R. P. C., Beyers, J., and Allbaugh, C. (1963) Diurnal Metabolism, Total Phospborus, Ohle Anomaly, and Zooplankton Diversity of Abnormal Marine Ecosystems of Texas. Publs. Inst. Mar. Sci., Univ. Texas, v. 9, pp. 404-53.

Ott, B., and Lewis, J. B. (1972) The Importance of the Gastropod *Corallophila abbreviata* (Lamarck) and the Polychaete *Hermodice carunculata* (Pallas) as Coral Reef Predators. *Canad. J. Zool.,* v. 50, pp. 1651-6.

Reeve, M. R. (1964) Studies on the Seasonal Variation of Zooplankton in a Marine Subtropical Inshore Environment. *Bul. Mar. Sci.,* v. 14, no. 1, pp. 102-22.

———. (1970) Seasonal Changes in the Zooplankton of South Biscayne Bay and Some Problems of Assessing the Effects on the Zooplankton of Natural and Artificial Thermal and Other Flucturations. *Bul. Mar. Sci.,* v. 20, no. 4, pp. 894-921.

Reeve, M. R., and Cosper, E. (1970) Acute Effects of Power Plant Entrainment on the Copopod *Acartia tonsa* from a Subtropical Bay and Some Problems of Assessment. FAO World Conf. on Mar. Pollution in the Ocean. Mimeo. Report, 8 pp.

Reish, D. J. (1955) The Relation of Polychaetous Annelids to Harbor Population. Publ. Health Rep., Wash., v. 70, pp. 1168-74.

———. (1957) Effect of Pollution on Marine Life. *Ind. Wastes,* v. 2, pp. 114-18.

———. (1959) An Ecologic Study of Pollution in Los Angeles-Long Beach Harbors, C.alifornia. Occ. Pap. Allan Hancock Fdn. no. 22, 119 pp.

———. (1960) The Use of Marine Invertebrates as Indicators of Water Quality. pp. 92-103 in *Waste Disposal in the Marine Environment,* E. A. Pearson (ed), New York, Pergamon Press, 569 pp.

———. *(1961) The Use of the Sediment Bottle Collector for Monitoring Polluted Marine Waters. Calif. Fish Game,* v. 47, no. 3, pp. 261-72.

Reish, D. J., and Barnard, J. L. (1960) Field Toxicity Tests in Marine Waters Utilizing the Polychaetous Annelid *Capitella capitata* (Fabricius). *Pacif. Nat.,* v. 1, no. 21, pp. 1-8.

Reish, D. J., and Winter, H. A. (1954) The Ecology of Alamitos Bay, California, with Special Reference to Pollution. *Calif. Fish Game,* v. 40, no. 2, pp. 105-21.

Robertson, R. (1970) Review of the Predators and Parasites of Stony Corals, With Special Reference to Symbiotic Prosobranch Gastropods. *Pac. Sci.,* v. 24, pp. 43-54.

Roessler, M. A. (1971a) Effects of a Steam Electric Station on a Subtropical Estuary in Florida. Presented NAE Conf. Power Plant Siting, Washington, D.C., March 17, Mimeo. Report.

———. (1971b) Environmental Changes Associated with the Turkey Point Power Plant on Biscayne Bay, Florida. *Mar. Poll. Bul.,* v. 2, no. 6, pp. 87-90.

Roessler, M. A., Rehrer, R. G., and Garcia, J. (1971) Studies of the Effects of Thermal Pollution in Biscayne Bay, Florida. Final Rep. to EPA Grant Nos. WP-01351-01-A and 18050 DFU Univ. of Miami, School of Marine and Atmospheric Science, Miami, Florida, mimeo. Report, 211 pp.

Roessler, M. A., and Tabb, D. C. (1972) Exclusion Temperatures for Subtropical Estuarine Organisms. *Quart. Jour. Fla. Acad. Sci.*

Roessler, M. A., and Zieman, J. C. (1970) Effects of Thermal Additions on the Biota of Southern Biscayne Bay, Florida. Proc. Gulf and Carib. Fish. Inst. 22nd Annual Session, pp. 136-45.

Rooth, C., and Lee, T. N. (1972) A Method for Estimating Thermal Anomaly Areas from Hot Discharges in Estuaries. *Sea Grant Spec. Bul,* v. 3, pp. 1-5.

Ryther, J. H. (1954) The Ecology of Phytoplankton Blooms in Moriches Bay and Great South Bay, Long Island, New York, *Biol. Bul. Mar. Biol. Lab.,* Woods Hole, Mass., v. 106, no. 2, pp. 198-209.

Segar, D. A. (1971) The Determination of Trace Metals in Saline Waters and Biological Tissues Using the Heated Graphite Atomizers. Conference on the Sensing of Environmental Pollutants, Palo Alto, Calif., AIAA Paper no. 71-1051.

Sherwood, C. B., and Klein, H. (1963) Saline Ground Water in Southern Florida. *Ground and Water,* v. 1, no. 2.

Shinn, E. A. (1972) Coral Reef Recovery in Florida and the Persian Gulf. Report dated May, 1972 of the Environmental Conservation Dept., Shell Oil Co., Houston, Texas. 9 pp.

Smith, F. G. W., Williams, R. H., and Davis, C. C. (1950) An Ecological Survey of the Subtropical Inshore Waters Adjacent to Miami. *Ecology,* v. 31, no. 1, pp. 119-46.

Tebo, L. B., Estes, R. L., and Lassiter, R. R. (1968) Temperature Studies, Lower Biscayne Bay, Florida. Federal Water Pollution Control Administration, Washington, D.C.

Thomas, L. P., Moore, D. R., and Work, R. C. (1961) Effects of Hurricane Donna on the Turtle Grass Beds of the Biscayne Bay, Florida. *Bul. Mar. Sci. Gulf & Carib.,* v. 11, no. 2, pp. 191-97.

Thorhaug, A. (1970) Temperature limits of five species of *Valonia. Jour. Phycol.,* v. 6, p. 27.

———. (1973) The Effect of Thermal Effluent in the Marine Biology of Southeastern Florida. 42 pp. in Thermal Ecology, J. W. Gibbons and R. R. Sharitz (eds), AEC Symp. Series (cerf. 730505).

———. (1974) Transplantation of the Sea Grass *Thalassia testudinum* Konig. Aquaculture, Elsevier Sci. Publishing Co., v. 4, 8 pp.

Thorhaug, A., Devany, T., Pepper, S. (1971) The Effect of Temperature on *Penicillus capitatus* Survival in Laboratory and Field Investigations. *Jour. Phycol.,* v. 7, p. 5-6.

Thorhaug, A., Devany, T., Murphy, B. (1971) Refining Shrimp Culture Methods: the Effect of Temperature. Proc. Gulf Carib. Fish. Instit. 23rd Annual Meeting, pp. 31-38.

Thorhaug, A., Drost-Hansen, W. (1967) Thermal Aspects of Membranes: Phase Transitions. Seventh Inter. Cong. of Biochemistry, Tokyo, Japan.

Thorhaug, A., Sterns, R. D. (1971) A Field Study of Marine Grasses in a Tropical Marine Estuary Before and After Heated Effluents. *Amer. Jour. Bot.* v. 68, no. 5, pp. 412-13.

Veri, A. R. (1971) An Environmental Land Planning Study, South Dade County, Florida. Division of Applied Ecology, Center for Urban Studies, Univ. of Miami.

Vernon, R. O. (1970) The Beneficial Uses of Zones of High Transmissivities in the Florida Subsurface for Water Storage and Waste Disposal. Florida Dept. of Natural Resources, Bureau of Geol., Info. Circular no. 70, Tallahassee, Florida.

Waldichuk, M., and Bousfield, E. L. (1962) Amphipods in Low-Oxygen Marine Waters Adjacent to a Sulphite Pulp Mill. *Jour. Fish. Res. Bd. Can.,* v. 19, no. 6, pp. 1162-5.

Weiss, C. M. (1948) The Seasonal Occurrence of Sedentary Marine Organisms in Biscayne Bay, Florida. *Ecology,* v. 29, no. 2, pp. 153-72.

Wells, J. W. (1957) Coral Reefs, *Mem. Geol. Soc. Amer.,* v. 67, no. 1, pp. 609-31.

Wilkinson, L. (1964) Nitrogen Transformations in Polluted Estuary. Pp. 405-20 in *Water Pollution Research,* E. A. Pearson (ed), v. 3, New York, Pergamon Press, vi + 437 pp.

Woodmansee, R. A. (1949) The zooplankton off Chicken Key in Biscayne Bay, Florida. Unpublished masters thesis, College of Arts and Sciences, Univ. of Miami, Coral Gables, Fla., 110 pp.

Wurster, C. F. (1968) DDT Reduces Photosynthesis by Marine Phytoplankton. *Science,* v. 159, pp. 1474-5.

Young, P. H. (1964) Some Effects of Sewer Effluent on Marine Life. Calif. *Fish Game,* v. 50, no. 1, pp. 33-41.

Zieman, J. C., Jr. (1970) The Effects of a Thermal Effluent Stress on the Sea Grasses and Macroalgae in the Vicinity of Turkey Point, Biscayne Bay, Florida. Ph.D. dissertation, Univ. of Miami, Coral Gables, Fla., 129 pp.

FIELD QUESTIONS
Environmental Quality

HEN AND CHICKENS REEF (About 6.5 km East of Islamorada):

1. Examine this reef which has been essentially dead since 1968.

    a.   Is there any indication of the type of stress(es) that caused death?

    b.   Is there a method by which one could determine the exact time of death of a particular part of this reef?

    c.   How would you evaluate if death could have been from an outside (nonreef) cause such as siltation, hurricane, pollution?

    d.   What has happened to the reef components (delicate branches, massive head, sand, etc.) since death?

    e.   Is there currently any reason to suppose that a vigorous reef will not develop again at this site?

    f.   What would happen if someone transplanted healthy reef organisms to this site? Could a "reforestation project" be successful?

2. *After* completing the above read Hudson *et al.,* (1976).

SAND KEY AND ROCK KEY (About 2 km Apart, Off Key West):

3. One of these reefs is flourishing—one is dead. For the dead reef ask yourself the same questions as noted above for Hen and Chickens. Compare the two dead reefs in as many ways as possible.

4. Visit the flourishing reef and do the following;

    a.   Look for two growth forms of the same coral species. Is the difference in growth form due to environmental stress of some type? Explain. Could one form really be a subspecies? Explain.

Sample No.

Photo No.

5. What would happen if you took corals from the flourishing reef and put them near the dead reef? put them near a tidal inlet from Florida Bay? put them near a sewage outlet? put them in shallower water nearshore during the winter months?

6. List the types of animal and plant predators you can observe actively working on both the dead and flourishing reef. Does your list compiled during the day vary from a list taken during a night dive? How? Why?

7. What causes "baldness" on reef surfaces? Do corals get "bald" only on their top sides? If so—why?

8. List all the various experiments you could run with an active "transplanting" program of various reef organisms. Which of these would be useful for isolating man-made stress vs natural stress?

FRENCH REEF AND KEY LARGO DRY ROCKS (Outer Reef—Northeast of Molasses Reef):

9. Visit French Reef. Is there a *reef* there? Examine the shape and composition of the rubble—what does it indicate? What probably caused the destruction of this reef?

10. Visit Key Largo Dry Rocks. How does this differ from French Reef (both are outer reefs along same platform edge)? Is there any danger of this reef dying off? List the various mechanisms which could cause the death of this reef. Are any of these active today here?

Data Relating to Water and Climatic Conditions in the Florida Keys

For data on tide amplitude, current velocities, surface water temperatures, dissolved oxygen content and salinity on outer reef areas, see Kissling's summary of the outer reef of the Lower Florida Keys written for this Guidebook on page 212.

The following paper by Warzeski (1976b), although on the topic of storm sedimentation in the Biscayne Bay region, summarizes the literature of the effect of climate and the various types of storm sedimentation processes applicable to a much wider area. See also pages 81 and 224 in this Guidebook.

STORM SEDIMENTATION IN THE BISCAYNE BAY REGION
by
E. R. Warzeski

Abstract

The Biscayne Bay region is a low energy environment, characterized by mild climatic conditions during most of the year. Three climatic energy levels are recognized: (1) prevailing mild southeast and east winds, (2) winter cold fronts, (3) rare major storms (hurricanes). Sediment erosion and transport occurs almost entirely during 2 and 3.

Dominant northerly winds of cold fronts cause southward longshore drift of sand along beaches, and southward transport of unstabilized sands on the bay bottom. Muddy sediment stirred into suspension by wave agitation tends to be redistributed (1) from the bay bottom as a whole to seagrass beds, (2) to seagrass covered tidal deltas and tidal bar belts, and (3) out of the bay. Waves breaking along unprotected shorelines contribute to erosion and shoreline retreat.

Wave agitation and storm surge during hurricanes are the major agents which modify sediment bodies in south Florida. Hurricane waves both winnow huge amounts of sediment from the bay bottom and reef tract and erode exposed shorelines. Breaking storm waves can shatter coastal mangroves and damage buildings in flooded areas. Storm surges may cause erosion where channeled through tidal passes or other gaps in barriers. Surge takes the form of a tidal wave under certain circumstances, in which case it may devestate buildings and vegetation in low-lying coastal areas. Receding storm surge deposits silt layers up to 10 cm or more in thickness on flooded land areas. Suspended sediment is also deposited on tidal deltas and tidal bar belts, and carried offshore by storm discharge and subsequent tides.

Man-made land features are generally more vulnerable to storm damage than natural sediment accumulations, the latter having been conditioned by many hurricanes over centuries. Bridge abutments, causeways, and land-fill islands that constrict the flow of storm surge are commonly washed away. "Natural" shorelines, where man's influence has hampered natural systems which prevent or repair storm erosion, are also vulnerable to storm erosion.

Introduction

South Florida and the Bahamas are "low energy" environments experiencing a mild, sunny climate throughout most of the year. Careful examination of Biscayne Bay on a typical summer day makes it clear that very little physical sedimentation—erosion, transport, and redeposition of sediment—takes place under these conditions. The water is clear except in the northern bay where man's influence is felt most. Swift currents in tidal passes carry little suspended sediment and merely shift channel bottom sands back and forth as the tides change. Yet the bay has ample evidence of sediment movement: tidal deltas, channels cut in sediment banks, accretionary beach ridges on Key Biscayne, and eroding mangrove peat exposed beyond the present shoreline.

The purpose of the present paper is to bring together existing literature, primarily from the past 10 to 15 years, which bears on the dynamics of sediment movement in the Biscayne Bay region, and to tie the sedimentary processes to the atmospheric conditions which cause them. The Biscayne Bay region is taken here as the bay itself, Card and Barnes Sounds, immediately adjacent low-lying land areas, and tidal passes connecting to the ocean. Much of this literature concerns specific areas other than the Biscayne Bay region but is relevant to sedimentary processes here.

Month	Mean[1] Precipitation (mm)	Mean[1] Air Temp. (°C)	Surface[2] Water Temp. (°C) Range (mean)	Surface[2] Water Salinity %° (Range (mean)
Jan.	44.0	20.8	13.8-25.0 (19.6)	28.0-37.9 (35.9)
Feb.	54.8	21.7	16.0-27.5 (22.2)	33.1-38.0 (36.0)
Mar.	36.9	23.2	18.2-28.1 (23.5)	33.2-38.2 (36.4)
April	40.7	25.3	21.5-29.7 (25.2)	33.5-38.8 (36.7)
May	85.6	27.0	23.7-30.8 (27.7)	33.3-38.6 (37.0)
June	93.5	28.3	25.9-31.9 (29.4)	32.1-38.8 (36.6)
July	91.8	29.1	27.0-32.5 (30.1)	31.5-38.8 (36.6)
Aug.	101.0	29.4	27.0-33.0 (30.3)	31.5-38.1 (36.6)
Sept.	165.8	28.5	27.0-32.5 (29.4)	33.7-37.6 (36.0)
Oct.	115.9	26.5	22.0-30.8 (27.3)	29.0-38.1 (35.8)
Nov.	99.8	24.1	18.7-28.1 (24.2)	32.5-38.8 (36.3)
Dec.	40.6	21.5	16.5-26.4 (22.1)	32.7-38.4 (36.2)

1. For 10 year period 1951 to 1960—World Weather Records, Vol. I, N. Am. U.S. Dept. Comm., Weather Bureau, Wash., D.C.

2. For 8 year period 1955 to 1962—Surface Water Temperature and Salinity, Atlantic Coast, N. and S. Am. Coast and Geodetic Survey Publ. 31-1, U.S. Dept. Comm., Wash., D.C.

Figure A.1. Monthly precipitation, air temperature, surface water temperature, and surface water salinity records for Key West, Florida. Data compiled by Zischke (1973); see page 23 of this Guidebook.

Ball (1967) recognized that high energy events—storms—play a major role in physical sedimentation in south Florida and the Bahamas. Earley and Goodell (1968) and Wanless (1969) concluded that intermediate to high energy events, including cold fronts and hurricanes, are important in sediment transport and deposition throughout the Biscayne Bay region. Warzeski (1975, 1976) and Wanless (1969) found that this has been the case ever since the postglacial rise of sea level first began to flood the bay between 6,000 and 7,000 and years ago.* Ball et al (1967), Craighead and Gilbert (1961), Perkins and Enos (1968) and Pray (1966) documented the major sedimentary significance of hurricanes. Ball et al., (1967) concluded that traces of these storms make up a disproportionately large part of the Holocene sedimentary record in south Florida when compared to the actual percentage of time that they affect the area.

Existing knowledge provides a good generalized picture of sedimentation within the Biscayne Bay region. A more detailed understanding which will allow prediction of the impact of man-made modifications awaits work on (1) water movements within the bay during storms, (2) concentration and composition of suspended sediments during storms, (3) rates and processes of shoreline erosion in different parts of the bay, and (4) natural systems which prevent or counteract shoreline erosion. Modeling techniques for predicting circulation within Biscayne Bay under various wind conditions have already been developed (Dean, this volume) and could be used to study effects of cold fronts. More complex models might conceivably predict movements during hurricanes of various strengths and paths. Turbidity studies await standardization of the technology for measuring turbidity, so as to allow comparison with results in other investigations as well as an actual research effort within the bay. Research which bears on 3 and 4 has been underway off Key Biscayne and Virginia Key for several years (Wanless and van de Kreeke, pers. comm. 1973). An expanded effort encompassing other parts of the Biscayne Bay region is necessary, however, to produce information usable in set-ting bulkhead lines or predicting the results of development in the region as a whole.

Discussion

Climatic Energy Levels in South Florida

Climatic conditions in south Florida are divisible into three energy levels or intensities, each of which has a distinct sedimentological significance. These are (1) prevailing mild southeast and east winds, (2) winter cold fronts and (3) rare major storms (hurricanes). Sedimentation during prevailing low energy conditions is essentially limited to in situ production of carbonate sediment by benthic organisms, and biological reworking of sediment (bioturbation) by the infauna. Significant erosion and transport of sediment takes place only during intermediate and high energy events—primarily cold fronts and hurricanes. Because the frequency, wind patterns and strengths of these two types of disturbances are radically different, their effects upon the sediments of south Florida and the Biscayne Bay region are also distinctly different.

Cold Fronts

Winter cold fronts pass over Biscayne Bay an average of once a week during the period from November to April. The following description of an average cold front is based on Mooers and Fernandez-Partagas (in preparation). An average cold front affects wind patterns in the Biscayne Bay region for 4 to 5 days, involving a slow 360° clockwise rotation of wind direction (direction from which the wind is blowing). Winds rise above ambient throughout this period, reaching maxima roughly half a day before and after passage of the front itself. Maximum winds ahead of the front are from the southwest and reach 8 m/sec. Maximum winds behind the front are from the northwest and

*These ages are, by convention, in radiocarbon years and are uncorrected. They are roughly equivalent to 7500 to 8500 "earth years" before present.

Date	Depth (meters)	Light (E.V.)	Temperature °C Air	Water Surface	Water Depth	Time	Sky
7/1	4.3		26.0	29.0	29.0	11:30	
7/8	2.5				29.5	11:30	sun
	15.8				25.6	11:30	
7/10			29.0	29.0		1:00	
7/14	1.0			32-33		4:00	
7/16	2.0		19.0	32.0		11:30	
7/17	7.3		30.0	29.4	29.4	1:20	
	1.0			33.3			
7/18	9.2		30.0		29.5	11:30	
7/21	1.8	9.4	28.0		30.2	11:30	cloud
	1.5	11.95				12:30	sun
	12.0	10.0				1:00	sun
7/22	4.3	12.0	29.0	29.9	29.9	11:30	sun
		12.4					
	1.0	13.1		30.0			sun
	4.3	11.1					cloud
7/23	1.6	11.2	28.9	30.5	30.5	12:00	cloud
	15.8				28.7	11:30	
7/27	.9	13.1		30.0		11:30	sun
7/30	16.0				28.7	11:30	
8/3	1.8	12.5				11:30	sun
	0.0	13.2					sun
8/4	2.5	11.5					sun
		12.0					
8/7	15.8	9.6				11:30	sun
8/14	7.6	10.9				11:00	sun
	3.6	9.3				11:20	cloud
	0.0	13.0				12:00	sun
	1.7	12.2				12:00	sun
	0.0	11.3				12:00	cloud
	23.0	7.2					cloud
1/6	1.5	12.6				12:00	sun
		11.9					cloud
	10.0	11.9				1:30	sun
1/7	4.5	11.2	27.0	25.0	24.8		sun
1/8	22.0	8.4		24.0	23.0		sun

Figure A.2. Values recorded for relative light intensities and temperatures in the environments of the outer reef of the lower Florida Keys during the summer of 1973 and January 1974 are listed with depth. The sky condition, whether overcast or clear, and the time of day are also given. For map of area involved and further discussion see page 202 of this Guidebook. Data from Landon (1975).

reach 10 to 15 m/sec. Northerly winds after frontal passage are of greater duration than southerly winds prior to passage. Maximum winds during an exceptional cold front can reach 20 to 26 m/sec (Fernandez-Partagas, pers. comm. 1976).

Wave agitation winnows surface sediments throughout the Biscayne Bay region during passage of a cold front. Breaking waves on shorelines and bars transport longshore sands and cause shoreline erosion.

Winnowing is most effective on open, seagrass-free bottoms. Seagrasses, particularly the turtlegrass *Thalassia testudinum,* act as a baffle to currents and wave surge, partially shielding the bottom. Large quantities of mud (silt and clay-size carbonate and organic detritus) are stirred into suspension. Much of this sediment is removed from Biscayne Bay, Card Sound and Barnes Sound by ebbing tides over a period of several days. Trapping of suspended mud

by seagrass carpets on tidal "deltas" and on the Safety Valve tidal bar belt contributes to the growth of these features.

Aerial photographs show that dominant northerly cold front winds cause a slow southward drift of turbid water within Biscayne Bay. Water is drawn in over the Safety Valve and through the northern tidal passes. In addition to noticeably raising the water level of Card and Barnes Sounds, this piling up of water to the south augments discharge of sediment-laden water from Caesars Creek, the southernmost tidal pass of Biscayne Bay. Partly as a result of this, the creek's ebb tidal "delta"Caesars Creek Bank—is larger than any other in south Florida (Warzeski, 1975, 1976).

Sediment which remains within the bay and sounds is preferentially redeposited on seagrass-covered bottoms due to lower turbulence within the mesh of *Thalassia*

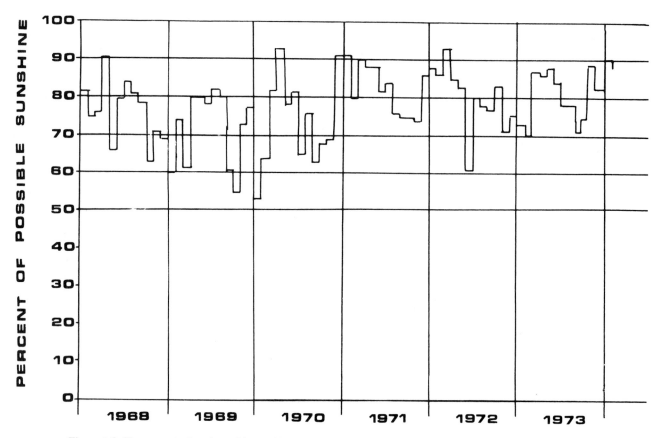

Figure A.3. The percent of total possible sunshine received at Key West for each month. From Landon (1975).

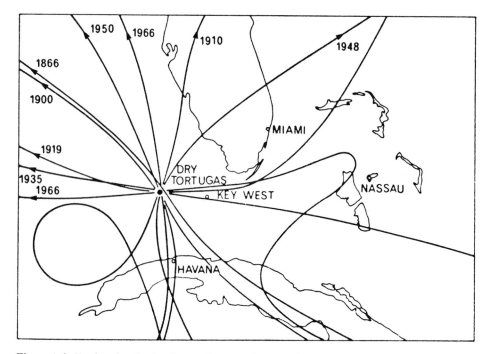

Figure A.4. Tracks of major hurricanes that passed across the Dry Tortugas since 1873. (Redrawn from Sugg *et al.*, 1971.) From Jindrich (1972).

blades. Storms thus yield a net transfer of mud (silt- and clay-size sediment) from the bay bottom as a whole to seagrass beds. This factor plus partial protection of the bottom from winnowing by *Thalassia* causes the higher percentages of mud found in seagrass stabilized substrates by Lynts (1966), Earley and Goodell (1968), and Wanless (1969).

Northerly winds from cold fronts also cause dominantly southward longshore drift of sand along beaches in the Biscayne Bay region and of unstabilized sand across the bay bottom. Southward longshore drift has caused southward accretion of barrier islands along the northern part of the bay over the past several thousand years (Wanless, 1969. Shepard and Wanless, 1971). It provides a continuing supply of sand to nourish beaches except where natural processes are cutoff by man. The distribution of fine-quartz sand in loose bottom sediments of Biscayne Bay, Card Sound and Little Card Sound documents southward movement of sand within the bay and sounds (Vaughan, 1910; Earley and Goodell, 1968; Wanless, 1969).

Shoreline erosion during cold fronts is difficult to separate from that during hurricanes.

Hurricanes

Sedimentary effects of rare major storms—hurricanes and very rare, extremely powerful hurricanes (100 or 500 year storms)—are more profound due to the far greater energy of hurricane winds. As a result hurricanes are disproportionately represented in the sedimentary record of south Florida (Ball *et al.*, 1967). Hurricane force winds affect the Biscayne Bay region at highly irregular intervals, averaging once every 7 years (Dunn and Miller, 1960).

Hurricanes are characterized by extreme wind velocities, from 33 m/sec to over 100 m/sec. They exhibit a counterclockwise wind circulation about an eye several 10's of kilometers or more in diameter. Winds of 30 m/sec or more pound a section of coastline extending 75 kilometers or more away from the eye (Gentry, 1974). The sequence of wind directions experienced by the Biscayne Bay region during a hurricane will depend upon the path of the storm: its direction of approach and whether the eye passes the north, south or over the bay. Hurricanes striking south Florida usually approach from the southeast to east, or southwest resulting in initial north and east winds followed by south and west winds (Ball *et al.*, 1967; Gentry, 1974). Storm surge and high energy wave action due to hurricane-force winds are important agents of sediment erosion, transport and redeposition.

Wave Action. Intense hurricane wave agitation reworks loose sediment throughout the Biscayne Bay region. *Thalassia*-stabilized muddy sediments are resistant to wave surge, and sustain only minor damage. Waves of Hurricanes Donna and Betsy, in 1960 and 1965 respectively, winnowed a few centimeters of sediment at most from seagrass beds in the inner reef tract, Florida Bay and Biscayne Bay (Pray, 1966; Ball *et al.*, 1967; Wanless, 1969). Huge quantities of silt- and clay-size sediment are stirred into suspension, to be carried landward by storm surge and seaward by ebbing storm discharge and tides.

More extensive damage may occur due to breaking of waves against the shoreline, and in flooded areas. Breaking storm waves generate short-term, high velocity currents capable of shattering shoreline mangroves, damaging

buildings, hurling large pieces of debris about, and eroding deeply into sediments along the shore (Craighead and Gilbert, 1962; Ball *et al.*, 1967). Craighead and Gilbert (1962) attribute excavation of "moats" commonly observed around mangrove islands in Florida Bay and elsewhere to the latter process. Ball *et al.*, (1967) found mangroves on the seaward side of Rodriguez Key, offshore from southern Key Largo, splintered and uprooted after Hurricane Donna.

Mangrove destruction during hurricanes and wave surge during both hurricanes and large cold fronts cause erosion of mangrove-stabilized shorelines in the Biscayne Bay region. Sediments overlying bedrock in the bay and sounds and along the present mainland shore record a history of shoreline retreat during the postglacial rise of sea level (Wanless, 1969). Retreat has been greatest where protection from waves of onshore storm winds is least—central and southern Biscayne Bay. An exposed platform of mangrove peat along parts of the mangrove fringe bordering this section of the bay indicates that erosion is still taking place (Wanless, 1969).

Erosion in natural areas, however, is generally significant only on a time scale of centuries or more. This is not the case where man has hampered natural systems which prevent or counteract storm erosion. Beaches where the longshore sand supply has been cutoff by jetties or groins are unable to recover from damage sustained during hurricanes and winter cold fronts. Shorelines from which protective shallow offshore seagrass beds have diminished may experience accelerated erosion. The northeast shore of Virginia Key is an example of both of these problems. The jetties of Government Cut shunt the zone of longshore and transport far offshore, preventing replenishment of the beach. Seagrass cover offshore from Virginia Key has decreased radically in area over the past decades (see Wanless, this volume). This seagrass bed loss may be related to cutoff of the longshore sediment supply. The net result of these factors has been substantial retreat of the northeast shore of the island. Mangrove peat originally deposited in the swamp behind the beach was exposed and eroding in places along the shore prior of the 1973 beach replenishment. In the absence of natural nourishment, future storms are likely to remove this beach as well.

Developed areas where bulkheads are placed too close to the shoreline are also vulnerable to shoreline erosion, particularly when combined with groins. Winter storms tend to erode beaches back, shifting sand to offshore bars. Where erosion reaches bulkheads, reflection of waves off the bulkhead excavates a "moat" along it. Groins, by keeping the zone of longshore drift offshore, prevent natural replenishment. Hurricane waves striking such an exposed bulkhead will undermine it. This occurred on parts of the beach on Key Biscayne during Hurricane Betsy in 1965 (Wanless, pers. comm., 1973).

Storm Surge. Storm tides and currents associated with them are the most damaging aspects of hurricanes in coastal areas. Hurricane winds generate mass flow of shallow coastal waters. Onshore winds pile storm tides of several meters or more against small sections of coastline, causing severe flooding in low-lying areas. Offshore winds tend to lower the coastal water level. Winds blowing along the axis of coastal bays may force water into or out of these water bodies. Storm tides of 2 to 3 meters have occurred within Biscayne Bay, Card and Barnes Sounds during Hurricanes Donna in 1960 and Betsy in 1965 (Perkins and Enos, 1968). The eyes of both of these storms skirted to the south of the Biscayne Bay region. Extreme tides of 4 meters or more were recorded within the bay during major hurricanes which passed directly over Biscayne Bay in 1926 and 1945 (Gentry, 1974).

Constriction of mass, wind-induced flow in narrow gaps between islands produces localized high energy currents. Currents are further enhanced if wind directions and coastal morphology combine to generate high storm tides on one side of the barrier while lowering the water level on the other. Hurricane Donna in 1960, provides an excellent example of this. Her effects are discussed in detail by Ball *et al.,* (1967) and Craighead and Gilbert (1962) and are summarized below.

Hurricane Donna passed northwestward over Florida Bay and Cape Sable. Initial northeast winds flushed water out of Florida Bay into the Gulf of Mexico, dropping the water levels to 0.5 m below MLW in the southern part of the bay and much lower in the northern part. At the same time, a storm tide of up to 4 meters above MLW built up along the southern side of the Florida Keys. Extremely powerful currents flowed bayward through passes between the keys. This situation was reversed after the eye passed and winds shifted 180° (Ball *et al.,* 1967). A 2 to 2.5 meter (6 to 8 feet) tidal wave swept over Cape Sable as water swept back into Florida Bay. The water reached 4 meters above MLW in the bay, and subsided in the reef tract.

The location and magnitude of storm tides and the strength of associated currents in the Biscayne Bay region depend upon the path and strength of the hurricane. The greatest effects result from a powerful storm moving directly onshore over Biscayne Bay (Ball *et al.,* 1967), as did the 1926 hurricane. Such a storm would generate a high storm tide in central and southern Biscayne Bay (4.0 meters at Dinner Key in the 1926 hurricane). Initial northerly winds and high water in the central bay would produce southward flow, raising the water level in southern Biscayne Bay, Card and Barnes Sounds as well. As the eye of the storm moves across the bay, offshore winds to the south of the eye would depress water levels in the reef tract to the south (Gentry, 1974), causing extremely high velocity discharge from Broad, Angelfish and Caesars Creeks.

Powerful storm discharge is potentially an important agent of erosion. Natural sediment bodies have been conditioned by many hurricanes, but still may, rarely, be affected. Norris Cut appears to have been formed when the Great Hurricane of 1835 passed directly over the bay (Chardon, this volume). Wanless (1969) concluded that new channels in the Safety Valve tidal bar belt are cut during major storms. Warzeski (1975a, 1976) found clear evidence of exceptionally powerful storm discharge from Caesars Creek within the past several hundred years. Cores through Caesars Creek Bank show that seagrass cover was ripped out and sediment eroded over much of the bank. A mud storm layer was deposited afterward, probably by waning currents of the same storm. This erosion is attributed to the 1926 and/or 1935 hurricanes on the basis of the storm layer's shallow burial, and aerial photographs showing incomplete recovery of seagrass beds by 1940.

Artificial obstructions to flow are not adapted to such currents and may experience severe erosion. Currents through tidal passes in the keys during Hurricane Donna removed hundreds of cubic meters of fill, undermining the overseas highway and nearby homes (Ball *et al.,* 1967). Water Flowing southward from Card and Barnes Sounds

into Florida Bay washed out two sections of U.S. 1 during Hurricane Betsy (Perkins and Enos, 1968). Portions of the Biscayne Bay region such as northern Biscayne Bay, where man-made features would severely obstruct the flow of storm waters, are particularly vulnerable to a direct strike by a major hurricane.

Hurricane "tidal waves" such as that which struck Cape Sable during Hurricane Donna can devestate developed areas along the coast. Greatest damage to the Biscayne Bay region would be inflicted on the keys and islands on the ocean side of the bay. Low- lying parts of the mainland coast would be hard hit as well. Waves of this type would result primarily from a 180° wind shift after initial winds had blown water away from the shore. A storm would have to approach from an odd direction, such as the north or northeast, in order to produce a mainland "tidal wave" in the Biscayne Bay region.

Much sediment stirred into suspension or eroded during a hurricane is redeposited on land by high storm tides or tidal waves. Hurricane Donna left a silt layer 0 to 13 cm in thickness extending up to 8 km into the Everglades and silt layers up to 6 cm in thickness on various keys in Florida Bay (Ball *et al.,* 1967). Hurricane Betsy generally deposited thinner and less extensive layers in the Florida Bay region (Pray, 1966; Perkins and Enos, 1968), but left up to 5 cm of silt on Key Biscayne (Wanless, 1969). Similar layers have been found in core borings through Key Biscayne, indicating a history of sporadic hurricane deposition over the last 3,000 to 4,000 years.

Storm silt deposits are an expensive and annoying side effect in developed areas, but are important geologically. Supratidal storm deposition is an important process of land-building in south Florida (Ball *et al.,* 1967; Craighead and Gilbert, 1962).

Silt layers are deposited on submerged tidal deltas and tidal bar belts as well. Deposition is due to a combination of (a) waning current energy as the storm passes, (b) decrease in velocity (and therefore sediment carrying capacity) of currents beyond narrow tidal passes, and (c) trapping of suspended sediment by seagrasses. Layers are rarely preserved in the manner of that found near the sediment surface in Caesars Creek Bank (Warzeski, 1975, 1976). Deposition in intact seagrass beds and subsequent burrowing of benthic organisms prevent formation or preservation of discrete layers. Despite this lack of preservation of direct evidence, the growth history of Caesars Creek Bank, (Warzeski, 1975, 1976) and inclusions of storm transported fine quartz sand in the Safety Valve tidal bar belt (Wanless, 1969) indicate that storms, including hurricanes, are a major source of fine sediment for those sediment bodies.

Summary

Sedimentation in the Biscayne Bay region occurs in discrete "events" associated with passage of cold fronts and hurricanes. Little or no erosion or transport of sediment occurs during prevailing mild climatic conditions.

Cold fronts, by virtue of the consistency of their wind patterns and their frequency (30 to 40 times per winter— Novermber-April) have a major role in surface sediment movement. Dominantly southward longshore drift of beach sands, and southward movement of sands on grass-free parts of the bay bottom are products of strong northerly winds after passage of a cold front. In addition, there is a net transfer of mud stirred into suspension by cold front winds (1) from the bay bottom as a whole to seagrass beds, (2) to seagrass-covered tidal "deltas" and tidal bar belts, and (3) out of the bay. Wave surge of hurricanes and cold fronts combine to erode exposed parts of natural mangrove shorelines, and may cause serious erosion of shorelines created or affected by man.

Hurricanes, through wind-induced large-scale flow and wave surge, are the agents of major modifications of sediment bodies, such as cutting of channels, rapid sediment accretion and erosion. Hurricane erosion occurs as a result of (1) high energy storm discharge through tidal passes and other gaps in barriers, (2) wave surge, and (3) "tidal waves." Storm discharge is a threat chiefly to artificial constrictions of flow such as causeways, artificial islands, and bridge abutments. In extreme cases, it may wash away or undermine buildings. Wave surge is a threat (a) to natural shorelines which lack bars, seagrass beds or nearby islands to protect them, (b) to seawalls and bulkheads placed too close to the shore, and (c) to boats, buildings and vegetation near the shoreline. "Tidal waves" would tend to occur where water was first blown away from shore and then blown back after the 180° shift in wind direction as the storm passes. The Biscayne Bay region could be subjected to such a wave only if a strong hurricane approached from an odd direction, such as the

northeast. The chief effects of a tidal wave from the east would be (1) destruction of vegetation and homes on the barrier islands and bedrock keys along the bay's ocean side and on the unprotected mainland shoreline of central Biscayne Bay, (2) landward transport of loose sediment—beach sand, vegetation and man-made debris.

Sediment deposition during hurricanes occurs (1) as surging waters lose energy and retreat, (2) as storm waters enter a much broader and/or deeper body of water. Deposits are rarely preserved except on land where they are not subject to subsequent biologic reworking. Silt layers left by retreating storm waters are usually 5 cm or less in thickness, and rarely exceed 10 cm. They are generally restricted to areas near the shoreline, but may reach several kilometers inland in low-lying areas such as southern Dade County.

Acknowledgment

The author gratefully acknowledges the patience, suggestions, and encouragement of Dr. Harold Wanless. Jose Fernandez-Partagas and Christopher Mooers kindly helped the author to understand the properties of cold fronts, and J. Fernandez-Partagas reviewed the meteorological descriptions of cold fronts and hurricanes herein. Finally, I gratefully acknowledge Sea Grant for publication of the symposium.

Data Relating
to Regional Stratigraphy

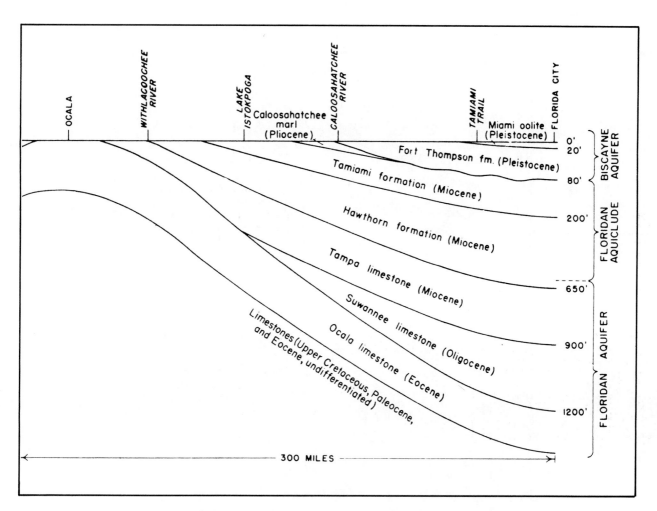

Figure A.5. The geology of south Florida (after Parker *et al.*, 1955).

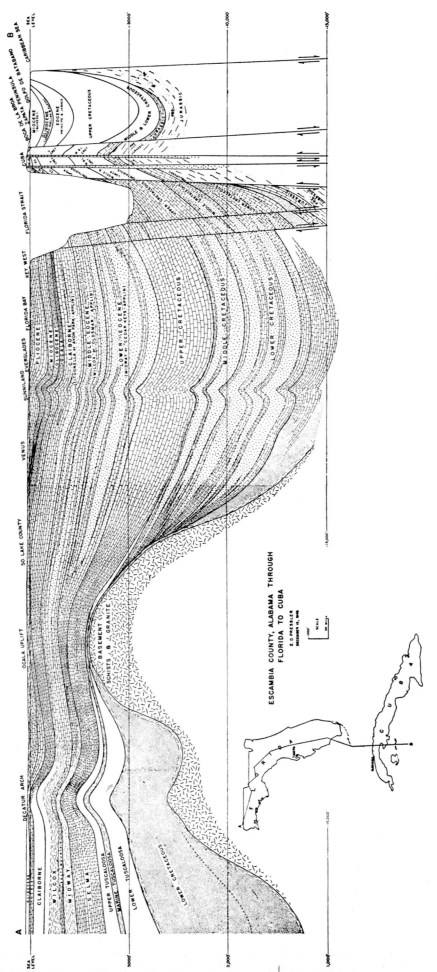

Figure A.6. A north-south cross section of the geology of Florida (after Presseler, 1947).

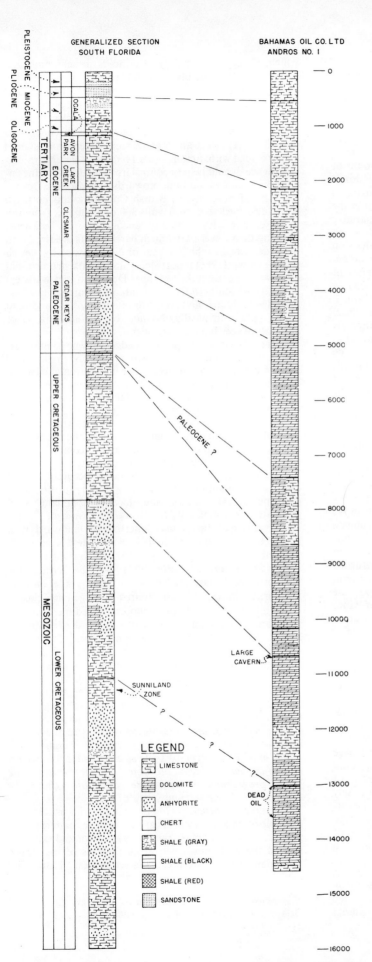

Figure A.7. Comparison of stratigraphic sections, Florida and Bahamas. Depths in feet. From Spencer (1967).

CARBONATE ANALOGS— MODERN AND ANCIENT*

by

J. C. Kraft

One of the most compelling reasons for studying modern carbonate environments of deposition is that models may be constructed for use in the interpretation of ancient carbonate environments. A very large amount of literature has developed on this subject. This discussion is to call your attention to a modern-ancient comparison which well illustrates the hypothesis—"the present is the key to the past." Figures A.8 and A.9 are comparative diagrams showing the areal distribution and cross-section interpretation of two large carbonate platform regions on a similar scale. Figure A.8 shows the Upper Devonian-Frasnian carbonate platform province of the subsurface of the western Canada sedimentary basin to be remarkably similar in areal distribution to that of the Recent through Cenozoic Florida-Bahama carbonate platform province. In both cases the areal geographic extent is approximately 250,000 sq miles. The cross-section interpretation (fig. A.9) shows one major difference. Much deeper "open marine" areas are indicated in the region of the Florida straits, Tongue of the Ocean, etc. Furthermore, the outer edge of the Bahama carbonate platform is located at the continental margin and borders on the deep Atlantic Ocean. However, it will also be noted that strong evidence exists, indicating a structural control for the "ocean deep" areas. A fairly close coincidence between negative gravity anomalies and the deeps has been demonstrated (Talwani 1960). Furthermore, examination of the Recent sediments of the Bahama-Florida area indicates that the areas of limestone deposition are in the shallow marine environment. Accordingly, very precise comparisons of facies change should be possible if one considers these carbonate provinces only in the light of deposition on shallow marine carbonate platforms. Other theories have been proposed in regard to the deep ocean separations of the platforms. Perhaps they have progressively deepened with long-term subsidence of the basin. The nearest sources of possible terrigenous clastic infill lie outside the carbonate province 800 + miles to the north. Accordingly, the deep ocean areas will continue to deepen because of a lack of available terrigenous sediment infill.

The sedimentary facies of the Florida-Bahama carbonate platform are thoroughly discussed elsewhere in this Guidebook and seen in the field trips to the Florida Keys and western Bahamas. Recent facies patterns may be compared with the following statement of carbonate facies change typical of the Upper Devonian of western Canada. The following discussion is a direct quote from Bassett and Stout (1967)—The Devonian of western Canada, page 728—in the publications of the International Symposium on the Devonian System by the Alberta Society of Petroleum Geologists. This statement very precisely describes the sedimentary facies patterns typical of the carbonate platforms of western Canada and calls attention to the fact that this type of sequence appears to be typical of nearly all carbonate platform provinces from Devonian to the present.

"The sedimentary facies adjacent to these margins and in their shelf equivalents have been studied by many workers. The carbonate rocks in front of a platform edge are commonly dark-coloured and rich in skeletal material which, if composed mainly of crinoids, brachiopods and solitary corals, probably represents deposition of what can be termed an 'open marine limestone' or 'limestone and shale' facies bordering the seaward edge of the platform. If the fauna consists of abundant colonial organisms such as corals, *Stromatopora* or calcareous algae, the deposit is thought to represent an 'organic reef' which, once started and provided with, among other factors, a tolerable rate of subsidence, would help to control its immediate marine environment and would build upwards. The Middle Devonian reefs of northeastern British Columbia and northern Alberta formed very steep walls 800 to 1,000 feet from top to base, but with contemporaneous basinal sediments flanking their lower parts. Such barrier reefs are similar to those described in the Permian of New Mexico by Newell (1953). Dooge (1966) reports talus slopes with strong initial dips occurring adjacent to Upper Devonian reef walls exposed in the southern Rocky Mountains. Some of the barrier edge buildups, however, are essentially lime and mud banks, not organically bound, but containing large amounts of reef-building organisms.

"Since the site of the platform edge was one of high carbonate production, the edge was invariably shallow and formed a barrier to effective water circulation shelfward. A typical facies progression shoreward from the barrier edge commences with a belt of variable width composed of light-coloured carbonates containing a limited variety of fossils such as *Amphipora,* ostracods, gastropods and calcispheres which may in places be very abundant. Brachiopods, solitary corals and crinoids are not common in this facies. Such an environment of deposition is termed 'semirestricted.' Adjacent to this facies occurs a belt of carbonates containing a sparse biota limited to ostracods, gastropods, calcispheres and locally abundant algal stromatolites. Commonly the strata are devoid of fossils. This 'restricted' facies may intertongue landward with finely textured and laminated carbonates, commonly dolomite of probably primary origin, which is interbedded with gypsum or anhydrite. This may be termed the 'evaporite carbonate' facies which may grade into an 'evaporite' facies. A varicoloured 'terrigenous clastic' facies may be present in these shoreward sediments if currents are strong enough to carry sand and silt into the basin. This lateral facies progression is repeated vertically in a series of carbonate-evaporite rhythms behind many of the barriers in response to minor changes in sea level combined with basin filling from the shorelines shelfward.

"This briefly-described shoreward facies progression can be traced from west to east in strata ranging in age from Cambrian to Triassic and has been substantiated in Recent carbonate provinces in many parts of the world. *Sedimentary structures, bedding types, carbonate grain types, and general type of skeletal assemblage, characteristic of the various facies, are seen in the same order in the facies progression of carbonate rocks throughout the Phanerozoic.* "(Bassett and Stout '67)

The modern-ancient analog presented in this article is an example of an area in which carbonate facies patterns have played a very large role in the exploration for and exploitation of the well-known petroleum reserves of the Upper Devonian of the western Canada sedimentary basin. The

*Submitted by John C. Kraft, Dept. of Geology, University of Delaware for the 1969 edition of this Guidebook.

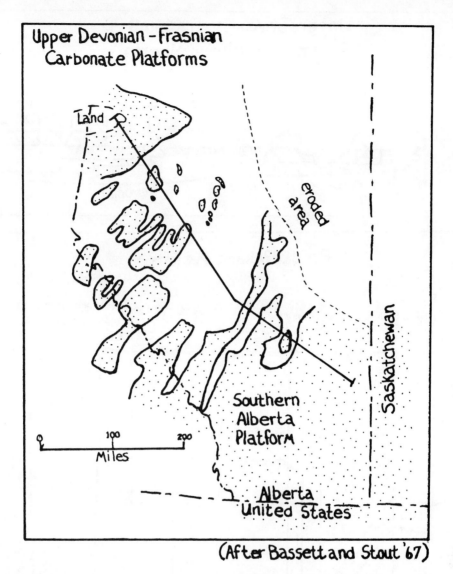

Upper Devonian - Frasnian
Carbonate Platforms

Land

eroded
area

Saskatchewan

Southern
Alberta
Platform

100   200
Miles

Alberta
United States

(After Bassett and Stout '67)

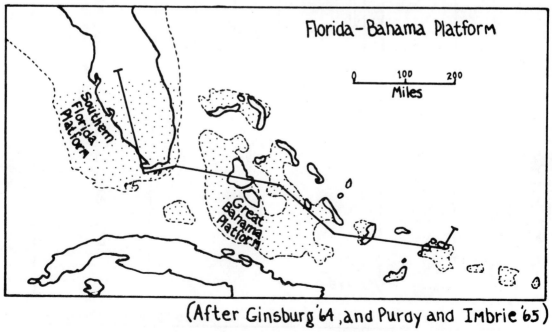

Florida-Bahama Platform

0   100   200
Miles

Southern
Florida
Platform

Great
Bahama
Platform

(After Ginsburg '64, and Purdy and Imbrie '65)

Figure A.8.

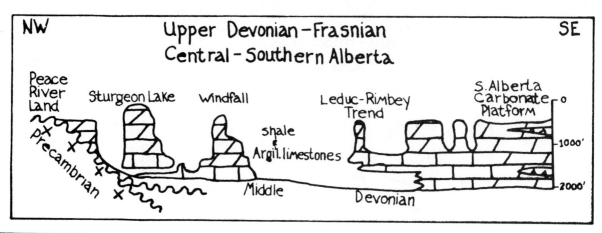

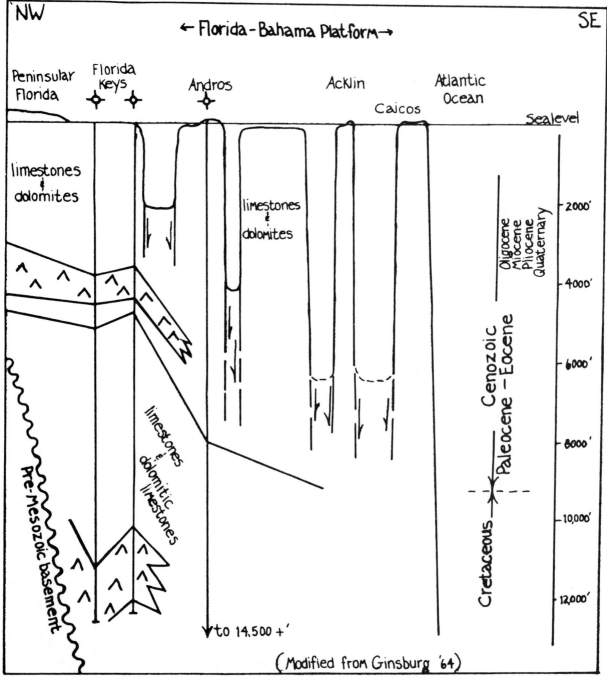

Figure A.9.

list of references are a guide to the very abundant literature on this subject in the Bulletin of Canadian Petroleum Geology and the Bulletin of the American Association of Petroleum Geologists. The list of references is by no means complete; however, it is representative of the literature available for comparisons of carbonate platform facies. No attempt has been made to list references for the many other carbonate platforms of the world which could be treated in a similar fashion.

Selected References to Ancient Carbonate Platform Environments and Sources of Data for the Diagrams

Bassett, H. G. and J. G. Stout (1967) Devonian of Western Canada, International Symposium on the Devonian System, pp. 717-52, Alberta Society of Petroleum Geologists, Calgary, Canada.

Belyea, H. R. (1960) Distribution of some reefs and banks of the Upper Devonian Woodbend and Fairholme Groups in Alberta and eastern British Columbia, Geological Survey of Canada, Paper 59-15.

Dooge, J. (1966) The Stratigraphy of an Upper Devonian carbonate shale transition between the North and South Ram Rivers of the Canadian Rocky Mountains, N. V. Drukkerij, J. J. Groenen ZoomLeiden, 53 p.

Ginsburg, R. N. (1964) South Florida Carbonate Sediments, Guidebook for Field Trip No. 1, Geological Society of America Convention, Miami, Florida, 71 pp.

Hoffmeister, J. F., J. I. Jones, J. D. Milliman, D. R. Moore, and H. G. Multer (1964) Living and Fossil Reef Types of South Florida, Guidebook for Field Trip No. 3, Geological Society of America Convention, Miami, Florida, 28 pp.

Klovan, J. E. (1964) Facies analysis of the Redwater reef complex, Alberta, Canada, *Bull. Canadian Petroleum Geology,* v. 12, p. 1-100.

McCrossan, R. G. and R. P. Glaister, Editors (1964) Geological History of Western Canada, 232 pp. Alberta Society of Petroleum Geologists, Calgary, Canada.

Murray, J. W. (1966) An oil producing reef-fringed carbonate bank in the Upper Devonian Swan Hills member, Judy Creek, Alberta, *Bull. Canadian Petroleum Geology,* v. 14, p. 1-103.

DATA RELATING TO SOME MARINE ORGANISMS OF THE SOUTH FLORIDA-BAHAMA AREAS

The following pages of Appendix 3 represent illustrations, identification and (in some cases) discussion concerning some of the more common organisms seen in the Holocene "carbonate world" of the Florida Keys-Bahama area. This section includes:

1. illustrations of some common corals,
2. illustrations of some other common invertebrates, courtesy of J. A. Zischke,
3. illustrations of some common marine algae,
4. an illustrated discussion concerning bryozoans in Floridian and Bahamian reefs by R. J. Cuffey,
5. an illustrated discussion of Florida/Bahamian Biofacies by A. H. Coogan, and
6. a partial listing of references to living organisms in the south Florida-Bahama area.

A-3

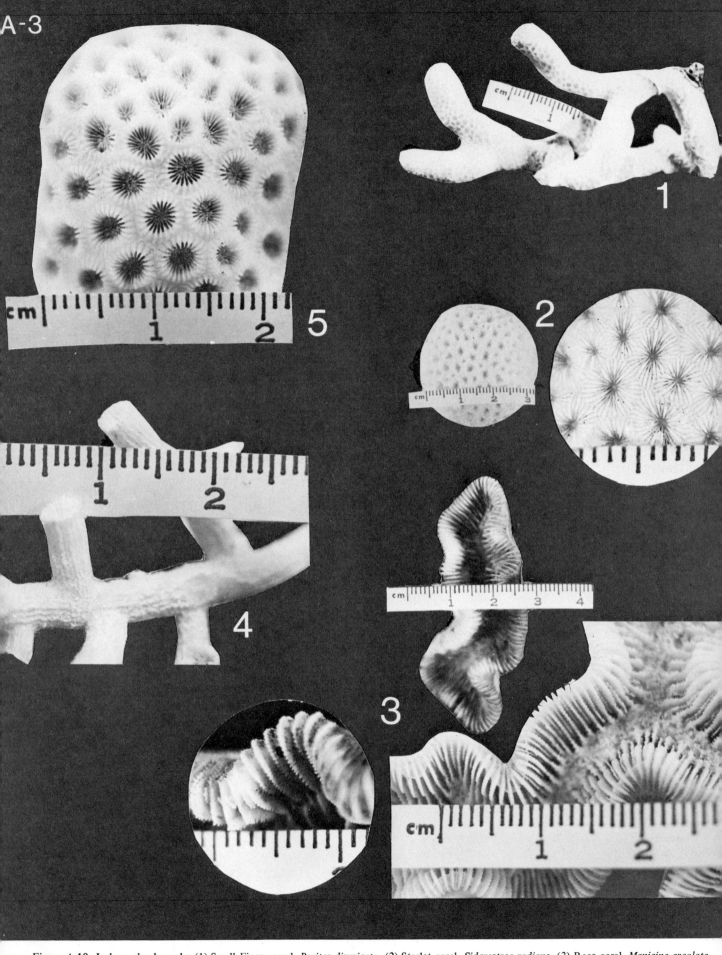

Figure A.10. Inshore shoal corals: (1) Small Finger coral, *Porites divaricata*. (2) Starlet coral, *Siderastrea radians*. (3) Rose coral, *Manicina areolata*. (4) Tube coral, *Cladocora arbuscula*. (5) Knobby Star coral, *Solenastrea hyades*.

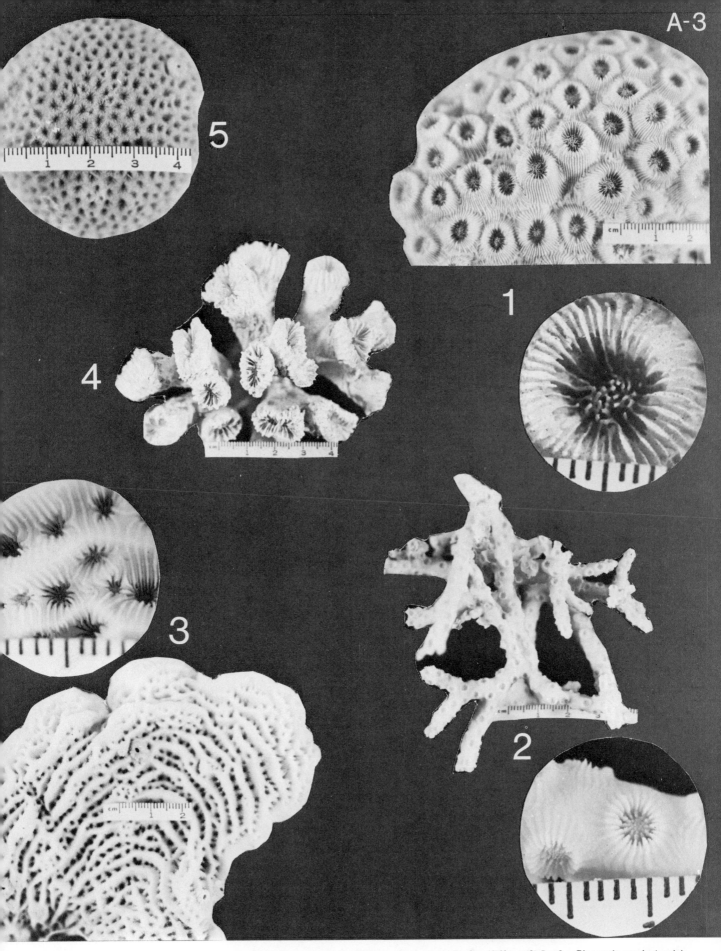

Figure A.11. Accessory corals: (1) Large Star coral, *Montastrea cavernosa.* (2) Ivory Bush coral, *Oculina diffusa.* (3) Leaf or Pineapple coral, *Agaricia agaricites.* (4) Flower coral, *Eusmilia fastigiata.* (5) Starlet coral, *Siderastrea siderea.*

Figure A.12. Massive and encrusting corals: (1) Common Brain coral, *Diploria strigosa*. (2) Encrusting Brain coral, *Diploria clivosa*. (3) Brain coral, *Diploria labrinthiformis*. (4) Common Star coral, *Montastrea annularis*. (5) Porous coral, *Porites asteroidies*.

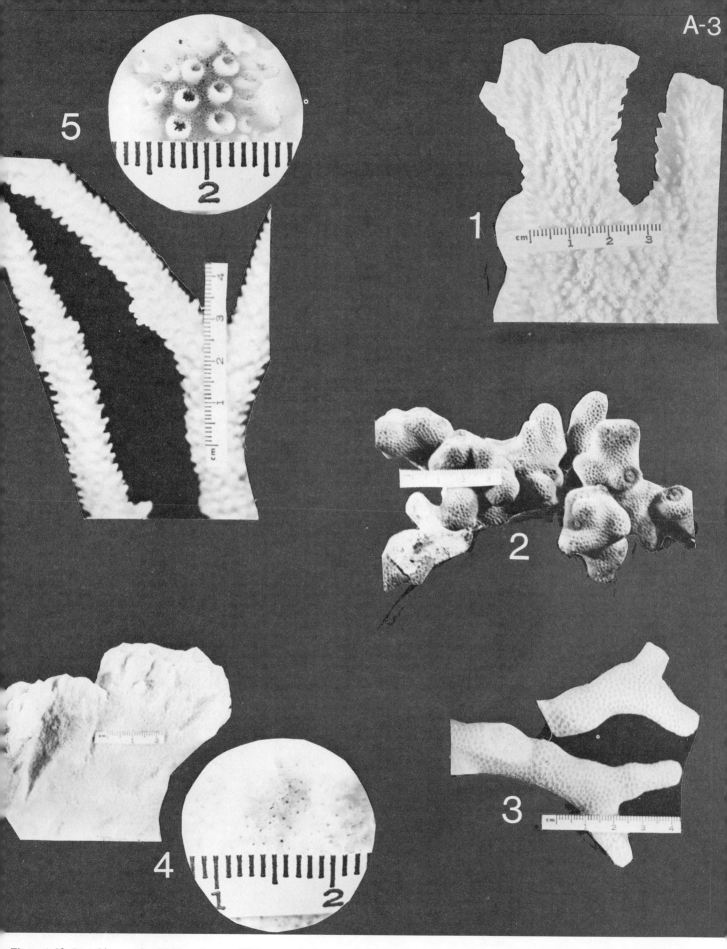

Figure A.13. Branching corals: (1) Moosehorn or Elkhorn coral, *Acropora palmata*. (2) Clubbed Finger coral, *Porites porites*. (3) Finger coral, *Porites porites,* var. *furcata.* (4) Stinging coral, *Millepora alcicornis* (**Note:** not a coral but a hydrozoan). (5) Staghorn or Deerhorn coral, *Acropora cervicornis.*

KEY

Number	Species	Number	Species
1.	*Ircinia campana*	46.	*Cypraea zebra*
2.	*Chondrilla nucula*	47.	*Conus mus*
3.	*Cinachyra cavernosa*	48.	*Columbella mercatoria*
4.	*Callyspongia vaginalis*	49.	*Batillaria minima*
5.	*Oreaster reticulata*	50.	*Fasciolaria tulipa*
6.	*Verongia longissima*	51.	*Astraea tecta americana*
7.	*Spheciospongia vesparia*	52.	*Cerithium litteratum*
8.	*Tethya* sp.	53.	*Nerita versicolor* (side view)
9.	*Porites furcata*	54.	*Nerita versicolor* (ventral view)
10.	*Favia fragum*	55.	*Crepidula fornicata*
11.	*Manicina areolata*	56.	*Crepidula fornicata*
12.	*Millepora alcicornis*	57.	*Siphonaria alternata*
13.	*Siderastrea siderea*	58.	*Diodora listeri*
14.	*Echinometra lucunter*	59.	*Arcopsis adamsi*
15.	*Tripneustes ventricosus*	60.	*Atrina rigida*
16.	*Ligia baudiniana*	61.	*Brachidontes exustus*
17.	*Eucidaris tribuloides*	62.	*Isognomon alatus*
18.	*Aiptasia pallida*	63.	*Lima scabra*
19.	*Eunicea knighti*	64.	*Barbatia cancellaria*
20.	*Pterogorgia anceps*	65.	*Anadara notabilis*
21.	*Pseudopterogorgia acerosa*	66.	*Chione cancellata*
22.	*Briareum asbestinum*	67.	*Antigona listeri*
23.	*Onuphis magna*	68.	*Chlamys sentis*
24.	*Onuphis magna* (anterior end)	69.	*Arca zebra*
25.	*Lumbrinereis maculata*	70.	*Codakia orbicularis*
26.	*Ischnochiton limaciformis*	71.	*Tetraclita squamosa*
27.	*Acanthopleura granulata*	72.	*Lepas anserifera*
28.	*Phascolosoma varians*	73.	*Paguristes tortugae*
29.	*Eurythoe complanata*	74.	*Pilumnus gemmatus*
30.	*Nereis dumerillii*	75.	*Pachygrapsus transversus*
31.	*Terebellides stroemi*	76.	*Clibanarius tricolor*
32.	*Amphitrite ornatus*	77.	*Pseudosquilla ciliata*
33.	*Nereis dumerillii* (anterior end)	78.	*Callinectes ornatus*
34.	*Eurythoe complanata* (anterior end)	79.	*Gonodactylus bredini*
35.	*Eunice longicirrata*		(= *G. oerstedii*)
36.	*Littorina angulifera*	80.	*Macrocoeloma trisponsum*
37.	*Tectarius muricatus*	81.	*Mithrax hispidus*
38.	*Modulus modulus*	82.	*Microphrys bicornatus*
39.	*Morula nodulosa*	83.	*Ophionereis reticulata*
40.	*Leucozonia nassa*	84.	*Echinaster sentus*
41.	*Thais deltoidea*	85.	*Ophioderma rubicundum*
42.	*Tegula fasciata*	86.	*Astrophyton muricatum*
43.	*Thais haemastoma*	87.	*Ophiothrix oerstedii*
44.	*Melampus coffeus*	88.	*Ophiocoma echinata*
45.	*Planaxis lineatus*		

Figure A.14. Some common invertebrates of Pigeon Key, Florida. By J. A. Zischke, St. Olaf College, Northfield, Minn.

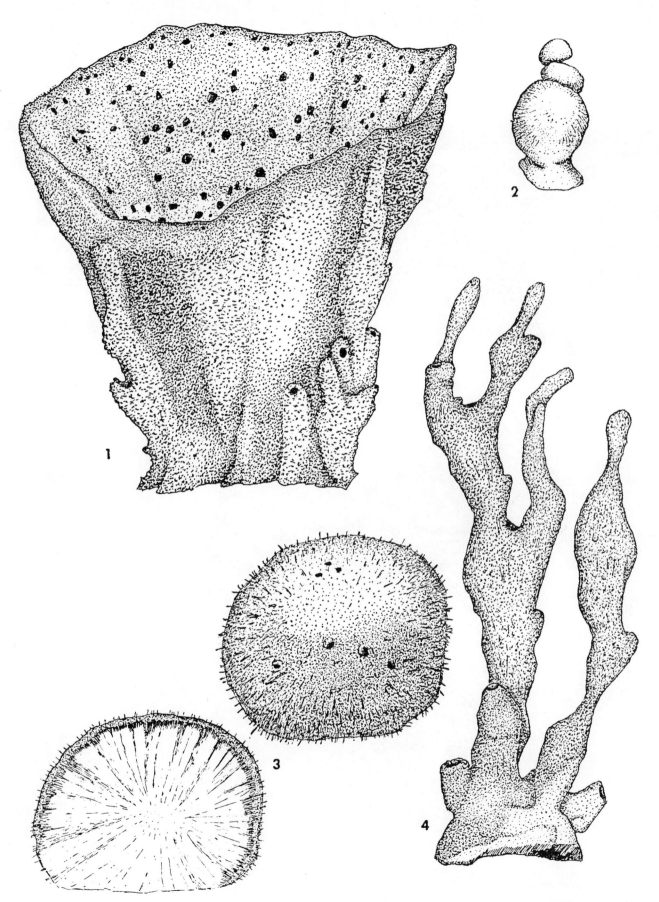

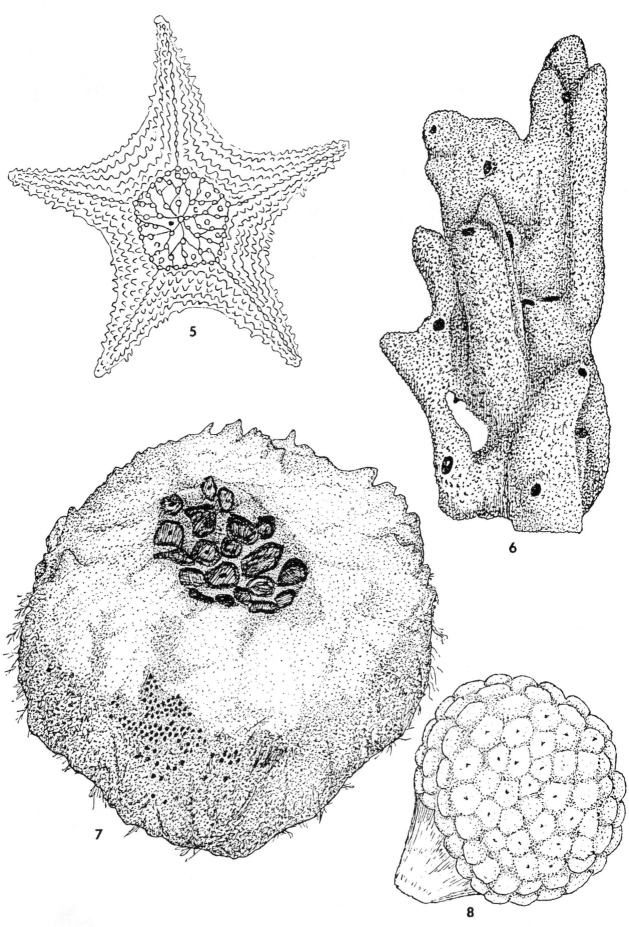

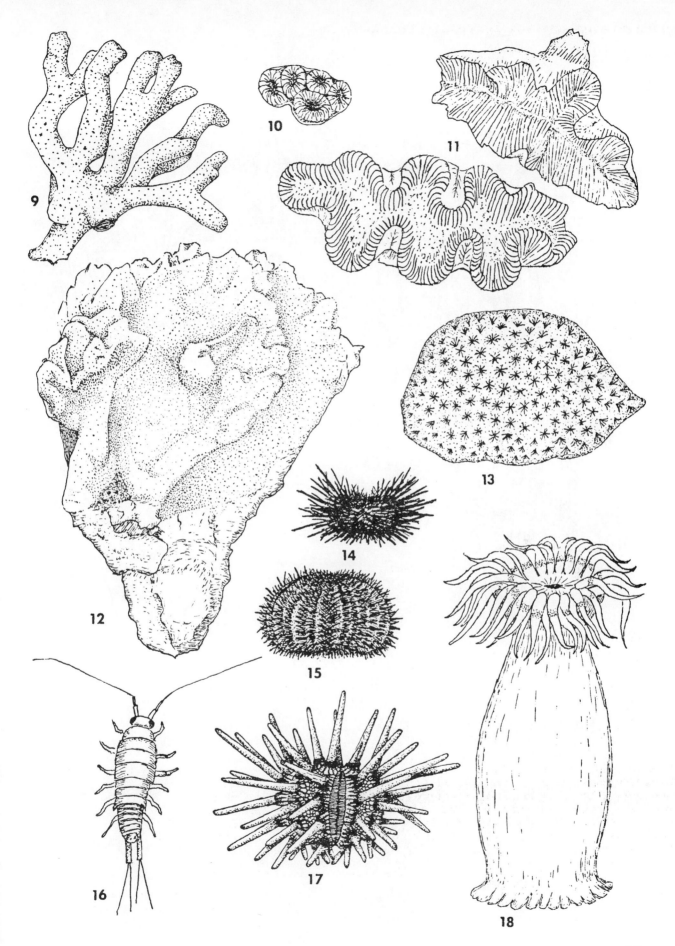

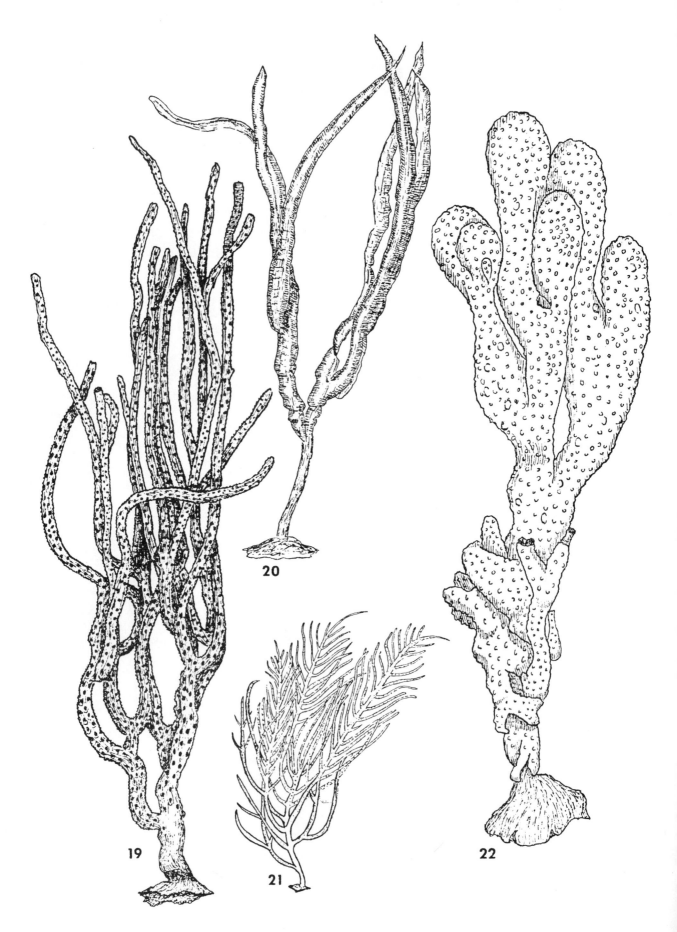

19

20

21

22

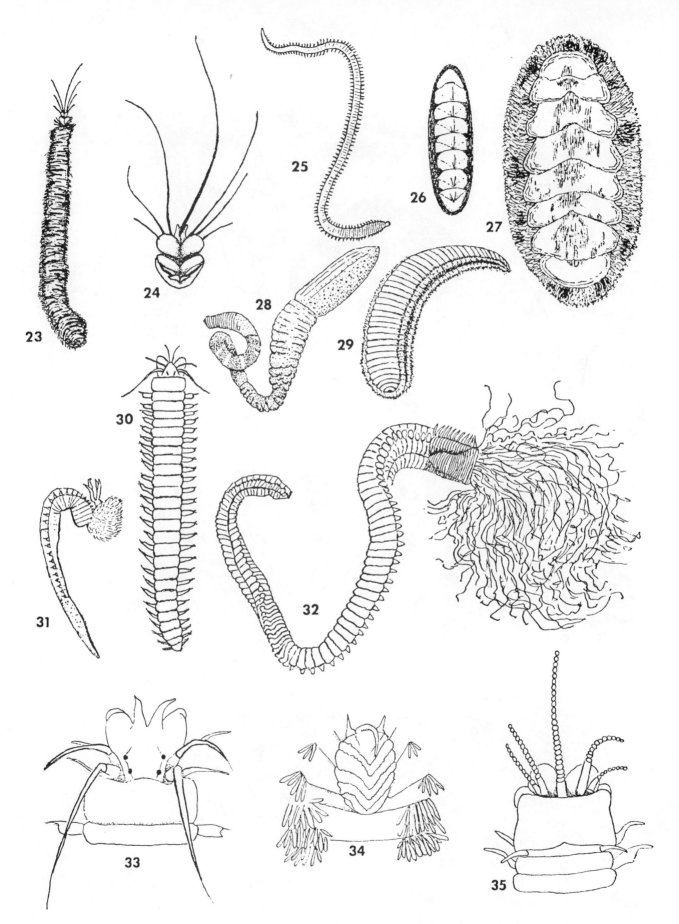

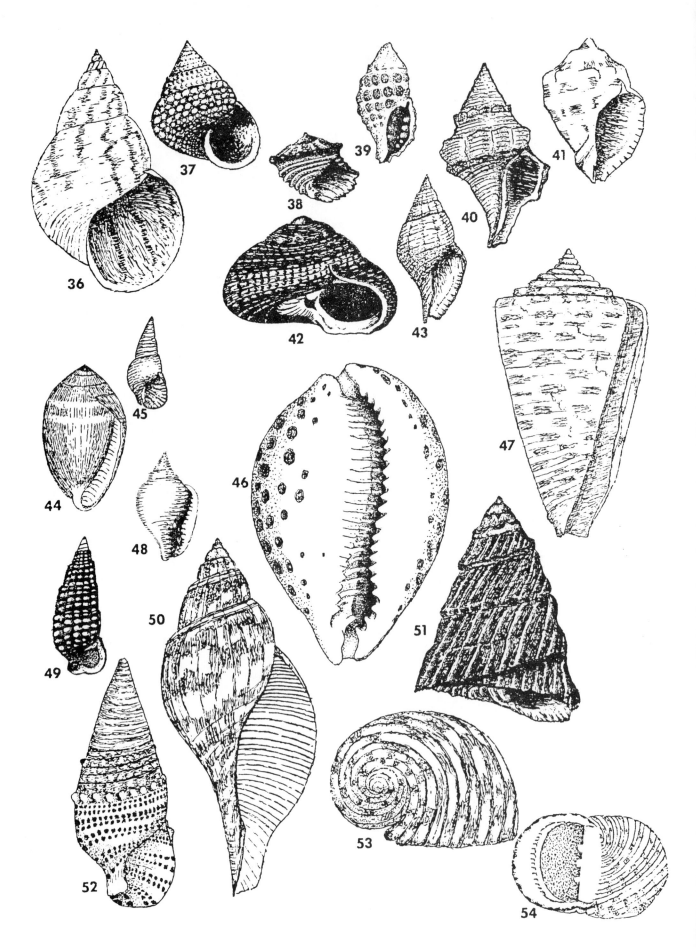

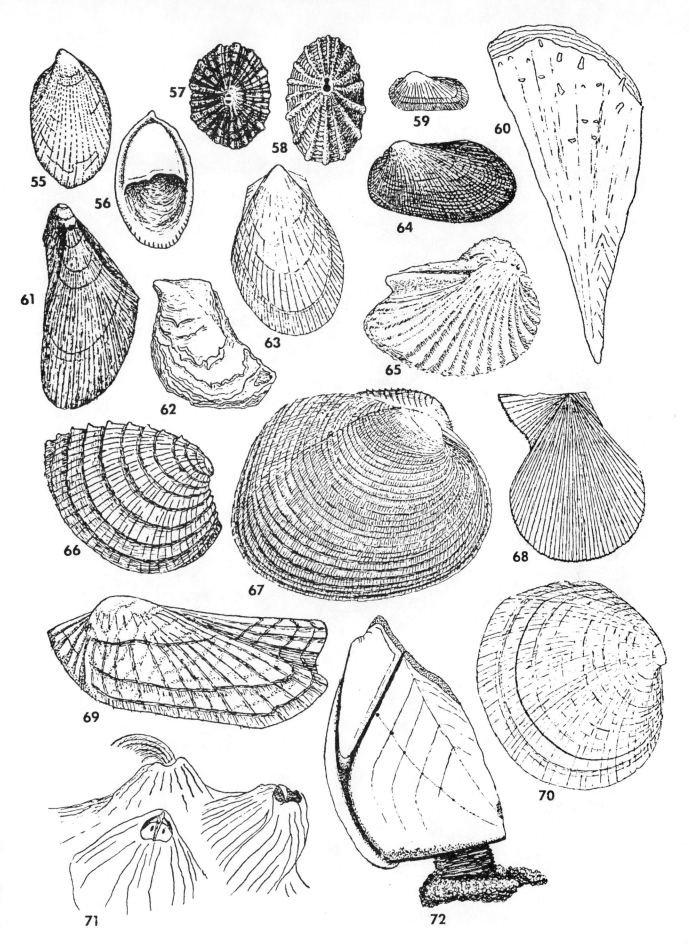

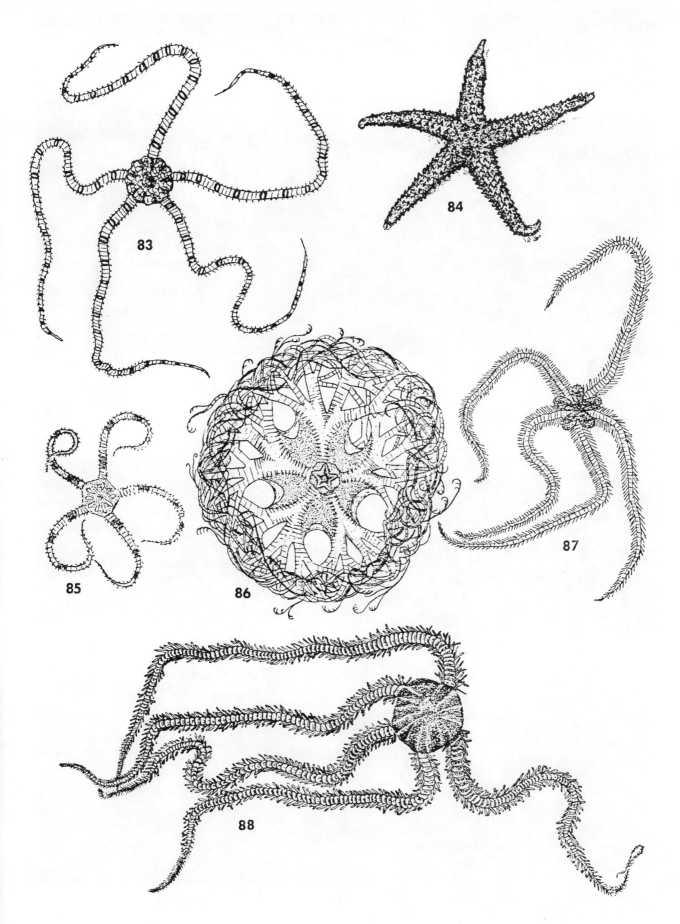

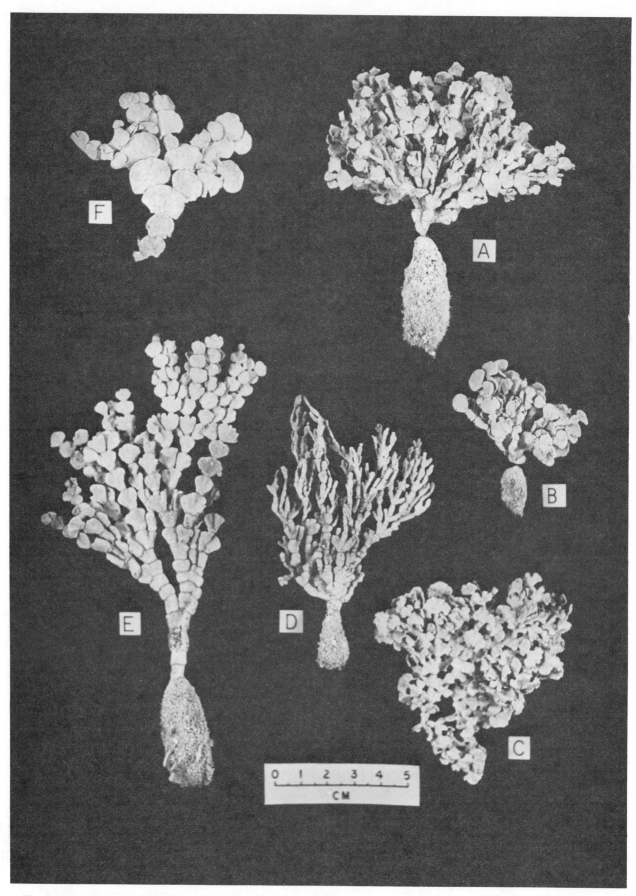

Figure A.15. Six species of calcareous green algae *Halimeda* found in the back-reef area of the Florida Keys: (A) *H. simulans*, (B) *H. scarba*, (C) *H. opuntia*, (D) *H. monile*, (E) *H. incrassata* (F) *H. tuna*.

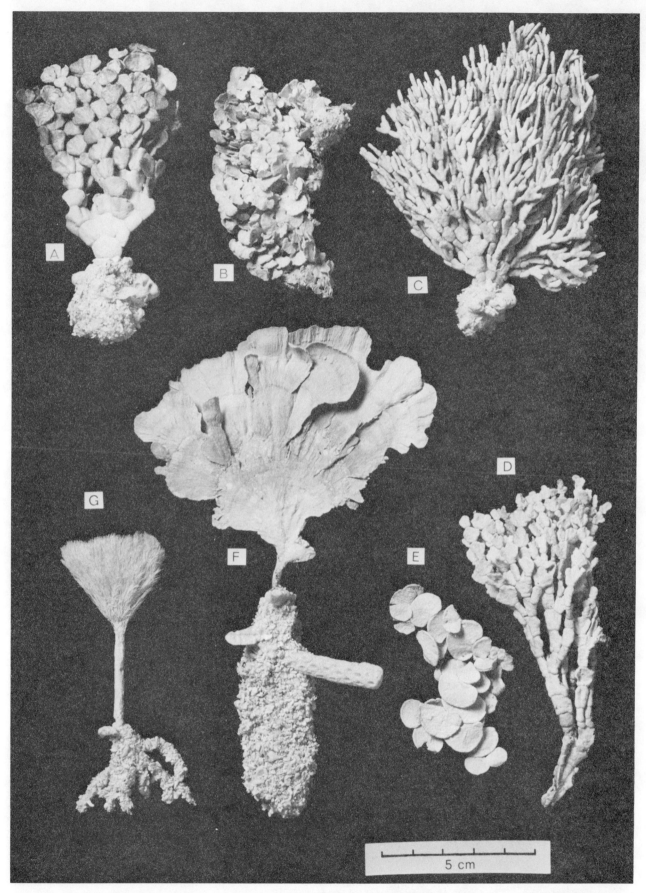

Figure A.16. Common species of green calcareous algae from the Dry Tortugas. (A) *Halimeda incrassata*. (B) *H. opunita*. (C) *H. monile*. (D) *H. incrassata*, forma *tripartita*. (E) *H. tuna*. (F) *Udotea flabellum*. (G) *Penicillus capitatus*. From Jindrich (1972).

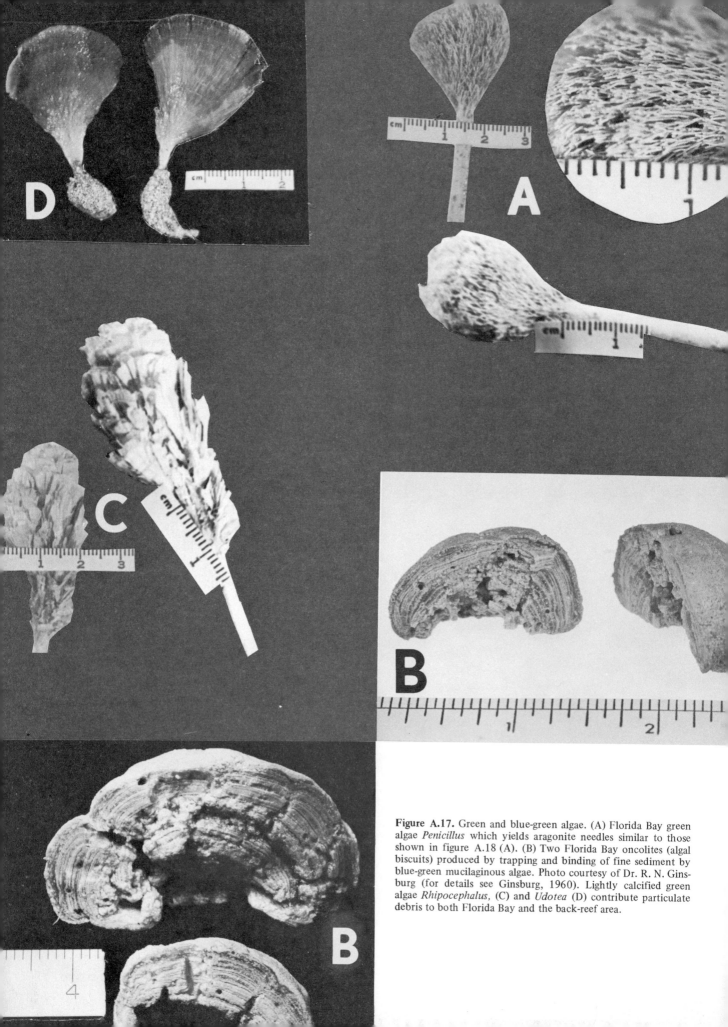

Figure A.17. Green and blue-green algae. (A) Florida Bay green algae *Penicillus* which yields aragonite needles similar to those shown in figure A.18 (A). (B) Two Florida Bay oncolites (algal biscuits) produced by trapping and binding of fine sediment by blue-green mucilaginous algae. Photo courtesy of Dr. R. N. Ginsburg (for details see Ginsburg, 1960). Lightly calcified green algae *Rhipocephalus,* (C) and *Udotea* (D) contribute particulate debris to both Florida Bay and the back-reef area.

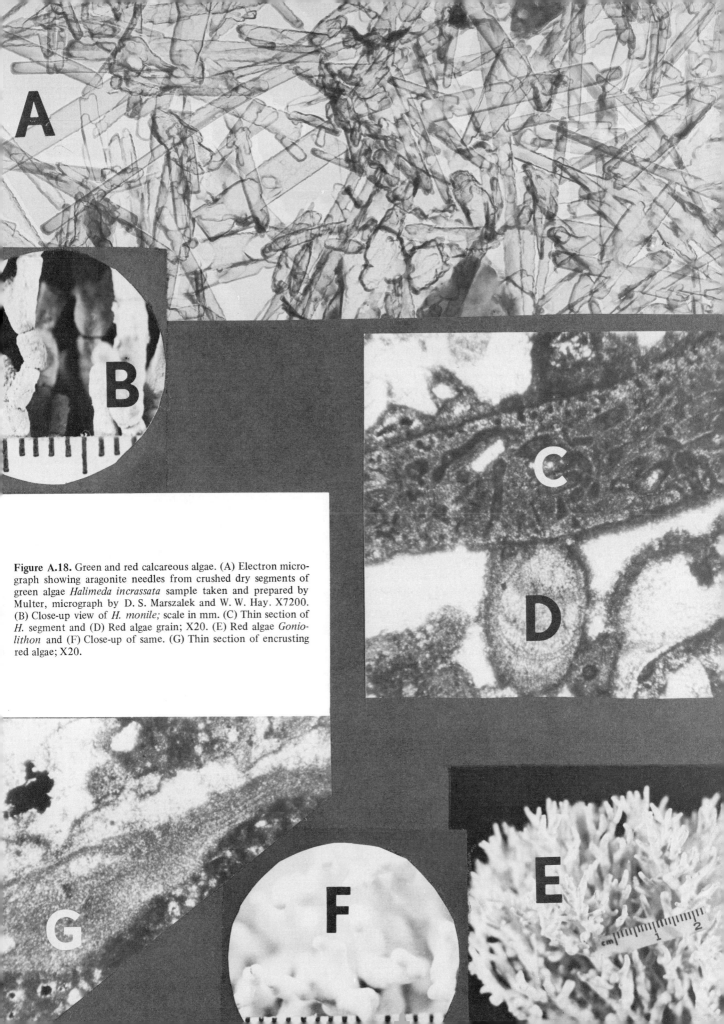

Figure A.18. Green and red calcareous algae. (A) Electron micrograph showing aragonite needles from crushed dry segments of green algae *Halimeda incrassata* sample taken and prepared by Multer, micrograph by D. S. Marszalek and W. W. Hay. X7200. (B) Close-up view of *H. monile;* scale in mm. (C) Thin section of *H.* segment and (D) Red algae grain; X20. (E) Red algae *Goniolithon* and (F) Close-up of same. (G) Thin section of encrusting red algae; X20.

BRYOZOANS IN FLORIDIAN
AND BAHAMIAN REEFS—
NOTES FOR FIELD OBSERVERS
by
R. J. Cuffey

Bryozoans have seldom been mentioned in the literature dealing with modern reefs, undoubtedly because their colonies are quite inconspicuous to visitors inspecting living reefs (Cuffey, 1972b). Actually, as recently realized, these animals are important elements in many modern coral reefs, including those in Florida and the Bahamas; consequently, detailed investigations are now elucidating bryozoans' roles in reefs (Cuffey, 1972b, 1974, in press b). However, in view of previous neglect of these interesting animals, the alert field observer can potentially add significantly to our understanding of reefal bryzoans. The following comments will thus help make visitors more aware of bryozoans which may be encountered on Floridian and Bahamian reef (figure A.19).

Several broad groupings can be distinguished readily among reef-dwelling bryozoans, although species-level identifications are frequently difficult; these major groups have therefore proven quite useful in reefal bryozoan investigations (Cuffey, 1973 and in press a). Those occurring commonly in reefs of this region are easily recognizable (fig. A.19). By far most common are *encrusting cheilostomes,* thin sheetlike crusts spreading (sometimes quite extensively) over the hard substrate, and composed of serially or irregularly arranged boxlike zooecia. Among the less common groups, three also adhere closely to their substrates but build much smaller colonies—*lichenoporid cyclostomes* (disk-shaped colonies consisting of radially arranged, thin, tubular zooecial), *idmoneid cyclostomes* (elongate series of tubular zooecia whose distalmost ends often stand erect), and *aeteid cheilostomes* (very fine, threadlike, encrusting networks with low erect tubular zooecoa). Colonies belonging to the remaining groups stand erect above the substrate. Very large, robustly branching or massively nodular colonies composed of boxlike zooecia are *schizoporellid cheilostomes,* superficially reminiscent of "finger corals" in gross form: these colonies are best referred to the species *Schizoporella errata,* at least tentatively (Cuffey and Fonda, 1976). Rather small, rigid, erect, perforated or fenestrated sheets represent *reteporid cheilostomes* (sometimes termed "lace corals"). Small, erect, many-branched, flexible or jointed colonies composed of serially arranged boxlike zooecia constitute *tuftlike cheilostomes;* similar but much more delicate colonies consisting of elongate tubular zooecia are *crisiid cyclostomes.*

Within Floridian and Bahamian reefs (figure A.19; Cuffey, 1971; Cuffey and Gebelein, 1975; Cuffey and Kissling, 1973), all these bryozoan groups (except robustly-branching schizoporellids) dwell primarily in sheltered niches—especially underneath dead coral fronds and heads, reef-rock ledges, detached rock fragments, and dead pelecypod shells—on coral knolls, patches, and rubble pavements. There, the bryozoans function as hidden-encrusters and cavity-dwellers, rather than as principal frame-builders (Cuffey, 1972b and in press b). On ocean-ward or barrier reefs, bryozoan abundance and diversity is greatest, with representatives from most groups present, as well as many species particularly of encrusting cheilostomes. Nearer shore, in back reef or lagoonal patch reefs and coral thickets, bryozoan abundance diminishes

somewhat, while diversity drops markedly, with encrusting cheilostomes (and sometimes even only one species of that group) being virtually the only bryozoans encountered. Between the reefs proper, marine grasses may bear a few encrusting cheilostomes, and lagoon-floor skeletal sands may include occasional fragments of such bryozoans; these forms thus are neither significant sediment-formers (unlike elsewhere) nor sediment-movement inhibitors. Finally, Pleistocene reefs such as the Key Largo Limestone (Cuffey, 1972a) display fossil bryozoans occurring in the same manner as in living reefs; these ancient reefal bryozoans— especially encrusting cheilostomes—are well shown in ornamental stone (outside walls of Coral Gabels City Hall; doorway of Comparative Sedimentology Laboratory of University of Miami)and weathered highway cuts (on Key Largo), but ironically are quite rare at certain classic localities (Windley Key quarry; Key Largo Canal).

Three exceptions to the above-outlined usual pattern of occurrence are noteworthy in considering bryozoans in Floridian and Bahamian reef complexes (fig. A-19). First, encrusting cheilostomes—possibly a species related to *Schizoporella*—build small patch reefs within tidal channels cutting through oolite shoals at Joulters Cays (Gebelein, 1974; Cuffey and Gebein, 1975). This occurrence is exceptionally interesting because, as one of the very few documented instances of bryozoans as frame builders in modern reefs, it provides unique opportunity for comparison with Paleozoic bryozoan reefs. Second, the robustly branching schizoporellid cheilostome *Schizoporella errata* forms extensive patches on the bank-interior shallows east of Cay Cays. Although this species' branches do not interlock to form a truly reefal frame, they do trap and stabilize (as well as contribute fragments to) loose sediment there, as they did during Pleistocene time in southern Florida (Hoffmeister, Stockman, and Multer, 1967; Multer, this Guidebook). Third, encrusting cheilostomes—including the widespread *Membranipora tuberculata*—occur as epiplanktonic drifters on rown algae floating in surface waters above reefs throughout the Florida-Bahama region (Cuffey, 1972b).

In conclusion, alert field observers can encounter various bryozoan groups on Floridian and Bahamian reefs (fig. A.19). Such observers can contribute new data concerning bryozoan distribution within these reefs and thus potentially help expand our insight into reefal bryozoans in general.

References Cited

Cuffey, R. J. (1971). Bryozoans in the modern reefs off northern Eleuthera Island (Bahamas): *Internat. Sedimentol. Cong. Abs.,* v. 8, p. 19.
———. (1972a). Discovery of and roles filled by bryozoans in Pleistocene reefs of southernmost Florida: *Geol. Soc. Amer. Abs. Prog.,* v. 4, p. 481-82.
———. (1972b). The roles of bryozoans in modern coral reefs: *Geol. Rdsch.,* v. 61, p. 542-50.
———. (1973). Bryozoan distribution in the modern reefs of Eniwetok Atoll and the Bermuda Platform: *Pacific Geol.,* v. 6, p. 25-50
———. (1974). Delineation of bryozoan constructional roles in reefs from comparison of fossil bioherms and living reefs: *Proc. 2nd Internat. Coral Reef Symp.,* v. 1, p. 357-64.

*Submitted by Roger J. Cuffey, Dept. of Geosciences, Penn. State Univ. University Park, Pa. 16802 for the 1977 Edition of this Guidebook.

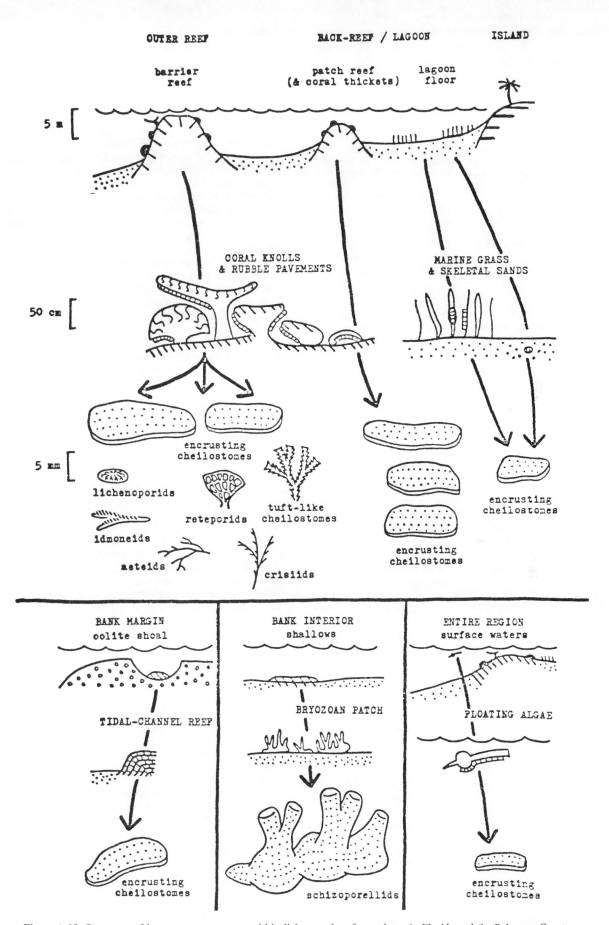

Figure A.19. Summary of bryozoan occurrences within living coral-reef complexes in Florida and the Bahamas. Courtesy of R. J. Cuffey.

———. (in press a). A note on bryozoans in the modern reefs of Eniwetok Atoll and the Australian Great Barrier Reef: Pacific Geol.

———. (in press b). Bryozoan contributions to reefs and bioherms through geologic time: Amer. Assoc. Petrol. Geol. Mem. (submitted).

Cuffey, R. J., and Fonda, S. S. (1976). "Giant" *Schizoporella*—Sedimentologically important cheilostome bryozoans in Pleistocene and Recent carbonate environments of Bermuda, the Bahamas, and Florida: *Geol. Soc. Amer. Abs. Prog.*, v. 8, p. 474-75.

Cuffey, R. J., and Gebelein, C. D. (1975). Reefal bryozoans within the modern platform-margin sedimentary complex off northern Andros Island (Bahamas): *Geol. Soc. Amer. Abs. Prog.*, v. 7, p. 743.

Cuffey, R. J., and Kissling, D. L. (1973). Ecologic roles and paleoecologic implications of bryozoans on modern coral reefs in the Florida Keys: *Geol. Soc. Amer. Abs. Prog.*, v. 5, p. 152-53.

Gebelein, C. D. (1974). Modern Bahaman platform environments: Geol. Soc. Amer. Field Trip Gdbk. (Miami), p. 1-96.

Hoffmeister, J. E., Stockman, K. W., and Multer, H. G. (1967). Miami Limestone of Florida and its Recent Bahamian counterpart: *Geol. Soc. Amer. Bull.*, v. 78, p. 175-90.

BAHAMIAN AND FLORIDIAN BIOFACIES
by
A. H. Coogan

Introduction

The main elements in organism bottom communities were outlined by Newell et al., in 1959 for the Andros Lobe of the Great Bahama Bank (fig. A.20). Their work stands as the only general statement of the distribution of Recent plants and animals in the Bahamian-Floridian region although many others have added important descriptions of parts of Recent faunas and floras. Notable among these contributions are the work of Storr (1964), Goreau (1959), Shinn (1963), Ginsburg (1964), on the main reef and inshore corals; Abbott (1955, 1962), Warmke and Abbott (1961), Turney (1964), Bayer and Work (1964) and Stephenson and Stephenson (1950) on molluscs and other invertebrates; Kier and Grant (1965) on echinoids; Howard (1965) on foraminifers and Shinn (1968) on burrowing organisms. Little attempt has been made, however, to apply the general approach of Newell *et al.,* (1959) to the Florida Keys biota with the exception of a short note by Turney (1964) in the Geological Society of America Guidebook on South Florida Carbonate Sediments (Ginsburg, 1964).

The purpose of this article is to review the general scheme presented by Newell *et al.,* (1959) for the Bahamas, to modify it slightly based on later work and to apply this modified scheme to the Florida Bay, Florida reef tract and southern Florida Keys with the objective of showing the importance of faunal distribution in the recognition of Recent environments. I have attempted this assignment without personally adding to the general or specific understanding of Recent ecology of the Bahamian-Floridian biota. This survey is based on the literature and my general interest and knowledge of the area.

Definition and Significance of Biofacies

The biofacies concept discussed in this Guidebook corresponds to the organism communities of Newell *et al.,* (1959), that is, the term biofacies is used to indicate any defined group of organisms living together in an area which can be mapped. This definition, of course, differs from the use of biofacies as part of a stratigraphic unit, insofar as Recent sediment and rocky areas do not represent stratigraphic units, and differs from the ecologist's use of the term community to indicate a self-contained steady state ecosystem (Kendeigh, 1961) in that few groupings of Recent marine organisms, with the exception of the coral reef biota, can be considered self-contained ecosystems. The advantage of using biofacies as defined groups of organisms is that this usage highlights the kinds of groupings commonly recognized by paleontologists, paleoecologists and stratigraphic geologists who are attempting to interpret ancient environments.

It is widely recognized that carbonate sediments and carbonate rocks are made principally of the remains of calcareous plants and animals. These sediments, although mineralogically simple, consist of diverse particles representing whole, disarticulated or fragmented plant and animal skeletons or are the direct or indirect result of biochemical or physiological activity of plants and animals. As a result, the carbonate lithofacies are more strongly related to those factors which control life and life zones than are clastic lithofacies. Factors such as water temperature, depth, light transmissibility, general circulation, salinity, nutrient supply and bottom type and stability exert control on the distribution of plants and animals and thus on the carbonate lithofacies as opposed to such factors as weathering rates, mineralogy of the source area, dispersal and stream patterns, point sources of sediment along coast lines etc., which control lithofacies in the clastic realm. Hence, carbonate lithofacies tend to closely parallel biofacies and both reflect the environment of deposition. Because of this parallelism, a paleoecological-textural approach to understanding carbonate depositional patterns is fundamental. This will become quite clear as you survey different carbonate environments in Florida and the Bahamas and the impression will be reinforced if you will distinguish the main faunal elements as well as the gross sediment type in each setting.

There is one disadvantage to the study of Recent carbonate sediments and biota as a basis for understanding ancient environments that is directly related to the organic origin of carbonate particles. Since the effect of organic evolution through geologic time is to change the total aspect of the biota, the more organisms control the sediment type the less Recent sediments resemble ancient limestones. The farther back in geologic time the geologist carries his investigation of environmental facies, the greater the likelihood he will encounter extinct facies and the less direct comparison of Recent sediments and faunas will aid him in an environmental interpretation. General principles apply, of course, in keeping with uniformitarianism.

*Submitted by Alan H. Coogan, Kent State University, Kent, Ohio for the 1969 Edition of the Guidebook.

General Biofacies Groupings

Twelve major biofacies are recognized in the Florida-Bahama area based on the predominance or characteristic occurrence of selected genera and species of calcareous invertebrates. These are the deep sea Planktonic-foraminiferal biofacies, three hard bottom reef and bank-edge coralline biofacies, seven sediment-bottom molluscan-echinoid biofacies and one hard bottom intertidal biofacies. Some of the groupings are based on nearly single group ecology, for example the Planktonic-foraminiferal biofacies; other facies are based on the recognition of a group of diverse species of many phyla and classes to which the name of one characteristic member has been assigned, for example the *Littorina* biofacies. In general, the approach has been to define the biofacies on broad groupings of animals and plants rather than on single groups except where the ecologic stress in the environment has restricted the variety of species to just a few types. Foraminifers are not considered in detail because of the difficulty in identifying them in the field. The illustrations (Plates I-VII) emphasize the molluscs and echinoids since the corals and algae are adequately illustrated in Ginsburg (1964) and this Guidebook.

Major Bahamian-Floridian Biofacies

Open marine, deep sea, Planktonic-foraminiferal biofacies: The areas of the deep sea adjacent to the Bahama Banks and Florida shelf are receiving sediments derived from the shallow shelf mixed with the planktonic rain of pteropods, foraminifers and coccoliths which thrive in the shallower, warm "upper strata" of the open subtropical sea. From a biotic standpoint, these areas which are known physiographically as the Florida Straits and the Tongue of the Ocean (fig. A.20, *1*) are part of the Planktonic-foraminiferal biofacies. Other elements of the biota include the usual nekton and soft-bodied plankton of subtropical seas and the sparse benthonic biota which includes brittle stars. The predominant, conspicuous faunal elements of the bottom sediment in the Tongue of the Ocean away from the bank edge are globogerinid planktonic foraminifers and pteropod molluscs with coccoliths making up a large part of the finer carbonate and mud matrix. This biota constitutes the main particle types in the relatively deep marine basins (500 to 1,200 fathoms) since early Tertiary time as indicated by dredging and coring as part of the JOIDES program (Bunce *et al.,* 1965). The same is probably true for the Florida Straits and the Tongue of the Ocean. Although you will not observe this biofacies as part of the field trip, it is useful to realize its great extent in the area for the sake of completeness.

Rocky bottom, shelf-edge (*Acropora palmata,*) Plexaurid and Millepore biofacies: The *Acropora palmata* or reef biofacies is characterized by an association of about thirty species of corals (Newell *et al.,* 1959; Goreau, 1959; Storr, 1964) including *Acropora palmata, A. cervicornis, Montastrea annularis, M. cavernosa, Siderastrea, Diploria labyrinthiformis* and *Porites porites.* These corals build the outer reef of the shelf margin. In the back-reef area, local patch reefs of the Plexaurid biofacies attach to rock bottom covered by a thin blanket of ephemeral sand. *Montastrea annularis* and *Diploria* are the main patch-reef builders which are associated with other corals such as *Porites porites, Agaricia, Siderastrea* and *Favia* (Ginsburg, 1964). The Plexaurid sea whip and gorgonacean "soft corals" are conspicuous but the associated molluscs (Plate I) such as *Chama, Arca,* the boring clam *Lithophaga, Aequipecten* and others are harder to find. Many are present in the Pleistocene reef exposed at the quarry on Windley Key (Stanley, 1966). These two biofacies represent the most exposed and slightly sheltered groups of intertidal and subtidal plants and animals on the wind and wave-swept edges of the bank. The biota is heavily dominated by strongly cemented and encrusting species which require a high quantity of nutrients and dissolved oxygen and calcium carbonate.

The millepore biofacies occupies subtidal rock ledges and rocky highs along the exposed shoreline and is transitional with the Plexaurid biofacies. Many of the coralline members of the other rocky-bottom, shelf-edge biofacies are present such as *Diploria, Siderastrea, Montastrea, Porites* and *Agaricia* as well as the millepore hydrozoan *Millepora,* the Plexaurids, gorgonaceans and rock dwelling urchin *Diadema* (spiny black urchin). The molluscan fauna (Plate I) is transitional between the reef molluscan fauna and the intertidal rocky shoreline fauna of the *Littorina* biofacies, and includes the rock boring *Lithophaga,* herbivorous gastropod *Livona pica,* and volcano limpets *Fissurella* and *Diodora.* Like the Plexaurid and *Acropora palmata* biofacies, the species of the millepore biofacies are mainly attached, cemented or encrusting species which can withstand turbulent water.

These three biofacies are mapped together (fig. A.20, *2*) since they occupy the narrow bank margin in the Bahamas and Florida. Characteristic foraminifers are *Archaias* and peneropolids. The area occupied by the three biofacies corresponds to the coralgal sand facies (Newell *et al.,* 1959) of the Andros Lobe of the Great Bahama Bank and the middle and outer reef track of the Florida Keys.

Inshore, rocky shoreline (*Littorina*) biofacies: An aerially restricted, distinct and strongly zoned rocky shoreline biota is widely recognized along the Atlantic coastline and has been designated the *Littorina* biofacies, generally following Newell *et al.,* (1959). Forty or more species of attached or cemented plants and animals constitute the biota including green algae, coralline algae, sponges, barnacles, chitons, gastropods, pelecypods, crabs, sipunculids, and echinoids. The abundance and occurrence of specific species is strongly controlled by the tidal range and a variety of niches are occupied in the subtidal, intertidal and supratidal (respectively, yellow, black and gray, and white zones) and in the tidal pools. The divisions are sharp to overlapping and occur within a few feet of one another. Typical examples are found on the rocky westward facing shores of Bimini and near Bahia Honda in the Lower Keys. The sparse abundance of pelecypods reflects the rock bottom. The zonation, a common intertidal effect, reflects the increased physiological and physical stress associated with dessication during low tide and wave attack. Common gastropods of this biofacies (Plate II) include *Diodora, Fissurella, Nodilittorina, Echininus, Livona, Littorina, Nerita, Acmaea* and *Batillaria.*

Subtidal, bank edge, unstable sand (*Strombus samba*) biofacies: The outer platform of the bank margin in the Bahamas and the immediate back-reef area of the Bahamas and Florida reef track is covered by patchy

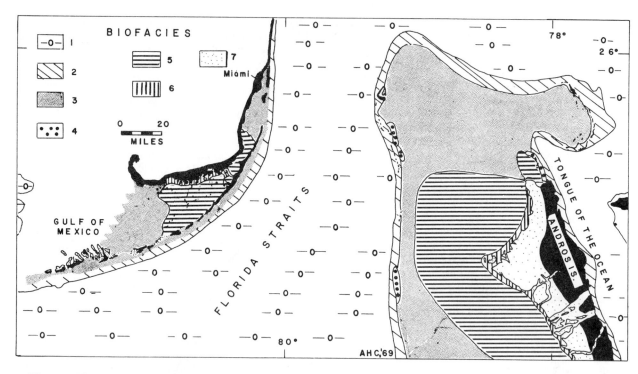

Figure A.20. Generalized distribution of seven of the Bahaman-Floridian biofacies discussed in this guidebook article and illustrated on Plates I-VII. It should be noted that the illustrations on the plates serve mainly to give the reader an overall view of the main faunal elements and do not substitute for better illustrated guides to the identification of the species such as Abbott (1955, 1962). Data and patterns compiled from Newell *et al.,* (1959), Purdy (1963), McCammon (1968) for the Bahamas and from Ginsburg (1964), Gorsline (1963), Kier and Grant (1965) and Turney (1964) for Florida. (1) Planktonic-foraminiferal biofacies, not illustrated. (2) Combined *Acropora palmata*, Plexaurid, Millepore and *Strombus samba* biofacies, Plates I, III. (3) *Strombus costatus* biofacies, Plates IV-VI. (4) *Tivela* biofacies, Plate II. (5) Transitional biofacies, Plate VI. (6) *Cerithidea costata* biofacies, Plate VII. (7) Combined *Fasciolaria-Batillaria* and *Crepidula* biofacies, Plate VII.

skeletal sand, only partly stabilized by sea grasses. This heavily rippled bottom in clear turbulent water represents a higher stress habitat than the more protected and shelfward stabilized sand bottom. Characteristic of the unstable sand in the Bahamian area is the black conch (*Strombus samba*) according to Newell *et al.,* (1959). Other species consist mainly of burrowing pelecypods and echinoids (Plate III), green algae, sea grasses and foraminifers. Although the gastropod *Strombus samba* may not be present in the Florida reef tract, the biofacies can be recognized there by the occurrence of many of the other species which also occur in the Bahamas. The *Strombus samba* biofacies occurs intermixed with the rocky bottom Plexaurid biofacies along the bank margins (fig. A.20, *1*), where, in Florida for example, it occupies the unstable coralgal sand areas of White Bank.

Echinoids are conspicuous including the burrowers *Encope* and *Leodia,* and the nonburrower *Clypeaster* (Kier and Grant, 1965). Burrowing pelecypods include *Laevicardium, Lucina, Chione, Macrocallista, Glycimeris Tellina* (Plate III) together with a few swimmers such as *Aequipecten* and *Chlamys* (Newell *et. al.,* 1959). Gastropods other than the large conch are sparse. Locally patches of green algae *Udotea, Halimeda* and *Acetabularia* and grasses *Cymodocea* and *Thalassia* mark the transition to the biofacies established on the grass-stabilized sand areas, discussed below.

Intertidal, bank-edge, unstable sand (*Tivela*) biofacies: Extreme mobility of grass-free, unstabilized sand bottom

is found in very small areas in the Bahamas coincident with the distribution of the actively growing intertidal oolite bars (fig. A.20, *4*). Examination of a sediment bottom map of the Bahamas shows that these bars occur within a broader oolitic belt along the western bank margin south of Bimini, in the Joulters Cay area, around the southern end of the Tongue of the Ocean and in the "hook" behind Eleuthera Island and the Berry Islands. These mobile sands are relatively devoid of living biota and the small numbers of species found there have been characterized as the *Tivela* biofacies by Newell *et al.,* (1959). The distribution of the *Tivela* biofacies does not coincide with the distribution of oolitic sand since away from the active bars the oolite sand tends to be stabilized by *Thalassia* and green algae and to be occupied by the *Strombus costatus* biofacies. The few species which constitute the *Tivela* biofacies are *Tivela abaconis* (Plate II), a burrowing pelecypod like the Pismo Beach clam, sparse patches of *Cymodocea* and *Thalassia* grass on the subtidal slopes of tidal channels, the echinoid *Leodia* and the starfish *Oreaster.* It might be expected that this biofacies would occupy the beach foreshore on islands at the edge of the bank platform. The only beach which fits this description in the area (fig. A.20) is on North Bimini where the *Tivela* biofacies has not been recognized and where the dead shells on the beach backshore represent mixing from the Plexaurid, *Strombus samba* and *Strombus costatus* biofacies.

Vegetation (*Thalassia* or algal) stabilized sand (*Strombus costatus*) biofacies: The most widespread and diverse biota is grouped into the *Strombus costatus* biofacies (Newell *et al.,* 1959). This rich biota lives in moderately heavy plant cover in the sheltered and relatively shallow waters of the back reef and open lagoonal areas adjacent to the bank edge in the Bahamas and Florida reef tract and in open parts of the Florida shelf not restricted by the mud shoals of Florida Bay or on the Lower Florida Keys. The stabilized-sand bottom and abundant vegetation provide many niches unavailable on the moving sand bottom and rocky bottom. On the Andros Lobe of the Great Bahama Bank (fig. A.20, *3*) the *Strombus costatus* biofacies occupies areas in which the sediment type ranges from cor-algal and grapestone sand to mixed sand and skeletal-mud or pellet-mud. In the Florida Keys, excellent samples of this biofacies may be obtained by dredging Hawk Channel, on the inshore shoals of Tavernier and Rodriquez Keys as well as along the front edges of the intertidal muddy flats on the seaward sides of Windley and Matecumbe Keys. Elements of this biota are also abundant in the channel close by the inlet to Bimini lagoon, but as Newell *et al.,* (1959) pointed out, the distribution of biota in Bimini lagoon is complex. The *Strombus costatus* biofacies (Plates IV, V and VI) has been recognized in Pleistocene rocks of the Florida Keys and southern Florida peninsula by Hoffmeister, Stockman and Multer (1967) as the bryozoan facies, which has mostly brozoans only, although this Pleistocene counterpart probably includes some of the Transitional biofacies described below.

Common algae are the green algae *Halimeda, Penicillus, Acetabularia* and the red alga *Goniolithon.* Corals typical of the inshore shoal coral faunas (Ginsburg, 1964) include *Porites furcatus, Manicina areolata. Porites divaricata, Solenastrea hyades* and *Siderastrea.* Common echinoderms are *Clypeaster rosaceus, Leodia, Meomia* (Plate IV) among the echinoids and the starfish *Oreaster.* Molluscs are abundant, including dozens of species of gastropods and pelecypods (Plates IV, V, VI). Typical are the burrowing pelecypods *Lucina, Laevicardium, Glycimeris* and *Codakia.* The gastropods include the conchs *Strombus gigas, Strombus rainus, Cerithium,* and other herbivores as well as the carnivorous *Natica* and *Olivella.* Peneropolid foraminifers are common in the grassy areas. Tentatively included in the *S. costatus* biofacies is the area of Florida Bay transitional to the Gulf of Mexico and away from the influence of the Lower Keys (fig. A.20, *3*) which was mapped as the "Gulf subenvironment" by Turney (1964) and for which he listed the burrowing pelecypods *Corbula, Tellina* and *Nucula* (Plate VI).

Muddy-sand, normal to hypersaline Transitional biofacies: Away from the shelf margin and transitional to the muddy shorelines and tidal flats of the land areas is a portion of the shallow bank characterized by fluctuating environmental conditions and occupied by a biota transitional between the diverse *Strombs costatus* biofacies and the strongly restricted salinity tolerant biofacies of the mangrove tidal flats. In the Bahamas, Newell *et al.,* (1959) designated this partly restricted biota the *Didemnum* community; it inhabits the sheltered, slow moving, hypersaline later in the lee of Andros Island. The distribution of this biofacies corresponds only generally to the distribution of the skeletal-mud and pellet-mud or to the area of hypersaline water. In southern Florida, a similar situation is represented by the molluscan biofacies of the "interior faunal assemblage" recognized by Turney (1964) as occu-

pying the interior of Florida Bay where mud banks cause the maximum restriction and where fluctuations in salinity are great. Salinity data for Florida Bay (Gorsline, 1963; Ginsburg, 1964) indicate fluctuations between 15 and 50 ppt while in the center of the Andros Lobe of the Great Bahama Bank, northwest of Williams Island, salinites as high as 43 ppt have been measured by Broecker and Takahashi (1966). These two investigators also give radiocarbon measurements showing that the water in this area moves so slowly that it spends as much as 200 days on the bank before exchanging with the sea.

Reconciliation of the Bahamian faunal lists with those from Florida Bay are difficult, hence the two areas, which were mapped separately by Newell *et al.,* and Turney using different criteria, are grouped into a Transitional biofacies (fig. A.20, *5*) recognizing that the differences in physiography, water circulation and biota may not justify detailed comparison.

The Bahamian *Didemnum* biofacies, named for a soft-bodied tunicate with calcareous spicules in the matrix of the body (perhaps a poor choice for the paleontologically oriented investigator) consists of the green algae *Halimeda, Penicillua, Acetabularia* and *Batophora,* the grasses *Thalassia* and *Cymodocea,* the bryozoan *Schizoporella* and a sparse surface bottom fauna of the coral *Manicina,* the echinoid *Clypeaster,* asteroid *Echinaster* and burrowing mollusk *Pitar.* In Florida Bay, some of these species may be absent, but in addition, the pelecypods *Chione cancellata, Transennella, Pinctada radiata* and the gastropods *Modulus* and *Cerithium muscarum* occur (Plate VI).

The Transitional biofacies passes locally into the biofacies of the intertidal and supratidal mangrove flats within Florida Bay and on the Lower Keys (too small to be shown on map, figure A.20). In the Bahamas, there is a salinity controlled biota, described next, which intervenes along the western Andros Island coast.

Subtidal, fluctuating high and low salinity, muddy bottom *Cerithidea costata,* biofacies: This nearshore, low-diversity, salinity-tolerant biofacies (fig. A.20, *6*), described and mapped by Newell *et al.,* (1959) as the *Cerithidea costata* community is characterized by only two molluscs—the burrowing pelecypod *Pseudocyrena colorata* and the gastropod *Cerithidea costata..* A small, non-calcareous green alga, *Batophora,* is especially abundant too. Newell had reservations concerning the distinctiveness of the biofacies which seemed limited largely to the shores of the western edge of Andros Island. It is listed here for completeness and included in it, for generalized mapping purposes, is the "northern faunal assemblage" of Turney (1964) consisting of one species of pelecypod, *Anomalocardia cuneimeris,* which he indicated lived where there is the most frequent and most intense dilution of Florida Bay by freshwater runoff from the Everglades. These two muddy shoreline areas receive freshwater runoff during high rain periods which accompany hurricanes, but may be hypersaline during the dry season. These fluctuating conditions impose high physiological stress on animals living in the area which probably accounts for the sparse biota. *Cerithidea costata,* a related Florida species *C. scaliformis,* and *Pseudocyrena* are figured (Plate VII) as part of the *Fasciolaria-Batillaria* biofacies since the species of these biofacies are commonly found mixed on tidal flats. Miliolid and peneropolid foraminifers are common.

Muddy-bottom, intertidal and supratidal, mangrove associated *Fasciolaria-Batillaria biofacies:* A distinct biota is found on the muddy intertidal shorelines and on the supratidal flats which are colonized by two readily recognizable floral associations, the intertidal red mangrove (*Rhizophora mangle*) and the supratidal black mangrove (*Avicennia nitida*) zones. These mangrove associations characterize the sheltered marshes and mud flats on the extensive lagoonal shores and intertidal areas of the western side of Andros Island (fig. A.20, 7), on the mud islands of Florida Bay and the tidal channels between the bedrock Upper Keys, and on the low-lying Lower Keys. Locally the mangrove zones develop around any protected island where restricted lagoonal waters occur such as the eastern and southern portions of Bimini and Pigeon Cay in Bimini lagoon. The red mangrove is easily recognized by its red wood and concave-upward, laterally extending prop root system. The black mangrove shoots are a distinctive single spikelike shoot, commonly found among the mud-cracked, stromatolitic algal covered supratidal muds.

The bottom sediments in and around the mangrove roots are rich in organic detritus and fecal pellets and have a high content of hydrogen sulfide. The biota, recognized by Newell *et al.,* (1959), and here called the *Fasciolaria-Batillaria* biofacies is characterized by abundant gastropods, few pelecypods, the mangrove bushes and the soft green alga *Batophora*. Especially common or distinctive are the gastropods *Batillaria, Cerithidea, Cerithium* and *Fasciolaria*. The only common pelecypods are the burrowing *Pseudocyrena colorata* and perhaps *Transennella* (Plate VII). Small windrows and beach ridges of these few but abundant species can be found ringing the many miles of intertidal shoreline on Andros Island and the Lower Florida Keys. Blackened and partially leached members of this biofacies are found well above the normal high tide marks on "dry land." Similar appearing associations of large numbers of a few species of gastropods have been recognized in carbonate intertidal deposits of Cretaceous and even older rocks.

Subtidal, muddy and gravelly-mud floored tidal creeks in mangrove areas (*Crepidula*) biofacies: Subtidal channels between the intertidal flats biotically differ from the *Fasciolaria-Batillaria* biofacies by having more mud-bottom dwelling pelecypods. This facies was recognized by Newell *et al.,* (1959) as a subdivision of the mangrove com-

munity and is designated here as the *Crepidula* biofacies (Plate VII) for the distinctive "slipper-shell" gastropod. The tidal creeks are commonly stabilized by patches of *Thalassia* grass and may have green algae growing on the bottom. The pelecypod fauna of *Codakia, Chione* and *Transennella* suggests a gradation with the Transition biofacies, while the presence of *Cerithium, Bittium* and *Cerithidea* links it with the *Fasciolaria-Batillaria* biofacies. It's worth noting, that in a similar physiographic setting, tidal channels emptying on the seaward side of the Upper Keys, near the inlet of Bimini lagoon, and between the oolite bars of Browns Cay carry a biota belonging to the *Strombus costatus* biofacies.

Summary

Twelve distinct biofacies can be recognized in the subtropical seas of the Florida-Bahama area which represents combinations of hard and soft bottom, salinity-stable and salinity-tolerant biotas. Comparison of the biofacies map (fig. A.20) with any of the many sediment bottom maps of the Bahamas (Newell *et al.,* 1959; Purdy and Imbrie, 1962; Purdy, 1963, McCammon, 1968) and Florida (Ginsburg, 1956, 1964; Gorsline, 1963) shows a strong parallelism between the distribution of carbonate biofacies and lithofacies as might be expected considering the biotic origin of carbonate particles. Exact correspondence is not found, nor expected, of course, because the sediment is more strongly affected by physical conditions while the biota "feels" the effect of water quality more strongly. Hence, the same local physical environment, for example a tidal channel, may be mapped as one or more different biofacies while substrates of different particle type are inhabited by a biota of the same biofacies.

Finally, one interesting aspect of the interplay between biofacies and lithofacies which has not been considered here may be left for your field investigations and discussions. A study of the role of competitive sedimentation will highlight many of the interrelationships between biota and sediment type in Florida and the Bahamas. You may want to compare, for example, the abundance of living species with the abundance of particles derived from these species in the sediment substrate.

FIELD NOTES

PLATE I

Molluscan species of the three Floridian-Bahamian normal marine, rocky-bottom, shelf-edge biofacies. Gastropods and pelecypods from the reef (*Acropora palmata*) and bank-edge (Plexaurid, including patch reef) biofacies are attached, borers, and swimmers (figs. 1-7). The rocky, subtidal, turbulent shoal (Millepore) biofacies is mainly of gastropods and boring pelecypods among the molluscan species (figs. 8-13).

Figure

1   Chama sp., two valves, X 0.5

2   Arca occidentalis, X 1.5

3   Aequipecten gibbus, X 1

4   "Lithophaga" sp., X 1

5   Cyphoma gibbosum, X 1.5

6   Cypraea sp., X 1

7   Olivella sp., X 1.5

8   Fissurella barbadensis, X 1.5

9   Diodora listeri, X 1.5

10   Livona pica, X 1

11   Cyphoma gibbosum, X 1.5

12   Astrea sp., X 1.5

13   Lithophaga sp., X 1

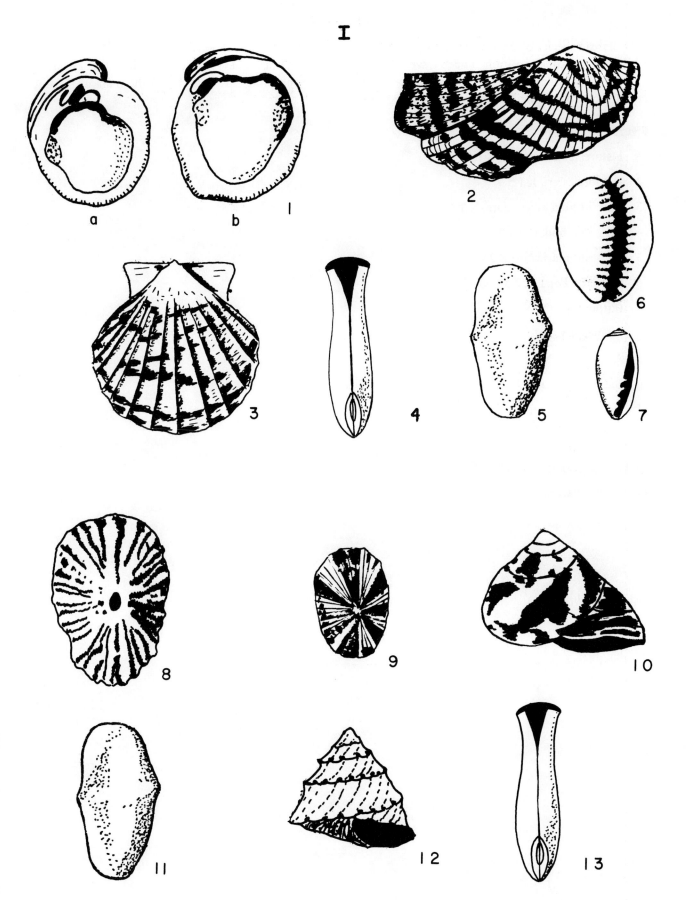

I

PLATE II

Molluscan species of the inshore, rocky shoreline are predominately gastropods which live in subtidal, intertidal and supratidal positions under wave attack and in tidal pools. These species constitute the *Littorina* biofacies (figs. 1-12). A very sparse fauna lives on the shifting, unstable bar-top sands of the Bahamian oolite bars and constitutes the *Tivela* biofacies (figs. 13-14).

Figure

1   Fissurella barbadensis, X 1.5

2   Diodora listeri, X 1.5

3   Nodilittorina tuberculata, X 3

4   Echininus nodulosus, X 2

5   Tectarius muricatus, X 2

6   Purperita pupa, X 2.5

7   Purpura patula, X 1

8   Livona pica, X 1

9   Acmaea sp., X 1.5

10   Batillaria minima, X 3

11   Nerita peloronta, X 1.5

12   Littorina angulifera, X 1

13   Leodia sexiesperforata, X 1

14   Tivela abaconis, X 0.5

II

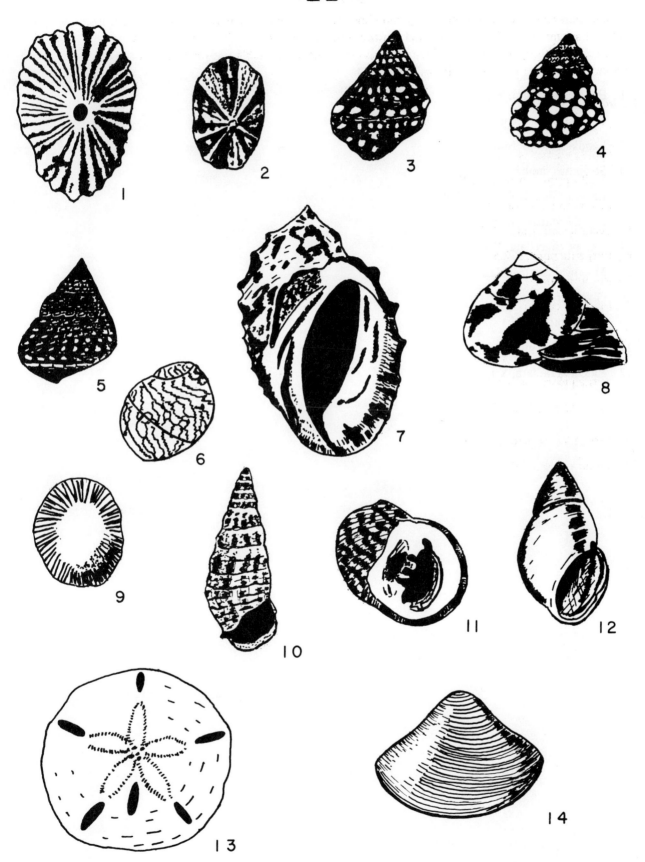

PLATE III

Molluscan and echinoid species which comprise the subtidal, bank-edge, unstable sand (*Strombus samba*) biofacies are mainly burrowing clams and sand dollars.

Figure

1   Encope michelini, X 1

2   Clypeaster rosaceus, X 0.5

3   Leodia sexiesperforata, X 1

4   Strombus samba, X 0.2

5   Laevicardium laevigatum, X 1

6   Lucina pennsylvanica, X 1

7   Chione cancellata, X 1.5

8   Macrocallista maculata, X 1

9   Glycimeris pectinata, X 1

10   Tellina radiata, X 1

11   Aequipecten gibbus, X 1

12   Pinctada radiata, X 1

III

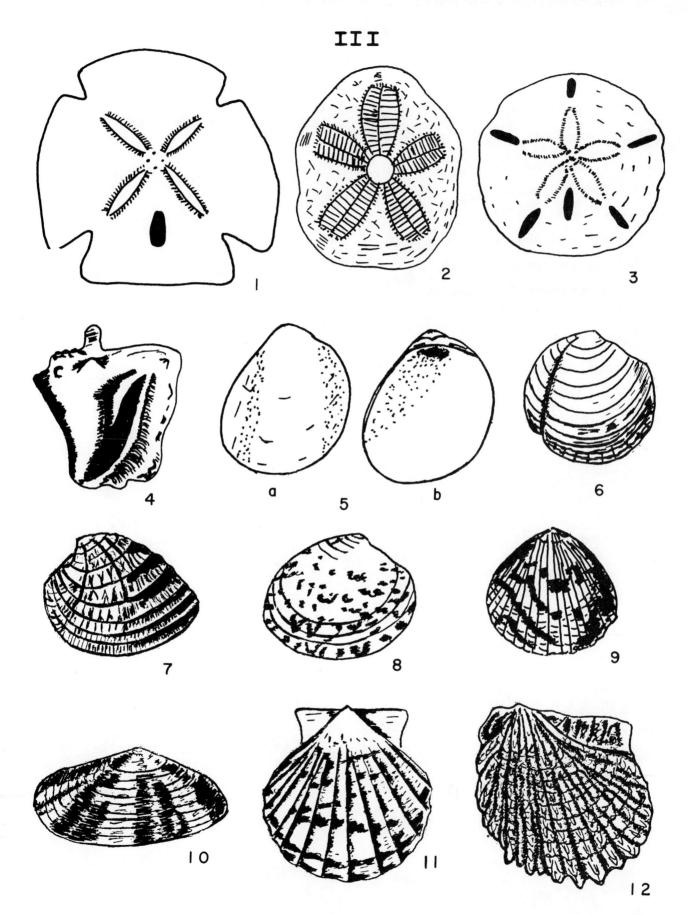

PLATE IV

Vegetation stabilized (*Thalassia* grass or algal) sand biofacies is known in the Bahamas as the *Strombus costatus* biofacies and in the Pleistocene rocks as the bryozoan *Schizoporella* biofacies. The fauna is characterized by abundant and varied species, probably representing the least stressed environment of the normal marine shelf seas on the Floridian and Bahamian platforms.

Figure

1   Clypeaster rosaceus, X 0.5

2   Leodia sexiesperforata, X 1

3   Meomia ventricosa, X 0.5

4   Strombus gigas, immature, X 1

5   Strombus gigas, adult with broad lip, X 0.25

6   Strombus raninus (like S. costatus), X 1

7   Cassis tuberosa, X 0.25

8   Xancus angulatus, X 0.5

9   Calliostoma jubjubinum, X 2

10   Olivella jaspidea, X 3

11   Conus spuricus, X 1.5

IV

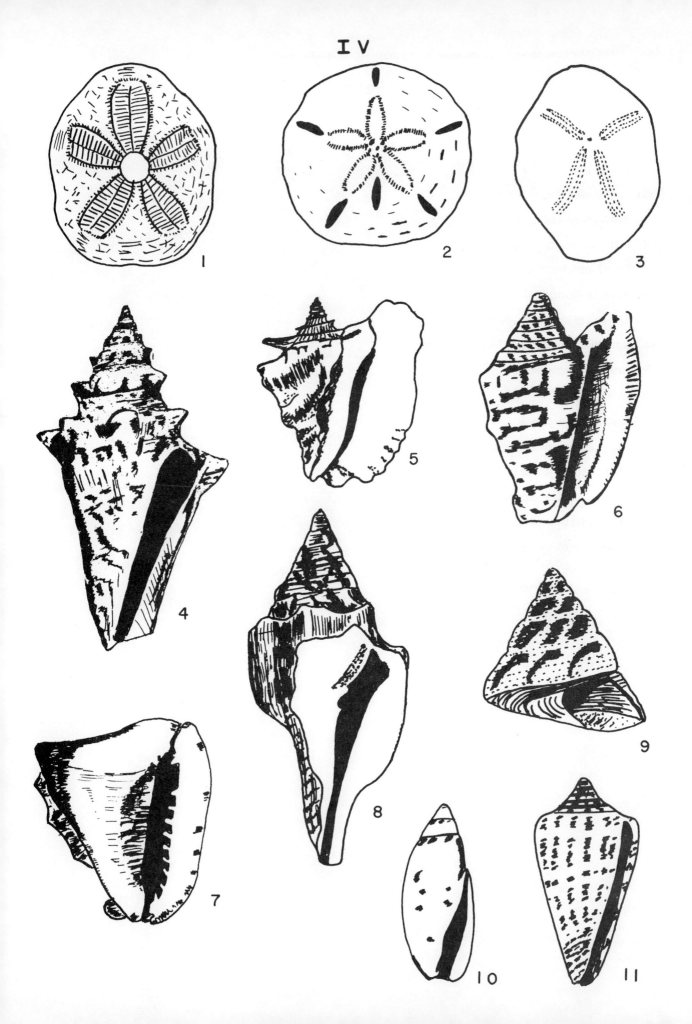

PLATE V

Molluscan species of the vegetation-stabilized sand (*Strombus costatus*) biofacies.

Figure

1   Astrea phoebia, X 1

2   Astrea longispina, X 1.5

3   Modulus modulus, X 3

4   Natica canrena, X 1.5

5   Cerithium eburneum, X 3

6   Terebra dislocata, X 1.5

7   Cerithium litteratum, X 1.5

8   Bulla occidentalis, X 1.5

9   Prunum apicinum, X 1

10 Olivella sp., X 3

11 Lucina pennsylvanica, X 1

12 Laevicardium laevigatum, X 1

13 Macrocallista maculata, X 1

14 Trigonocardia medium, X 1

15 Glycimeris pectinata, X 1

16 Codakia orbicularis, X 1

V

PLATE VI

Vegetation-stabilized sand (*S. costatus*) biofacies (figs. 1-7) and the Transitional biofacies between the normal marine and highly saline marine faunas (figs. 8-13). The Transitional biofacies has mainly sandy mud-bottom burrowers.

Figure

1   Pinna carnea, X 0.5

2   Aequipecten gibbus, X 1.5

3   Pinctada radiata, X 1

4   Arca occidentalis, X 1.5

5   Tellina radiata, X 1

6   Corbula sp., X 7

7   Nucula proxima, X 4

8   Clypeaster rosaecus, X 0.5

9   Modulus modulus, X 3

10   Cerithium muscarum, X 1.5

11   Chione cancellata, X 1.5

12   Transennella sp., X 3

13   Pinctada radita, X 1

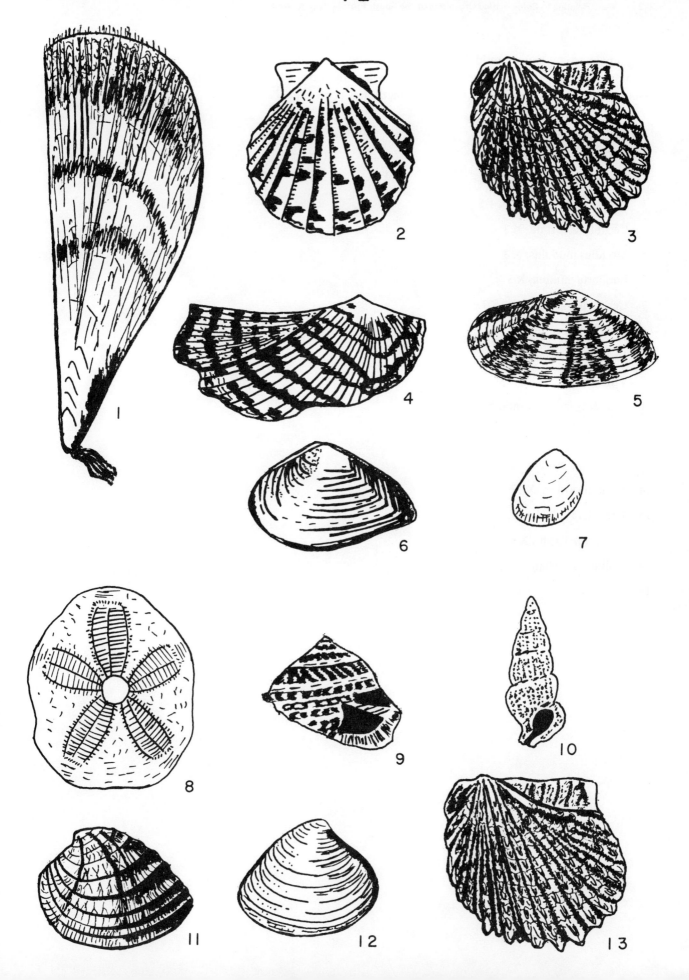

VI

PLATE VII

Muddy intertidal and supratidal mangrove-associated (*Fasciolaria-Batillaria*) biofacies is found on tidal flats of Andros Island and on the keys of Florida Bay and the Lower Keys (figs. 1-11). The tidal creek mangrove associated (*Crepidula*) biofacies is found in adjacent low areas (figs. 12-18).

Figure

1   Fasciolaria tulipa, X 1

2   Fasciolaria gigantea, X 0.5

3   Cerithium litteratum, X 1.5

4   Cerithidea costata, X 3

5   Modulus modulus, X 3

6   Batillaria minima, X 3

7   Cerithidea scaliformis, X 1.5

8   Cerithium floridensis, X 2

9   Melampus coffeus, X 2

10   Transennella sp., X 3

11   Pseudocyrena colorata, X 3

12   Cerithium eburneum, X 3

13   Cerithidea costata, X 3

14   Bittium varium, X 12

15   Crepidula sp., X 1.5

16   Transennella sp., X 1

17   Chione cancellata, X 1

18   Codakia orbicularis, X 1

VII

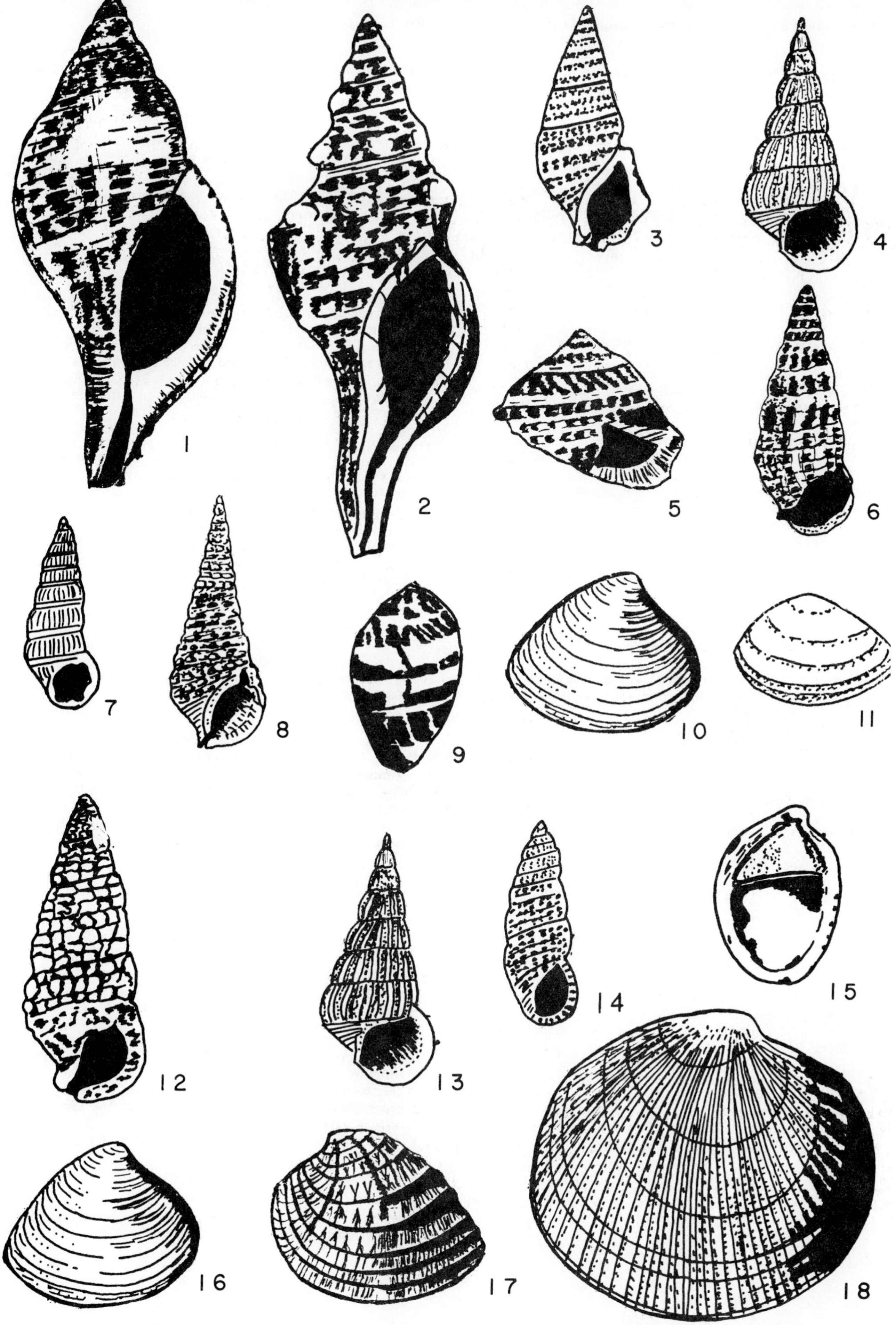

Partial Listing of References to Living Organisms in the South Florida— Western Bahama Area

The following list is intended for field and class/research use for those interested in identification and ecological investigations of local organisms. The author is indebted to Don Kissling and James Zischke for helping compile this list of useful references.

Algae

Chapman, V. J., 1946, Algal zonation in the West Indies: *Ecology,* v. 27, p. 91-93.

Ginsburg, R. N. and others, 1954, Laminated algal sediments of south Florida and their recognition in the fossil record: unpub. rept. no. 54-21, Univ. Miami Marine Lab, Coral Gables, Fla., 33p.

Humm, Harold J., 1963, Some new records and range extensions of Florida marine algae. *Bull. Mar. Sci. Gulf & Carib.,* 13 (4): 516-26.

Taylor, W. R., 1960. Marine algae of the eastern tropical and subtropical coasts of the Americas: Ann Arbor, University of Michigan Press, 870 p.

Marine grasses

Phillips, R. C., 1960, Observations on the ecology and distribution of the Florida sea grasses: Professional Papers Series, *Fla. Bd. Conserv.,* v. 2, p. 1-72.

Humm, H. J., 1964, Epiphytes of the sea grass *Thalassia testudinum,* in Florida: *Bull. Mar. Sci. Gulf & Carib.,* v. 14, p. 306-41.

Ginsburg, R. N. and Lowenstam, H. A., 1958, Influence of communities on the depositional environment of sediments: *Jour. Geol.,* v. 66, p. 310-18.

Moore, D. R., 1963, Distribution of the sea grass, *Thalassia,* in the United States: *Bull. Mar. Sci. Gulf & Carib.,* v. 13, p. 329-42.

Bernatowicz, A. J., 1952, Marine Monocothyledonous plants of Bermuda: *Bull. Mar. Sci. Gulf & Carib.,* v. 2, p. 338-45.

Foraminifers

Bock, Wayne, 1976, Monthly variation in the Foraminiferal Biofacies on *Thalassia* and sediment in the Big Pine Bay area, Florida, (Jan. 1967) Doctoral dissertation at the Univ. of Miami, see p. 81 and 84 this Guidebook.

Howard, J. F., 1965, Shallow-water foraminiferal distribution near Big Pine Key, southern Florida Keys: Compass, Sigma Gamma Epsilon, v. 42, p. 265-80.

Lynts, G. W., 1966, Variation of the Foraminiferal standing crop over short lateral distances in Buttonwood Sound, Florida Bay: *Limn. & Oceanog.,* v. 11, p. 562-66.

———., 1962, Distribution of Recent foraminifera in upper Florida Bay and adjacent sounds: Contrib. Cushman Found. Foram Research, v. 13, pt. 4, p. 127-44.

———., 1965, Observations on the Recent Florida Bay Foraminifera: Contrib. from the Cushman Found. for Foraminiferal Research, v. 16, pt. 2, p. 67-69.

Moore, W. E., 1957, Ecology of Recent Foraminifera in northern Florida Keys: *Bull. Am. Assoc. Petro. Geol.,* v. 41, p. 727-41.

Sponges

De Laubenfels, M. W. 1953, A Guide to the Sponges of Eastern North America, Spec. Publ. Mar. Lab. Univ. Miami, Univ. Miami Press, Miami, Fla., 33 p.

De Laubenfels, M. W., and John F. Storr, 1958, The Taxonomy of American Commercial Sponges. *Bull. Mar. Sci. Gulf & Caribbean* v. 8, no. 2.

Tierney, J. Q., 1954, Porifera, Gulf of Mexico Bibliography, *Fish. Bull., U.S.* v. 55, Mo. 89.

Corals

Bayer, F. M., 1961, The Shallow-water Octocorallia of the West Indian Region. Martinus Nijhoff, The Hague, Netherlands, 373 p.

Cary, L. R., 1918, The Gorgonaceae as a factor in the formation of coral reefs: Pap. from Dept. Biol. Carn. Inst. Wash., v. 9, p. 341-62.

Hedgepeth, J. W., 1954, Anthozoa: The anemones, In Galtsoff, P. L. (ed.), Gulf of Mexico its origin, waters, and marine life. *Fish, Bull. U.S.,* v. 55, Bull. 89: 285-90

Jones, J. A., 1963, Ecological studies of the southeastern Florida patch reefs. Part I. Diurnal and seasonal changes in the environment: *Bull, Mar. Sci. Gulf & Caribbean* v. 13, p. 282-307.

Kissling, D. L., 1965, Coral distribution on a shoal in Spanish Harbor, Florida Keys: *Bull. Mar. Sci.* v. 15, no. 3, p. 599-611.

Newell, N. D., 1958, American coral seas (West Indies): N.Y. Acad. Sci. Trans., ser. 2, v. 21, no. 2, p. 125-27.

Shinn, Eugene, 1963, Spur and groove formation on the Florida Reef Tract: *Jour. Sed. Petrology,* v. 33, p. 291-303.

Smith, F. G. Walton, 1943, Littoral fauna of the Miami area: I. The Madreporaria: *Fla. Acad. Sci. Proc.,* v. 6, no. 1, p. 41-48.

———., 1948, Atlantic reef corals: Univ. Miami Press, Coral Gables, 112 p.

———., 1954, Corals and Gorgonians. Gulf of Mexico Bibliography, *Fish. Bull., U.S.,* v. 55, no. 89.

Squires, D. F., 1958, Stony corals from the vicinity of Bimini, Bahamas, West Indies: *Bull. Amer. Mus. Nat. Hist.,* v. 115, art. 4, p. 221-62.

Vaughan, T. W., 1919, Fossil corals from Central America, Cuba and Puerto Rico, with an account of American Tertiary, Pleistocene and Recent coral reefs: *U.S. Nat. Mus. Bull.,* no. 103, p. 189-524.

Molluscs

Abbott, R. Tucker, 1954, American seashells: New York, Van Nostrand Co., 541 p.

———., 1968, Seashells of North America. Golden Press, New York, N.Y., 280 p.

Dubar, J. R., 1958, Stratigraphy and paleontology of the Late Neogene strata of the Caloosahatchie River area of southern Florida: *Florida Geol. Survey Bull.* 40.

Lloyd, R. M., 1964, Variation in the oxygen and carbon isotope ratios of Florida Bay mollusks and their environmental significance: *Jour. Geology,* v. 72, p. 84-111.

Lenderking, Ruth, 1954, Some recent observations on the biology of *Littorina angulifera* Lam. of Biscayne and Virginia Keys, Florida. *Bull. Mar. Sci. Gulf & Caribbean,* v. 3, no. 4.

Marcus, E. and E. Marcus, 1960, Opisthobranchs from American Atlantic warm waters. *Bull. Mar Sci. Gulf & Carib.* 10: 129-203.

———., 1967, American Opisthobranch Mollusks. *Stud. Trop. Oceanogr. Miami.* no. 6, 256 p.

Perry, L. M. and Schwengel, J. S., 1955, Marine shells of the western coast of Florida: New York, Paleontological Research Institute.

Wade, B. A., 1967, Studies on the biology of the West Indian beach clam, *Donax denticulatus* Linne: *Bull. Mar. Sci.,* v. 17, no. 1, p. 149-74.

Warmke, G. L. and Abbott, R. T., 1962, Caribbean seashells: Narberth, Penn. Livingston Publishing Co.

Arthropods

Barnard, J. L., 1969, The families and genera of marine gammaridean Amphipoda. *Bull. U.S. Nat. Mus.* 271: 1-535

Benson, R. H., and Coleman, G. L., 1963, Recent marine ostracodes from the eastern Gulf of Mexico: Univ. Kans. Paleont. Contr., Arthropoda, Art. 2, p. 1-52.

Chace, F. A., 1972, The Shrimps of the Smithsonian-Bredin Caribbean Expeditions with a Summary of the West Indian Shallow-water Species (Crustacea: Decapoda: Natantia). Smithsonian Contributions to Zoology, no. 98, 179 p.

Manning, R. B., 1969, Stomatopod Crustacea of the Western Atlantic. *Stud. Trop. Oceanogr. Miami.* no. 8, 380 p.

Pilsbry, H. A, 1916, The sessile barnacles (Cirripedia) contained in the collections of the U.S. National Museum: including a monograph of the American species. *Bull. U.S. Nat. Mus.* 93: 1-357.

Provenzano, Jr. Anthony J., 1959, The shallow water hermit crabs of Florida. *Bull. Mar. Sci. Gulf & Caribbean,* v. 9, no. 4.

Puri, H. S. and Hulings, N. C., 1957, Recent ostracode facies from Panama City to Florida Bay area: Gulf Coast Assoc. *Geol. Soc. Trans,* v. 7, p. 167-90.

Puri, H. S., 1960, Recent Ostracoda from the west coast of Florida: Gulf Coast Assoc. *Geol. Soc. Trans.,* v. 10, p. 107-49.

Rathbun, M. J., 1918, The grapsoid crabs of America. *Bull, U.S. Nat. Mus.* 97: 1-461.

———., 1925, The spider crabs of America. *Bull. U.S. Nat. Mus.* 129: 1-613.

———., 1930, The cancroid crabs of America, *Bull. U.S. Nat. Mus.* 152: 1-609.

———., 1933, Brachyuran crabs of Porto Rico and the Virgin Islands. N.Y. Acad. Sci. Scientific Survey of Porto Rico and the Virgin Islands. 15: 1-121.

———., 1937, The oxystomatous and allied crabs of America. *Bull. U.S. Nat. Mus.* 166: 1-278.

Schmitt, W. L., 1935, Crustacea. Macrura and Anomura of Porto Rico and the Virgin Islands. N.Y. Acad. Sci. Scientific Survey of Porto Rico and the Virgin Islands. 16: 123-227.

Schultz, G. A. 1969, the Marine Isopod Crustaceans. Wm. C. Brown Company Publishers, Dubuque, Iowa, 359 p.

Shinn, E. A., 1968, Burrowing in recent line sediments in Florida and the Bahamas: *Jour. Paleontology,* v. 42, p. 879-94.

Shoemaker, C. R., 1935, The amphipods of Porto Rico and the Virgin Islands. N.Y. Acad. Sci. Scientific Survey of Porto Rico and the Virgin Islands. 15: 229-62.

Voss, G. L., 1955, A key to the commercial and potentially commercial shrimp of the family Penaeidae of the western North Atlantic and Gulf of Mexico. Fla. St. Bd. of Cons., Tech. Ser. 14: 1-23.

Echinoderms

Clark, H. L., 1933, A handbook of the littoral echinoderms of Porto Rico and the other West Indian Islands. N.Y. Acad. Sci. Scientific Survey of Porto Rico and the Virgin Islands. 16: 1-147.

Deichmann, E., 1930, The holothurians of the western part of the Atlantic Ocean. *Bull. Mus. Comp. Zool. Harv.* 71:43-226.

———., 1954, The holothurians of the Gulf of Mexico. In Galtsoff . L. (ed.), Gulf of Mexico its origin, waters, and marine life. *Fish. Bull. U.S.,* v. 55, Bull. 89: 381-410.

Thomas, Lowell P., 1962, The shallow water amphiurid brittle stars (Echinodermata, Ophiuroidae) of Florida. *Bull. Mar. Sci. Gulf & Carib.,* 12(4): 623-94.

General

Greenberg, J. and Greenberg, I., 1972, The Living Coral Reef: Corals, and Fishes of Florida and the Bahamas, Bermuda and the Caribbean. Seahawk Books, Miami, Fla., 110 p.

Voss, G. L., 1976-in press, Guide to the Commoner Shallow Water Invertebrates of Florida and the Caribbean. Seeman Press, Miami.

Field Trips

GENERAL STATEMENT: The Florida Keys-Bahamas represents the only easily accessible area in North America for viewing aspects of the modern and ancient "carbonate world." As such it is the focus of a great many field trips of all sizes and with various objectives.

In some cases successful "expeditions" have been made by several departments (i.e. geology and biology) from one or several campuses joining forces. One interdisciplinary technique is described by Steinker and Floyd (1976) in the abstract given below.

AN INTERDISCIPLINARY FIELD COURSE IN MARINE SCIENCE
by
D. C. Steinker and J. C. Floyd

Many field courses in geology and biology tend to suffer from the effects of compartmentalization in science. For several years we have taught an interdisciplinary field course in shallow water marine environments in the Florida Keys, where diverse environments are easily accessible. Students are drawn primarily from geology and biology, but also have included majors from other areas of science. Lectures, discussions, field excursions, and group research projects serve to integrate various biological, geological, and chemical aspects of the marine environment. Students with backgrounds in different areas work toegether on field-oriented research problems. The main objectives of the course are: (1) to give the student a better understanding of the marine environment through first-hand experience, and (2) to instill a better appreciation of the fundamental unity of science through problem solving.

In addition to individual or campus organized and led trips as noted above there are professionally-led "Package Tours" of from 2 to 10 day duration for various groups (these are described on page 383 of this Guidebook).

Whatever the size or objectives of various visiting groups, however, there are some common basic necessities and a series of "classic field stops" usually required. The purpose of this last chapter is to help in planning and implementing your visit by noting some of these basic necessities and providing various length itineraries with some of these "classic field stops." Have a good trip!

CHARTS: Charts for the Florida area are available from:

> Director, Coast and Geodetic Survey
> ESSA
> Washington Science Center
> Rockville, Maryland 20852

For map coverage between Miami and Marathon and in Florida Bay use the latest edition of chart 11451 (formerly 141-SC).

Chart 1251 covers the area below Marathon to Key West. Other charts—same scale as 1251—covering the Keys from Marathon to Miami are 1250 and 1249.

For the western Bahaman area persons should send to the same above address for chart 1112; chart 1002 for all Bahamas and south Florida area. From the United States Navy Hydrographic Office, Washington, D. C., charts H. O. 26-A and H. O. 1854 are useful for more detailed work in the western Bahamas.

For those interested in details of Florida Bay there is the "Boaters Guide to the Upper Florida Keys: Jewfish Creek to Long Key" by John O'Reilly, 64 pages, available from the University of Miami Press, Coral Gables, Florida. This is a descriptive (nongeological) account of the Florida Bay area with detailed maps of boat routes to be used in conjunction with standard charts. **Note:** "channels" and "banks" in Florida Bay are subject to constant change (without notice!).

GENERAL SAFETY RULES: For work out of boats in the south Florida—Bahama area one should observe certain rules at all times: (1) have one good life preserver aboard for each person, (2) have a good ladder for getting in and out of boat, (3) use "buddy system" when in water, (4) have a good anchor, anchor chain and line (use grapnel when anchoring in coral), (5) leave one's destination and estimated time of return with some reliable person onshore, (6) take flashlight (useful if caught out after dark—i.e., motor trouble) and sun protective gear (lotion, hats, etc.). See note on page 199 about offshore boating.

FIELD CONDITIONS (. . .or "fun in the sun"?): Despite rumors circulated to the contrary, the weather in Florida is always unpredictable and can be nasty without warning. Precipitation and wind data as well as local advice indicates that June is usually a good month for fieldwork with rain and hurricane potentialities increasing towards September-October. Strong winds disturb the bottom sediment and often make the water opaque for several days and small boat handling difficult. The best weather (low winds) should be used for outer reef observations; the more protected Florida Bay sometimes can be observed even on windy days.

EQUIPMENT: Besides rain and wind, Florida has lots of sun for which the field trip participant should be properly prepared. The following items are listed to make the initial venture more comfortable and hence more productive!

- hat with brim
- mask, snorkel, flippers, weight belt
- suntan oil and/or lotion, sun glasses
- sun preventative paste (white, opaque)
- lightweight, long-sleeved shirt to wear in water
- lightweight, long pair of slacks to wear in boat
- one pair of old canvas tennis sneakers to wear in the water (when flippers are not appropriate)
- seasickness pills if believed possibly necessary
- rubber wetsuit if swimming in the winter months
Other accessory items which night prove valuable would be:
- Plastic bucket, plastic sample bags, waterproof black felt marker pens, gloves for working in coral areas
- underwater camera

A simple "home made" size-grade analysis device for field use is described in figure A.21 with an example of a data sheet for tabulating results given on page 381 of this Guidebook. Box cores can be made out of one gallon rectangular tin cans with bottoms cut out. A more sophisticated box core is shown in use in figure 3.4. Core samplers come in a variety of sizes (see figure A.22) and can be a very useful tool for obtaining samples from which impregnated slabs, such as shown in figure 6.14, can be made.

PROTECTED AREAS: Visitors should note the boundaries of various Protected Areas in the Florida Keys and abide by rules and regulations governing their use. These areas include the Everglades National Park (Crane Key is in this part), Ft. Jefferson National Monument (Dry Tortugas) and John Pennekamp Coral Reef State Park. In such areas the removal or destruction of natural features and

marine life is prohibited. Collecting of scientific specimens is by permit only. For further information concerning the Everglades Park and Ft. Jefferson areas write to: Superintendent, Everglades National Park, P. O. 279. Homestead, Florida 33030. **Note:** There are abundant opportunities for sampling Key Largo/Miami limestones from huge quanities of dredged rock piles (often of perfect specimen size!) along canals and construction sites. Visitors are urged to use these sources of supply and not to deface vertical walls of bedrock in quarries or along canals (i.e., Windley Key quarry or Key Largo Canal cut).

FIELD ACCOMMODATIONS: Motels and camp sites are available in the Florida Keys and persons wishing to use these facilities should check the various directories available. Persons making their initial trip should, if at all possible, check by phone or, better yet, with a reliable local source of information as to availability of boats for rent. Weekends are usually a more difficult time to charter due to increased demands for boats by Florida resident fishermen. If one is leading a large group for the first time it is highly advisable to make a "recon" trip beforehand or discuss details with someone who has recently returned from a similar field trip. The latter advice is even more imperative if plans are being made to visit the Bahamas where availability of accommodations and boat rentals (with a few exceptions) can vary with amazing regularity!

FIELD TRIP ITINERARIES

A. One-Half Day (Miami Area)

Mangrove Reef Rock (Northern Key Biscayne): Map figure 2.3; photo figures 2.12 to 2.16. Text page 11. Arrive at 12:30 p.m.; park car at city park northeast of where bridge comes onto Northern Key Biscayne and walk to outcrop. Also present are thin layers of *worm reef rock* as seen in figure 2.16. See Bayer (1964) for littoral marine organisms along this beach, (p. 14 this Guidebook). Return to car by 2:30 p.m.

Worm Reef Rock (Northeast Virginia Key): Map figure 2.3, figure 2.4 to figure 2.11. Text pages 7-11. Drive to northeast end of Virginia Key and park as close to shoreline as possible. Walk the beach and observe encrusting worm reef deposits. Offshore in two to four foot depths can be found additional worm reef accumulations. Return to car by 4:00 p.m.

Miami Formation-Oolitic Facies (Coconut Grove): Index map 1; text chapter 9. Return to mainland and drive southwest for about a mile on south Bayshore Drive. You will be driving on top of the Silver Bluff

Terrace. As the road comes off the terrace you will observe on the north side of the road several excellent exposures of the oolitic facies of the Miami Limestone (see figure 9.1). *Please* do not walk on lawns nor chip rock from these exposures—this is private property.

B. One Day (Florida Keys)

Lagoon Area (Florida Bay): Index map 3. Leave Islamorada (Upper Matecumbe Key) by 7:30 a.m. in hired boat and motor north of *mud mounds "78"* (fig. 2.39) and pages 76-80. Leave mud mound and go directly north until you intercept Cross Bank, a *mud bar.* Examine bar (pp. 71-73). Continue westward on south side of bar until you reach channel cut through bar just east of the largest of the Crane Keys. Go ashore on the northwest shore of the larger of the Crane Keys. Figures 2.40 to 2.44 and pages 40-42. Return to Islamorada area via same route or use channel about 1/2 mile west of smaller of Crane Keys, west of the larger of the Crane Keys; whichever return route is selected, use Route U.S. 1 bridge at Islamorada as bearing marker for return.

Back Reef (Hens and Chickens Patch Reef): Pass through Islamorada bridge by 1:00 p.m. and proceed to Hens and Chickens *back reef.* Use questions given on pages 193 and 315. Return to Islamorada by 5:00 p.m.

C. One and One-Half Days (Miami Area Plus Florida Keys)

One-half day spent doing trip A noted above; drive to Tavernier at night for early start the next morning.

One Day

Florida Bay (Leaving from Tavernier): Map figure 2.39 and 3.1; text pages 40-42 and pages 65-75. Leave Tavernier by 7:30 a.m. in rented boat and go through Cross Bank at cut about 1/2 mile west of where Tavernier Creek empties into Florida Bay (fig. 2.39) at the flashing light "73". Proceed west to post "78A" which is a *mud mound* (figure 2.39 and page 76). Leave mud mound and go directly north until you intercept Cross Bank, a *mud bar.* Examine bar (pp. 71-73). Continue westward on south side of bar until you reach channel cut through bar just east of the larger of the Crane Keys. Go ashore on the north western shore of the larger of the Crane Keys, figures 2.40 to 2.44 and pages 40-42. Return to Tavernier Creek by 12:30 p.m.

Back Reef (Rodiguez Key Area): Map figure 6.1 and figures 6.8 to 6.14. Leave Tavernier Creek by 12:30 p.m. and motor to Rodriguez Key anchoring off the southeast corner. Swim and wade ashore to mangrove island. Answer questions page 194. Return to Tavernier Creek by 5:30 p.m.

D. Two Days (Florida Keys)

One-Half Day

Lagoon Area (Coupon Bight, Big Pine Key): Maps figure 3.27 to figure 3.29, text p. 94-97. Use Coast and Geodetic chart 1251. Charter boat from southeast Big Pine Key or Newfound Harbor Channel area. Arrive at Coupon Bight by 8:30 a.m. Observe and sample flora and fauna; note sediment and bottom topography as described pages 94-97. Leave Coupon Bight by 12:30 p.m.

One-Half Day

Back Reef (Patch Reef off South Shore of Big Pine Key): Map figure 3.27 and Coast and Geodetic chart 1215. Motor to small patch reefs which lie in offshore band parallel to the shore starting about 1/2 mile southeast of flashing light "2". Use text pages 176-182 and note figures 6.16 and 6.18 to 6.20 which were taken underwater at this Big Pine Key site. Answer questions pages 193-194. Return to dock by 5:30 p.m.

One Day

Outer Reef (Looe Key): Map Coast and Geodetic chart 1215. Leave dock by 8:00 a.m. and motor to Looe Key taking compass bearing and looking for single spar on Looe Key. **Note:** safety statement for outer reef work on page 199. Anchor in southwestern corner of protected sand area in back of outer reef. Zonation is not as distinct as that found at Key Largo Dry Rocks. Read pages 199-215. Answer questions pages 235-236. Leave Looe Key by at least 4:00 p.m. for mainland or sooner if weather becomes questionable.

E. One Day Reef Trip (Florida Keys)

Stops all in the park—so no sampling allowed. See page 199 for safety note on outer reef trips.

Half Day

Back Reef (Mosquito Banks): Map figure 6.1. Coast and Geodetic chart 11451. Text pages 176-182. Leave Tavernier by 8:00 a.m. and motor up Hawk Channel to Mosquito Bank reef. Answer questions pages 193-194. Leave for Key Largo Dry Rocks by 11:00 a.m. taking an easterly bearing directly toward the nun "4DS" on edge of outer reef platform. Enroute stop at one of the shallow White Banks sand patches shown on chart 11451.

Blanket Sediments (White Banks): Anchor and observe bottom sediment. Read pages 182-191 and figures 6.23 to 6.26. Answer questions page 196.

Half Day

Outer Reef (Key Largo Dry Rocks): Leave White Banks by 12:15 p.m. for Key Largo Dry Rocks by proceeding on an easterly course to edge of shallow

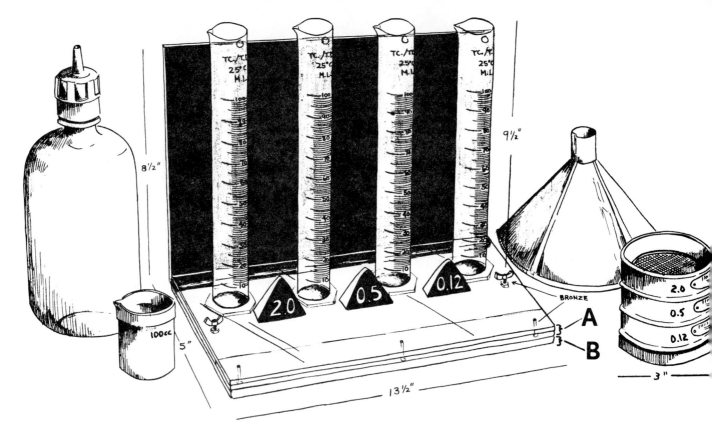

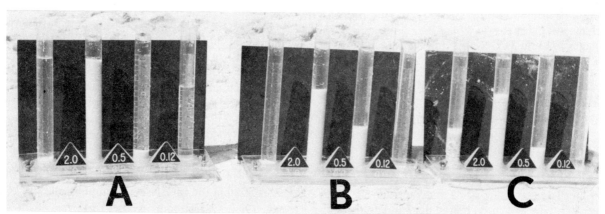

Figure A.21. Portable noncorrosive size-grade analysis equipment for quick field measurements. (*top*) Components are all made of plastic except stainless sieves and bronze wing-nut bolts. Basal plastic sheet "A" has four holes to receive plastic graduates. Wing-nuts secure plastic sheet "A" to sheet "B" and hold graduates in place. Protruding edge of black backboard slips between sheet "A" and "B" so light-colored sediment can be photographed (see below). **Procedure:** Take 100 cc sample using plastic cup, dump into top sieve, wash sample through screens using plastic squirt bottle, dump and wash contents of each screen into appropriate graduate using plastic funnel.

(*above*) Results in N. Bimini beach analysis (A) 100 feet offshore, 10 feet depth; (B) 60 feet offshore, 7 feet depth; (C) littoral midtide sample. (*left*) Field analysis underway. Note: see page 381 for data recording sheet.

BOTTOM SAMPLE DESCRIPTION SHEET

LOCATION

| 2 | .5 | .12 |

DESCRIPTION

TRACT	OUTER REEF					
REEF	BACK REEF					
LAGOON						
OTHER						

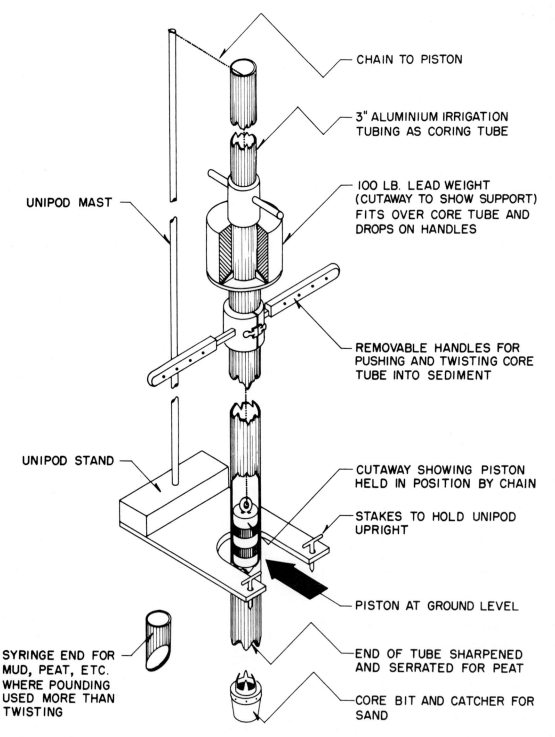

CHAIN TO PISTON

3" ALUMINIUM IRRIGATION TUBING AS CORING TUBE

100 LB. LEAD WEIGHT (CUTAWAY TO SHOW SUPPORT) FITS OVER CORE TUBE AND DROPS ON HANDLES

REMOVABLE HANDLES FOR PUSHING AND TWISTING CORE TUBE INTO SEDIMENT

UNIPOD MAST

UNIPOD STAND

CUTAWAY SHOWING PISTON HELD IN POSITION BY CHAIN

STAKES TO HOLD UNIPOD UPRIGHT

PISTON AT GROUND LEVEL

END OF TUBE SHARPENED AND SERRATED FOR PEAT

SYRINGE END FOR MUD, PEAT, ETC. WHERE POUNDING USED MORE THAN TWISTING

CORE BIT AND CATCHER FOR SAND

Figure A.22. Sketch of drop-weight coring sampler for various types of sediment by Wanless (1969). For other papers on coring devices see Ginsburg and Lloyd (1956), Ginsburg *et al.*, (1966) and Enos (1977 – in press).

outer reef platform (nun "4DS") and then northeast along edge of platform to Key Largo Dry Rocks (or Grecian Rocks) areas. A second procedure would be to estimate position at White Banks and plot a direct compass course to Key Largo Dry Rocks. Anchor using chain at anchor end of line and examine reef. Read pages 199-215 and figures 7.8 to 7.13. Answer questions pages 235-236. Leave Key Largo Dry Rocks by 3:45 p.m. for mainland.

F. One Day Lagoon and Beach Trip (Bimini, Bahamas)

Half Day

Lagoon (Bimini): Maps figures 3.30 to 3.32; U.S.N. H.O. charts 26-A and H.O. 1854. Also figures 3.33 to 3.43. Leave South Bimini dock at 8 a.m. in skiff and motor to Pigeon Cay. Examine bottom sediment and flora/fauna along a northwest-southeast traverse from southeast corner of Cay to the middle of deepest portion or rock-floored channel running along west shore of south Bimini. At 10 a.m. motor to a point off northeast corner of Mosquito Point and compare bottom sediment and fauna/flora with that just seen off Pigeon Cay. Read text pages 100-109 and answer questions page 127. Return to north Bimini dock by 12 noon. Lunch in Alice Town.

Half Day

Sand Beach (North Bimini): Walk to beach west of Alice Town. Examine beachrock (text page 33 and figures 2.33 and 2.34), examine beach sand and cliffs of indurated sand. See questions page 58 and page 59. Return to south Bimini by ferry by 3:15 p.m.

Rocky Shore (South Bimini): Taxi from south Bimini dock to Nixsons Harbor (map figure 3.30). Figures 2.21 to 2.25 text pages 20-22. Walk (at low tide) to Rabbit Cays and examine erosion of rocky shore. Answer questions page 57.

G. One Day—Oolitic Sand Belt and Supratidal Flats (Cat Cay and Andros Island)

Half Day

Oolitic Sand Belt (Cat Cay): Map U.S.N. H.O. charts 26-A and H.O. 1854; figure 5.15. Leave in chartered amphibian plane from north Bimini at 8:30 a.m. (or from Miami at 8 a.m.) and fly to Cat Cay area landing in one of the channels traversing the oolitic sand belt (fig. 5.15); anchor plane on one of the bars and examine bar and adjacent channel bottom. Text pages 151-155 and answer questions page

163. Fly out of area by 10:30 a.m. Chart course towards Williams Island. Circle Williams Island (fig. 2.47).

Half Day

Supratidal Flats (Williams Island and Palm Hammock): Land near old fish camp and dock at Williams Island. See page 63 for typical cross section for running traverse and answer questions on page 62. **Note:** if time is short skip stop at Williams Island and go directly to Palm Island shown by arrow in figure 2.48 and in photos of figures 2.49 and 2.50. Land in adjacent channel and make traverse from channel across tidal flats to palm hammock. Examine dolomite crust band ringing hammock (fig. 2.50). See also figures 2.51 and 2.52 and text pages 45-55.

H. "Package Tours"—2 to 10 Days—Professionally Led (Florida/Bahamas)

"Package tours" can be readily arranged for viewing whatever aspects of the modern carbonate world of the Florida-Bahamas one wishes to see. Time, pocketbook, and availability of an authoritative guide are the only limitations. For example, the author has (with M. P. Weiss) led ten day continuous field trips six consecutive summers for twenty-two college professors (NSF supported Carbonate Short Course) covering the Pliocence-Pleistocene bedrock of southeast Florida and the modern reef environments of Florida Keys and western Bahamas. On-site lectures and evening seminars were used extensively. Logistics involved coordinating charted buses, boats, rental cars, amphibian planes, motels and restaurants. Another technique (used by the writer for five consecutive summers for an average of fifteen petroleum geologists from a major oil company) involves having an advance "prep session" (at the company's main office) for three days with intensive lecturing on various aspects of modern carbonate sedimentation followed immediately by four days of continuous field trips in the Florida Keys and one day overflight with stops in the Bahamas. Other "package tours" could involve working out of an available boat with sleeping and working facilities off eastern Andros. If well planned and executed, these types of trips can be extremely beneficial.

As an example of the above types of trips, the following published account from the September 1970 issue of the Shale Shaker magazine is reproduced as an illustration of what one group of (petroleum) geologists chose to see on a typical

Florida-Bahama field trip. Index maps II and III are keyed to each day's stops.

THE BAHAMAS AND SOUTHERN FLORIDA: A MODEL FOR CARBONATE DEPOSITION
by
Gerald M. Friedman[*]

Introduction

In petroleum exploration the emphasis has shifted in recent years from structural to stratigraphic trap exploration. In stratigraphic trap exploration in a carbonate terrain, the geologist searches for favorable lithologies which represent environments that at the time of depsotion had primary interparticle porosity or those in which secondary porosity was prone to develop after deposition. These environments include (1) the high energy zone of strong currents or waves where intergranular winnowing deposits well-sorted calcareous particles devoid of lime mud (micrite) and (2) these environments in the supratidal zone where secondary porosity develops during dolomitization (Friedman, in press).

Both of these favorable environments can be studied in the Bahamas and Florida together with a spectrum of other environments which are less favorable for exploration. Acquaintance with the entire spectrum of environments is necessary for successful exploration.

Most major oil companies send their exploration geologists to visit the Bahamas and Florida to learn recognition of carbonate environments and prediction of the favorable interfingering facies. These carbonate schools are an essential part of the exploration business. Most geologists attended college prior to the development of modern carbonate terminology and concepts (Folk 1959). Hence they have misconceptions about carbonate sediments and rocks and about carbonate depositional environments. Geological societies are now beginning to follow the example of the major companies, particularly for the benefit of independents, consultants, and the smaller companies. Yet personnel from the major companies cannot always be accommodated on the company field trips because the size of the group has to be kept small. Geological societies have an educational obligation toward all their members, irrespective of company affiliation.

The Oklahoma City Geological Society has pioneered in the petroleum business in being one of the first to organize a society-sponsored field trip to Florida and the Bahamas. This field trip was prompted only in part as an educational experience; the major inducement was economic. When you drill a well you do not know the kinds of facies that may interfinger with those observed in the well. In the Bahamas the sharp facies boundaries are startling. An understanding of the sequence of facies changes together with the marine processes responsible for these changes is a necessary prerequisite for an intelligent interpretation of a well drilled in carbonate terrain. Current trends are toward drilling deeper wells which cost from one hundred thou-

sand dollars to several million dollars; thus money spent on a well must be coupled with sound interpretation. A field trip to the Bahamas and Florida is inexpensive in comparison with the cost of a well. The Oklahoma City Geological Society trip cost each participant about $600, cheap in relation to the cost of a well.

At the outset we should mention two of the most startling impressions that are apparent on any field trip to the carbonate environments in the Bahamas and Florida: (1) the sharp facies boundaries and (2) the great amount of organic material as revealed by the abundance of "critters[**]" in the sediments and waters. Thus the source for hydrocarbons is found in the sediments themselves.

One aspect of caution needs to be stressed: the purpose of field work in the Bahamas-Florida tract is to teach principles, such as the conditions under which the various facies are deposited, the favorable facies for exploration, or the conditions of dolomitizations. The field area does not teach scale. Sea level has reached its present position only 4,000 years ago. The beginning of modern carbonate deposition in the area does not exceed 4,000 years and the thickness of the various facies must be measured in feet, not in hundreds or thousands of feet. This observations has to be kept in mind when applying the principles learned in the Bahamas and Florida to the subsurface. Nevertheless we know from the Superior Oil Company well drilled on Andros Island in the Bahamas that the same shallow water facies extends to a minimum depth of 14,587 feet at which the well bottoms. Therefore, although we cannot apply the scale of *modern* sediments in the Bahamas to our subsurface problems, if we include in the Bahamas premodern (that is, Pleistocence, Tertiary, and Cretaceous) carbonate sediments, we come up with thickness analogous to those found in subsurface exploration.

The following sections discuss the Oklahoma City Geological Society field trip conducted between June 8 and 13, 1970. The sequence in which the stops are described follows the field trip sequence.

Pleistocene Outcrops
June 8

Stop 1. Windley Key Quarry located along the Florida Keys. This quarry which is in the Pleistocene Key Largo limestone allows the observer to place himself within the reef and observe along the reef walls the reef facies in three-dimensions. This stop makes an ideal beginning for a field trip to the area because it is the only opportunity for him to see in three-dimensions the anatomy of a reef. All later stops concern themselves only with two dimensions, as seen from the air or by swimming or wading in the sediment. This quarry has been described by Stanley (1966), brief descriptions are given by Multer (1969) and the diagenetic changes are explained by Friedman (1964, p. 807).

[*]Department of Geology, Rensselaer Polytechnic Institute, Troy, New York.
[**]A descriptive term derived by Friedman to eliminate the generic names of numerous gastropods, pelecypods, worms, and shallow water sea life.

This quarry is in the back-reef facies; the framework builder of the reef is a coral *Montastrea annularis*. Between the corals is abundant reef rubble as well as interstitial skeletal sand both of which were derived by the degradation of the reef builders and of the tests of other organisms that lived in and around the reef. Stanley (1966) made a statistical analysis of the amount of framework builders versus skeletal debris and arrived at a 30:70 ratio. Estimates made by our field party in looking at the quarry walls came closer to a 10:90 ratio. Irrespective of which figure is more correct, these measurements and estimates indicate that it would be difficult or even impossible to recognize in subsurface cuttings that the reef has been penetrated. A core would be necessary to identify the facies correctly as a reef; yet even in cores careful examination would be essential.

The reef is well lithified; most of the original aragonite in reef rubble and skeletal debris, but not in framework builders, has gone over to calcite (Friedman, 1964, p. 807). Yet porosity though variable is high; we have not measured porosity in the rocks of this quarry, but estimate the porosity to exceed 25%. The permeability is impressive and could amount to 500 or more millidarcies. A similar Pleistocene reef facies along the Red Sea coast studied by G. M. Friedman, G. Gvirtzman, and A. Yvneh gave porosity values of 25.9 to 68.3% and permeability values of 9.5 to > 10,000 md. Most of the permeability measurements in these Pleistocene Red Sea reefs exceed 1,000 md.

From a petroleum geologist's viewpoint, the lessons to be learned from this quarry are (1) an appreciation of the amount of reef rubble and reef sand actually present in a reef, and (2) the high primary constructional porosity and permeability still existing in reef facies.

Modern Carbonate Sediments in the Bahamas
June 9

Our Bahamas traverse originated in Miami, Florida, where we charted an amphibious plane and flew across the Gulf Stream (Florida Straits) to the island of North Bimini which lies on the west edge of the Great Bahama Bank. The plane landed in Bimini Lagoon before it taxied on to the island and parked at "Bimini International Airport". Before splashdown in the lagoon, sharks can almost invariably be seen from the air and stingingrays are often visible. Hence geologists on arrival in the Bahamas should be careful in watching out for these fish and try to avoid them. During stops the pilot is usually requested to watch from the vantage point of the wing while geologists are in the water. Despite numerous field trips by companies, universities, and geological societies no geologist has yet fallen victim to an attack by a predatory fish.

Stop 1. Bimini Island. On the island of Bimini we examined skeletal carbonate sands in the tidal zone along the west shore facing the Florida Straits. This skeletal sand consisting of fragments of corals, coralline algae, sea urchin spines, *Halimeda* (green algae) and molluscan shells is a high energy facies with a high interparticle porosity. It is the kind of facies that makes a good reservoir rock. The lime mud or lime silt offshore below wave base is full of organic material which is derived as an "organic rain" from the high energy zone or is locally derived from planktonic or benthonic organisms. This organic matter is preserved in the offshore reducing sediments, and is available as a source rock. This combination of source rock in the offshore lime mud or lime silt and skeletal sand in the high energy zone illustrates the making of a stratigraphic trap. Fine-grained low-energy sediments to the east are necessary to provide a seal.

Recent beachrock is exposed along with the skeletal sand in the tidal zone. This beachrock consists of skeletal sand particles cemented together by a calcium carbonate cement. Introduction of the cement is related to chemical and possibly biological processes in the tidal environments and shows that cementation can take place in a marginal marine environment. Pleistocene limestone is also exposed on the beach immediately east of the recent beachrock. The dips on the Pleistocene are to the east while the Recent dips west.

Many geologists have studied the Bimini Islands and Bimini Lagoon; these include Bathurst (1967), Hay (1967), Kornicker (1960), Newell and Imbrie (1955), Newell and Rigby (1957), Newell et al., (1959), Purdy (1963), Scoffin (1969), Squires (1958), Turekian (1957), and Voss and Voss (1960). Mutler (1969, pp. 20, 23-24, 62-66) has summarized some of this data.

Stop 2. Oolite shoal between South Cat Key and Sandy Key. This stop, about 12 to 14 miles south of Bimini Island along the west edge of the Great Bahama Bank, demonstrates a high energy environment in which the carbonate sediments consist almost entirely of oolite. The distribution of the oolite facies is practically coincident with that of the shoals facing the ocean where optimum conditions for oolite formation are obtained (Purdy, 1961). By and large, the oolites are confined to the intertidal position with tidal channels dissecting the tidal flats. Both channels and flats are underlain by oolite. Our plane landed in a tidal channel and we moved from there by swimming and wading to the tidal flats. We saw the oolite shoals at high tide and appreciated the fast tidal currents flowing across the shoals. This high energy of strong currents and waves causes intergranular winnowing leading to well-sorting oolites devoid of lime mud and hence high interparticle porosity. This oolite shoal which faces the deep water and towards the platform interfingers with lime mud would make a good reservoir rock. In the subsurface it would have high interparticle porosity, the lime mud towards the platform would form a seal and the offshore lime mud and lime silts with their ample organic matter and reducing conditions would serve as a source for hydrocarbons.

In this area a chemical company is dredging the oolite as a source for aragonite. In an environment that is charted on bathymetric charts as part of the ocean, the presence of these large dredging machines provides a scale for the broad lateral expanse of this shallow oolite belt. According to Purdy (1963) the length of this oolite belt facing the deep water of the Florida Straits is about 100 miles.

Observations from the air indicate that grasses and green algae are absent from the oolite shoals contiguous to the ocean. Oolite sands are washed over onto the adjacent lime mud as tidal or washover fans, and near these washover fans, grasses, and green algae begin to settle on the oolite shoals. Broad ripple marks attesting to the action of the

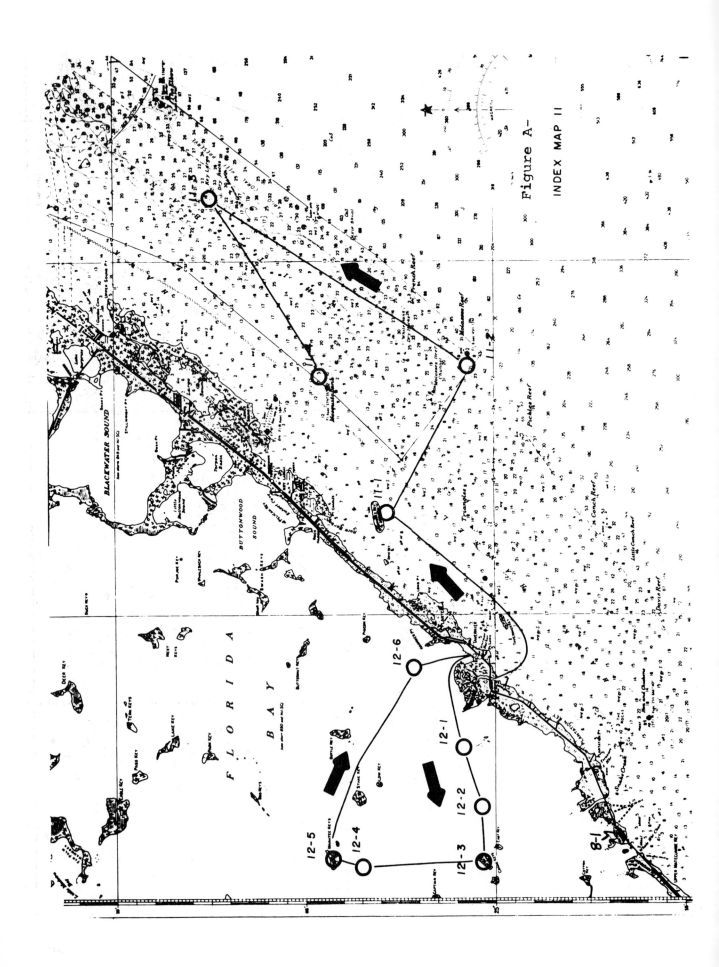

Figure A—

INDEX MAP II

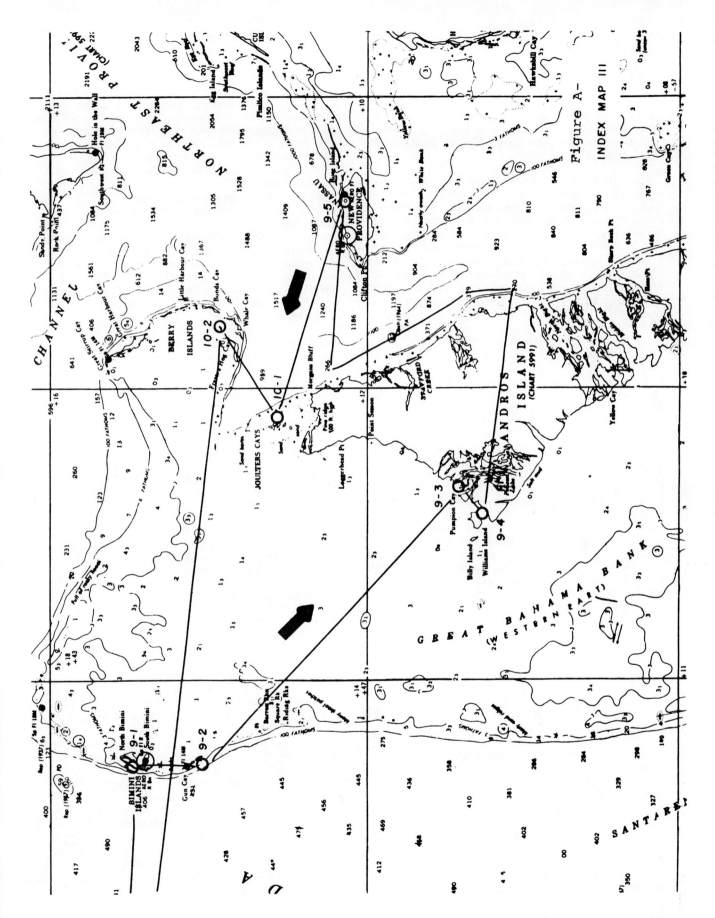

Figure A— INDEX MAP III

tidal currents which sweep across these shoals are evident from observations in flight.

On the seaward side as well as where the tidal or washover fans have migrated across the lime mud, crossbedding with steep avalanche faces is prominent. On the seaward side the crossbeds face seaward and on the washover fans they point toward the platform.

The main lesson at this stop is to become acquainted with the high energy oolite shoal which commonly serves as a reservoir in the subsurface.

Stop 3. West shore of Andros Island approximately one to two miles northeast of Williams Island. At this stop we landed in a tidal channel where we examined the pellet mud facies. We gained an appreciation of the turbid nature of the sediment when it is stirred up and the lime mud particles became suspended. Green algae are abundant in the sediment, including *Penicillus, Acetabularia, Rhipocephalus,* and *Halimeda.* This pelletal mud sediment is full of organic matter and exudes an odor of H₂S, indicating strong reducing conditions. Locally in the tidal channel the lime mud is lithified showing that marine lithification takes place in the intertidal zone, an observation already made in the high energy zone of Bimini Islands.

From the tidal channel we waded ashore to study dolomite crusts in the supratidal zone (zone above normal high tide level). The sediment in this zone consists of pellet mud that has become dolomitized. The dolomite content varies from 0 to 80%. The best exposures of dolomitized sediment are near palm hammocks where dolomite crusts have the appearance of an old asphalt road coating. The pellet muds are finely laminated, a characteristic of low energy environments resulting from the baffling or trapping action of blue-green algae. The algae trap the lime mud and their mucilage binds it. These finely-laminated algal mat structures are known as stromatolites and are diagnostic of uppermost intertidal to supratidal environments. The black dolomitized pellet mud crusts are commonly broken up by desiccation or "mud" cracks. Dolomite chips of these desiccation or "mud" cracks commonly separate and can be moved away by storm surges into the subtidal environment. These chips are analogous to the intraformational conglomerate so common in ancient rocks. Small voids, 1 to 3 mm in diameter, known as birds-eye structures are common in these sediments. These voids are likely to be the result of gas bubbles and are diagnostic of tidal sediments, especially those of the supratidal zone, although they are preserved sometimes in intertidal sediments (Shinn, 1968a). Additional details on the seidments of this area are given by Shinn and Ginsburg (1964) and Shinn et al., (1965a, b; 1969).

The supratidal sediments, such as those examined at this stop, are characterized by finely-laminated dolomite or dolomitic limestones, stromatolitic (undulating) structures, birdseye textures, desiccation cracks, chips derived from desiccation cracks, burrow mottles, and trace fossils (tracks or organisms) (Friedman, 1969; Friedman and Sanders, 1967).

Dolomites from this type of environment form reservoirs in the subsurface. Examination of the dolomite on the west side of Andros Island seems to belie this fact because the dolomite here is not very porous. However, this sediment is only partially dolomitized and porosity in this type of rock is created by the dissolution of the unreplaced calcium carbonate (Murray, 1960).

This dolomitized sediment is part of the pellet mud facies of the Great Bahama Bank. The pellet mud facies makes up the single most abundant facies west of Andros Island. The pellet mud facies shows reducing conditions and an abundance of organic matter. In the subsurface it could serve as a source of hydrocarbons. Dolomitized pellet mud with its high porosity after the removal of unreplaced calcium carbonate could evolve into a reservoir rock. Much of the Arbuckle in southern Oklahoma and west Texas (Ellenberger) consists of this type of rock.

Before leaving this area we made a brief stop between Andros Island and Williams Island to see more pellet muds. We sank into the mud to our waists.

Flying along the reef. On our traverse we flew across Andros Island and then along the reef which faces the Tongue-of-the-Ocean. Observations of this reef from the air showed its patchiness. The reef seems to consist of many patch reefs which locally coalesce, yet for the most part the reef is discontinuous. The width of the reef is variable but small, to be measured at most in hundreds of feet. Skeletal sand forms a facies behind the reef as a result of degradation of the reef. In a subsurface play you would have to pinpoint the reef. A reef is not necessarily something big, yet its porosity and permeability are high. However the reef rubble and interfingering skeletal sands may be part of the same reservoir. Reefs, reef rubble, and skeletal sands are part of the same high energy zone in which high interparticle porosity prevails.

On our traverse after flying along the reef we crossed the Tongue-of-the-Ocean and headed to Nassau for the night.

June 10th

Stop 1. Joulter's Cay. From Nassau we flew across the Tongue-of-the-Ocean toward Joulter's Cay. Approaching the Great Bahama Bank we crossed the reef where the shoaling waters break. Here we saw from the air the reef in the highest energy zone facing the ocean. Behind it, toward the platform occurs the skeletal facies and then the sharp boundary of the oolite facies. The oolite facies of Joulter's Cay like that between South Cat Cay and Sandy Cay is intertidal with tidal channels cutting across the oolite tidal flats. We landed in a tidal channel and waded across to the tidal flats. The sediment here consists of oolites, but the oolites are smaller than those observed between South Cat Cay and Sandy Cay. No measurements were made by us of the size of the oolites in these two separate areas, but qualitatively the size differences are readily apparent in the field. The explanation for the smaller size of the grains is that the conditions at Joulter's Cay are not as favorable for the precipitation of calcium carbonate as in the area between South Cat Cay and Sandy Cay. Between South Cat Cay and Sandy Cay the oolite facies is located along the shoals where waves and current break in the highest energy zone. By contrast, at Joulter's Cay, the highest energy shoaling zones are coincident with the reef which is followed toward the platform by the skeletal facies resulting from reef degradation, and only then by the oolite shoals. Moreover islands of Pleistocene rock (the "cays" for which Joulter's Cay has been named) partially protect the oolite shoals. Although these shoals are located in a high energy tidal zone, they are not in the highest energy environment, hence less calcium carbonate precipitation and smaller oolite particles.

On the oolite tidal shoals horizontal tracks of moving

organisms (known in ancient rocks as "trace fossils") are abundant as are truncated ripples.

The lesson pertinent to petroleum exploration that this stop offers is essentially the same as that for the area between South Cat Cay and Sandy Cay. However, this stop shows that the oolite shoals do not necessarily face the deep water but may be separated from it by reef and skeletal facies. In subsurface a Joulter's Cay oolite shoal equivalent could be part of the same reservoir as the reef and skeletal facies.

From the oolite shoal we flew about 1/4 to 1/2 mile westward to the lee side of the shoal where we examined volcanolike mounds made by *callianassid* shrimps. These mounds represent burrows made by the shrimps in the intertidal environment. The size of the mounds was variable but some were up to 1 ft high and up to 1 1/2 ft in diameter at the base. Shinn (1968b) who has studied these mounds by making polyester plastic casts of their burrow networks has shown that 1 to 2 ft away from the mounds are funnel-shaped entrances that lead down into passages having diameters as large as 1 in. At the top of the volcanolike mound is a small crater rimming the burrow opening from which the sediment issues which the shrimp pumps up from below. These mounds are most abundant in the intertidal zone and are exposed at low tide. Shinn (1968b) has found them down to water depth of 35 ft in Florida and the Bahamas, but notes that they are less numerous there than in the intertidal zone.

Burrows of this type are abundant in ancient rocks and are generally known as *Ophiomorpha*. *Ophomorpha* burrows have been described from rocks ranging in age from Upper Jurassic to Pleistocene (Kennedy and Seelwood, 1970) and are credited to *callianassid* shrimps. Hence these burrows, if found in ancient rocks, should be considered as tidal, most likely intertidal. Burrows similar to those made by *callianassids* extends down to the Cambrian and may suggest a tidal origin for the rocks in which these burrows are found.

Stop 2. Cockroach Cay in the Berry Islands. From Joulter's Cay we flew to Cockroach Cay in the Berry Islands chain. Here we landed on the lee side of Cockroach Cay to examine the grapestone facies. Grapestones are cemented aggregates of particles that resemble detrital limestone fragments. In fact though recognized from the Bahamas as early as 1879 (Sorby, 1879, p. 76; Purdy, 1963, pp. 343-344) it was not until 1954 that Illing (pp. 44-45) showed that these aggregates were not derived from the breakdown of older limestones, but rather were forming in place on the Great Bahama Bank. He used the term "lumps", a term still used by some of the companies (Leighton and Pendexter, 1962). Under the binocular microscope the grapestones resemble clusters of grapes, hence the name.

Near Cockroach Cay we found both grapestones and "crusts". The crusts are flat aggregates of particles that are much larger than grapestones, commonly up to an inch or more in length. As the term crusts indicates they probably formed as a crust parallel to the sediment floor.

The significance of the grapestone lies in their position in the energy spectrum between the oolite and mud facies. The oolite facies is in a zone of maximum, or close to maximum, turbulence, whereas the mud facies is in the zone of minimum turbulence. The grapestone facies occupies a position between these two. Currents and waves do not dislodge contiguous grains constantly, otherwise cement would not be able to form between them.

In a well it may be difficult to recognize grapestones in samples or even in cores. However, careful examination of cores should show the difference between grapestones and detrital limestone particles. In grapestones, the grains which make up the aggregates tend to be mostly of similar size and are whole. By contrast, in detrital limestone particles, the edges are commonly rounded from wear during transportation and individual grains within the detrital limestone fragments show the effect of breakage with individual grains along the margin sharply truncated.

This stop was the last of our traverses in the Bahamas and we flew across the Great Bahama Bank and Florida Straits back to Miami.

Modern Carbonate Sediments in South Florida
June 11

On June 11th a traverse was made by boat from Tavenier on Key Largo to the outer reef tract. The purpose of this trip was to examine first a mound of lime mud in the back-reef area, then the reef itself, and finally a patch reef behind the reef. The first stop was the mud mound of Rodriguez Key.

Stop 1. Rodriguez Key. Rodriguez Key is analogous to Pennsylvanian mud mounds which form reservoirs in the southwestern United States. A good example of such a reservoir is Aneth Field. Both the modern mud mounds and their ancient counterparts exhibit the principle of "competitive sedimentation". According to this principle, primary porosity is formed if sedimentation of skeletal particles exceeds that of lime mud which can be interstially entrapped. Contrarywise, if the production of lime mud is dominant, all pore space between the skeletal particles is taken up by lime mud resulting in zero porosity.

In mud mounds, both modern and ancient, plants act as a baffle and trap fine sediment from suspension. The resultant accumulation of the fine sediment results in the formation of the mounds. At Rodriguez Key the baffles are principally *Thalassia* (turtle) grass, and green algae. The mound has no rigid organic framework itself. It forms an elongate, flat-topped shoal and is surrounded by deeper water ranging from 2 to 10 ft in depth. The central part of the mound, where the mangroves grow, is above mean low water. Ginsburg and others (1964) provide details on Rodriguez Key and Multer (1969) summarizes data from various sources including those of Turmel and Swanson (publication pending).

Rodriguez Key, located about 1 mile east of Key Largo, shows windward zonation of plants and animals. The dominant wind direction in Florida is from the east and southeast, hence on the windward side of the mound, several zones follow each other. In our study our boat was anchored on the windward side and we swam, floated, or waded across these zones.

The most windward zone is that of *Porites,* a small branching finger coral. This coral, together with interspersed coralline algae (*Goniolithon*), turtle grasses, and green algae forms an interlocking wave-resistant ridge.

The second zone is that of *Goniolithon,* a branching coralline (red) algae. This algae covers the bottom; its branches crunch under the feet of the waders. *Porites,* green algae, and *Thalassia* grass are interspersed between *Goniolithon*.

Both *Porites* and *Goniolithon* zones are very narrow, ranging from less than 100 ft to 250 ft in width. These two zones are known collectively as the windward bank margin.

Beyond the windward bank margin is the bank environment or the grass algae zone. In this zone *Thalassia* and green algae carpet the floor. Among green algae *Halimeda*, *Penicillus*, *Rhipocephalus*, and *Udotea* are most abundant. The *Thalassia* leaves are coated by an abundant epifauna as well as by find grained sediment. The grass is the most important baffle in the bank environment and its rhizomes stabilize the lime mud. The lime mud is in places very soft and we sank into it to out knees and even to our waists.

The lime mud probably derives from green algae, especially *Penicillus*. Stockman, Ginsburg, and Shinn (1967) have computed that in Florida Bay and the near-shore part of the Florida Reef Tract, *Penicillus* is a major contributor of lime mud. The skeletal fragments buried in the lime mud are principally derived from the breakdown of *Halimeda*, *Porites*, and *Goniolithon*, although other organisms contribute skeletal fragments as well.

The central part of the mound is made up by the subaerial part of Rodriguez Key in which a dense jungle of mangroves baffles the waves and currents during storms to catch the lime mud.

Stop 2. Molasses Reef. From Rodriguez Key we went by boat to Molasses Reef to examine part of the Florida Reef Tract. Although this reef faces the ocean (Gulf Stream; Florida Straits) and the water drops off from less than 10 feet to more than 100 feet within a few hundred feet lateral distance, the surprising feature is that most of the reef consists of rubble only. Dead blocks of coral and coralline algae are everywhere with only small solitary corals sitting on these blocks. Sea whips and sea fans (Gorgonians) which move gracefully with each passing wave and which do not have rigid calcium carbonate skeletons are the only impressive life organisms of Molasses Reef. Already Agassiz in 1852 (p. 152) reported on the dead boulders of coral rock at Molasses Key which he recorded "lie scattered about in great profusion—flung pell-mell among the fragments of more delicate forms." Hurricanes are responsible for these large boulder-sized coral blocks and transport them to the leeward of the reef. Ball, Shinn and Stockman (1967) studied the effect of the passage of Hurricane Donna (September 9 to 10th, 1960) across south Florida and noted that the amount of boulder-size rubble formed by hurricane surf far exceeds the amount produced by day to day processes of death and deterioration. Ball, Shinn and Stockman (1967) documented the leeward movement of reef rubble some 200 to 500 feet distance behind Molasses Reef during Hurricane Donna.

Below Molasses Reef a former reef exists which is now covered by a veneer of rubble. Sea level fluctuations may have destroyed the former reef by drowning it. Shinn (1963) exhumed part of this dead reef by blasting it open with dynamite and found well-developed colonies of *Acropora* coral in growth position encrusted by coralline algae. At present living colonies of *Acropora* are absent on this reef.

In a subsurface analogue to Molasses Reef, rubble would overlie true reef. Both would be highly porous and interfinger with organic-rock offshore lime mud and lime silt which might serve as source rock. Landward lime muds, especially those that would be analogous to the lime muds of Florida Bay (studied on the following day), would be a seal to the porous zone of reef and reef rubble.

Banks (1959) reported that the reef rubble of Molasses Reef would make an excellent reservoir. He compared it with the Cretaceous (Comanche) Sunniland pay zone of the Forty Mile Bend oil field in southern Florida.

Stop 3. Key Largo Dry Rocks. From Molasses Reef the boat took us to Key Largo Dry Rocks, located about 7 1/2 miles northeast of Molasses Reef and also a part of the Florida Reef Tract. Key Largo Dry Rocks in contrast with Molasses Reef, which consists of reef rubble, is composed of flourishing reefs with well-developed ecologic zonation. The zones at Key Largo Dry Rocks consist of back reef, reef flat and fore reef. The fore reef is further subdivided into *Acropora*, *Millepora* and rubble zones (Shinn 1963).

On our field trip we stopped first in the back-reef zone at Key Largo Dry Rocks and anchored our boat there. Coral life in this zone is very abundant with both solitary and colonial corals. Large heads of *Montastrea* are dispersed over a wide area and a concentration of *Acropora palmata* (moose horn coral) occurs near the reef flat which is the next zone. Sea whips and sea fans (Gorgonians) are interspersed between the coral life. Skeletal sand and rubble carpet patches of the bottom not covered by life growth.

After studying the back reef, the boat took us to the fore reef side of the reef. In the fore reef contiguous to the reef flat, massive colonies of *Acropora palmata* spread out across the zone which forms a seaward sloping terrace and is known as *Acropora* zone. Most of the *Acropora palmata* corals of this zone are oriented in the direction of wave movement. This orientation allows them to thrive without competition in the most turbulent zone of the reef. This orientation contrasts sharply with the *Acropora palmata* colonies of the back-reef environment in which the corals are unoriented. Landward of the *Acropora* zone is the reef flat where the corals are for the most part dead and the coralline algae cover and entomb the corals. In fact the corals in the reef flat are already in the subsurface in the sense that they are completely covered by coralline algal growth and can only be exhumed by drastic measures such as a dynamite blast. Swimming over the reef flat shows a relatively flat surface which is caused by the low tide (above low tide corals and coralline algae die). Below this flat the corals are entombed, yet the surface of the flat offers no evidence that below it massive colonies of corals were once thriving.

Seaward of the reef flat is the *Acropora* zone already described which forms a terrace, and seaward of the *Acropora* zone is a second terrace covered with the fire coral *Millepora*. This *Millepora* zone is followed seaward by the rubble zone in which reef rubble carpets the bottom.

Stop 4. Mosquito Banks. From Key Largo Dry Rocks we went to Mosquito Banks to look at a patch reef. Mosquito Banks has a solid framework of corals and abundant sea fans and sea whips (Gorgonians). Coral rubble and skeletal sand underlie the area contiguous to the patch reef.

From Mosquito Banks we returned to Tavenier for the night.

June 12

On June 12th we studied Florida Bay starting out from Tavenier on Key Largo. By and large Florida Bay is floored by lime mud. Florida Bay has been described in several papers by Ginsburg (1956, 1957) and Ginsburg and others (1964). A review of the literature is given by Mutler (1969).

Stop 1. "Lake" Near Cross Bank. Our first stop was about 2 1/2 miles west of Tavenier in one of the "mud lakes" in water about 7 feet deep. Here we noted lime mud in suspension. Details of the bottom were difficult to make out because of water turbidity. Diving to the bottom showed that *Thalassis* grass was abundant at the bottom of this "lake".

Stop 2. Cross Bank. Cross Bank, where studied, is part of a bank of the "mud lakes" in Florida Bay. The sediment of Cross Bank consists of lime mud with much included coarse shelly material. *Thalassia* grass is dense and acts as a baffle trapping lime mud. The sediment is soft and easy to sink into. A strong odor of H_2S escapes the sediment. Mangroves have begun tenuously to hold on where the lime mud is shallowest but they are still very sparse.

This sediment would make a poor reservoir but a good source rock. The underlying Pleistocene Key Largo reef limestone, with its high porosity, would make a suitable reservoir with the lime mud acting as both source rock and seal.

Stop 3. Crane Key. From Cross Bank we went to Crane Key where we landed to see a supratidal mud flat. Here we observed finely-laminated lime mud. The fine undulating laminations result from the trapping action of blue-green algae (stromatolites). Stromatolites, as pointed out in the discussion on the west side of Andros Island (June 9, stop 3), are indicative to uppermost intertidal to supratidal environments. The algal mats which occur as crusts on the surface are commonly broken up by desiccation or "mud" cracks. Chips curl up and spall off and can be moved by storm surges into the subtidal environment. Such chips are recognized in the subsurface as intraformational conglomerate. Birdseye structures similar to those found on the west side of Andros Island are common in the lime mud of Crane Key. Some dolomitization of the lime mud has locally occurred in the algal mats of Crane Key. In one example 23% dolomite was found (G. M. Friedman in Friedman and Sanders, 1967, p. 287).

Stromatolites are widespread in the stratigraphic record, including the Arbuckle (Cambro-Ordovician) Hunton, (Siluro-Devonian) Lansing and Kansas City (Pennsylvanian) formations.

Stop 4. Locality near Manatee Key. After Crane Key we visited a locality on Cross Bank about half a mile south of Manatee Key. The purpose of this stop was to find living bryozoans (*Schizoporella floridana*) which in this shallow water commonly encrust *Thalassia* grass. Despite an intensive search we found none. A possible explanation offered by Herb Alley was that the previous winter's unusually hard freeze may have killed them off in this area as it did the parrot fish. The sediments at this stop consisted of lime mud with abundant *Thalassia* grass acting as a baffle.

Stop 5. Manatee Key. Manatee Key is another typical Florida Bay supratidal mud flat. On wading ashore from the boat was found a storm layer of coquina consisting of dominantly pebble size shells at the edge of the key. This storm layer records a passing high energy pulse in a low energy mud environment. In subsurface such as coquina may be misleading in that it may suggest more widespread high energy conditions than the passing of a storm.

The key is populated by a dense jungle of mangroves. Decaying mangrove roots leave doughnutlike rings in the lime mud. Algal mats similar to those on Crane Key are found on Manatee Key. Some of these, as at Crane Key, are broken up by desiccation. Working on this supratidal mud flat, sinking into the mud or disturbing the mud with a stick or a shovel leads to escape from the sediment. The odor is strong indicating highly reducing conditions.

Stop 6. Spoil Bank. At this stop we observed trails of *Batillaria*. This gastropod has been shown to be responsible for making pellets in carbonate depositional environment. Kornicker and Purdy (1957) collected individuals of this species in a glass bottle and found that their excreted pellets are identical to those for the northern extremity of Bimini Lagoon.

Thalassia is abundant in the lime mud at this stop. The effectiveness of *Thalassia* as a stabilizer of lime mud as well as a trapper or baffler of sediment became apparent here as its deep roots were examined. Mud mounds made by *callianassid* shrimps similar to those found at Joulter's Cays in the Bahamas abound in this area.

References Cited

Abbott, B. M. (1975) Implications for the Fossil Record of Modern Carbonate Bank Corals. *Bul. Geol Soc. Amer.,* v. 86, p. 203-04.

Abbott, R. T. (1954) American Seashells. New York, Van Nostrand Co. Inc., 541 p.

———. (1962) Seashells of the World. New York, Golden Press, 160 p.

———. (1968) Seashells of North America. New York, Golden Press, 280 p.

Agassiz, L. (1852) Extracts from the Report of Professor Agassiz to the Supt. of the Coast Survey, on the Examination of the Florida Reef, Keys and Coast. *Exec. Doc. U.S. Senate,* v. 5, no. 3, p. 145-60.

———. (1880) Report on the Florida Reefs. *Mem. Mus. Comp. Zool., Harvard College,* v. 7, no. 1, 61 p.

Alexander, T. R. (1974) Evidence of Recent Sea-Level Rise Derived from Ecological Studies on Key Largo, Fla. p. 219-22. in Environments of South Florida: Present and Past, P. J. Gleason (ed), *Memoir 2, Miami Geol. Soc.,* 452 p.

Alexandersson, T. (1972) Micritization of Carbonate Particles: Processes of Precipitation and Dissolution in Modern Shallow-Marine Sediments. *Bul. Geol. Instn. Univ. Upsala N. S. 3,* v. 7, p. 201-36.

Andrews, J. E., Shepard, F. P., and Hurley, R. J. (1970) Great Bahama Canyon. *Geol. Soc. Amer. Bul.,* v. 81, p. 1061-78.

Andrews, P. B. (1964) Serpulid Reefs, Baffin Bay, Southeast Texas *in* Depositional Environements, South Central Texas Coast, Gulf Coast Assoc., *Geol. Soc.,* Field Trip Guidebook for Annual Meeting.

Antonius, A. (1973) New Observations on Coral Destruction in Reefs. Prepared for the Tenth Meeting of the Assoc. of Island Marine Laboratories of the Caribbean in Mayquez, Puerto Rico.

Atwood, D. K., and Bubb, J. N. (1968) Distribution of Dolomite in a Tidal Flat Environment, Sugarloaf Key, Florida (abst). *Geol. Soc.* Amer. Annual Meeting, Mexico City, November.

Baars, D. L. (1963) Petrology of Carbonate Rocks. Four Corners Geol. Soc., p. 101-29.

Bader, R. G. (1971) Testimony for the State of Florida Department of Pollution Control, Miami Florida, October 13.

Bader, R. G., and Roessler, M. A. (eds) (1972) An Ecological Study of South Biscayne Bay and Card Sound, Florida. Prog. Rep. to U.S. A.E.C. (AT (40-1)-3801-4) and FPl, Univ. of Miami, School of Marine and Atmospheric Science, Miami, Florida. Mimeo. Rep. ML 72060.

Bader, R. G., Roessler, M. A., Voss, G. L., and deSylva, D. P. (1971) Testimony for the State of Florida Department of Pollution Control, Tallahassee, Florida, September 8.

Bain, R. J., and Teeter, J. W. (1975) Previously Undescribed Carbonate Deposits on Key Largo, Fla. *Geology,* v. 3, no. 3, p. 137-39.

Ball, M. M. (1967a) Carbonate Sand Bodies of Florida and the Bahamas. *Jour. Sed. Pet.,* v. 37, no. 2, p. 556-91.

———. (1967b) Tectonic Control of the Configuration of the Bahama Banks. *Trans. Gulf Coast Geol. Soc.,* v. XVII, p. 265-67.

Ball, M. M., Shinn, E. A., and Stockman, K. W. (1967) The Geologic Effects of Hurricane Donna in South Florida. *Jour. Geol.* v. 75, no. 5, p. 583-97.

Ballard, R. D., and Uchupi, E. (eds) (1970) Morphology and Quaternary History of the Continental Shelf of the Gulf Coast of the U.S. *Bul. Mar. Sci.,* v. 20, no. 3, p. 547-59.

Banks, J. E. (1959) Limestone Conglomerates (Recent and Cretaceous) in Southern Florida. *Bul. Amer. Assoc. Petrol. Geol.* v. 43, no. 9, p. 2237-43.

———. (1964) The Geologic Setting of Southern Florida. p. 3-13 in South Florida Carbonate Sediments: Guidebook for Field Trip No. 1, R.N. Ginsburg (ed), Geol. Soc. Amer. Annual Meeting, Miami

———. (1967) Geologic History of the Florida-Bahama Platform. Trans. Gulf Coast Asoc. of Geological Soc., v. XVII, p. 261-64.

Barnard, J. L. (1969) The Families and Genera of Marine Gammaridean Amphipoda. *Bul. U.S. Nat. Mus.,* v. 271, p. 1-535.

Barnard, L. A. Macintyre, I. G., and Pierce, J. W. (1964) Possible Environmental Index in Tropical Reef Corals. *Nature,* v. 252, no. 5480, p. 219-20.

Barron, E. J. (1976) Source and Dynamics of Suspended Sediments in Nearshore Regions, Southwestern Florida (abst). *Bul. Amer. Assoc. Petrol. Geol.,* v. 60, no. 4, p. 448.

Basan, P. B. (1970) Aspects of Sedimentation and Development of a Carbonate Bank in the Barracuda Keys, Florida (abst). *Geol Soc. Amer. Abstracts,* v. 2, no. 3, p. 193.

———. (1973) Aspects of Sedimentation and Development of a Carbonate Bank in the Barracuda Keys, South Florida. *Jour. Sed. Pet.,* v. 43, no. 1, p. 42-53.

Bass, R. O., and Sharps, S. L. (eds) (1963) Shelf Carbonates of the Paradox Basin (includes papers on Fla-Bahamas): 4th Field Conference, *Four Corners Geol. Soc.,* 273 p.

Bassett, H. G. and Stout, J. G. (1967) Devonian of Western Canada, International Symposium on the Devonian System, p. 717-52. Alberta Society of Petroleum Geologists, Calgary, Canada.

Bathurst, R. G. C. (1967) Oölitic Films on Low Energy Carbonate Sand Grains, Bimini Lagoon, Bahamas. *Marine Geol.,* v. 5, p. 89-109.

———. (1971) Carbonate Sediments and Their Diagenesis. New York, Elsevier Publishing Co., 620 p.

Bavendamm, W. (1932) Die Mikrobiologische Kalkföllung in der Tropischen See. Arkiv für Microbiologie, *Zeitschr Erforschung Pflanzlichen Mikroorganismen,* v. 3, no. 2, p. 205-76.

Bayer, F. M. (1961) The Shallow-Water Octocorallia of the West Indies Region. The Hague, Netherlands, Martinus Nijhoff, 373 p.

Bayer, F. M., and Work, R. C. (1964) Littoral Marine Life of South Florida: Guidebook for Field Trip No. 7, Geol. Soc. Amer. Annual Meeting, Miami, Florida.

Beales, F. W. (1958) Ancient Sediments of Bahamian Type. *Bul. Amer. Assoc. Petrol. Geol.,* v. 42, p. 1845-80.

Belyea, H. R. (1960) Distribution of Some Reefs and Banks of the Upper Devonian. Woodbend and Fairholme Groups in Alberta and Eastern British Columbia. Geological Survey of Canada, Paper 59-15.

Benham, S. R., Boyce, R. L., Drexler, W. W., and Hamilton, M. M. (1970) A Carbonate Sand Beach, Bahia Honda Key, Fla. (abst) Geol. Soc. Amer. Abstracts, v. 2, no. 3, p. 194.

Benson, R. H., and Coleman, G. L. (1963) Recent Marine Ostracodes from the Eastern Gulf of Mexico. Univ. Kans. Paleont. Contr., Arthropoda, art. 2, p. 1-52.

Bernatowicz, A. J. (1952) Marine Monocotyledonous Plants of Bermuda. Bul. Mar. Sci. Gulf & Carib., v. 2, p. 338-45.

Bishop, W. F. (1968) Petrology of Upper Smackover Limestone in Northern Haynesville Field, Claiborne Parish, La. Bul. Amer. Assoc. Petrol. Geol., v. 52, no. 1, p. 92-128.

Black, M. (1933a) The Algal Sediments of Andros Island, Bahamas. Phil. Trans Roy. Soc. London, Series B, v. 222, p. 165-92.

———. (1933b) The Precipitation of Calcium Carbonate on the Great Bahama Bank. Geol. Mag., v. 70, no. 10, p. 455-66.

Boardman, M. (1976) Holocene Depositional History of the Bight of Abaco, Bahamas. Unpublished Masters Thesis, Univ. of North Carolina, Chapel Hill.

Bock, W. D. (1967) Monthly variation in the Foraminiferal Biofacies on Thalassia and Sediment in the Big Pine Bay Area, Florida. Doctoral Dissertation, Univ. of Miami, January.

———. (1971) A Handbook of the Bethonic Foraminifera of Florida Bay and Ajacent Waters in A Symposium of Recent South Florida Foraminifera, W. D. Bock et al., Memoir 1, p. 245, Miami Geol. Soc., 10 Rickenbacker Causeway, Miami, Fla. 33149.

Bock, W. D., Hay, W. W., Jones, J. I., Lynts, G. W., Smith, S. L., and Wright, R. C. (1971) A Symposium of Recent South Florida Foraminifera. Memoir 1, 245 p., Miami Geol. Soc., 10 Rickenbacker Causeway, Miami, Fla. 33149.

Bourrouilh, F. (1974) Donnees Geomorphologieques sur la Region de Fresh Creek, Ile Andros, Bahama. Mar. Geol., v. 16, no. 4, p. 213-35.

Bowman, H. H. M. (1917) Physiology of the Red Mangrove. Amer. Philos. Soc. Proc., v. 56, p. 589-672.

Boyer, B. W. (1972) Grain Accretion and Related Phenomena in Unconsolidated Surface Sediments of the Florida Reef Tract. Jour. Sed. Pet., v. 42, no. 1, p. 205-10.

Braun, M., and Friedman, G. M. (1968) Carbonate Microfacies and Environments of the Tribes Hill Formation (Lower Ordovician) of the Mohawk Valley, New York (abst). Geol. Soc. Amer., Program, Annual Meeting, Northeastern Section, p. 18.

Bretz, J. H. (1940) Solution Cavities in the Joliet Limestone of Northeastern Illinois. Jour. Geol., v. XLVIII, no. 4, p. 337-84.

Bricker, O. P. (ed) (1971) Carbonate Cements: The Johns Hopkins University Studies of Geology, no. 19. The Johns Hopkins Press, 376 p.

Broecker, W. S., and Takahashi, T. (1960) Calcium Carbonate Precipitation on the Bahama Banks. Jour. Geophysical Resch., v. 71, p. 1575-1602.

———. (1966) Calcium Carbonate Precipitation on the Bahama Banks. Jour. Geophysical Resch., v. 71, p. 1575-1602.

Broecker, W. T., Takahashi, T., and Thurber, D. L. (1965) Uranium-Series Dating of Corals and Oolites from Bahaman and Florida Key Limestones. Science, v. 149.

Brooks, H. K. (1962a) Observations on the Florida Middle Ground. Geol. Soc. Amer. Spec. Paper 68, p. 65-66.

———. (1962b) Reefs and Bioclastic Sediments of the Dry Tortugas (abst). Geol. Soc. Amer. Spec. Paper 73, p. 1-2.

———. (1968) The Plio-Pleistocene of Florida, with Special Reference to the Strata Outcropping on the Caloosahatchee River. p. 3-42 in Late Cenozoic Stratigraphy of Southern Florida, R. D. Perkins (ed), Miami Geol. Soc. Annual Field Trip.

———. (1974) Lake Okeechobee. p. 256-86 in Environments of South Florida: Present and Past, R. J. Gleason (ed), Memoir 2, Miami Geol. Soc., 452 p.

Bunce, E. T., Emery, K. O., Gernad, R. D., Knott, S. T., Lidz, L., Saito, T., and Schlee, J. (1965) Ocean Drilling on the Continental Margin. Science, v. 150, p. 709-16.

Busby. R. F. (1962) Submarine Geology of the Tongue of the Ocean, Bahamas. Tech. Report TR-108, Naval Oceanographic Office, Washington, D.C., 20390. 84 p.

Busby, R. F., and Dick, G. F. (1964) Oceanography of the Eastern Great Bahama Bank. Part 1. Temperature-Salinity Distribution. U.S. Naval Oceanographic Office, Tech. Report TR-174, 42 p.

Busby, R. F., and Merrifield, R. (1967) Undersea Studies with the D.S.R. Alvin, Tongue of the Ocean, Bahamas. U.S. Navy Ocenographic Offic, I.R. no. 67-51, 54 p.

Bush, J. (1951) Rock from the Straits of Florida. Geol. Soc. Amer. Bul., v. 35, p. 102-07.

Carlton, J. M. (1976) Land-Building and Stabilization by Mangroves. Envir. Conservation, v. 1, no. 4, p. 285-94.

Carter, L. J. (1976) Dade County: The Politics of Managing Urban Growth. Science, v. 192. p. 982-85.

Cary, L. R. (1918) The Gorgonaceae as a Factor in the Formation of Coral Reefs. Pap. from Dept. Biol., Carn. Inst. Wash., v. 9 p. 341-62.

Chace, F. A. (1972) The Shrimps of the Smithsonian-Bredin Caribbean Expeditions with a Summary of the West Indian Shallow-Water Species (Crustacea: Decapoda: Natantia). Smithsonian Contributions to Zoology, no. 98, 179 p.

Chapman, V. J. (1946) Algal Zonation in the West Indies. Ecology, v. 27, p. 91-3.

Chave, K. E. (1967) Recent Carbonate Sediments—An Unconventional View. Jour. Geol. Ed., v. XV, no. 5, p. 200-204.

Chen, Chih Shan (1965) The Regional Lithostratigraphic Analysis of Paleocene and Eocene Rocks of Florida. Fla. Geol. Survey Bul., no. 45, 105 p.

Chesher, R. H. (1971) Biological Impact of a Larger-Scale Desalination Plant at Key West. Environmental Protection Agency, Water Pollution Control Research Series 18080 GBX, 150 p.

———. (1973) Environmental Analysis, Canals and Quarries, Lower Florida Keys. Unpublished final report prepared for Charley Toppino and Sons, Inc., Rockland Key, Florida, March 162 p.

———. (1974) Canal Survey Florida Keys. Marine Research Foundation, Inc., March, 173 p.

Chilingar, G. V., Bissel, H. J., and Fairbridge, R. W. (eds) (1967) Carbonate Rocks. New York, Elsevier Publishing Co., 417 p.

Clark, E. E. (1972) Environmental Assessment Study, Punta Gorda Area. Punta Gorda Isles, Inc., Florida, 136 p.

Clark, H. L. (1933) A Handbook of the Littoral Echinoderms of Porto Rico and the Other West Indian Islands. N.Y. Acad. Sci. Scientific Survey of Porto Rico and the Virgin Islands, v. 16, p. 1-147.

Cloud, P. E., Jr. (1959) U.S. Geol. Survey Profess. Papers no. 280-K, p. 398.

———. (1962) Environments of Calcium Carbonate Deposition West of Andros Island, Bahamas. U.S. Geol. Survey Prof. Paper 350, 138 p.

Cohen, A. D. (1970) An Allochthonous Peat Deposit from Southern Florida. Geol. Soc. Amer. Bul., v. 81, p. 2477-82.

Cohen, A. D., and Spackman, W. (1972) Methods in Peat Petrology and Their Application to Reconstruction of Paleoenvironments. Geol. Soc. Amer. Bul., v. 83, p. 129-42.

Cole, S. A. (1974) The Effect of Thermal Stress Conditions on Benthic Foraminifera in Biscayne Bay, Fla. Masters Thesis, Univ of Illinois.

Collier, C. A. (1975) Seawall and Revetment Effectiveness, Cost, and Construction. Rept. No. 6, Florida Sea Grant Program and Florida Coastal Coordinating Council, Reproduced by National Tech. Info. Service (COM-75-11175) U.S. Dept. Commerce, Springfield, Va. 22151.

Cooke, C. W., and Mossom, S. (1929) Geology of Florida. Florida Geol. Survey Annual Report 20, p. 29-227.

Craighead, F. C. (1964) Land, Mangroves and Hurricanes. Fairchild Tropical Garden Bul., v. 19, no. 4, p. 5-32.

Cuffey, R. J., and Fonda, S. S. (1976) Giant Schizoporella-Sedimentologically Important Cheilostome Bryozoans in

Pleistocene and Recent Carbonate Environments of Bermuda, Bahamas and Florida (abst). *Geol. Soc. Amer.*, v. 8, no. 4, p. 474-75.

Cuffey, R. J., and Gebelein, C. D. (1975) Reefal Bryozoans within the Modern Platform-Margin Sedimentary Complex off Northern Andros Island (Bahamas) (abst). *Geol. Soc. Amer.*, v. 7, no. 6, p. 743.

Cuffey, R. J., and Kissling, D. L. (1973) Ecologic Roles and Paleoecologic Implications of Bryozoans on Modern Coral Reefs in the Florida Keys. *Geol. Soc. Amer. Abstracts with Programs*, v. 5, no. 2, p. 152-3.

D'Argenio, B., De Castro, P., Emiliani, C., and Simone, L. (1975) Bahamian and Apenninic Limestones of Identical Lithofacies and Age. *Bul. Amer. Assoc. Petrol. Geol.*, v. 59, no. 3, p. 524-30.

Davis, J. H. (1940) The Ecology and Geology Role of Mangroves in Florida. Carnegie Inst., Washington, Pub. 517, v. 32, p. 307-409.

Deichmann, E. (1939) The Holothurians of the Western Part of the Atlantic Ocean. *Bul. Mus. Comp. Zool. Harv.*, v. 71, p. 43-226.

———(1954) The Holothurians of the Gulf of Mexico. *in* Gulf of Mexico its Origin, Waters, and Marine Life, P. L. Galtsoff (ed), *Fish Bul. U.S.*, v. 55, bul. 89, p. 381-410.

De Laubenfels, M. W. (1953) A Guide to the Sponges of Eastern North America. *Spec. Publ. Mar. Lab. Univ. Miami, Miami, Fla., Univ. of Miami Press*, 33 p.

De Laubenfels, M. W., and Storr, J. F. (1958) The Taxonomy of American Commercial Sponges, *Bull. Mar. Sci. Gulf & Caribbean*, v. 8, no. 2.

Dietz, R. S., and Holden, J. C. (1973) Geotectonic Evolution and Subsidence of Bahama Platform: A Reply. *Bul. Geol. Soc. Amer.*, v. 84, p. 3477-82.

Dietz, R. S., Holden, J. C., and Sproll, W. P. (1970) Geotectonic Evolution and Subsidence of Bahama Platform. *Geol. Soc. Amer. Bul.*, v. 81, p. 1915-28.

Dodd, J. R., Hatten, D. E., and Liebe, R. M. (1973) Possible Living Analog of the Pleistocene Key Largo Reefs of Florida. *Geol. Soc. Amer. Bul.*, v. 84, p. 3995-4000.

Dodd, J. R., and Siemers, C. T. (1971) Effect of Late Pleistocene Karst Topography Holocene Sedimentation and Biota, Lower Florida Keys. *Geol. Soc. Amer. Bul.*, v. 82, p. 211-18.

Dollfus, R. P. (1960) Sur un Recif Actuel, le Banc des Hermelles de Baie du Mont-Saint-Michel; Quelques Rensignments Documentaries. *Soc. Geol. France*, B.S.7, no. 1, p. 133-40.

Donaldson, J. A. (1962) Stromatolites in the Denault Formation, Marion Lake, Coast of Labrador, Newfoundland. Dept. of Mines and Tech. Surveys, *Bul. 102, Geol. Survey of Canada*, 33 p.

Dooge, J. (1966) The Stratigraphy of an Upper Devonian Carbonate Shale Between the North and South Ram Rivers of the Canadian Rocky Mountains, N. V. Drukkerij, J. J. Groenen Zoom-Leiden, 53 p.

Dooge, J., and Raasch, G. O. (1966) Faunal and Lithological Aspects of the Lower Paleozoic at Royal Creek, Yukon Territory. Canadian Petrol. *Geol. Bul.*, v. 14, no. 4, p. 601.

Dubar, J. R. (1958a) Neocene Stratigraphy of Southwestern Florida. *Trans. Gulf Coast Assoc., Geol. Soc.*, v. VIII, p. 129-55.

———. (1958b) Stratigraphy and Paleontology of the Late Neogene Strata of the Caloosahatchie River Area of Southern Florida. *Florida Geol. Survey Bul.* 40.

———. (1974) Summary of the Neogene Stratigraphy of Southern Florida. *in* Post-Miocene Stratigraphy, Central and Southern Atlantic Coastal Plain, R. Q. Oaks and J. R. Dubar (eds), Logan, Utah, Utah State Univ. Press.

Duke, T. W., Willis, J. N., and Price, T. J., (1966) Cycling of Trace Elements in the Estuarine Environments. 1. Movement and Distribution of Zinc 65 and Stable Zinc in Experimental Ponds. *Chesapeake Science*, v. 7, no. 1, p. 1-10.

Earley, C. F., and Goodell, H. G. (1968) The Sediments of Card Sound, Florida. *Jour. Sed. Pet.*, v. 38, no. 4, p. 985-99.

Ebanks, W. J., and Bubb, J. N. (1975) Holocene Carbonate Sedimentation Matecumbe Keys Tidal Bank, South Florida. *Jour. Sed. Pet.*, v. 45, no. 2, p. 422-39.

Emery, K. O. (1956) Marine Solution Basins. *Jour. Geol.*, v. LIV, no. 4, p. 209-28.

Enos, Paul (1974a) A Carbonate Shelf-Margin, The Third Dimension (abst) *Bul. Geol. Soc. Amer.*, v. 6, no. 7, p. 723.

———. (1974b) Carbonate Sedimentation Patterns of South Florida Shelf Margins (abst). *Amer. Assoc. Petrol. Geol. Abstracts*, v. 1, p. 31.

Enos, Paul, and Perkins, R. D. (1976) Evolution of Florida Bay Interpreted from Island Stratigraphy (abst). S.E.P.M. Annual Meeting, New Orleans.

———. (1977—in press) Quaternary Depositional Framework of South Florida. Part I: Carbonate Sediment Accumulations of the South Florida Shelf Margin, by P. Enos, Part II: Discontinuity Surfaces in Stratigraphic Analysis: The Pleistocene of South Florida, by R. D. Perkins. *Memoir 147, Geol. Soc. Amer.*

Eutsler, R. L. (1970) Algal Borings, Homogenization, and Porosity in Recent Limestone of the Lower Florida Keys (abst). *Geol. Soc. Amer. Abstracts*, v. 2, no. 3, p. 207-8.

Fairbridge, R. W. (1952) Marine Erosion. Seventh Pacific Sci. Congr. Proc. New Zealand, 1949, v. 3, p. 347-59.

———. (1974) The Holocene Sea-Level Record in South Florida p. 223-32 *in* Environments of South Florida: Present and Past, P. J. Gleason (ed), Memoir 2, Miami Geol. Soc. 452 p.

Fischer, A. G. (1964) The Lofer Cyclothems of the Alpine Triassic. Kansas Geol. Survey Bul. 169, p. 107-49.

Fischer, P. H. (1950) Vie et Moeures des Mollusques. Payot, Paris, 312 p.

Fleece, J B. (1962) The Carbonate Geochemistry and Sedimentology of the Keys of Florida Bay, Florida. Contribution No. 5, Sed. Resch. Lab., Dept. Geol., Fla. State Univ. Tallahassee, Fla.

Folk, R. L. (1959) Practical Petrographic Classification of Limestone. *Bul. Amer. Assoc. Petrol. Geol.*, v. 43, no. 1, p. 1-38.

Foster, H. L. (1956) Annotated Bibliography of Geologic and Soils Literature of Western North Pacific Islands. Intel. Div. Office of the Engineer, U.S. Army and U.S.G.S., 884 p.

Fowler, M. L., Atkinson, L. C., Gill D., Keer, F. R., and Raring, A. N. (1970) Biota and Sediment in a Small Lagoon on Big Pine Key, Florida (abst). *Geol. Soc. Amer.*, v. 2, no. 3, p. 210.

Friedman, G. M. (1964) Early Diagenesis and Lithification in Carbonate Sediments. *Jour. Sed. Pet.*, v. 34, p. 777-813.

———. (ed) (1969a) Depositional Environments in Carbonate Rocks. S.E.P.M. Spec. Publ. No. 14, 209 p.

———. (1969b) Recognizing Tidal Environments in Carbonate Rocks with Particular Reference to Those of the Lower Paleozoics in the Northern Appalachians (abst). *Geol. Soc. Amer.*, N.E. Sec., p. 20-21 March.

———. (1970) The Bahamas and Southern Florida: A Model of Carbonate Deposition, Shale Shaker, September.

———. (in press) Petroleum Geology—Criteria for Recognition of Depositional Environments in Carbonate Rocks. Encyclopedia of Science and Technology, New York, McGraw-Hill Book Co.

Friedman, G. M., and Sanders, J. E. (1967) Origin and Occurrence of Dolostones. p. 267-348 *in* Carbonate Rocks: Origin, Occurrence and Classification, Developments in Sedimentology 9A, G. V. Chilingar, H. J. Bissell, R. W. Fairbridge (eds), New York, Elsevier, 471 p.

Frost, J. G. (1974) Subtidal Algal Stromatolites from the Florida Backreef Environment. *Jour. Sed. Pet.*, v. 44, no. 2, p. 532-37.

Frye, J. C., and Swineford, A. (1947) Solution Features on Cretaceous Sandstone in Central Kansas. *Amer. Jour. Sci.*, v. 245, p. 366-79.

Galaine, C., and Houlbert, C. (1916) Les Recifs d'Hermelles et l'Assechement de la Baie du Mt-Saint-Michel. *C.R. Acad. Sci. Paris*, v. 163, p. 613-16.

Gassaway, J. D. (1970) Mineral and Chemical Composition of Sediments from the Straits of Florida. *Jour. Sed. Pet.,* v. 40, n. 4, p. 1136-46.

Gebelein, C. D. (1969) Distribution, Morphology, and Accretion Rate of Recent Subtidal Algal Stromatolites, Bermuda. *Jour. Sed. Pet.,* v. 39, no. 1. p. 49-69.

———. (1973) Sedimentology and Stratigraphy of Recent Shallow-Marine and Tidal-Flat Sediments, Southwest Andros Island, Bahamas. *Amer. Assoc. Petrol. Geol. Bul.,* v. 57, no. 4, p. 780-81.

———. (1974) Guidebook for Modern Bahama Platform Environments. Field Trip Guidebook for Geol. Soc. Amer. Annual Meeting, Miami, Nov., 97 p. Obtainable from C. D. Gebelein, Dept. Geol. Sciences, Univ. of Calif. at Santa Barbara, Santa Barbara, Calif. 93106 ($4.40).

Gentry, R. C. (1974) Hurricanes in South Florida. p. 73-81. *in* Environments of South Florida: Present and Past, P. J. Gleason (ed), Memoir 2, Miami Geological Society, 452 p.

Gibson, R. G., and Schlee, J. (1967) Sediments and Fossiliferous Rocks from the Eastern Side of the Tongue of the Ocean, Bahamas, Deep-Sea Research, v. 14, p. 691-702.

Gillis, W. T., Byrne, R., and Harrison, W. (1975) Bibliography of the Natural History of the Bahama Islands. Atoll Research Bul. no. 191, 123 p.

Ginsburg, R. N. (1953a) Beachrock in South Florida. *Jour. Sed. Pet.,* v. 23, no. 2, p. 85-92.

———. (1953b) Intertidal Erosion on the Florida Keys. *Bul. Mar. Sci.,* v. 3, no. 1. p. 54-69.

———. (1953c) Lithification and Alteration Processes in South Florida Carbonate Deposits. Unpublished Ph.D. Dissertation, *Dept. Geol. Univ. of Chicago, Chicago, Ill.* 104 p.

———. (1955) Recent Stromatolitic Sediments from South Florida (abst). *Jour. Paleo.,* v. 29, no. 4, p. 723-24.

———. (1956) Environmental Relationships of Grain Size and Constituent Particles in Some South Florida Carbonate Sediments. *Bul. Amer. Assoc. Petrol. Geol.,* v. 40, no. 10, p. 2384-2427.

———. (1957) Early Diagenesis and Lithification of Shallow Water Carbonate Sediments in South Florida. p. 80-99 *in* Regional Aspects of Carbonate Deposition, *S.E.P.M. Spec. Publ.* No. 5.

———. (1960) Ancient Analogues of Recent Stromatolites. *Inter. Geol. Congress XXI Session,* Part XXII, p. 26-35.

———. (ed) (1964) South Florida Carbonate Sediments: Guidebook for Field Trip No. 1, *Geol. Soc. Amer. Annual Meeting, Miami,* 72 p.

———. (1967) Stromatolites. *Science,* v. 157, no. 3786, p. 339-40.

Ginsburg, R. N., Bernard, H. A., Moody, R. A., and Daigle, E. E. (1966) The Shell Method of Impregnating Cores of Unconsolidated Sediment. *Jour. Sed. Pet.,* v. 36, no. 4, p. 1118-25.

Ginsburg, R. N., Bricker, O. P., Wanless, H. R., and Garret, P. (1970) Exposure Index and Sedimentary Structures of a Bahama Tidal Flat. *Geol Soc. Amer. Abstacts,* v. no. 7, p. 744.

Ginsburg, R. N., and Hardie, L. A. (1975) Tidal and Storm Deposits, Northwestern Andros Island, Bahamas. p. 201-08. *in* Tidal Deposits: A Casebook of Recent Examples and Fossil Counterparts, R. N. Ginsburg (ed), New York, Springer-Verlag.

Ginsburg, R. N., Isham, L. B., Bein, S. J., and Kuperberg, J. (1954) Laminated Algal Sediments of South Florida and their Recognition in the Fossil Record. *Unpublished Final Report (ML No. 8034) to the Nat. Science Foundation (Inst. of Mar. Sci.) Miami, Florida.*

Ginsburg, R. N., and James, N. P. (1974a) Holocene Carbonate Sediments of Continental Shelves, p. 137-55 *in* The Geology of Continental Margins, C. A. Burk and C. L. Drake (eds), New York, Springer-Verlag, 1009 p.

———. (1974b) Spectrum of Holocene Reef-Building Communities in the Western Atlantic. p. 7.2-7.21 *in* Principles of Benthic Community Analysis, Sedimenta IV, Comparative Sedimentology Laboratory, Univ. of Miami.

Ginsburg, R. N., and Lloyd, R. M. (1956) A Manual Piston Coring Device for Use in Shallow Water. *Jour. Sed. Pet.,* v. 26, no. 1, p. 64-66.

Ginsburg, R. N., Lloyd, R. M., Stockman, K. W., and McCallum, J. S. (1963) Shallow-Water Carbonate Sediments, p. 554-82 *in* The Sea, New York, Inter Science Publishers, John Wiley.

Ginsburg, R. N., and Lowenstam, H. A. (1958) The Influence of Marine Bottom Communities on the Depositional Environment of Sediments. *Jour. Geol.,* v. 66, p. 310-18.

Ginsburg, R. N., Rēzak, R., and Wray, J. L. (1971) Geology of Calcareous Algae. Sedimenta I, 40 p., Comparative Sedimentology Lab., Fisher Island Station, Miami Beach, Fla. 33139.

Ginsburg, R. N., Shinn, E. A., and Schroeder, J. H., (1967) Submarine Cementation and Internal Sedimentation within Bermuda Reefs (abst). *Geol. Soc. Amer. Annual Meeting.*

Glockhoff, C. (1973) Geotectonic Evolution and Subsidence of Bahama Platform: Discussion. *Bul. Geol. Soc. Amer.,* v. 84, p. 3473-76.

Glynn, P. W. (1970) Smithsonian Contribution Zool., v. 66, p. 1.

———. (1973) Aspects of the Ecology of Coral Reefs in the W. Atlantic Region. p. 271-319 *in* Biology and Geology of Coral Reefs, v. II, O. A. Jones and R. Endean (eds), Academic Press, 480 p.

Gomberg, D. (1973) Drowning of the Floridian Platform and Formation of a Condensed Sedimentary Sequence (abst). *Geol. Soc. Amer.,* v. 5, p. 640.

———. (1974) Geology of the Pourtales Terrace (abst). *Fla. Sci.,* v. 37, Suppl. 1, p. 15.

Goreau, T. F. (1959) The Ecology of Jamaican Coral Reefs, I. Species Composition and Zonation. *Ecology,* v. 40, p. 67-90.

Goreau, T. F., and Wells, J. W. (1967) The Shallow- Water *Scleractinia* of Jamacia; Revised List of Species and their Vertical Distribution Range. *Bul. Mar. Sci.,* v. 17, no. 2, p. 442-53.

Gorsline, D. S. (1963) Environments of Carbonate Deposition, Florida Bay and Florida Straits. p. 130-43 *in* Shelf Carbonates of the Paradox Basin, Four Corners Geol. Soc. 4th Annual Field Conference.

Gough, D. I. (1967) Magnetic Anomalies and Crustal Structure in Eastern Gulf of Mexico. *Bul. Amer. Assoc. Petrol. Geol.,* v. 51, no. 2, p. 200-11.

Grant, K., Hoare, T. B., Ferrall, K. W., and Steinker, D. C. (1973) Some Habitats of Foraminifera, Coupon Bight, Florida. Compass, v. 50, no. 4, p. 11-16.

Graus, R. R., and Macintyre, I. G. (1975) Light-Adapted Growth of Massive Reef Corals: Computer Simulation (abst). *Geol. Soc. Amer.* Abstracts, v. 7, no. 7, p. 1090.

Gray, G. W. (1974) The Origin and Evolution of the Sediments of Butch Key, An Emergent Mud Mound in Florida Bay. Umpublished Senior Thesis, Dept. Geology, College of Wooster, Wooster, Ohio.

Greenberg, J., and Greenberg, I. (1972) The Living Coral Reef: Corals and Fishes of Florida and The Bahamas, Bermuda and the Caribbean. Miami, Fla., Seahawk Books, 110 p.

Greenfield, L. J. (1963) Metabolism and Concentration of Calcium and Magnesium and Precipitation of Calcium Carbonate by a Marine Bacterium. Annals of N.Y. Acad. Science, v. 109, no. 1, p. 23-45.

Greenfield, L. J., and Buck, J. D. (1964) Calcification in Marine-Occurring Yeasts. *Bul. Mar. Sci. Gulf & Carib.,* v. 14, no. 2, p. 239-45.

Greenfield, L. J., Carroll, J. J., and Johnson, R. F. (1965) Mechanisms of Calcium and Magnesium Uptake from Sea Water by a Marine Bacterium. *Jour. Cellular and Compar. Physiol.,* v. 66, no. 1, p. 109-18.

Greenfield, L. J., and Rich, E. R. (1972) Biological Study of the Waters of the Key Haven Associated Enterprises, Inc., Key Haven, Florida. 18 p.

Greenfield, L. J., Weiner, C., and Goodman, H. (1967) Enumeration of Bacteria in Calcareous Marine Sediments. Limnol. and Oceanog.

Griffin G. M. (1974a) Natural Variations in Suspended Sediment Concentration, Diver's Horizontal Visibility, and Color Recognition Limits Upper Florida Keys—1972-1974. Harbor Branch Foundation Laboratory, Publ. no. 025.

Griffin, G. M., and Antonious A. (1974b) Turbidity and Coral Reef Health in Waters of Pennekamp Park, Upper Keys (abst). *Florida Scientist*, v. 37, supplement 1, p. 15.

Griffin, G. M., and Manker, J. P. (1974c) Effects of Dredging on Water Clarity Around a Key Largo Dredge-Fill Site, with Recommendations for Dredgers and Regulatory Agencies. (Oral presentation)—presented at the Florida Keys Coral Reef Workshop, Oct. 21-22, 1974 sponsored by the Coastal Coordinating Council, Dept. of Nat. Resources, State of Fla., Tallahassee, Fla. The complete text is available from G. M. Griffin, Geol. Dept., Univ. of Fla., Gainesville, Fla. 32611

Gross, M. G., and Tracey, J. I. (1966) Oxygen and Carbon Isotopic Composition of Limestones and Dolomites, Bikini and Eniwetok Atolls. Science, v. 151, no. 3714, p. 1082-84.

Gutstadt, A. M. (1968) Petrology and Depositional Environments of the Beck Spring Dolomite (Precambrian) Kingston Range, Calif. *Jour. Sed. Pet.,* v. 38, p. 1280-89.

Gygi, R. A. (1975) *Sparisoma viride* (Bonnaterre), the Stoplight Parrotfish, a Major Sediment Producer on Coral Reefs of Bermuda?, Eclogae Geologicae Helvetiae, v. 68, no. 2, p. 327-59.

Halley, R. B. (1976) Pleistocene Barrier Bar Seaward of the Ooid Shoal Complex Near Miami, Florida (abst). *Fla. Sci.,* v. 39, suppl. 1, p. 11.

Halley, R. B., Lukas, K. J., and Harris, P. M. (1976) A Comparison of Endolith Floras From Holocene-Pleistocene (Bahama-Florida) Ooids (abst). *Geol. Soc. Amer. Annual Meeting,* Denver.

Ham, W. E. (ed) (1962) Classification of Carbonate Rocks: A Symposium. *Amer. Assoc. Petrol. Geol. Memoir 1,* 279 p.

Harris, P. M. (1976) Holocene Carbonate Sediments, Joulters Cays Area Great Bahama Bank (abst). *Fla. Sci.* v. 39, suppl. 1, p. 11.

Harris, R. W. (1976) The Distribution of Whitings, Bahama Banks. *Florida Scientist,* v. 39, Suppl. 1, p. 14.

Hartman, Olga (1944-47) Polychaetous Annelids; Part VI. Paraonidae, Magelonidae, Longosomidae, Ctenodrilidae, and Sabellariidae: Allan Hancock Pacific Expeditions. *Los Angeles, Univ. Soc. Calif. Press,* p. 311-89.

Hastings, S. C. (1970) Size-Grade and Constituent Particle Analysis of Some Carbonate Sands Across Florida Lagoon and Reef Tract Environments. Unpublished senior thesis, College of Wooster, Wooster, Ohio, 86 p.

Hay, W. W. (1967) Bimini Lagoon: Model Carbonate Epeiric Sea (abst). *Bul. Amer. Assoc. Petrol. Geol.,* v. 51, p. 468-69.

Hay, W. W., Wiedenmayer, F., and Marszalek, D. S. (1970) Modern Organism Communities of the Bimini Lagoon and their Relation to the Sediments. p. 19-30 *in* Sedimentary Environments and Carbonate Rocks, Bimini, Bahamas, Supko, Marszalek, and Bock (eds), 4th Annual Field Trip of Miami Geol. Soc., 30 p.

Heckel, P. H. (1974) Carbonate Buildups in the Geologic Record: a Review. p. 90-154 *in* Reefs in Time and Space, L. F. La Porte (ed), Soc. Econ. Paleon. and Mineral., Special Publ., no. 18.

Hedgepeth, J. W. (1954) Anthozoa: The Anemones. p. 285-90 *in* Gulf of Mexico, Its Origin, Waters, and Marine Life, P. L. Galtsoff (ed), *Fish. Bul. U.S.,* v. 55. bul. 89.

Herdman, W. A. (1920) The Marine Biological Station at Port Erin (Isle of Man). 34th Annual Rept., former Liverpool Mar. Biol. Comm., p. 1-32.

Hine, A. C. (1975) Shallow Bank Margin Structure and Depositional Processes: N. W. Little Bahama Bank. Ph.D. Thesis, Univ. of South Carolina, Columbia, S. C.

Hoffman, Paul (1967) Algal Stromatolites: Use in Stratigraphic Correlation and Paleo Current Determination. Science, v. 157, p. 1043-45.

Hoffmeister, J. E. (1974) Land from the Sea. Univ. of Miami Press, 143 p.

Hoffmeister, J. E., Jones, J. I., Milliman, J. D., Moore, D. R., and Multer, H. G. (1964) Living and Fossil Reef Types of South Florida: Guidebook for Field Trip No. 3. *Geol. Soc. Amer. Annual Meeting, Miami, Florida,* 26 p.

Hoffmeister, J. E., and Mutler, H. G. (1961) The Key Largo Limestone of Florida (abst). Geol. Soc. Amer. Special Paper 68, p. 199-200.

———. (1964a) Growth-Rate Estimates of a Pleistocene Coral Reef of Florida. *Geol. Soc. Amer. Bul.,* v. 75, p. 353-58.

———. (1964b) Pleistocene Limestones of the Florida Keys. p. 57-61 *in* Guidebook for Field Trip No. 1, *Geol. Soc. Amer. Annual Meeting, Miami, Florida.*

———. (1965) Fossil Mangrove Reef of Key Biscayne Florida. *Geol. Soc. Amer. Bul.,* v. 76, p. 845-52.

———. (1968) Geology and Origin of the Florida Keys. *Geol. Soc. Amer. Bul,* v. 79, p. 1487-1502.

Homeister, J. E., Stockman, K. W., and Multer, H. G. (1967) Miami Limestone of Florida and Its Recent Bahamian Counterpart. *Geol. Soc. Amer. Bul.,* v. 78, p. 175-90.

Hohlt, R. B. (1948) The Nature and Origin of Limestone Porosity. Colorado School of Mines Quart., v. 43, p. 5-51.

Howard, J. F. (1965) Shallow-Water Foraminiferal Distribution Near Big Pine Key, Southern Florida Keys. Compass, v. 42, p. 265-80.

Howard, J. F., and Faulk, K. L. (1968) Microfaunal Distribution and Controls, Coupon Bight, Southern Florida Keys (abst). Ohio Acad. Science Meeting, Toledo, Ohio, April 19.

Howard, J. F., Kissling, D. L., and Lineback, J. A. (1970) Sedimentary Facies and Distribution of Biota in Coupon Bight, Lower Florida Keys. *Geol. Soc. Amer. Bul.,* v. 81, p. 1929-46.

Hubbard, D. K., Ward, L. G., and Fitzgerald, D. M. (1976) Reef Morphology and Sediment Transport, Lucaya, Grand Bahama Island (abst). *Bul. Amer. Assoc. Petrol. Geol.,* v. 60, no. 4, p. 682.

Hudson, J. H., Shinn, E. A., Halley, R. B., and Lidz, B. (1976) Sclerochronology: A Tool for Interpreting Past Environments. *Geology,* v. 4, p. 361-64.

Huffman, S. F., Anderson, G. G., Orosco, C. A. (1970) A Carbonate Sandbar Near Bahia Honda Key, Florida (abst). *Geol. Soc. Amer. Abstracts,* v. 2, no. 3, p. 219.

Humm, H. J. (1963) Some New Records and Range Extensions of Florida Marine Algae. *Bul. Mar. Sci. Gulf & Carib.,* v. 13, no. 4, p. 516-26.

———. (1964) Epiphytes of the Sea Grass *Thalassia testudinum,* in Florida. *Bul. Mar. Sci. Gulf and Carib.,* v. 14, p. 306-41.

Hunt, M. (1969) A Preliminary Investigation of the Habits and Habitat of the Rock Boring Urchin *Echinometra lucunter* Near Devonshire Bay, Bermuda. p. 35-40. *in* Seminar on Organism-Sediment Inter-Relationships, R. N. Ginsburg and P. Garrett (eds), Bermuda Biological Station, Special Publ. 2, p. 153.

Illing, L. V. (1954) Bahamian Calcareous Sands. *Bul. Amer. Assoc. Petrol. Geol.,* v. 38, no. 1. p. 1-95.

———. (1959) Deposition and Diagenesis of Some Upper Paleozoic Carbonate Sediments in Western Canada. Fifth World Petroleum Congress Proceedings, Section 1, Paper 2.

Imbrie, J., and Buchanan, H. (1965) Sedimentary Structures in Modern Carbonate Sands of the Bahamas. S.E.P.M. Spec. Publ. 12, p. 149-72.

Imbrie, J., and Purdy, E. G. (1962) Classification of Modern Bahamian Carbonate Sediments. *Amer. Assoc. Petrol. Geol., Memoir* 1, Tulsa, Okla., p. 253-72.

Jindrich, V. (1969) Recent Carbonate Sedimentation by Tidal Channels in the Lower Florida Keys. *Jour. Sed. Pet.,* v. 39, no. 2, p. 531-53.

———. (1972) Biogenic Buildups and Carbonate Sedimentation, Dry Tortugas Reef Complex, Florida. Unpublished dissertation, State Univ. of New York at Binghamton, 96 p.

Johnson, J. H. (1951) An Introduction to the Study of Organic Limestones. *Quart. Col. Sch. Mines.,* v. 46, no. 2. 185 p.

Jones, J. A. (1963) Ecological Studies of the Southeastern Florida Patch Reefs. Part I. Diurnal and Seasonal Changes in the Environment. *Bul. Mar. Sci. Gulf & Carib.*, v. 13, p. 282-307.

Jones, J. I. (1971) The Ecology and Distribution of Living Planktonic Foraminifera in the Straits of Florida. p. 175-245 *in* A Symposium of Recent South Florida Foraminifera by Jones and Bock (eds), Memoir 1, Miami Geol. Soc., 245 p.

Jordan, G. F. (1954) Large Sink Holes in Straits of Florida. *Bul. Amer. Assoc. Petrol. Geol.*, v. 38, no. 8, p. 1810-17.

———. (1962) Submarine Physiography of the U.S. Continental Margins. *Tech. Bul. No. 18, U.S. Dept. of Commerce, Coast and Geodetic Survey March*, 28 p.

Jordan, G. F., Malloy, R. J., and Kofoed, J. W. (1964) Bathymetry and Geology of Pourtales Terrace, Florida. *Marine Geol.*, v. 1, p. 259-87.

Jordan, G. F. and Steward, J. B., Jr. (1961) Submarine Topography of the Western Straits of Florida. *Geol. Soc. Amer. Bul.*, v. 72, p. 1051-58.

Kahle, C. F. (1968) Petrology and Structure of a Salina (Silurian) Dolomitized Algal Stromatolite Complex, Northwestern Ohio (abst). Amer. Assoc. Petrol. Geol. Annual Meeting, April.

———. (1974) Algae, Fungi, and Transformation of Sparry Calcite to Micrite (abst) *Amer. Assoc. Petrol. Geol. Abstracts*, v. 1, p. 51.

Keck, W. G. (1969) Aquifer Performance Test Analysis, Stock Island Steam Power Plant, Utility Board of the City of Key West. East Lansing, Mich., W. G. Keck and Assoc.

Kelkowsky, E. (1908) Oolith und Stromatololith im Norddeutschen Buntsandstein. Deutsch. *Geol. Gesell. Zeitschr.*, Bd. 60, p. 68-125.

Kellerman, K. F., and Smith, N. R. (1914) Bacterial Precipitation of Calcium Carbonate. *Jour. Wash. Acad. Sci.*, v. 4, p. 400-2.

Kelly, M. G., and Conrad, A. (1969) Aerial Photographic Studies of Shallow Water Benthic Ecology. *in* Remote Sensing in Ecology, P. L. Johnson (ed), Athens, Georgia, Univ. of Georgia Press.

Kendeigh, S. C. (1961) Animal Ecology. Englewood Cliffs, N. J., Prentice Hall, 468 p.

Kennedy, W. J., and Sellwood, B. W. (1970) *Ophiomorpha nodosa* Lundgren, a Marine Indicator from the Sparnacian of South East England. *Proc. Geol. Assoc. London*, v. 81, p. 99-110..

Kier, P. M., and Grant, R. E. (1965) Echinoid Distribution and Habits, Key Largo Coral Reef Preserve, Florida. *Smithsonian Miscell. Coll.*, v. 149, no. 6, (Publ. 4649), 68 p.

Kirtley, D. W. (1966) Intertidal Reefs of Sabellariidae (Annelida Polychaeta) Along the Coast of Florida. Unpublished Masters Thesis, The Florida State, University, Tallahassee, August.

———. (1974) Guidebook for Sabellariid Reefs, Beach Erosion and Environmental Problems of the Barrier Island-Lagoon System of the Southeast Florida Coast. *Guidebook for Geol. Soc. Amer. Annual Meeting*, Miami, Nov., Harbor Branch Found. Lab. Rt. 1, Box 196, Ft. Pierce, Fla. 33450.

Kirtley, D. W., and Tanner, W. F. (1968) Sabellariid Worms Builders of a Major Reef Type. *Jour. Sed. Pet.*, v. 38, no. 1, p. 73-78.

Kissling, D. L. (1965) Coral Distribution on a Shoal in Spanish Harbor, Florida Keys. *Bul. Mar. Sci.*, v. 15, no. 3, p. 599-611.

———. (1968) Sedimentary Controls and Facies in Coupon Bight Lower Florida Keys (abst). Ohio Acad. Sci. Meeting, Toledo, Ohio, April 19.

Kissling, D. L., and Lineback, J. A. (1967) Paleoeclogical Analysis of Corals and Stromatoporoids in a Devonian Biostrome, Falls of the Ohio, Kentucky-Indiana. *Geol. Soc. Amer. Bul.*, v. 78, p. 157-74.

Klovan, J. E. (1964) Facies Analysis of the Redwater Reef Complex, Alberta, Canada. *Bul. Canadian Petrol. Geol.*, v. 12, p. 1-100.

Knight, S. H. (1968) Precambrian Stromatolites, Bioherms and Reefs in the Lower Half of the Nash Formation, Medicine Bow Mountains, Wyoming. *Contributions to Geology, Univ. of Wyoming*, v. 7, no. 2, p. 73-116.

Kohout, F. A. (1960) Cyclic Flow of Salt Water in the Biscayne Aquifer of Southeastern Florida. *Jour. Geophyisical Resch.*, v. 65, no. 7, p. 2133-41.

Kornicker, L. S. (1958) Bahamian Limestone Crusts. *Transactions Gulf Coast Assoc. of Geological Societies*, v. VIII, p. 167-70.

———. (1960) Ecology and Taxonomy of Recent Ostracodes in the Bimini Area, Great Bahama Bank. *Bul. Inst. Mar. Sci.*

Kornicker, L. S., and Purdy, E. G. (1957) A Bahamian Faecal-Pellet Sediment. *Jour. Sed. Pet.*, v. 27, no. 2, p. 126-28.

Kotila, D. A., and Harris, F. W. (1975) Nature, Distribution, and Constituent-Particle Analyses of Modern Carbonate Sediments, Long Key, Fla. (abst). *Geol. Soc. Amer. Abstracts*, v. 7, no. 4, p. 506.

Krawiec, W. (1963) Solution Pits and Solution Breccias of the Florida Keys. Unpublished Masters Thesis, Dept. Geology, Univ. of Rochester.

Landon, S. M. (1975) Environmental Controls on Growth Rates in Hermatypic Corals from the Lower Florida Keys. Unpublished M. A. Thesis, SUNY, Binghamton, N.Y. **Note:** publication pending.

Lang, J. (1971) Interspecific Aggression by Scleractinian Corals. *Bul. Mar. Sci.*, v. 21, no. 4, p. 952-59.

La Porte, L. F. (1967) Carbonate Deposition Near Mean Sea-Level and Resultant Facies Mosaic: Manlius Formation (Lower Devonian) of N.Y. State. *Bul. Amer. Assoc. Petrol. Geol.*, v. 51, no. 1, p. 73-101.

———. (ed) (1974) Reefs in Time and Space. S.E.P.M. Spec. Publ. No. 18, 256 p.

Larsen, P. W., and Michel, J. F. (1972) A Hydrographic Study of the Proposed Development of Sea Air Estates, Vaca Key, Monroe County, Florida. Marine Univ. of Miami, 17 p.

Le Blanc, R. J., and Breeding, J. G. (eds) (1957) Regional Aspects of Carbonate Deposition. S.E.P.M., Spec. Publ. No. 5, 178 p.

Leighton, M. W., and Pendexter, C. (1962) Carbonate Rock Types. p. 33-62 *in* Classification of Carbonate Rocks, W. E. Ham (ed) *Bul. Amer. Assoc. Petrol. Geol.*, Mem. 1, 279 p.

Lenderking, R. (1954) Some Recent Observations on the Biology of *Littorina angulifera* Lam. of Biscayne and Virginia Keys, Florida. *Bul. Mar. Sci. Gulf and Carib.*, v. 3, no. 4.

Lineback, J. A. (1968) Macrofaunal and Floral Distributions and Controls in Coupon Bight, Lower Florida (abst). Ohio Acad. Sci. Meeting, Toledo, Ohio, April 19.

Lloyd, R. M. (1964) Variation in the Oxygen and Caron Isotope Ratios of Florida Bay Mollusks and Their Environmental Significance. *Jour. Geol.*, v. 72, p. 84-111.

Logan, B. W., Rezak, R., and Ginsburg, R. N. (1964) Classification and Environmental Significance of Algal Stromatolites. *Jour. Geol.*, v. 72, p. 68-83.

Lowenstam, H. A. (1955) Aragonite Needles Secreted by Algae and Some Sedimentary Implications. *Jour. Sed. Pet.*, v. 25, p. 270-272.

———. (1963) Biologic Problems Relating to the Composition and Diagenesis of Sediments. p. 137-95 *in* The Earth Sciences, Univ. of Chicago Press.

Lowenstam, H. A., and Epstein, S. (1957) On the Origin of Sedimentary Aragonite Needles of the Great Bahama Bank. *Jour. Geol.*, v. 65, p. 364-75.

Lynts, G. W. (1962) Distribution of Recent Foraminifera in Upper Florida Bay and Adjacent Sounds. Contrib. Cushman Found. Foram. Resch., v. 13, pt. 4, p. 127-44.

———. (1965) Observations on the Recent Florida Bay Foraminifera. Contrib. Cushman Found. Foram. Resch., v. 16, pt. 2, p. 67-69.

———. (1966a) Relationship of Sediment-Size Distribution Through Ecologic Factors in Buttonwood Sound, Florida Bay. *Jour. Sed. Pet.*, v. 36, no. 1, p. 66-74.

———. (1966b) Variation of the Foraminiferal Standing Crop over Short Lateral Distances in Buttonwood Sound, Florida Bay. *Limnol. Oceanog.*, v. 11, p. 562-66.

———. (1971) Distribution and Model Studies on Foraminifera Living in Buttonwood Sound, Florida Bay. p. 74-120 *in* A Symposium of Recent South Florida Foraminifera, Memoir 1, *Miami Geol. Soc.*, 245 p.

Lynts, G. W., Judd, J. B., and Stehman, C. F. (1973) Late Pleistocene History of Tongue of the Ocean, Bahamas. *Geol. Soc. Amer. Bul.,* v. 84, p. 2665-84.

Manker, J. P. (1975) Distribution and Concentration of Mercury, Lead, Cobalt, Zinc and Chromium in Suspended Particulates and Bottom Sediments—Upper Florida Keys, Florida Bay and Biscayne Bay. Unpublished Ph.D. Dissertation, Rice University, Houston, Texas.

Macintyre, I. G., and Smith, S. V. (1974) X-Radiographic Studies of Skeletal Development. 21 p. *in* Proceedings, v. II, Second Coral Reef Symposium, Brisbane, Great Barrier Reef Committees.

Malloy, R. J., and Hurley, R. J. (1970) Geomorphology and Geologic Structure Straits of Florida. *Bul. Geol. Soc. Amer.,* v. 81, p. 1947-72.

Manker, J. P., and Griffin, G. M. (1971) Source and Mixing of Insoluble Clay Minerals in a Shallow Water Carbonate Environment—Florida Bay. *Jour. Sed. Pet.,* v. 41, p. 302-6.

Manning, R. B. (1969) Stomatopod Crustacea of the Western Atlantic. Stud. Trop. Oceanogr. Miami, no. 8, 380 p.

Marcus, E. (1967) American Opisthobranch Mollusks. Stud. Trop. Oceanogr. Miami, no. 6, 256 p.

Marcus, E., and Marcus, E. (1960) Opisthobranchs from American Atlantic Warm Waters. *Bul. Mar. Sci. Gulf & Carib.,* v. 10, p. 129-203.

Markel, A. L. (1968) What Has the Manned Submersible Accomplished? *Oceanology International,* v. 3, no. 5, p. 35-38.

Matter, Albert (1967) Tidal Flat Deposits in the Ordovician of Western Maryland. *Jour. Sed. Pet.,* v. 37, p. 601-9.

Matthews, R. K. (1966) Genesis of Recent Lime Mud in Southern British Honduras. *Jour. Sed. Pet.,* v. 36, no. 2, p. 428-54.

McCammon, R. B. (1968) Multiple Component Analysis and Its Application in Classification of Environments. *Amer. Assoc. Petrol. Geol. Bul.,* v. 52, p. 2178-96.

McCrossan, R. G., and Glaister, R. P. (eds) (1964) Geological History of Western Canada. Alberta Soc. Petrol. Geol., Calgary, Canada, 232 p.

McLean, R. F. (1967) Measurements of Beachrock Erosion by Some Tropical Marine Gastropods. *Bul. Mar. Sci.,* v. 17, p. 551-61.

McNulty, J. K. (1970) Effects of Abatement of Domestic Sewage Pollution on the Benthos Volumes of Zooplankton, and the Fouling Organisms of Biscayne Bay, Florida. Univ. of Miami Press, Studies in Tropical Oceanography, no. 9, 107 p.

Michel, J. F. (1972a) A Study of the Hydrology of Proposed Canals at Largo Key Subdivision. Univ. of Miami.

———. (1972b) A Study of the Hydrodynamic Effects of Additional Canal Construction at Port Largo Estates. Univ. of Miami.

———. (1973) A Study of the Hydrodynamic Effects of Additional Canal Construction at Venetian Shores. Univ. of Miami, June, 19 p.

Milliman, J. D. (1965) An Annotated Bibliography of Recent Papers on Corals and Coral Reefs. *Atoll Resch. Bul.,* no. 111, 58 p.

———. (1966) The Marine Geology of Hogsty Reef, A Bahamian Atoll. Ph.D. Dissertation, University of Miami, Fla., 292 p.

———. (1967a) Carbonate Sedimentation of Hogsty Reef, A Bahamian Atoll. *Jour. Sed. Pet.,* v. 37, p. 658-76.

———. (1967b) The Geomorphology and History of Hogsty Reef, A Bahamian Atoll. *Bul. Mar. Sci. Gulf & Carib.,* v. 17, p. 519-43.

———. (1973) Caribbean Coral Reefs. *in* Biology and Geology of Coral Reefs, O. A. Jones and R. Endean (eds), New York, *Academic Press,* v. 1, 410 p.

———. (1974) Marine Carbonates. New York, Springer Verlag, 375 p.

Milliman, J. D., and Emery, K. O. (1968) Sea Levels During the Past 35,000 Years. *Science,* v. 162, no. 3850, p. 1121-23.

Minter, L. L., Keller, G. H., and Pyle, T. E. (1974) Morphology and Sedimentary Processes in and around Tortugas and Agassiz Sea Valleys, Southern Straits of Florida (abst). *Fla. Sci.,* v. 37, suppl. 1, p. 11.

Missimer, T. M.. (1976) Structural Stability, Coastal Subsidence and the Late Holocene Eustatic Sea Level Rise in South Florida (abst). *Fla. Sci.,* v. 39, suppl. 1, p. 14.

Mitterer, R. M. (1974) Pleistocene Stratigraphy in Southern Florida Based on Amino Acid Diagenesis in Fossil *Mercenaria. Geology,* v. 2, no. 9, p. 425-28.

———. (1975) Ages and Diagenetic Temperatures of Pleistocene Deposits of Florida Based on Isoleucine Epimerization in *Mercenaria.* p. 275-82, *in* Earth and Planetary Science Letters, v. 28, Elsevier Scientific Publ. Co.

Monty, C. (1965) Recent Algal Stromatolites in the Windward Lagoon, Andros Island, Bahamas. Extrait des Annales de la Societe Geologique de Belgique, t 88, 1964-65, *Bul. no. 5-6,* p. 270-78.

Moore, C. H. (1973) Intertidal Carbonate Cementation Grand Cayman, West Indies. *Jour. Sed. Pet.,* v. 43, no. 3, p. 591-602.

Moore, H. B. (1965) Bottom Temperatures between Miami and Bimini. Unpublished Tech. Report, No. ML65-215, Inst. Mar. Sci., Univ. of Miami, 3 p.

Moore, W. E. (1957) Ecology of Recent Foraminifera in Northern Florida Keys. *Amer. Assoc. Petrol. Geol. Bul.,* v. 41, p. 727-41.

Müller, G., and Müller, J. (1967) Mineralogisch-Sediment-petrographische und Chemische Untersuchungen an Einem Bank-Sediment (Cross-Bank) der Florida Bay, U.S.A. *Jb. Miner, Abh.,* v. !06, no. 3, p. 257-86.

Multer, H. G. (1969) Field Guide to Some Carbonate Rock Environments. Published by the author, Fairleigh Dickinson Univ., Madison, N.J., 312 p.

———. (1971a) Reef Tracts and Carbonate Sedimentation in the Southeastern United States. *Geol. Soc. Amer. Abstracts,* v. 3, no. 5, p. 335.

———. (1971b) Drusy Mosaic Growth in the Miami Limestone. p. 137-38 *in* Carbonate Cements, O. P. Bricker, ed. The Johns Hopkins Press, Baltimore, 376 p.

———. (1971c) Cementation and Recrystalization of Miami Limestone Ooids. p. 139-40 *in* Carbonate Cements, O. P. Bricker, ed. The Johns Hopkins Press, Baltimore, 376 p.

———. (1971d) Holocene Cementation of Skeletal Grains into Beachrock, Dry Tortugas, Florida. p. 25-26 *in* Carbonate Cements, O. P. Bricker, ed., The Johns Hopkins Press, Baltimore, 376 p.

———. (1975) Field Guide to Some Carbonate Rock Environments, Florida Keys-Western Bahamas. Guidebook for Geol. Soc. Amer. Annual Meeting, Miami, November 1974, published by the author, Fairleigh Dickinson Univ., Madison, N.J., 430 p.

Multer, H. G., and Hoffmeister, J. E. (1968a) Subaerial Laminated Crusts of the Florida Keys. *Geol. Soc. Amer. Bul,* v. 79, p. 183-92.

———. (1968b) Petrology and Significance of the Key Largo (Pleistocene) Limestone, Florida Keys (abst). Geol. Soc. Amer. Annual Meeting, Mexico City, November.

Multer, H. G., and Milliman, J.D. (1967) Geologic Aspects of Sabellarian Reefs, Southeast Florida. *Bul. Mar. Sci.,* v.. 17, no. 2, p. 257-67.

Multer, J. D., and Multer, H. G. (1966) The Mangrove World. American Forests, August, p. 34-37.

Murray, J. W. (1966) An Oil Producing Reef-Fringed Carbonate Bank in the Upper Devonian Swan Hills Member, Judy Creek, Alberta. *Bul. Canadian Petrol. Geol.,* v. 14, p. 1-103.

Murray, R. C. (1960) Origin of Porosity in Carbonate Rocks. *Jour. Sed. Pet.,* v. 30, p. 59-84.

Nesteroff, W. D. (1956) De l'Origine des Oolithes. *Compt. Rend.,* v. 242, p. 1047-49.

Neumann, A. C. (1966) Observations on Coastal Erosion in Bermuda and Measurements of the Boring Rate of the Sponge, *Cliona lampa, Limnol. Oceanogr.,* v. 11, no. 1, p. 92-108.

———. (1974a) Cementation, Sedimentation and Structure on the Flanks of a Carbonate Platform, NW Bahamas. p. 26-30 in Recent Advances in Carbonate Studies, Special Publ. no. 6, Gerhard and Multer (eds), West Indies Lab, Fairleigh Dickinson Univ., St. Croix, U.S. Virgin Islands 00820.

———. (1974b) Shallow and Deep Bank Margin Structure and Sedimentation: Little Bahama Bank (abst). *Bul. Geol. Soc. Amer.*, v. 6, no. 7, p. 888.

Neumann, A. C., and Ball, M. M. (1970) Submersible Observations in the Straits of Florida: Geology and Bottom Currents. *Geol. Soc. Amer. Bul*, v. 81, p. 2861-74.

Neumann, A. C., and Moore, W. S. (1975) Sea Level Events and Pleistocene Coral Ages in the Northern Bahamas. *Quat. Resch.*, v. 5, p. 215-24.

Neumann, A. C., Gebelein, D. C., and Scoffin, T. P. (1969) Composition, Structure, and Erodability of Subtidal Mats, Abaco, Bahamas. *Jour. Sed. Pet.*, v. 40, p. 274-98.

Neumann, A. C., and Land, L. S. (1975) Lime Mud Deposition and Calcareous Algae in the Bight of Abaco, Bahamas: A Budget. *Jour. Sed. Pet.*, v. 45, no. 4, p. 763-86.

Newell, N. D. (1958a) American Coral Seas. XVth Intern. Cong. Zool., Section III, Paper 43.

———. (1958b) American Coral Seas (West Indies). N.Y. Acad. Sci. Trans., ser. 2, v. 21, no. 2, p. 125-27.

———. (1963) Recent Terraces of Tropical Limestone Shores. Zeitschrift fur Geomorphologie, v. 3, p. 87-106.

Newell, N. D., and Imbrie, J. (1955) Biogeological Reconnaissance in the Bimini Area, Great Bahama Bank. Trans. N.Y. Acad. Sci., ser. 2, v. 18, no. 1, p. 3-14.

Newell, N. D., Imbrie, J., Purdy, E. G., and Thurber, D. L. (1959) Organism Communities and Bottom Facies, Great Bahama Bank. *Bul. Amer. Mus. Nat. His.*, v. 117, p. 180-228.

Newell, N. D., Purdy, E. G., and Imbrie, J. (1960) Bahamian Oolitic Sand. *Jour. Geol.*, v. 68, p. 481-97.

Newell, N. D., and Rigby, J. K. (1957) Geologic Studies in the Great Bahama Bank. p. 15-79 *in* Regional Aspects of Carbonate Deposition, Soc. Eco. Paleo. and Min., no. 5.

Newell, N. D., Rigby. J. K.,Fischer, A. G., Whiteman, A. J., Hickox, J. E., Bradley, J. S. (1953) The Permian Reef Complex of the Guadalupe Mountains Region, Texas and Mexico. San Francisco, W. H. Freeman & Co., 236 p.

Newell, N. D., Rigby, J. K., Purdy, E. G., and Thurber, D. L. (1959) Organism Communities and Bottom Facies, Great Bahama Bank. *Bul. Amer. Mus. Nat. His.*, v. 117, p. 180-228.

Newell, N. D., Rigby, J. K., Whiteman, A. J., and Bradley, J. S. (1951) Shoal-Water Geology and Environments, Eastern Andros Island, Bahamas. *Amer. Mus. Nat. His.*, v. 97, art. 1.

Oaks, R. Q., and Du Bar, J. R. (1974) Post-Miocene Stratigraphy, Central and Southern Atlantic Coastal Plain. Utah State Univ. Press, 275 p.

Odum, W. E. (1967) The Influence of the Seagrass Community on the Depositional Environment of Sediments. Rep. to Organism-Sediment Interaction Seminar, Bermuda Biol. Sta., 7 p.

Ogden, J. C. (1977—in press) Estimates of Carbonate Sediment Production by Some Common Fish and Echinoid Grazers in Reef Areas of the Caribbean. *in* Caribbean Reef Systems: Holocene and Ancient., S. Frost and M. Weiss (eds). Amer. Assoc. Petrol. Geol. Special Paper no. 4.

———. (Publication pending) The Colonization of an Artificial Reef by Coral Reef Fishes in the Caribbean.

O'Neill, C. (1976) Sedimentology of East Key, Dry Tortugas (abst). *Fla. Sci.*, v. 39, suppl. 1, p. 10.

Oppenheimer, C. H., and Kornicker, L., (1958) Effect of the Microbial Production of Hydrogen Sulphide and Carbon Dioxide on the pH of Recent Sediments. Pub. Inst. Marine Sci., Univ. of Miami, v. 5, p. 5-15.

Oppenheimer, C. H., and Master (1965) On the Solution of Quartz and Precipitation of Dolomite in Sea Water During Photosynthesis and Respiration. Zeitschrift fur Allg., Microbiologie, Bd. 5, Heft 1, p. 48-51. (Akademie-Verlag, Berlin).

Otter, G. W. (1937) Rock-Destroying Organisms in Relation to Coral Reefs. Rep. Gt. Barrier Reef Comm., v. 1, no. 12, p. 324-50.

Parker, G. G., Ferguson, G. E., Love, S. K. *et al.,* (1955) Water Resources of Southeastern Florida with Special Reference to the Geology and Ground Water of the Miami Area: U.S. Geol. Survey Water Supply Paper 1255, 903 p.

Partington, W. M., and Barada, W. R. (1973) Florida's Endangered Coral Reefs. January Info. Newsletter, Env. Info. Center, Winter Park, Florida.

Pasley D. (1972) Field Guide Windley's Key Quarry. Amer. Quat. Assoc. Second National Conference, Miami, Florida, Meeting no. 1, 14 p.

Perkins, R. D. (1963) Petrology of the Jeffersonville Limestone (Middle Devonian) of Southeastern Indiana. *Geol. Soc. Amer. Bul.*, v. 74, p. 1335-54.

———. (1974) Discontinuity Surfaces as a Stratigraphic Tool. The Pleistocene of South Florida (abst). *Geol. Soc. Amer. Abst.*, v. 6, no. 7, p. 908.

Perkins, R. D., and Enos P. (1968) Hurricane Betsy in the Florida-Bahama Area—Geologic Effects and Comparison with Hurricane Donna. *Jour. Geol.*, v. 76, p. 710-17.

Perkins, R. D., *et al.,* (1968) Late Cenozoic Stratigraphy of Southern Florida—A Reappraisal. Miami Geol. Soc. 2nd Annual Meeting Guidebook, Feb., 110 p.

Perry, L. M., and Schwengel, J. S. (1955) Marine Shells of the Western Coast of Florida, New York, Paleontological Research Institute.

Pettijohn, F. J. (1949) Sedimentary Rocks. Harper, New York, 526 p.

Phillips, R. C. (1960) Observations on the Ecology and Distribution of the Florida Sea Grasses. Professional Papers Series, Fla. BD. Conserv., v. 2, p. 1-72.

Pilsbry, H. A. (1916) The Sessile Barnacles (Cirripedia) Contained in the Collections of the U.S. National Museum; Including a Monograph of the American Species. *Bul. U.S. Nat. Mus.*, v. 93, p. 1-357.

Pitt, W. A. (1972) Effects of Septic Tank Effluent on Ground-Water Quality, Dade County, Florida. U.S. Geol. Survey, October.

Pray, L. C. (1966) Hurricane Betsy (1965) and Near Shore Carbonate Sediments of the Florida Keys. Geol. Soc. Amer. Annual Meeting, San Francisco, Calif., p. 168-69.

Pray, L. C., and Murray, R. C. (eds). (1965) Dolomitization and Limestone Diagenesis. S.E.P.M. Spec. Publ. no. 13, 180 p.

Pressler, E. D. (1947) Geology and Occurrence of Oil in Florida. *Bul. Amer. Assoc. Pet. Geol.*, v. 31, p. 1851-62.

Price, W. A. (1967) Development of the Basin-in-Basin Honeycomb of Florida Bay and Northeastern Cuban Lagoon. *Trans. Gulf Coast Assoc.*, v. 17, p. 368-99.

Provenzano, A. J., Jr. (1959) The Shallow Water Hermit Crabs of Florida. *Bul. Mar. Sci. Gulf & Caribbean,* v. 9, no. 4.

Purdy, E. G. (1961) Bahamian Oolite Shoals., p. 53-62 in Geometry of Sandstone Bodies, Spec. Vol., *Bul. American Assoc. Pet. Geol.*

———. (1963) Recent Calcium Carbonate Facies of the Great Bahama Bank. 1. Petrography and Reaction Groups. 2. Sedimentary Facies. *Jour. Geol.* v. 71, p. 334-35, 472-97.

Purdy, E. G., and Imbrie, J. (1964) Carbonate Sediments, Great Bahama Bank. Guidebook for Meeting no. 2, Geol. Soc. Amer. Annual Meeting, Miami, Fla., 66 p.

Puri, H. S. (1960) Recent Ostracoda From the West Coast of Florida. Gulf Coast Assoc. *Geol. Soc. Trans.*, v. 10, p. 107-49.

Puri, H. S., and Hulings, N. C. (1957) Recent Ostracode facies from Panama City to Florida Bay area. Gulf Coast Assoc. *Geol. Soc. Trans.*, v. 7, p. 167-90.

Puri, H. S., and Vernon, R. O. (1964) Summary of the Geology of Florida and a Guide to the Classic Exposures. Florida Geol. Survey Spec. Publ. no. 5, 312 p.

Raphael, C. N. (1975) Coastal Morphology. Southwest Great Abaco Island, Bahamas. *Geoform*, v. 6, p. 237-46. Pergamon Press.

Rathbun, M. J. (1918) The Grapsoid Crabs of America. *Bul. U.S. Nat. Mus.*, v. 97, p. 1-461.

———. (1925) The Spider Crabs of America. *Bul. U.S. Nat. Mus.*, v. 129, p. 1-613.

———. (1930) The Cancroid Crabs of America. *Bul. U.S. Nat. Mus.*, v. 152, p. 1-609.

———. (1933) Brachyuran Crabs of Porto Rico and the Virgin Islands. N.Y. Acad. Sci. Scientific Survey of Porto Rico and the Virgin Islands., v. 15, p. 1-121.

———. (1937) The Oxystomatous and allied Crabs of America. *Bul. U.S. Nat. Mus.* v. 166, p. 1-278.

Rehm, A. E., and Humm, H. J. (1973) *Sphaeroma terebrans:* A Threat to the Mangroves of Southwestern Florida. *Science,* v. 182, p. 173-74.

———. (1976) The Effects of the Wood-Boring Isopod *Sphaeroma terebrans* on the Mangrove Communities in Florida. *Envir. Conservation,* v. 3, no. 1, p. 47-57.

Revelle, R., and Emery, K. O. (1957) Chemical Erosion of Beach Rock and Exposed Reef Rock. U.S. Geol. Sur. Prof. Paper 260-T, p. 699-709.

Richter, R. (1920) Ein Devonisher "Pfeifenquarzit" vergleichen mit der Heutigen "Sandkoralle" (Sabellaira, Annelidae). *Senckenbergiana,* v. 2, p. 215-35.

———. (1921) Scolithus, Sabelithus. Sabellarifex und Geflecht-quartzite. *Senckenbergiana* v. 3, p. 49-52.

———. (1927) "Sandkorallen" Riffe in der Nordsee. *Natur und Volk,* v. 57, no. 2, p. 49-62.

———. (1928) Observations on Organisms and Sediments on Shallow Sea Bottoms. *Amer. Midland Nat.,* v. 2, p. 236-42.

Robertson, P. B. (1963) A Survey of the Marine Rock-Boring Fauna of Southeast Florida. Unpubl. Masters Thesis, Inst. Mar. Sci., Univ. Miami, June.

Robinson, R. B. (1967) Diagenesis and Porosity Development in Recent and Pleistocene Oolites from Southern Florida and the Bahamas. *Jour. Sed. Pet.,* v. 37, p. 355-64.

Rodis, H. G., Klein, H., and Sherwood, C. B. (1974) Hydrology of the Miami Area. Field Trip Guidebook for Geol. Soc. Amer. Annual Meeting, Miami, Nov., c/o H. G. Rodis, U.S.G.S., 801 S. Miami Ave., Miami, Fla. 33130.

Roebeck, G. A., Clarke, N. A., and Dostal, K. A. (1967) Effectiveness of Water Treatment Process in Virus Removal. *Jour. Amer. Water Works Assn.,* v. 54, p. 1275-89.

Roehl, P. O. (1967) Stony Mountain (Ordovician) and Interlake (Silurian) Facies Analogs of Recent Low-Energy Marine and Subaerial Carbonates, Bahamas, *Amer. Assoc. Petrol. Geol. Bul.,* v. 51, no. 10, p. 1979-2032.

Roessler, M. A. (1971) Testimony for the State of Florida Department of Pollution Control, Orlando, Florida, November 22.

Rubin, M., and Suess, H. E. (1955) U.S. Geological Survey Radiocarbon Dates 22 *Sci.,* v. 121, no. 3145, p. 481-88.

Russell, R. J. (1962) Origin of Beachrock. Annals Geomorph. Neue Folge, Heft 1, Band 6, Juni., p. 1-21.

Ryther, J. H. (1971) Recycling Human Wastes to Enhance Food Production From the Sea. Environmental Letters, v. 1, no. 2, p. 79-87.

Sander, Bruno (1936) Contributions to the Study of Depositional Fabrics. Published in English (1951) by Amer. Asso. Petrol. Geol., Translated by Knopf, E. B., 207 p.

Sandford, S. (1909) The Topography and Geology of Southern Florida. Florida Geol. Survey 2nd Ann. Rept., p. 175-231.

Schlager, W., Hooke, R. L., and James, N. P. (1976) Episodic Erosion and Deposition in the Tongue of the Ocean (Bahamas). *Geol. Soc. Amer. Bul.,* v. 87.

Schlanger, S. O. (1963) (ed), U.S. Geol. Survey Profess. Papers 260-BB, p. 991-1066.

Schmalz, R. R., and Swanson, F. J. (1969) Diurnal Variations in the Carbonate Saturation of Seawater. *Jour. Sed. Pet.,* v. 39, no. 1, p. 255-67.

Scholl, D. W. (1958) Effects of an Arenaceous Tube-Building Polychaete upon the Sorting of a Beach Sand at Abalone Cove, Calif. Compass Mag. 35, p. 276-83.

———. (1964) Recent Sedimentary Record in Mangrove Swamps and Rise in Sea Level Over the Southwestern Coast of Florida, Part I and II. Mar. Geol. 1, p. 344-66, 2, p. 343-64.

———. (1966) Florida Bay. A Modern Site of Limestone Formation, p. 282-88 *in* Fairbridge, R. W. (ed)., The Encyclopedia of Oceanography., Reinhold Publishing Corp., New York, 1021 p.

Scholl, D. W., Craighead, F. C., and Stuiver, M. (1969) Florida Submergence Curve Revised; Its Relation to Coastal Sedimentation Rates. *Science,* v. 163, p. 562-64.

Schmitt, W. L. (1935) Crustacea, Macrura and Anomura of Porto Rico and the Virgin Islands. New York Acad. Sci. Scientific Survey of Porto Rico and the Virgin Islands, v. 16, p. 123-227.

Schroeder, J. H., and Zankl, H. (1974) Dynamic Reef Formation: A Sedimentological Concept Based on Studies of Recent Bermuda Reefs. p. 413-28 in Proceedings of the Second International Coral Reef Symposium, v. 2, Great Barrier Reef Committee, Brisbane, Australia.

Schultz, G. A. (1969) The Marien Isopod Crustaceans. Dubuque, Iowa, Wm. C. Brown, Co. Publishers, 359 p.

Scoffin, T. P. (1970a) The Trapping and Binding of Subtidal Carbonate Sediments by Marine Vegetation in Bimini Lagoon, Bahamas, *Jour. Sed. Pet.,* v. 40, p. 249-73.

———. (1970b) A Conglomeratic Beachrock in Bimini, Bahamas. *Jour. Sed. Pet.,* v. 40, no. 2, p. 756-58.

Segar, D. A., Pellenbarg, R., Gilio, J. (1972) Observations on the Distribution of Ag, Cu, Co, Ni, Cd, Zn, Pb, Fe, and V in a Coastal Ecosystem, (abst)., Trans. EOS Amer. *Geophysical Union,* v. 53, no. 11, p. 1030.

———. (1973) Trace Metals in Carbonate and Organic Rich Sediments. *Marine Pollution Bul.,* v. 4, p. 138-42.

Seibold, E. (1962) Untersuchungen zur Kalkfällung und Kalklösung am Westrand der Great Bahama Bank. *Sedimentology,* v. 1, p. 50-74.

———. (1964) Beobachtungen zur Schichtung in Sedimenten am Westrand der Grand Bahama Bank. N. Jb. Geol. Palaonti, Abh. 120, 3, p. 233-52.

Shepard, F.P., (1963) Thirty-five thousand Years of Sea Level. Essays in Marine Geology, Univ. of S. Calif. Press.

Sheridan, R. E., Drake, C. L., Nafe, J. E., and Hennion, J. (1966) Seismic-refraction Study of Continental Margin East of Florida. *Bul. Amer. Assoc. Petrol. Geol.,* v. 50, no. 9, p. 1972-91.

Shier, D. E. (1969) Vermetid Reefs and Coastal Development in the Ten Thousand Islands, Southwest Florida. *Geol. Soc. of Amer. Bul,* v. 80, p. 485-508.

Shinn, E. A. (1963) Spur and Grove Formation on the Florida Reef Track. *Jour. of Sedimentary Petrology,* v. 33, no. 2, p. 291-303.

———. (1964) Recent Dolomite, Sugarloaf Key, p. 62-67 in Guidebook for GSA Field Trip No. 1, South Florida Carbonate Sediments, R. N. Ginsburg (ed), 72 p.

———. (1966) Coral Growth-Rate, an Environmental Indicator. *Jour. Paleo.,* v. 40, no. 2, p. 233-40.

———. (1968a) Selective Dolomitization of Recent Sedimentary Structures. *Jour. Sed. Pet.,* v. 38, no. 2, p. 612-16.

———. (1968b) Practical Significance of Birdseye Structures in Carbonate Rocks. *Jour. Sed. Pet.,* v. 38, p. 215-23.

———. (1968c) Burrowing in Recent Lime Sediments of Florida and the Bahamas. *Jour. Paleo.,* v. 42, p. 879-94.

Shinn, E. A., Ball, M. M., and Stockman, K. W. (1967) The Geologic Effects of Hurricane Donna on South Florida. *Jour. Geol.,* v. 75, no. 5, p. 583-97.

Shinn, E. A., and Ginsburg, R. N. (1964) Formation of Recent Dolomite in Florida and the Bahamas (abst). *Bul. Amer. Assoc. Petrol. Geol.,* v. 48, p. 547.

Shinn, E. A., Ginsburg, R. N., and Lloyd, R. M. (1965) Recent Supratidal Dolomite from Andros Island, Bahamas, p. 112-23 *in* Dolomitization and Limestone Diagenesis, a Symposium, L. C. Pray and R. D. Murray (eds), Soc. Econ. Paleontologists and Mineralogists, Spec. Publ. no. 13, 180 p.

Shinn, E. A., and Lloyd, R. M. (1965) Recent Supratidal Dolomitization in Florida and the Bahamas (abst). *Bul. Amer. Assoc. Petrol. Geol.,* v. 48, p. 547.

Shinn, E. A., Lloyd, R. M., and Ginsburg, R. N. (1969) Anatomy of a Modern Carbonate Tidal-Flat, Andros Island, Bahamas, *Jour. Sed. Pet.,* v. 39, p. 1202-28.

Shoemaker, C. R. (1935) The Amphipods of Porto Rico and the Virgin Islands, N.Y. Acad. Sci. Scientific Survey of Porto Rico and the Virgin Islands, v. 15, p. 229-62.

Shrock, R. R. (1945) Surficial Breccias Produced from Chemical Weathering of Eocene Limestone in Haiti, West Indies, Proced. Ind. Acad. Sci., v. 55, p. 107-10.

Siemers, C. T., and Dodd, J. R. (1969) Buried Karst Topography on the Florida Keys and its Influence on Recent Sedimentation and Biotic Distribution. p. 204-206 in Absts. Annual Geol. Soc. Meeting, Atlantic City.

Sisler, F. D. (1962) Microbiology and Biochemistry of the Sediments and Overlying Water. U.S. Geol. Survey Prof. Paper 350, p. 64-69.

Smith, F. G. W. (1943) Littoral Fauna of the Miami Area: I. The Madreporaria. *Proc. Fla. Acad. Sci.,* v. 6, no. 1, p. 41-48.

———. (1948) Atlantic Reef Corals. Coral Gables, Univ. of Miami Press, 112 p.

———. (1954) Corals and Gorganians. Gulf of Mexico Bibliography, *Fish. Bul., U.S.,* v. 55, no. 89.

Smith, S. L. (1971) Distribution of Recent Foraminifera in Lower Florida Bay. p. 116-120 in A Symposium of Recent South Florida Foraminifera, Bock *et al.,* Memoir 1, 243 p., Miami Geol. Soc., 10 Rickenbacker Causeway, Miami, Fla. 33149.

Sorby, H. C. (1879) The Structure and Origin of Limestones. *Geol. Soc. London Proc.,* v. 25, p. 56-95.

Spencer, M. (1967) Bahama Deep Test. *Bul. Amer. Assoc. Pet. Geol.,* v. 51, no. 2, p. 263-68.

Squires, D. F. (1958) Stony Corals from the Vicinity of Bimini, Bahamas, B. W. I. *Bul. Amer. Mus. Nat. His.,* v. 115, p. 215-62.

Stanley, S. M. (1966) Paleoecology and Diagenesis of Key Largo Limestone, Florida. *Bul. Amer. Assoc. Petrol. Geol.,* v. 50, p. 1927-47.

Starck, W. A., II (1966) Marvels of a Coral Realm. *Nat. Geog. Mag.,* v. 130, no. 5, p. 710-38.

———. (1968) A List of Fishes of Alligator Reef, Florida with Comments on the Nature of the Florida Reef Fish Fauna. *Undersea Biology,* v. 1, no. 1, p. 5-36.

Starck, W. A., and Davis, W. P. (1966) Night Habits of Fishes of Alligator Reef, Florida. *Ichthyologica,* v. 38, p. 313-56.

Stehli, F. G., and Hower, J. (1961) Mineralogy and Early Diagenesis of Carbonate Sediments. *Jour. Sed. Pet.,* v. 31, p. 358-71.

Steinker, D. C. (1974) Foraminiferal Studies in Tropical Carbonate Environments. p. 37-39 in Recent Advances in Carbonate Studies, Special Publ. no. 6, West Indies Laboratory, St. Croix, U.S. Virgin Islands.

Steinker, D. C., and Floyd, J. C. (1975) Previously Undescribed Carbonate Deposits on Key Largo, Florida: Comment. *Geology,* v. 3, no. 7, p. 360.

———. (1976) An Interdisciplinary Field Course in Marine Science. (abst) Fla. Sci. v. 39, suppl. no. 1, p. 26.

Steinker, P. J., and Steinker, D. C. (1972) The Meaning of Facies in Stratigraphy. *Compass,* v. 49, p. 45-53.

Stephenson, T. A., and Stephenson, A. (1950) Life Between Tide Marks in North America, 1. The Florida Keys. *Jour. Ecology,* v. 38, no. 2, p. 354-402.

Stockman, K. W., Ginsburg, R. N., and Shinn, E. A. (1967) The Production of Lime Mud by Algae in South Florida. *Jour. Sed. Pet.,* v. 37, no. 2, p. 633-48.

Storr, J. F. (1964) Ecology and Oceanography of the Coral Reef Tract, Abaco Island, Bahamas. Geol. Soc. Amer. Special Paper no. 79, 98 p.

Supko, P. (1970) Some Aspects of the Geology of Bimini, Bahamas. p. 5-14c, *in* Sedimentary Environments and Carbonate Rocks of Bimini, Bahamas, Supko *et al.,* (ed.) 4th Annual Meeting of the Miami Geol. Soc., 30 p.

Supko, P., Marszalek, D., and Bock, W. (1970) Sedimentary Environments and Carbonate Rocks, Bimini, Bahamas. 4th Annual Meeting of the Miami Geol. Soc., 30 p.

Swinchatt, J. P. (1965) Significance of Constituent Composition,

Texture, and Skeletal Breakdown in Some Recent Carbonate Sediments. *Jour. Sed. Pet.,* v. 35, no. 1, p. 71-90.

Tabb, D. C., and Jones, A. C. (1962) Effect of Hurricane Donna on the Aquatic Fauna of North Florida Bay. Trans. of the *Amer. Fish. Soc.,* v. 91, no. 4, p. 375-78.

Taft, W. H., Arrington, F., Haimovitz, A., Macdonald, C., and Woolheater, C. (1968) Lithification of Modern Carbonate Sediments at Yellow Bank, Bahamas. *Bul. Mar. Sci.,* v. 18, p. 762-828.

Taft, W. H., and Harbaugh, J. W. (1964) Modern Carbonate Sediments of Southern Florida, Bahamas, and Espiritu Santo Island, Baja, California: A Comparison of their Mineralogy and Chemistry. Stanford Univ. Publ. Geol. Sci., v. 8, no. 2, 133 p.

Talwani, M. (1960) Gravity Anomalies of the Bahamas and their Interpretations. Unpub. Ph.D. Thesis, Columbia Univ., 89 p.

Taylor, W. R. (1960) Marine Algae of the Eastern Tropical and Subtropical Coasts of the Americas. Ann Arbor, Univ. of Michigan Press, 870 p.

Teichert, C. (1958) Cold and Deep-water Coral Banks. *Amer. Assoc. Petrol. Geol. Bul.,* v. 42, no. 5, p. 1064-82.

Textoris, D. A. (1968) Petrology of Supratidal, Intertidal, and Shallow Subtidal Carbonates, Black River Group, Middle Ordovician, New York, XXIII International Geol. Congress, v. 8, p. 227-48.

Thomas, L. P. (1962) The Shallow Water Amphiurid Brittle Stars (Echinodermata, Ophiuroidae) of Florida. *Bul. Mar. Sci. Gulf & Carib.,* v. 12, no. 4, p. 623-94.

Thomas, T. M. (1970) A Detailed Analysis of Climatological and Hydrological Records of South Florida with Reference to Man's Influence upon Ecosystem Evolution. Rep. DI-NPS-14-10-1-160-18 to U.S. Nat. Park. Ser., Univ. of Miami, 89 p.

———. (1974) A Detailed Analysis of Climatological and Hydrological Records of South Florida with Reference to Man's Influence upon Ecosystem Evolution. p. 82-122 *in* Environments of South Florida: Present and Past, P. J. Gleason (ed), Memoir 2, Miami Geol. Soc., 452 p.

Thorhaug, A. (1971) Testimony for the State of Florida Department of Pollution Control, Miami, Florida, October 13.

Tierney, J. Q. (1954) Porifera. Gulf of Mexico Bibliography, *Fish. Bul., U.S.* v. 55, no. 89.

Till, R. (1970) The Relationship Between Environment and Sediment Composition (Geochemistry and Petrology) in the Bimini Lagoon, Bahamas. *Jour. Sed. Pet.,* v. 40, no. 1, p. 367-85.

Traverse, A., and Ginsburg, R. N. (1966) Palynology of the Surface Sediments of Great Bahama Bank, as Related to Water Movement and Sedimentation. *Marine Geol.* v. 4, p. 417-59.

Turekian, K. H. (1957) Salinity Variations in Sea Water in the Vicinity of Bimini, Bahamas, British West Indies. *Am. Mus. Novitates,* v. 1822, p. 1-12.

Turmel, R., and Swanson, R. (1964) Rodriguez Bank. p. 26-33 in South Florida Carbonate Sediments, R. N. Ginsburg (ed), Guidebook for Field Trip No. 1, Geol. Soc. Amer. Annual Meeting, Miami, Nov.

———. (pub. pending) The Evolution of Rodriguez Bank, a Modern Carbonate Mound.

Turney, W. J. (1964) Molluscan Fauna. p. 15 in South Florida Carbonate Sediments, R. N. Ginsburg (ed), 72 p.

Turney, W. J., and Perkins, R. F. (1972) Molluscan Distribution in Florida Bay. Sedimenta III, Comparative Sedimentology Lab., Univ. of Miami, Miami, Florida, 36 p.

Uchupi, E. (1968) Atlantic Continental Shelf and Slope of The United States—Physiography. U.S. Geol. Survey Prof. Paper no. 529-C, 30 p.

Uchupi, E., and Emery, K. O. (1967) Structure of Continental Margin off Atlantic Coast of United States. *Bul. Amer. Assoc. Petrol. Geol.,* v. 51, no. 2, p. 223-34.

U.S. Dept. of the Interior (1974) Resource and Land Information for South Dade County, Florida. U.S. Geol. Survey Invest., 1-850, 65 p.

Van Houten, F. B. (1948) Origin of Red-banded Early Cenezoic Deposits in Rocky Mountain Region. *Amer. Assoc. Pet. Geol. Bul.,* v. 32, p. 2083-2126.

Van Tuyl, F. M. (1914) The Origin of Dolomite. *Iowa Geol. Sur. Ann. Brochure,* v. 25, p. 251-422.

Vaughan, T. W. (1909) Geology of the Florida Keys and the Marine Bottom Deposits and Recent Corals of Southern Florida. Carnegie Inst. Washington, Yearbook, no. 7, p. 131-36.

———. (1910a) The Geologic Work of Mangroves in Southern Florida. *Smithsonian Misc. Coll.,* v. 52, *Quart. Issue,* v. 5, p. 461-64.

———. (1910b) Geology of the Keys, the Marine Bottom Deposits, and Recent Corals of Southern Florida. Carnegie Inst. Washington, Yearbook no. 8, p. 140-44.

———. (1911) The Recent Madreporaria of Southern Florida. Carnegie Inst. Washington, Yearbook no. 9, p. 135-44.

———. (1912) The Madreporaria and Marine Bottom Deposits of Southern Florida. Carnegie Inst. Washington, Yearbook no. 10, p. 147-56.

———. (1913) Studies of the Geology and of the Madreporaria of the Bahamas and of Southern Florida. Carnegie Inst. Washington, Yearbook no. 11, p. 153-62.

———. (1914) Reef Corals of Southern Florida. Carnegie Inst. Washington, Yearbook no. 12, p. 181-83.

———. (1915a) The Reef Corals of the Bahamas and of Southern Florida. Carnegie Inst. Washington, Yearbook no. 13, p. 222-26.

———. (1915b) The Geologic Significance of the Growth Rate of the Floridian and Bahamian Shoal Water Corals. *Jour. Washington Acad. Sci.,* v. 5, p. 591-600.

———. (1915c) Growth Rate of the Floridian and Bahamian Shoal Water Corals. Carnegie Inst. Washington, Yearbook p. 221-31.

———. (1916a) On Recent Madreporaria of Florida, the Bahamas, and the West Indies, and on Collections from Murray Island, Australia. Carnegie Inst. Washington, Yearbook no. 14, p. 220-31.

———. (1916b) The Results of Investigations of the Ecology of the Floridian and Bahamian Shoal Water Corals. *Proc. Natl. Acad. Sci.,* v. 2, p. 95-100.

———. (1918) The Temperature of the Florida Coral Reef Tract. Carnegie Inst. Washington, Pub. 213. p. 319-39.

———. (1919) Fossils Corals from Central America, Cuba and Puerto Rico, with an Account of American Tertiary, Pleistocene and Recent Coral Reefs. *U.S. Nat. Mus. Bul.* no. 103, p. 189-524.

Voss, G. L. (1955) A Key to the Commercial and Potentially Commercial Shrimp of the Family Penaeidae of the Western North Atlantic and Gulf of Mexico. Fla. St. Bd. of Cons., Tech. Ser. no. 14, p. 1-23.

———. (1971) Testimony for the State of Florida Department of Pollution Control, Miami, Florida, October 13.

———. (1976-in press) Guide to the Commoner Shallow Water Invertebrates of Florida and the Caribbean. Miami, Seeman Press. (does not include mollusks)

Voss, G. L., and Voss, N. A. (1955) An Ecological Survey of Soldier Key, Biscayne Bay, Florida. *Bul. Mar. Sci.,* v. 5, no. 3, p. 203-29.

———. (1960) An Ecological Survey of the Marine Invertebrates of Bimini, Bahamas, with a Consideration of their Zoogeographical Relationships. *Bul. Mar. Sci. Gulf & Carib.,* v. 10, no. 1, p. 96-116.

Wade, B. A. (1967) Studies on the Biology of the West Indian Beach Clam, *Donax denticulatus* Linne. *Bul. Mar. Sci.,* v. 17, no. 1, p. 149-74.

Walker, K. R., and La Porte, L. F. (1968) Mutually Congruent Carbonate Lithofacies and Biofacies from the Middle Ordovician and Early Devonian of New York. Geol. Soc. America 1968 Annual Meetings, (abst), p. 310-11.

Walper, J. L. (1974) The Origin of the Bahama Platform. Trans. Gulf Coast Assoc. *Geol. Soc.,* v. 24, p. 25-30.

Wanless, H. R. (1969) Sediments of Biscayne Bay—Distribution and Depositional History. Tech. Rept. 69-2, Inst. of Marine Sci., Miami, Florida, 260 p.

———. (1974a) Fining-Upwards Sequence Generated by Sea-Grass Beds. (abst)., *Bul. Geol. Soc. Amer.,* v. 6, no. 7, p. 999.

———. (1974b) Mangrove Sedimentation in Geologic Perspective, p. 190-200 in Environments of South Florida: Present and Past. P. J. Gleason (ed)., Memoir 2, Miami Geol. Soc., 252 p.

Warmke, G., and Abbott, R. T. (1961) Caribbean Seashells. Livingston Publ. Co. Naberth, Penn., 348 p.

Warzeski, E. R. (1976a) Growth History and Sedimentary Dynamics of Caesars Creek Bank. Unpublished Masters Thesis, Univ. of Maine.

———. (1976b) Storm Sedimentation in the Biscayne Bay Region, p. 33-38 in Biscayne Bay Symposium I, April 2-3, 1976 Published as part of Univ. of Miami Sea Grant Special Report no. 5.

Wayne, C. D. (1974) Effect of Artificial Sea-Grass on Wave Energy and Near-Shore Sand Transport. Trans. Gulf Coast Assoc. Geol. Soc., v. 24, p. 279-88.

Weber, J. N., White, E. W., and Weber, P. H. (1975) Correlation of Density Banding in Reef Coral Skeletons with Environmental Parameters: The Basis for Interpretation of Chronological Records Preserved in the Coralla of Corals. *Paleobiology,* v. 1, p. 137-49.

Wells, J. W. (1932) Study of the Reef Corals of the Tortugas. Carnegie Inst. Washington, Yearbook no. 31, p. 290-91.

Wilber, R. J., and Neumann, A. C. (1976) Petrology of Subsea Cemented Carbonate Mounds, Lithoherms, in Northern Straits of Florida (abst). *Amer. Assoc. Petrol. Geol. Bul.,* v. 60, no. 4, p. 733.

Wilson, J. L. (1974) Characteristics of Carbonate-Platform Margins. *Amer. Assoc. Petrol. Geol. Bul.* v. 58, no. 5, p. 810-24.

Wiman, S. K., and McKendree, W. G. (1975) Distribution of *Halimeda* Plants and Sediments on and around a Patch Reef near Old Rhodes Key, Florida. *Jour. Sed. Pet.,* v. 45, no. 2, p. 415-21.

Winland, H. D., and Matthews, R. K. (1974) Origin and Significance of Grapestone, Bahama Islands. *Jour. Sed. Pet.,* v. 44, no. 3, p. 921-27.

Woelkerling, W. J. (1976) South Florida Benthic Marine Algae. Sedimenta V, 148 p., Comparative Sedimentology Lab., Univ. Miami, Fisher Island Station, Fla. 33139.

Wolf, K. H. (1965) Littoral Environment Indicated by Open-Space Structures in Algal Limestones. *Paleogeography, paleoclimatology, paleoecology,* v. 1, p. 183-223.

Woodring, W. P., Brown, J. S., and Burbank W. S. (1924) Geology of the Republic of Haiti. Geol. Survey of the Republic of Haiti, p. 109-34.

Wright, R. C., and Hay, W. W. (1971) The Abundance and Distribution of Foraminifers in a Back-Reef Environment, Molasses Reef, Florida. p. 121-274 in A Symposium of Recent South Florida Foraminifera. Jones & Bock (eds), Memoir 1, Miami *Geol. Soc.,* 245 p.

Young, P. H. (1964) Some Effects of Sewer Effluent on Marine Life. *Calif. Fish and Game,* v. 50, January.

Zahl, P. A. (1952) Man-of-War Fleet Attacks Bimini. Nat. Geog., Feb., p. 185-212.

Zankl, H., and Schroeder, J. H. (1972) Interaction of Genetic Processes in Holocene Reefs off North Eleuthera Island, Bahamas. *Geol. Rundschau,* v. 61, no. 2, p. 520-41.

Ziegler, A. M., Walker, K. R., Anderson, E. J., Kaufmann, E. G., Ginsburg, R. N., and James, N. P. (1974) Principles of Benthic Community Analysis, Sedimenta IV, Comparative Sedimentology Laboratory, Univ. of Miami, Fla.

Zischke, J. A. (1973) An Ecological Guide to the Shallow-Water Marine Communities of Pigeon Key, Florida. Published by the author and supported in part by N.S.F. Sci. Fac. Fellowship no. 61213. St. Olaf College, Northfield, Minnesota.

Author Index

Subject Index

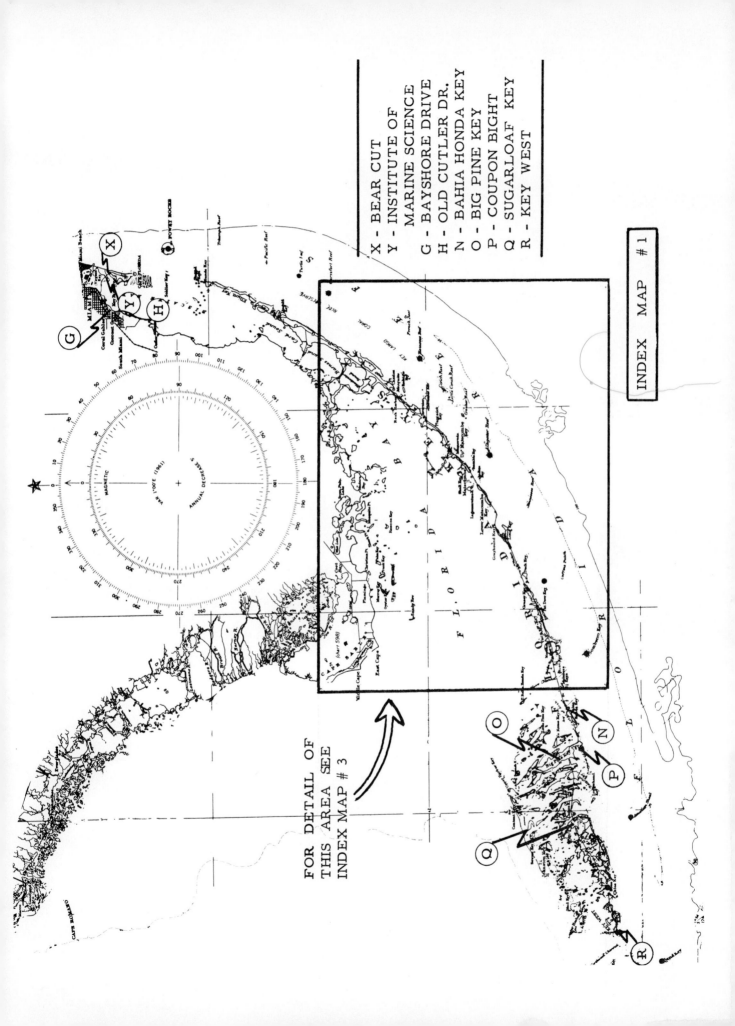

INDEX MAP #1

X – BEAR CUT
Y – INSTITUTE OF MARINE SCIENCE
G – BAYSHORE DRIVE
H – OLD CUTLER DR.
N – BAHIA HONDA KEY
O – BIG PINE KEY
P – COUPON BIGHT
Q – SUGARLOAF KEY
R – KEY WEST

FOR DETAIL OF THIS AREA SEE INDEX MAP #3

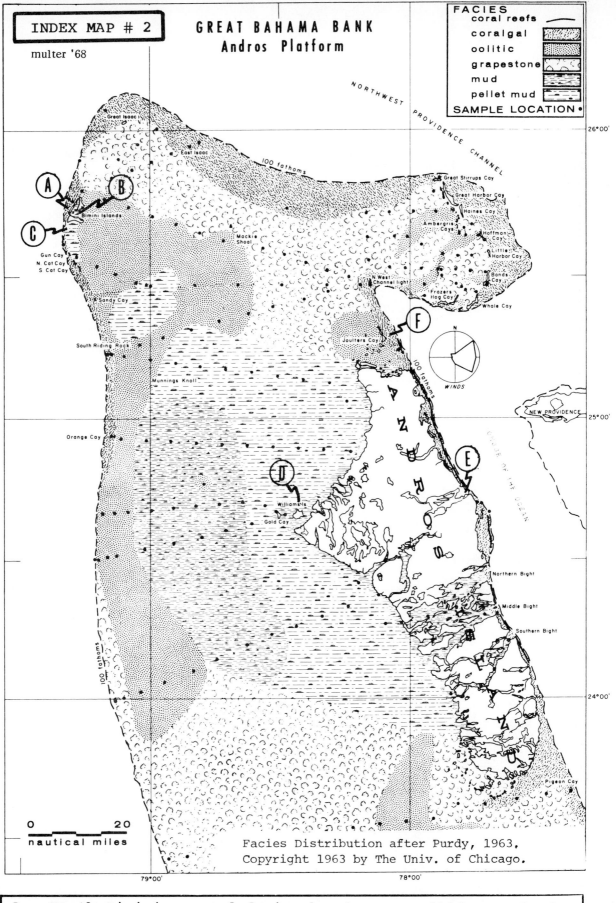

INDEX MAP # 2

multer '68

GREAT BAHAMA BANK
Andros Platform

FACIES
coral reefs
coralgal
oolitic
grapestone
mud
pellet mud
SAMPLE LOCATION •

Facies Distribution after Purdy, 1963,
Copyright 1963 by The Univ. of Chicago.

0   20
nautical miles

A - North Bimini seaward-facing beach   D - Williams Island
B - Bimini lagoon   E - Fresh Creek
C - Open water - platform edge   F - Joulters Cay

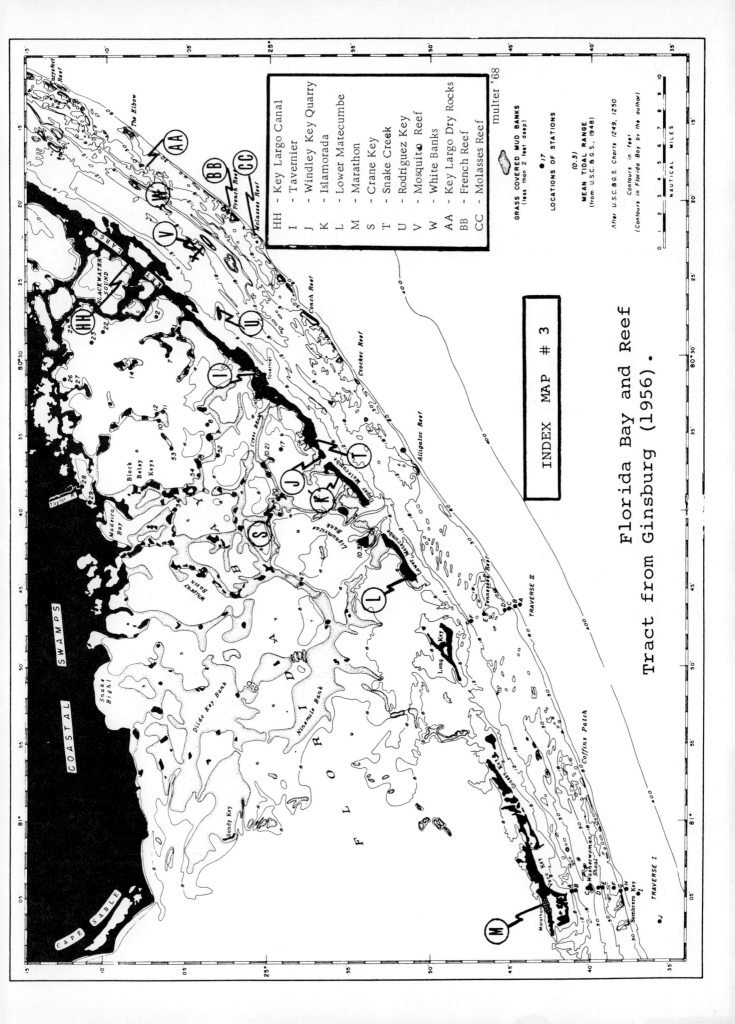

Florida Bay and Reef
Tract from Ginsburg (1956).

INDEX MAP # 3

HH - Key Largo Canal
I - Tavernier
J - Windley Key Quarry
K - Islamorada
L - Lower Matecumbe
M - Marathon
S - Crane Key
T - Snake Creek
U - Rodriguez Key
V - Mosquito Reef
W - White Banks
AA - Key Largo Dry Rocks
BB - French Reef
CC - Molasses Reef

multer '68

⬭ GRASS COVERED MUD BANKS
(less than 2 feet deep)

● 17 LOCATIONS OF STATIONS

(0.3) MEAN TIDAL RANGE
(from U.S.C.&G.S., 1948)

After U.S.C.&G.S. Charts 1249, 1250
Contours in feet
(Contours in Florida Bay by the author)

0 1 2 3 4 5 6 7 8 9 10
NAUTICAL MILES

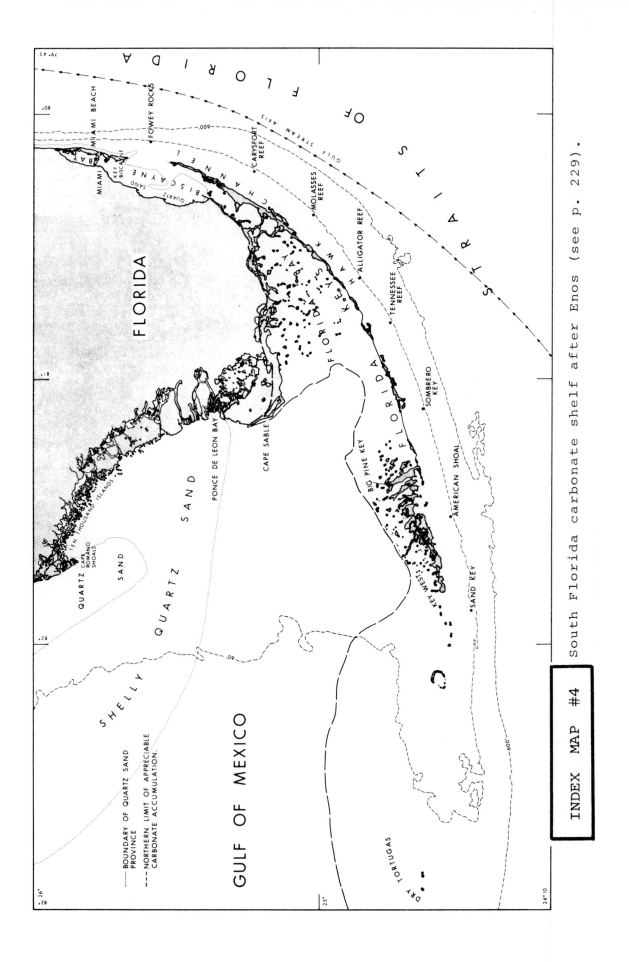

INDEX MAP #4   South Florida carbonate shelf after Enos (see p. 229).

INDEX MAP # 5: Key Largo Dry Rocks –Molasses Reef and central Key Largo area. From NOAA Nautical Chart 11451 (formerly 141-SC).

INDEX MAP # 6 : Triangles reef–Hen and Chickens reef, Tavernier and Florida Bay. From NOAA Nautical Chart 11451 (formerly 141–SC).

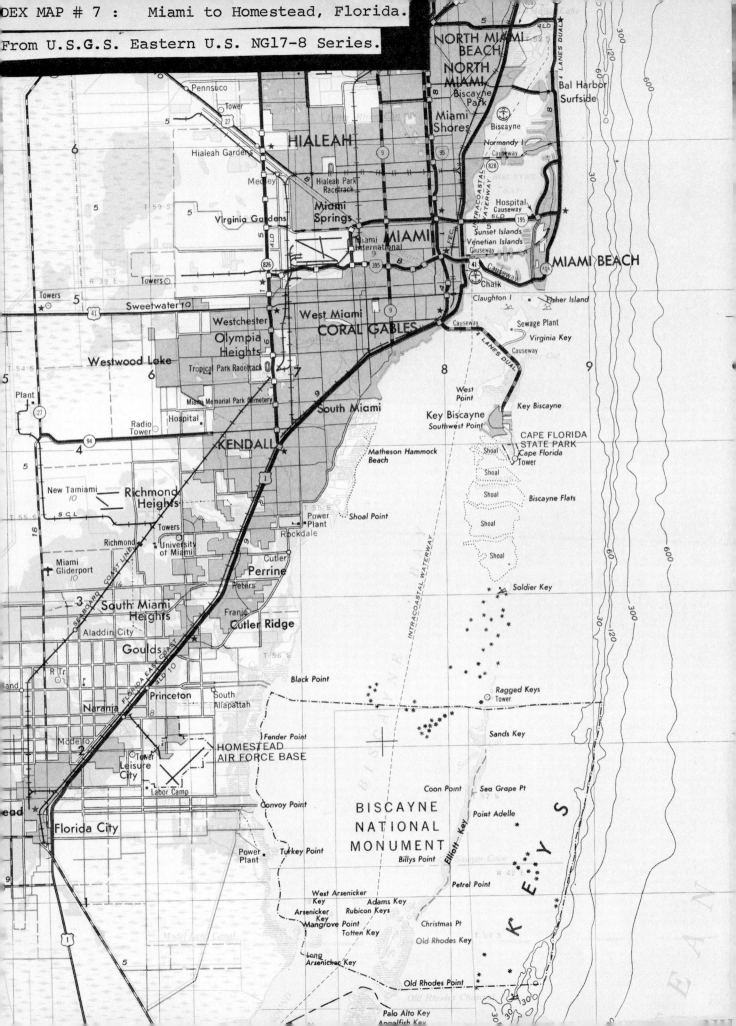

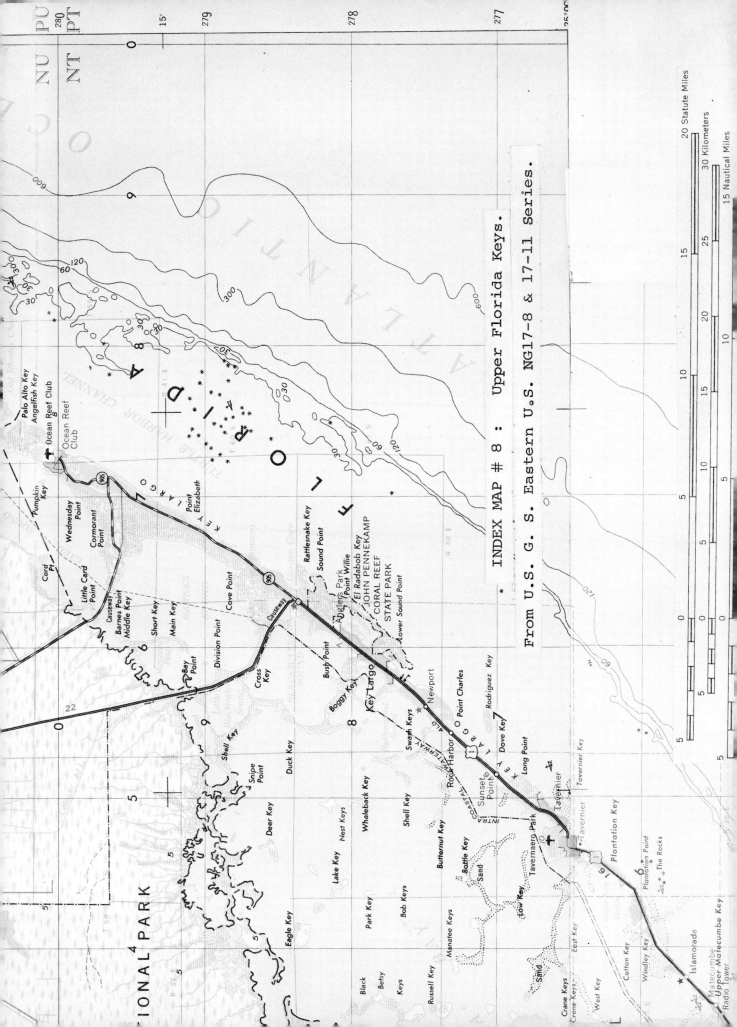

INDEX MAP # 8 : Upper Florida Keys.

* From U.S.G.S. Eastern U.S. NG17-8 & 17-11 Series.

INDEX MAP # 9 : Middle Florida Keys.

From U.S. G. S. Eastern U.S. NG17-8 & 17-11 Series.

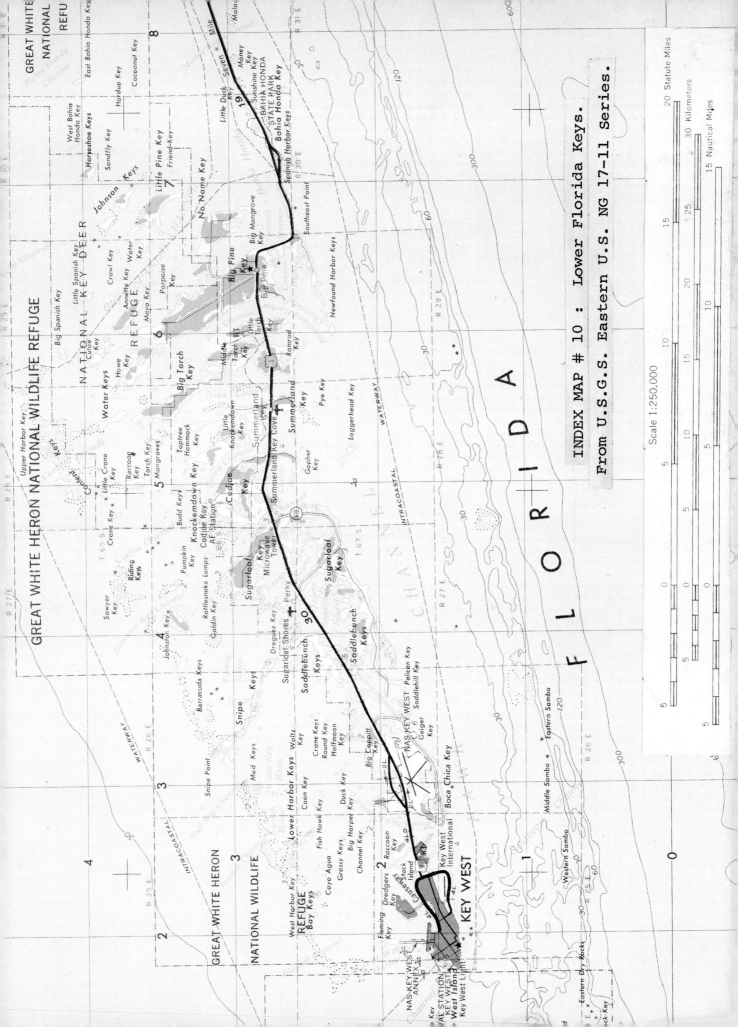

INDEX MAP # 10 : Lower Florida Keys.

From U.S.G.S. Eastern U.S. NG 17-11 Series.

Scale 1:250,000